The series "Studies in Computational Intelligence" (SCI) publishes new developments and advances in the various areas of computational intelligence—quickly and with a high quality. The intent is to cover the theory, applications, and design methods of computational intelligence, as embedded in the fields of engineering, computer science, physics and life sciences, as well as the methodologies behind them. The series contains monographs, lecture notes and edited volumes in computational intelligence spanning the areas of neural networks, connectionist systems, genetic algorithms, evolutionary computation, artificial intelligence, cellular automata, self-organizing systems, soft computing, fuzzy systems, and hybrid intelligent systems. Of particular value to both the contributors and the readership are the short publication timeframe and the world-wide distribution, which enable both wide and rapid dissemination of research output.

Indexed by SCOPUS, DBLP, WTI Frankfurt eG, zbMATH, SCImago.

All books published in the series are submitted for consideration in Web of Science.

Studies in Computational Intelligence

Volume 1107

Series Editor

Janusz Kacprzyk, Polish Academy of Sciences, Warsaw, Poland

Michael Zgurovsky · Nataliya Pankratova
Editors

System Analysis and Artificial Intelligence

Springer

Editors
Michael Zgurovsky
Research Supervisor of the Institute for Applied Systems Analysis
National Technical University of Ukraine "Igor Sikorsky Kyiv Polytechnic Institute"
Kyiv, Ukraine

Nataliya Pankratova
Deputy Director for Research of the Institute for Applied Systems Analysis
National Technical University of Ukraine "Igor Sikorsky Kyiv Polytechnic Institute"
Kyiv, Ukraine

ISSN 1860-949X ISSN 1860-9503 (electronic)
Studies in Computational Intelligence
ISBN 978-3-031-37449-4 ISBN 978-3-031-37450-0 (eBook)
https://doi.org/10.1007/978-3-031-37450-0

This Springer imprint is published by the registered company Springer Nature Switzerland AG
The registered company address is: Gewerbestrasse 11, 6330 Cham, Switzerland

Preface

The collection of scientific works System Analysis and Artificial Intelligence brings together the latest scientific works of Ukrainian scientists and their colleagues from other countries of the world in three interrelated areas: system analysis, artificial intelligence, and intellectual data analysis. The main goal of this collection is an attempt to combine these three actual areas of research to solve complex problems and improve decision-making processes.

The collection is divided into three parts that reflect different aspects of the book's subject:

(1) System Analysis of Processes and Phenomena of Different Nature.
In particular, the authors of the articles in the collection show that system analysis involves the analysis of complex systems of various nature to identify areas for improvement and optimization of their performance. Examples of the use of systems analysis for the study of different complex systems, from computer networks to production processes and organizational structures, are given.

(2) Artificial Intelligence Systems: Mathematical Foundations and Applications.
According to the authors of the scientific works of this collection, artificial intelligence involves the development of intelligent algorithms that can learn from large amounts of data and be used to make decisions or predict various processes based on the use of this data. The works of the authors of the articles in this collection show how artificial intelligence can be used to automate processes, optimize decision-making, and increase the efficiency of human activity.

(3) Intelligent Data Analysis for Complex Systems and Processes.
The authors of the collection show that intelligent data analysis involves the analysis of data to reveal hidden patterns, trends, and connections in complex systems, which can provide a better understanding of the nature of their functioning. Specific applications illustrate the applications of the analysis of data

obtained from a wide range of sources, including social networks, mass media, and financial databases for decisions making.

By combining these three areas, the authors of the scientific works of this collection try to convey the idea of how to get a deep understanding of complex systems of various nature and develop the most effective solutions to complex problems.

The general narrative of the collection of scientific works is a combination of systems analysis, artificial intelligence, and intelligent data analysis with the aim of providing large companies, corporations, and other commercial and government structures with powerful tools to optimize operations, make better decisions, and stay ahead of competitors and improving the functioning of large social systems.

We would like to express our sincere appreciation to all authors for their contributions. We certainly look forward to working with all contributors again in nearby future.

Kyiv, Ukraine
May 2023

Michael Zgurovsky
Nataliya Pankratova

Contents

System Analysis of Processes and Phenomena of Different Nature

Approach to Development of Digital Twin Model for Cyber-Physical System in Conditions of Conceptual Uncertainty 3
Nataliya Pankratova and Igor Golinko

Constructing Mathematical Models for Actuarial Processes 27
Petro Bidyuk, Yoshio Matsuki, and Vira Huskova

Functional Planning Optimization of Exploiting Underground Space in Large Cities Using System Methodology 43
Hennadii Haiko, Illia Savchenko, and Yurii Haiko

Analysis and Modelling of the Underground Tunnel Planning in Uncertainty Conditions 63
Nataliya Pankratova, Vladymyr Pankratov, and Danylo Musiienko

Stabilization of Impulse Processes of the Cognitive Map of Cryptocurrency Usage with Multirate Sampling and Coordination Between Some Nodes Parameters 83
Viktor Romanenko, Yurii Miliavskyi, and Heorhii Kantsedal

Wireless Sensor Networks for Healthcare on SoA 101
Anatolii Petrenko and Oleksii Petrenko

Improving Predictive Models in the Financial Sector Using Fractal Analysis 117
Alexey Malishevsky

K-LR Modeling with Neural Economy and Its Utilization in Unclear Data 133
Glib Mazhara and Kateryna Boiarynova

Artificial Intelligence Systems: Mathematical Foundations and Applications

Formalization and Development of Autonomous Artificial Intelligence Systems 153
Pavlo Kasyanov and Liudmyla Borysivna Levenchuk

Attractors for Differential Inclusions and Their Applications for Evolution Algorithms 165
Nataliia Gorban, Oleksiy Kapustyan, Pavlo Kasyanov, and Bruno Rubino

System Analysis and Method of Ensuring Functional Sustainability of the Information System of a Critical Infrastructure Object 177
Oleg Barabash, Valentyn Sobchuk, Andrii Musienko, Oleksandr Laptiev, Volodymyr Bohomia, and Serhii Kopytko

Intellectual Data Analysis and Machine Learning Approaches for Car Insurance Rates Reconstruction 193
Pavlo Pustovoit and Olga Kupenko

On Generation of Daily Cloud-Free Satellite Images at High Resolution Level 203
Natalya Ivanchuk, Peter Kogut, and Petro Martyniuk

Computational Intelligence for Digital Healthcare Informatics 233
Abdel-Badeeh M. Salem

Continuous and Convex Extensions Approaches in Combinatorial Optimization 257
Sergiy Yakovlev and Oksana Pichugina

Action Encoding in Algorithms for Learning Controllable Environment 271
Andrii Tytarenko

Comparison of Constrained Bayesian and Classical Methods of Testing Statistical Hypotheses in Sequential Experiments 289
Kartlos Kachiashvili, Vakhtang Kvaratskhelia, and Archil Prangishvili

Investigation of Artificial Intelligence Methods in the Short-Term and Middle-Term Forecasting in Financial Sphere 307
Yuriy Zaychenko, Helen Zaichenko, and Oleksii Kuzmenko

Neo-Fuzzy Radial-Basis Function Neural Network and Its Combined Learning 323
Yevgeniy Bodyanskiy, Olha Chala, Valentin Filatov, and Iryna Pliss

Investigations of Different Classes Hybrid Deep Learning Networks and Analysis of Their Efficiency in Forecasting 341
Galib Hamidov and Yuriy Zaychenko

Generalized Models of Logistics Problems and Approaches to Their Solution Based on the Synthesis of the Theory of Optimal Partitioning and Neuro-Fuzzy Technologies 355
Anatolii Bulat, Elena Kiseleva, Liudmyla Hart, and Olga Prytomanova

Intelligent Data Analysis for Complex Systems and Processes

Technological Principles of Using Media Content for Evaluating Social Opinion 379
Michael Zgurovsky, Dmytro Lande, Oleh Dmytrenko, Kostiantyn Yefremov, Andriy Boldak, and Artem Soboliev

Scenario Modelling in the Context of Foresight Studies 397
Serhiy Nayev, Iryna Dzhygyrey, Kostiantyn Yefremov, Ivan Pyshnograiev, Andriy Boldak, and Sergii Gapon

Assessing the Development of Energy Innovations and Its Impact on the Sustainable Development of Countries 419
Maryna Kravchenko, Olena Trofymenko, Kateryna Kopishynska, and Ivan Pyshnograiev

Studies of the Intercivilization Fault Level Dynamics 439
Michael Zgurovsky, Maryna Kravchenko, and Ivan Pyshnograiev

Exploring the Vulnerability of Social Media for Crowdsourced Intelligence Under a False Flag 455
Illia Varzhanskyi

System Analysis of Processes and Phenomena of Different Nature

Approach to Development of Digital Twin Model for Cyber-Physical System in Conditions of Conceptual Uncertainty

Nataliya Pankratova and Igor Golinko

Abstract The approach to the digital twin model development for cyber-physical system as a physical process for which there is an analytical mathematical description in conditions of conceptual uncertainty is presented. Several models of digital twins for the process of air heating on the electric heater are given. The proposed approach takes into account the conceptual uncertain parameters of the analytical model with subsequent passive identification of the model by controlling the dynamic characteristics of the physical process. The algorithm of passive identification with the quadratic deviation minimization of the mathematical model results from the dynamics of the functioning physical process results is proposed. It is shown that the analytical model adaptation of physical process is a one-extreme optimization problem for which numerical methods of uncertain parameters search are applicable. Simulation results confirmed the effectiveness of the proposed methodology for developing a digital twin model of a physical process for enterprise cyber-physical systems.

Keywords Mathematical model · Uncertain parameters · State space · Digital twin · Identification · Quality criterion · Cyber-physical system · Electric heater

1 Introduction

Today, production management needs Industry 4.0 competencies, which have appeared relatively recently and largely relate to IT technologies. The classical pyramidal methodology of management system design is being superseded by modern approaches of direct interaction between system components based on the concept

N. Pankratova (✉) · I. Golinko
Institute for Applied System Analysis, Igor Sikorsky Kyiv Polytechnic Institute, Peremogy avenue 37, Kyiv 03056, Ukraine
e-mail: natalidmp@gmail.com

I. Golinko
e-mail: conis@ukr.net

M. Zgurovsky and N. Pankratova (eds.), *System Analysis and Artificial Intelligence*, Studies in Computational Intelligence 1107,
https://doi.org/10.1007/978-3-031-37450-0_1

of Industry 4.0. The digital twin refers to an innovative new toolkit that helps exploit advanced scenarios of the Internet of Things (IoT) and other technologies. This toolkit is used to create digital models of physical objects. These physical objects can be buildings, factories, cities, power grids, transportation systems, and more.

The term digital twin is used in a variety of contexts and it is expanding as companies and technologies evolve. With the availability of cloud technology, innovations in modeling, IoT platform capabilities, and improved interoperability with sensors, companies have begun to invest heavily in the development of digital twin tools.

According to the Industrial Internet Consortium (IIC) [1], a digital twin is a formal digital representation of some asset, process or system that captures attributes and behaviors of that entity suitable for communication, storage, interpretation or processing within a certain context. Designing digital twins is based on the use of simulation modeling techniques that provide the most realistic representation of the physical environment or object in the virtual world. Mathematical description of digital twins can be obtained using statistical modeling, machine learning, or analytical modeling techniques.

Methods of statistical modeling can be divided into three groups [2]: models of regression analysis, models of classification and models of anomalies detection. The choice of method depends essentially on the size, quality and data nature, as well as on the type of problems to be solved and the knowledge of the process being modeled. For man-made processes in cyber-physical systems (CPS), analytical models are often used, which have valuable properties in engineering [3]. In [4], the development of CPS using deterministic models, which have proven to be extremely useful, is discussed. Deterministic mathematical models of CPS are based on differential equations and include synchronous digital logic and single-threaded imperative programs. However, CPS combines these models in such a way that determinism is not preserved.

One of the fundamental works on the digital twin's standardization is the Industrial Internet Reference Architecture (IIRA) reference model proposed by IIC [5]. The document describes guidelines for the development of systems, solutions and applications using IoT in infrastructure solutions and industry. This architecture is abstract, and provides general definitions for various stakeholders, system decomposition order, design patterns, and terms glossary. The architectural framework of the model under consideration is shown in Fig. 1.

The IIRA model identifies at least four stakeholder strands: business; usage; operation; implementation. Each strand focuses on the implementation of the digital twin functional model, the structure, interfaces and interactions between the digital twin components, and the interaction of the digital twin model's system with external elements of the environment to support CPS functioning. Based on the IIRA model, digital twin information includes (but is not limited to) a combination of categories: physical model and data; analytical model and data; temporal variable archives; transactional data; master data; and visual models and calculations. Digital twins are created for widespread use with simulation and optimization methods, databases, etc. to implement rational management of a physical process in real time and support strategic and operational decision making. Thus, the digital twin concept has

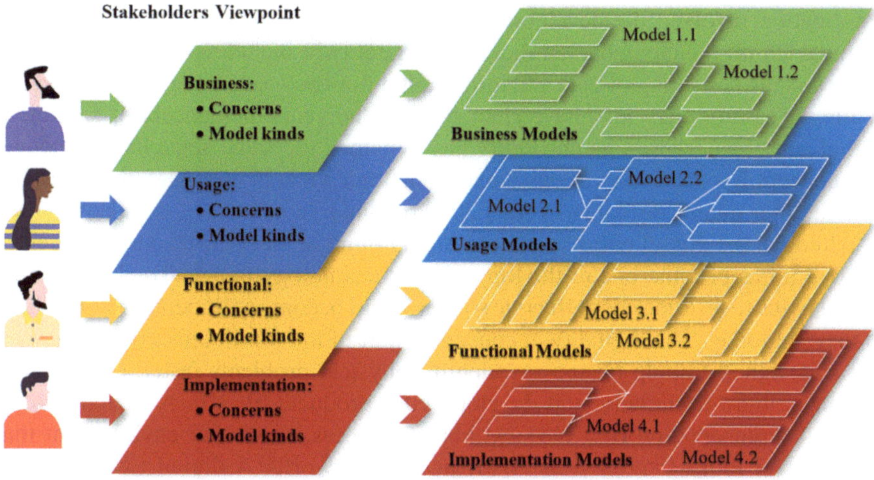

Fig. 1 The digital twin model architecture according to IIC version

a multifaceted architecture and correspondingly complex mathematical support for implementation.

Considering the above in this paper is proposed to develop a digital twin model for cyber-physical systems by applying the developed analytical model of air heating on an electric heater in conditions of conceptual uncertainty. The problem of conceptual uncertainty disclosure in its contentive statement is reduced to a problem of system-coordinated disclosure of a set of diverse uncertainties on the basis of unified principles, techniques, and criteria. This set includes the uncertainty of parameters for each type of electric heater, uncertainty of its physical and mechanical characteristics, situational uncertainty of risks in the process of development and its operation. Such an uncertainty refers to the conceptual one in the sense that, as distinct from information uncertainty, it represents a unified complex of the lack of information, ambiguity, and contradictoriness of interconnected and interdependent elements of a specified set of polytypic uncertainties [6].

2 Related Papers

The first digital twin in history can be considered a program used by NASA in the 1960s to design the Apollo 13 mission. It was created to test how a future object would behave in the physical world. Later, NASA engineers discovered that the same twin could be used to control existing equipment and predict what would happen to it take real readings from sensors. This later led to modern digital twin [7]. The digital twin concept gained recognition in 2002 after Challenge Advisory has hosted a presentation for Michael Grieves [8] in the University of Michigan [7]. The presentation involved the development of a product lifecycle management center. It contained all the elements familiar for a digital twin including real space, virtual space and the spreading of data and information flow between real and virtual space. While the terminology may have changed over the years the concept of creating a digital and physical twin as one entity has remained the same since its emergence. The concept has received various names, but John Vickers of NASA proposed the now-familiar name "digital twin" in a 2010 roadmap report.

Although the digital twins have been highly familiar since 2002, only as recently as 2017 it has become one of the top strategic technology trends. The IoT enabled digital twins to become cost-effective so they could become as imperative to business as they are today. The digital twin concept is part of the fourth industrial revolution and is designed to help businesses identify physical problems faster, predict their outcomes more accurately, and develop better products [9, 10]. A digital twin is a digital representation of a physical object, process, service or environment that behaves and looks like its counterpart in the real-world. A digital twin can be a digital replica of an object in the physical world, such as a jet engine or wind farms, or even larger items such as buildings or even whole cities, alternatively digital twin technology can be used to replicate processes in order to collect data to predict how they will perform [11]. In more detail we can say so: it is a software analogue of the physical device, simulating the internal processes, technical characteristics and behavior of the real object under the influence of interference and environment. An important feature of a digital twin is that it uses information from the sensors of a real device operating in parallel to set input influences on it.

What is meant is that a digital twin is a set of parameters that exhaustively describe a physical object or process. Often it looks very clear, because for convenience the digital twin is visualized on the screen. We get a full-fledged digital copy of a real object to work in special software. A single item, an entire technological process, an enterprise or even an entire industry can get its "twin".

Until recently, digital modeling was static. To accurately simulate processes, you had to capture and adjust data manually. Today, things have changed: IoT tools, open APIs, artificial intelligence, and Big Data tools make digital twins as responsive as possible - the model can automatically update based on constantly incoming data. This makes modeling much easier and more accurate. Through a systematic literature review and a thematic analysis of 92 digital twin publications from the last ten years, the paper [12] provides a characterization of the digital twin, identification of gaps in knowledge, and required areas of future research. In characterizing the digital twin, the state of the concept, key terminology, and associated processes are identified, discussed, and consolidated to produce 13 characteristics (Physical Entity/Twin; Virtual Entity/Twin; Physical Environment; Virtual Environment; State; Realization; Metrology; Twinning; Twinning Rate; Physical-to-Virtual Connection/Twinning; Virtual-to-Physical Connection/Twinning; Physical Processes; and Virtual Processes) and a complete framework of the digital twin and its process of operation. Following this characterization, seven knowledge gaps and topics for future research focus are identified: Perceived Benefits; Digital Twin across the Product Life-Cycle; Use-Cases;

Technical Implementations; Levels of Fidelity; Data Ownership; and Integration between Virtual Entities; each of which are required to realize the digital twin.

Digital twin is currently a term applied in a wide variety of ways. Some differences are variations from sector to sector, but definitions within a sector can also very significantly. Within engineering, claims are made regarding the benefits of using digital twinning for design, optimization, process control, virtual testing, predictive maintenance, and lifetime estimation. In many of its usages, the distinction between a model and a digital twin is not made clear. However, many interesting open questions exist, some connected with the volume and speed of data, some connected with reliability and uncertainty, and some to do with dynamic model updating. In paper [13] are considered the essential differences between a model and a digital twin, outline some of the key benefits of using digital twins, and suggest directions for further research to fully exploit the potential of the approach.

The paper [14] reviews the latest state of the art in methodologies and techniques related to the development of digital twins, primarily from a modeling perspective. Current challenges and best practices are also covered in detail, along with recommendations and considerations for various stakeholders.

Digital twin can facilitate interaction between the physical and the cyber worlds and achieve smart manufacturing. However, the digital twin's development in the industry remains vague. This study investigates the global patent databases of digital twin patents and summarizes related technologies, effects, and applications. Patent map analysis is used to uncover the patent development trajectory of digital twin in the patent databases of the USA, China, and the World Intellectual Property Organization among European nations [15].

In 2023 year, the digital twin concept is evolving into something much slicker and incredibly practical the executable digital twin (xDT). In simple terms, the xDT is the digital twin on a chip. The xDT uses data from a (relatively) small number of sensors embedded in the physical product to perform real-time simulations using reduced-order models. From those small numbers of sensors, it can predict the physical state at any point on the object (even in places where it would be impossible to place sensors) [16].

3 Approach to the Development of Digital Twin Model for Cyber Physical System Heater in Conditions of Conceptual Uncertainty

The approach to the development of digital twin mathematical model to predict and support the functioning of the real physical process is proposed. This takes into account the parametric uncertainty of the physical process mathematical description. As an example, the digital twin model development is given on the example of the electric heater analytical model. Generalized structural scheme of digital twin

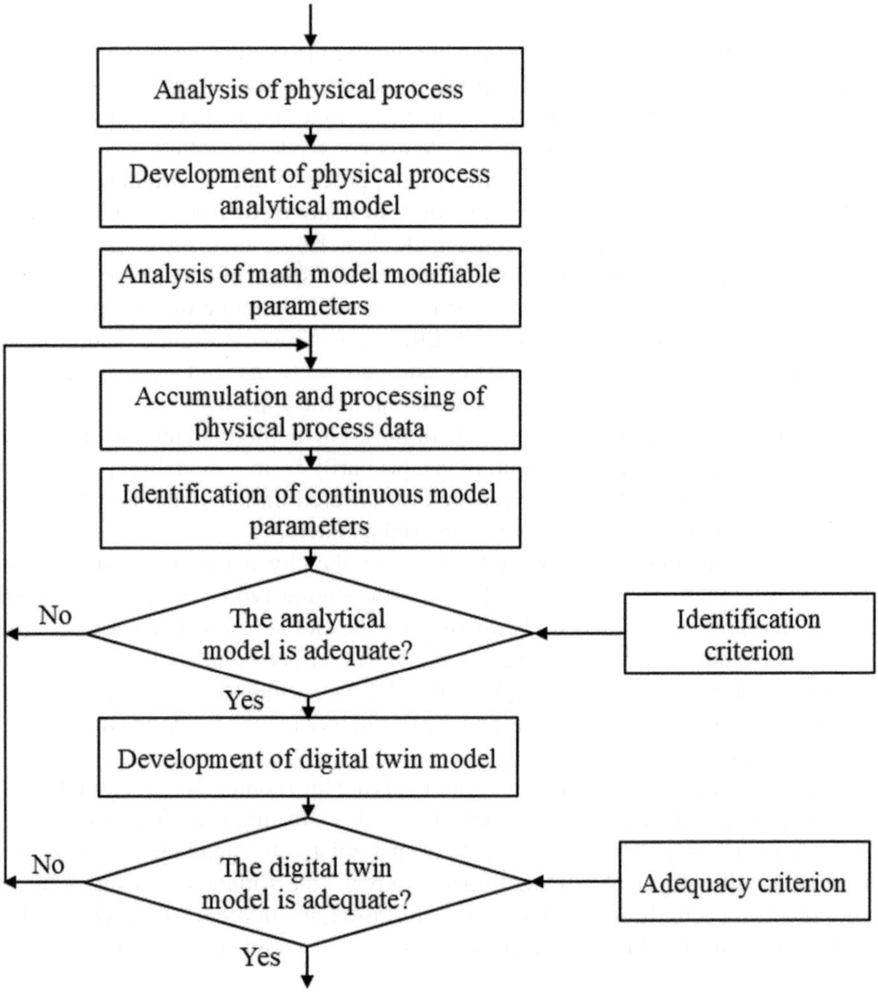

Fig. 2 The structural scheme of the digital twin model development procedure

model development procedure based on analytical model of physical process [17] is presented in Fig. 2.

In the first stage of development, it is necessary to conduct a literature analysis in the applied field of research for the physical process. This will help to determine the model structure, existing advantages and disadvantages. As a rule, the analytical model of the studied process has the system form of differential, difference or algebraic equations.

The practice of using analytical models shows that ready-to-use models are very rare. Even tested models require adjustment of parameters in order to adapt them to specific conditions of use. Thus, when a digital twin model is developed for a particular physical process, the researcher needs to determine the uncertainty "physical limits of that process" in the numerical values form of mathematical model parameters. To do this, the researcher needs to perform passive identification of the mathematical model parameters. A very important role at this stage is played by the data quality for the model identification, so the formation of the database should be guided by the known requirements of informativeness, synchroneity and correctness [2].

The last step in the digital twin model development is the identified model discretization. Here it is necessary to set correctly the sampling time for the mathematical model. On the one hand, the sampling time should not be small in order to ensure the information distribution over the CPS network. On the other hand, a large sampling time will lead to the loss of intermediate information for short-term forecasts. The obtained numerical model, even a sufficiently adequacy high degree, does not yet guarantee a prediction estimates high quality if the basic uncertainties for the mathematical model of the physical process are not taken into account. Therefore, after designing the digital twin model, it is necessary to check the possibility of using it for solving the assigned forecasting tasks.

An important characteristic for a digital twin model is to determine the received prediction quality. Often, the quality of forecast estimates is determined with the help of least squares method (LSM). However, LSM is one of many possible statistics that depends on the data scale. Therefore, only this characteristic is not enough for the analysis of a qualitative prediction. The quality of linear and pseudo-linear models is assessed using several statistical quality criteria [18], since each criterion has its own specific purpose and characterizes one property of a prediction evaluation. Therefore, a digital twin model developer must comprehensively study a physical process, existing mathematical models, possible perturbing effects on a physical process and justify a use of adequacy criteria for a digital twin in conditions of conceptual uncertainty.

4 Models and Methods

4.1 The Analytical Model of Air Heating on an Electric Heater. Evaluation of the Model Parameters Influence on the Calculation's Accuracy

The analytical model of air heating on an electric heater is considered, which is proposed in [17]:

$$\begin{cases} T_E \dfrac{d\Delta\theta_E}{dt} + \Delta\theta_E &= k_0 \Delta N_E + k_1 \Delta\theta_A, \\ T_A \dfrac{d\Delta\theta_A}{dt} + \Delta\theta_A &= k_2 \Delta\theta_E + k_3 \Delta\theta_{A_0} + k_4 \Delta G_A, \\ T_d \dfrac{d\Delta d_A}{dt} + \Delta d_A &= k_5 \Delta d_{A_0} + k_6 \Delta G_A; \end{cases} \tag{1}$$

here $T_E = \dfrac{c_E M_E}{K_E}$, $K_E = \alpha_0 F_0$, $k_0 = \dfrac{1}{K_E}$, $k_1 = 1$; $T_A = \dfrac{c_A M_A}{K_A}$, $K_A = c_A G_A + \alpha_0 F_0$, $k_2 = \dfrac{\alpha_0 F_0}{K_A}$, $k_3 = 1 - k_2$, $k_4 = \dfrac{c_A \left(\theta_{A_0} - \theta_A\right)}{K_A}$; $T_d = \dfrac{\omega V_A}{G_A}$, $k_5 = 1$, $k_6 = \dfrac{d_{A_0} - d_A}{G_A}$. To solve the system of differential equations (1), the state space can be used:

$$\mathbf{X}' = \mathbf{AX} + \mathbf{BU},$$
$$\mathbf{X} = \begin{bmatrix} \Delta\theta_A \\ \Delta d_A \\ \Delta\theta_E \end{bmatrix}, \mathbf{A} = \begin{bmatrix} -1/T_A & 0 & k_2/T_A \\ 0 & -1/T_d & 0 \\ k_1/T_E & 0 & -1/T_E \end{bmatrix},$$
$$\mathbf{B} = \begin{bmatrix} k_3/T_A & 0 & k_4/T_A & 0 \\ 0 & k_5/T_d & k_6/T_d & 0 \\ 0 & 0 & 0 & k_0/T_E \end{bmatrix}, \mathbf{U} = \begin{bmatrix} \Delta\theta_{A_0} \\ \Delta d_{A_0} \\ \Delta G_A \\ \Delta N_E \end{bmatrix}; \tag{2}$$

or the Laplace transform

$$\mathbf{Y}(p) = \mathbf{W}(p)\mathbf{U}(p),$$
$$\mathbf{Y} = \begin{bmatrix} \Delta\theta_A \\ \Delta d_A \end{bmatrix}, \mathbf{W} = \begin{bmatrix} W_{11} & 0 & W_{13} & W_{14} \\ 0 & W_{22} & W_{23} & 0 \end{bmatrix}, \mathbf{U} = \left[\Delta\theta_{A_0}\ \Delta d_{A_0}\ \Delta G_A\ \Delta N_E\right]^{\mathrm{T}}, \tag{3}$$

here $W_{11} = \dfrac{b_1 p + b_0}{a_2 p^2 + a_1 p + 1}$, $W_{13} = \dfrac{b_3 p + b_2}{a_2 p^2 + a_1 p + 1}$, $W_{14} = \dfrac{b_4}{a_2 p^2 + a_1 p + 1}$,

$W_{22} = \dfrac{k_5}{T_d p + 1}$, $W_{23} = \dfrac{k_6}{T_d p + 1}$; $a_2 = \dfrac{T_E T_A}{1 - k_1 k_2}$, $a_1 = \dfrac{T_E + T_A}{1 - k_1 k_2}$;

$b_0 = \dfrac{k_3}{1 - k_1 k_2}$, $b_1 = \dfrac{k_3 T_E}{1 - k_1 k_2}$, $b_2 = \dfrac{k_4}{1 - k_1 k_2}$, $b_3 = \dfrac{k_4 T_E}{1 - k_1 k_2}$, $b_4 = \dfrac{k_0 k_2}{1 - k_1 k_2}$.

The parameters classification of the mathematical model (1) is shown in Fig. 3. The numerical values analysis of the model parameters (1) allows to conclude that the thermophysical quantities of material flows and structural materials of the electric heater are determined with high accuracy from handbooks on the thermophysical properties of substances and materials, these parameters refer to blocks 1–3 in Fig. 3.

The modifiable in general case uncertain parameters of the model there are in Block 4. Formally for model (1) there are six changing parameters α_0, G_A, θ_{A_0}, θ_A,

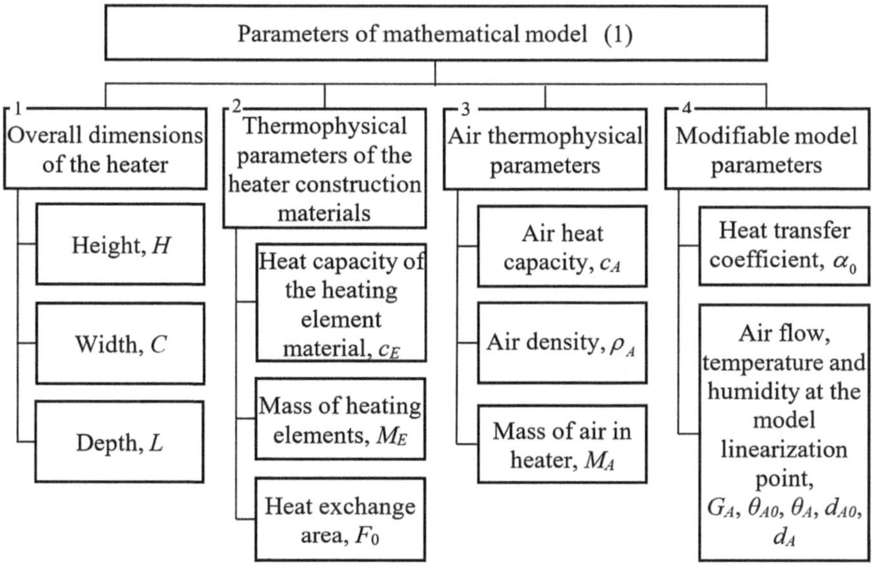

Fig. 3 The analytical model parameters classification of electric heater

d_{A_0}, d_A from which all coefficients of mathematical models (1), (2), (3) depend. The task of finding the numerical values of the parameters α_0, G_A, θ_{A_0}, θ_A, d_{A_0}, d_A should be solved in conditions of conceptual uncertainty [6], because these parameters are linked into a single complex of interrelated parameters of the mathematical model. For example, an increase in air flow G_A leads to an increase in the heat transfer coefficient α_0, exactly the same effect can be achieved by increasing the air humidity d_{A_0}. Thus, the same numerical value α_0 has an infinite number of combinations of the considered parameters. To solve such problems, it is necessary to formulate a strategy for developing an analytical model, which will later be used to design a digital twin model. In these conditions, a choice of an analytical model from the available set, the uncertain parameters analysis of the model and their justification, the method of identifying uncertain parameters is a nonformalizable procedure, and only the researcher can carry it out. The result depends on the competence, skill, experience, intuition, and other individual qualities of an actual researcher who is carrying out the given procedure.

For example, for models (1), (2), (3), the list of uncertain parameters is six, which complicates the search. However, the parameters G_A, θ_{A_0}, θ_A, d_{A_0}, d_A can be measured using CPS sensors and thus solve the problem of the uncertainty of these quantities. In this case, for models (1), (2), (3) it is necessary to reveal the uncertainty of only one parameter α_0, which greatly simplifies the search task.

Note that heat transfer coefficient α_0 depends on many factors and there is no sensor to measure this parameter. Numerical value α_0 depends on: temperature of heating element θ_E; temperature θ_A, humidity d_A and air flow G_A; design features of heat exchange surface and many other factors. This parameter is a subject of research in thermal engineering, which are based on experimental studies and similarity theory [19].

Let's consider the influence analysis of heat transfer coefficient α_0 on dynamic properties of electric heater HE 36/2. In modeling, let's assume that the basic static mode parameters of electric heater G_A, θ_{A_0}, θ_A, d_{A_0}, d_A are determined, using CPS sensors. In calculations for model (2), let's assume that by reducing the flow rate of heated air from $G_A = 0.43kg/s$ to $\overline{G}_A = 0.2kg/s$ (the new nominal operating mode of the heater, all other conditions being equal), the heat transfer coefficient has changed from $\alpha_0 = 161W/\left(m^{2\circ}C\right)$ to $\overline{\alpha}_0 = 100W/\left(m^{2\circ}C\right)$. In this case, numerical values of model matrices (2) will change:

$$\text{from } \mathbf{A} = \begin{bmatrix} -2.631 & 0 & 0.268 \\ 0 & -2.357 & 0 \\ 0.129 & 0 & -0.178 \end{bmatrix}, \mathbf{B} = \begin{bmatrix} 2.362 & 0 & -21.98 & 0 \\ 0 & 2.358 & 0 & 0 \\ 0 & 0 & 0 & 0.0036 \end{bmatrix}; \tag{4}$$

$$\text{to } \overline{\mathbf{A}} = \begin{bmatrix} -1.265 & 0 & 0.167 \\ 0 & -1.097 & 0 \\ 0.111 & 0 & -0.111 \end{bmatrix}, \overline{\mathbf{B}} = \begin{bmatrix} 1.099 & 0 & -21.98 & 0 \\ 0 & 1.097 & 0 & 0 \\ 0 & 0 & 0 & 0.0036 \end{bmatrix}. \tag{5}$$

The simulation results are shown in Fig. 4. In this figure and further the following designations are used: the output vector $\mathbf{X}$ for the reference mathematical model (4), for calculation of which the initial values of parameters α_0 and G_A were used; the output vector $\overline{\mathbf{X}}$ for the model (2), where new values of parameters $\overline{\alpha}_0$ and $\overline{G}_A$ were used. The output vectors $\mathbf{X}$ and $\overline{\mathbf{X}}$ contain the variables: $\Delta\theta_A$ and $\Delta\overline{\theta}_A$ are temperature of heated air; Δd_A and $\Delta\overline{d}_A$ are moisture content of heated air; $\Delta\theta_E$ and $\Delta\overline{\theta}_E$ are temperature of heating elements.

When modeling the transients for the electric heater HE 36/2, the control action $\mathbf{U}$ is taken into account by a step change in the input variables. The input vector $\mathbf{U}$ contains the variables: $\Delta\theta_{A_0}$, Δd_{A_0} are temperature and moisture content of the air at the electric heater inlet; ΔG_A is air flow through the electric heater; ΔN_E is electric power supplied to the electric heater elements.

Similar simulation results can be obtained for model (3). Based on the simulation results, it can be concluded that the heat transfer coefficient α_0 affects the numerical values of the matrices $\mathbf{A}$ and $\mathbf{B}$ of the model (2). Dynamic characteristics of the air heating process significantly depend on this parameter, which is demonstrated in the simulation graphs (see Fig. 3). Specialists in the modeling area of heat exchange processes should accurately determine this parameter, since its numerical value significantly affects the properties of the studied process.

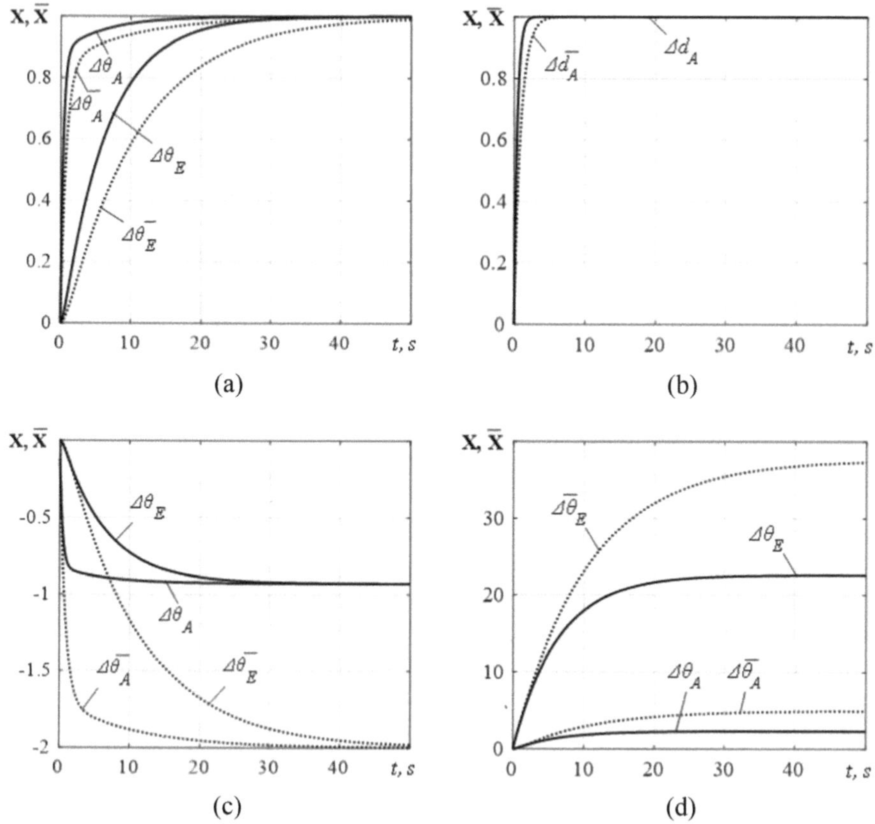

Fig. 4 Influence of the heat transfer coefficient α_0 on the dynamic characteristics of the heater HE 36/2 by the influence channels $\mathbf{U} \to \mathbf{X}$ (the graphs shows the output variables that have received increments as the result of changing α_0): **a** $\Delta\theta_{A_0} \to \mathbf{X}$, $\Delta\theta_{A_0} = 1°C$; **b** $\Delta d_{A_0} \to \mathbf{X}$, $\Delta d_{A_0} = 1$ g/kg; **c** $\Delta G_{A_0} \to \mathbf{X}$, $\Delta G_{A_0} = 0.1$ kg/s; **d** $\Delta N_E \to \mathbf{X}$, $\Delta N_E = 1$ kW

4.2 Identification of Mathematical Model Parameters

Analytical model (1) is obtained using the studied laws of heat and mass transfer. For adapting the model to the concrete process of air heating, specification of its parameters is required. Identification of model parameters can be carried out using active or passive experiment on the operating equipment. Passive identification methods are most often used, since there is no need to expend additional production resources in the course of the experiment and this approach is justified by its cost-effectiveness. To identify the model in the state space (2), the Kalman filter is most often used, and for the model (3), the LSM and its modifications are used [18]. In our case we will apply LSM, taking into account its universality.

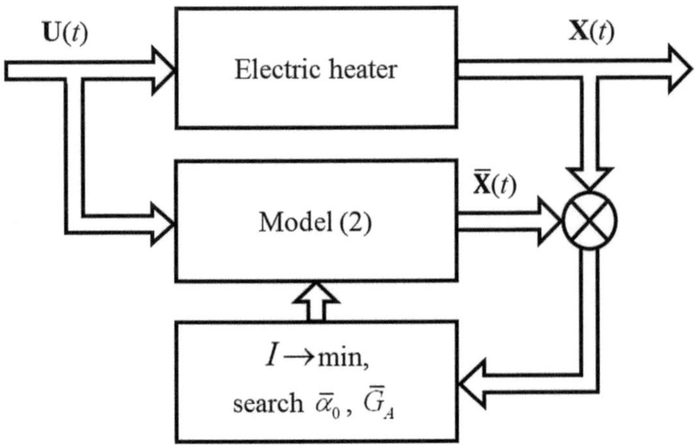

Fig. 5 The schematic diagram for identification of mathematical model

Formally, model (1) has six changing parameters α_0, G_A, θ_{A_0}, θ_A, d_{A_0}, d_A (see Fig. 3). In the passive identification process, it is sufficient to specify α_0, G_A. Other uncertain parameters of the electric heater model θ_{A_0}, θ_A, d_{A_0}, d_A can be estimated using the CPS sensors. As an identification criterion, we use a dependence that minimizes the square of the error of the state variables measured values of the physical process $\mathbf{X}$ and the output vector $\overline{\mathbf{X}}$ estimates of the model being identified (2) at the same input $\mathbf{U}$

$$I = M\left\{\int_{t_0}^{t_0+t_f} \left(\mathbf{X}-\overline{\mathbf{X}}\right)^{\mathrm{T}} \mathbf{Q}\left(\mathbf{X}-\overline{\mathbf{X}}\right) dt\right\} \to \min, \tag{6}$$

here t_0 is the initial time of the trend, t_f is the duration of the trend sampling, $\mathbf{Q}$ is the unit square matrix, T is the matrix transpose operator, M is the mathematical expectation operator, which takes into account industrial perturbations. A generalized structural diagram of the passive identification process is shown in Fig. 5. The search criterion (6) is calculated during the implementation of the identification algorithm. Therefore, it is recommended to use numerical zero-order optimization methods to identify model parameters [20]. Numerical methods require significant computational resources, so the identification algorithm can be implemented within a decision support system (DSS) [21].

The algorithm for parametric identification of mathematical model consists of the following steps:

(1) the CPS sensors monitor vector of input influence $\mathbf{U}(t)$ and output state $\mathbf{X}(t)$ of the physical process in real time, the initial data are processed in the DSS and the database of input and output states for the physical process is formed;
(2) the output state $\overline{\mathbf{X}}(t)$ of the mathematical model with the initial values of the parameters $\overline{\alpha}_0$ and $\overline{G}_A$ is estimated by the time trends of the input action $\mathbf{U}(t)$;

(3) the identification criterion (6) is defined, using vectors of output $\mathbf{X}(t)$ and identifiable $\overline{\mathbf{X}}(t)$ the physical process states;
(4) the parameters $\overline{\alpha}_0$ and $\overline{G}_A$ of the identified model (2) are numerically optimized using criterion (6);
(5) if the minimum of criterion (6) is found then proceed to the design of the digital twin, else continue to minimize the identification criterion and go to step 2.

In the identification process it is necessary to take into account the numerical method peculiarities for minimization of criterion (6). Also, for qualitative identification the time trends t_f of input and output states should be several times longer than the physical process transient's duration. To eliminate overflow effects of the numerical minimization algorithm of criterion (6), it is necessary to set search limits for identifiable parameters $\overline{\alpha}_0$ and $\overline{G}_A$, based on mathematical model physical feasibility.

4.3 Results Estimation of Identification of Mathematical Model Parameters

The proposed numerical identification algorithm was investigated using MatLAB. The reference model (2) with numerical values of matrices **A** and **B** (4) was used to form the time trends of the operating electric heater $\mathbf{X}(t)$. To simulate production disturbances, a random signal with an amplitude of ± 0.2 was mixed into the reference output variables of the vector $\mathbf{X}(t)$. In the identified model, initial values of the parameters $\overline{\alpha}_0$, $\overline{G}_A$ differed significantly from the reference values α_0, G_A, and numerical values of matrices $\overline{\mathbf{A}}$ and $\overline{\mathbf{B}}$ were used (5). For the numerical identification of $\overline{\alpha}_0$ and $\overline{G}_A$, the MatLAB function fminsearch(...) was used, where the simplex Nelder-Mead optimization method is applied. The main results of the numerical study are presented in Figs. 6 and 7.

Figure 6 shows the case when the initial conditions of the reference model $\mathbf{X}(0) = [2, 1, 15]^{\mathrm{T}}$ and the identified model $\overline{\mathbf{X}}(0) = [0, 0, 0]^{\mathrm{T}}$ are significantly different. Also, the identification parameters are significantly different: $\alpha_0 = 161$, $G_A = 0.43$, $\overline{\alpha}_0 = 110$, $\overline{G}_A = 0.15$. With the step influence of the input vector were obtain the transients shown in Fig. 6a. The difference between the output values of the reference $\mathbf{X}(t)$ and the identifiable model $\overline{\mathbf{X}}(t)$ is quite significant. In Fig. 6b shows the isolines surface of criterion (6) and its minimization trajectory. It is seen that the functional has one extremum in the area of the model physical practicability, so any numerical optimization method can be used as an optimization method. According to the proposed identification algorithm $\overline{\alpha}_0 = 160.54$, $\overline{G}_A = 0.4$ were determined. The found parameter values are quite close to the reference ones $\alpha_0 = 161$, $G_A = 0.43$. Figure 6c shows the time characteristics of the state variables of the reference $\mathbf{X}(t)$ and the identified $\overline{\mathbf{X}}(t)$ model after identification under the input influence $\mathbf{U}(t) = [1, 1, -0.2, 1]^{\mathrm{T}}$ on both models. Based on the simulation results, it can be concluded that the proposed algorithm for passive identification of the electric heater

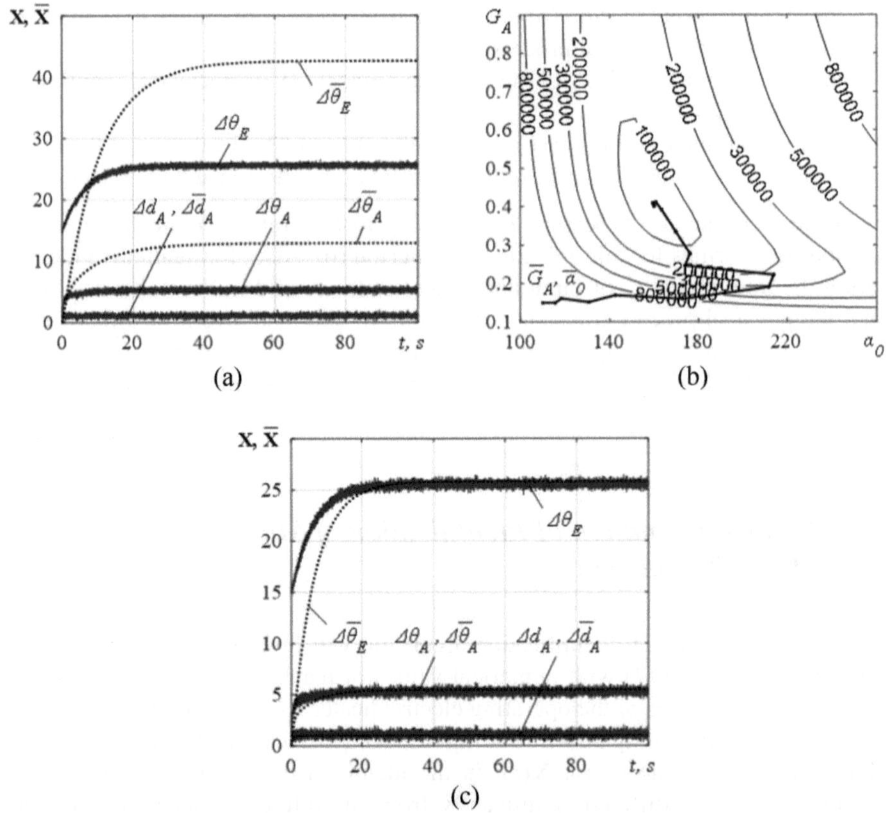

Fig. 6 The parametric identification of $\overline{\alpha}_0$ and $\overline{G}_A$ with step influence $\mathbf{U}(t) = [1, 1, -0.2, 1]^{\mathrm{T}}$: **a** the simulation of transients before identification; **b** the identification trajectory of parameters $\overline{\alpha}_0$ and $\overline{G}_A$ by criterion (6); **c** the simulation of transients after identification parameters $\overline{\alpha}_0$ and $\overline{G}_A$

has good convergence in the case of different initial conditions and the presence of random perturbations.

Figure 7a shows the case when the reference model is in the stationary state $\mathbf{X}(0) = [2.3, 0.22.5]^{\mathrm{T}}$ during the presence of random perturbations. This state is provided by the input influence vector $\mathbf{U}(t) = [0, 0, 0, 1]^{\mathrm{T}}$. The initial conditions of the identifiable model are zero $\overline{\mathbf{X}}(0) = [0, 0, 0]^{\mathrm{T}}$. According to the simulation condition, the parameters of the identified model $\overline{\alpha}_0 = 250$, $\overline{G}_A = 0.8$ are significantly different from the reference model $\alpha_0 = 161$, $G_A = 0.43$. Figure 7b shows the surface isolines of criterion (6) and its minimization trajectory. During the identification process, the values of the parameters $\overline{\alpha}_0 = 157.28$, $\overline{G}_A = 0.42$ are optimized. As in the first study, the found values of the parameters are quite close to the reference ones. Figure 7c depicts the temporal characteristics of the state variables of the reference $\mathbf{X}(t)$ and identifiable $\overline{\mathbf{X}}(t)$ model after identification under the input influence $\mathbf{U}(t) = [0, 0, 0, 1]^{\mathrm{T}}$. According to the modeling results can conclude that the pro-

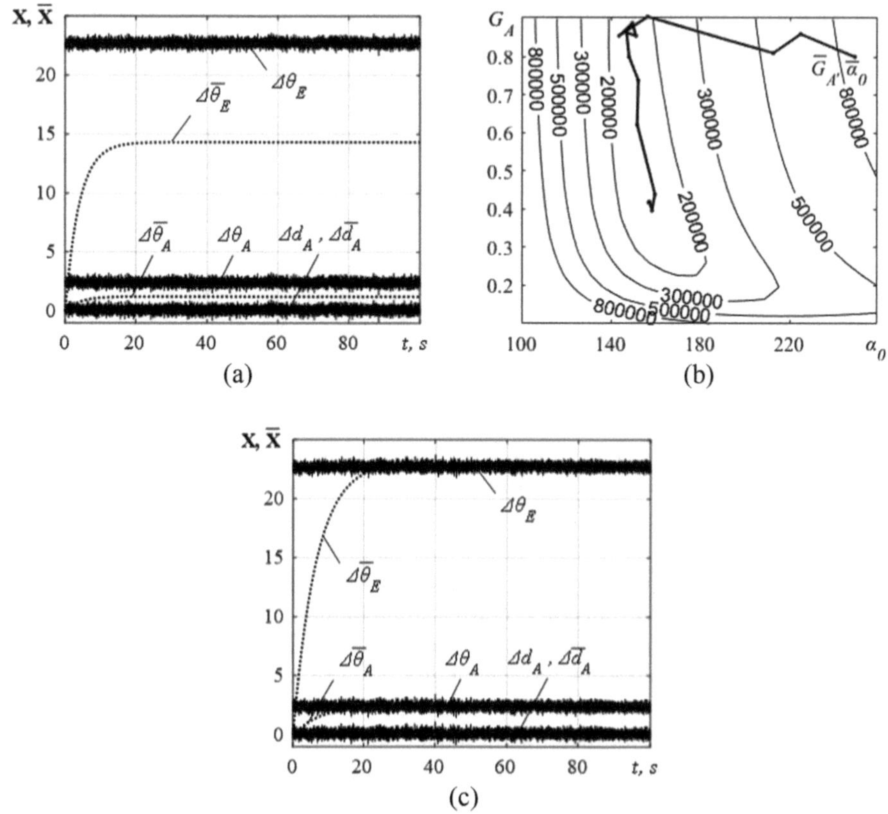

Fig. 7 The parametric identification of $\overline{\alpha}_0$ and $\overline{G}_A$ with step influence $\mathbf{U}(t) = [0, 0, 0, 1]^{\mathrm{T}}$: **a** the simulation of transients before identification; **b** the identification trajectory of parameters $\overline{\alpha}_0$ and $\overline{G}_A$ by criterion (6); **c** the simulation of transients after identification parameters $\overline{\alpha}_0$ and $\overline{G}_A$

posed passive identification algorithm has good convergence in the case of the object stationary and the random perturbations presence.

The identification algorithm considered above is universal. It is recommended to use it for identification of mathematical model (3). For this case it is necessary to take into account that the output state vector **Y** has two state variables and the identification criterion (6) will look like

$$I = M\left\{\int_{t_0}^{t_0+t_f} \left(\mathbf{Y} - \overline{\mathbf{Y}}\right)^{\mathrm{T}} \mathbf{Q}\left(\mathbf{Y} - \overline{\mathbf{Y}}\right) dt\right\} \to \min, \quad (7)$$

here **Y** is the vector of the output signal for the physical process, $\overline{\mathbf{Y}}$ is the vector of the output signal estimates for the identified model (3). Below is the example of parameter identification for a mathematical model (3). To form the time trends $\mathbf{Y}(t)$

of the functioning electric heater HE 36/2 the reference model (3) with the transfer matrix

$$\mathbf{W}(p) = \begin{bmatrix} \frac{5.6p+1}{2.4p^2+6.7p+1} & 0 & -\frac{52.1p+9.3}{2.4p^2+6.7p+1} & \frac{0.002}{2.4p^2+6.7p+1} \\ 0 & \frac{1}{0.42p+1} & 0 & 0 \end{bmatrix}. \quad (8)$$

$$\overline{\mathbf{W}}(p) = \begin{bmatrix} \frac{9p+1}{8.2p^2+11.3p+1} & 0 & -\frac{180.4p+20}{8.2p^2+11.3p+1} & \frac{0.005}{8.2p^2+11.3p+1} \\ 0 & \frac{1}{0.91p+1} & 0 & 0 \end{bmatrix}. \quad (9)$$

In the example the input signal $\mathbf{U}(t)$ has a harmonic component. Here, both models were subjected to the input influence $\mathbf{U}(t) = [1, 1 + 0.2\sin(0.2t), -0.2 + 0.1\sin(0.3t), 0.5 + 0.5\sin(0.1t)]^{\mathrm{T}}$. Figure 8a depicts the simulation case when the parameters of the identified model $\overline{\alpha}_0 = 300$, $\overline{G}_A = 0.25$ differ significantly from the reference ones $\alpha_0 = 161$, $G_A = 0.43$. Figure 8b shows surface isolines of criterion (7) and its minimization trajectory. During the identification process, the values of the parameters $\overline{\alpha}_0 = 161.9$, $\overline{G}_A = 0.426$ are optimized. The found numerical values of identification parameters are close enough to the reference ones. Figure 8c shows the time characteristics of the output vector variables of the reference $\mathbf{Y}(t)$ and the identified $\overline{\mathbf{Y}}(t)$ model after identification under the harmonic change $\mathbf{U}(t)$. According to the simulation results can conclude that the proposed passive identification algorithm has good convergence in the harmonic component presence of the vector $\mathbf{U}(t)$ and the random perturbations presence.

5 Digital Twin Development for the Air Heating Process by an Electric Heater

The digital twin development will be carried out on the basis of analytical models (2) and (3). The choice of mathematical model (2) or (3) depends on the management strategy applied methods at production. In order to obtain adequate computational data at the first stage of digital twin synthesis, it is necessary to identify the continuous model of electric heater, using the passive identification algorithm discussed above. Also, it is highly desirable to reduce the computational resources of the digital twin model. It is known from modeling theory that computational resources for modeling differential equations are more than for their discrete analogs based on difference

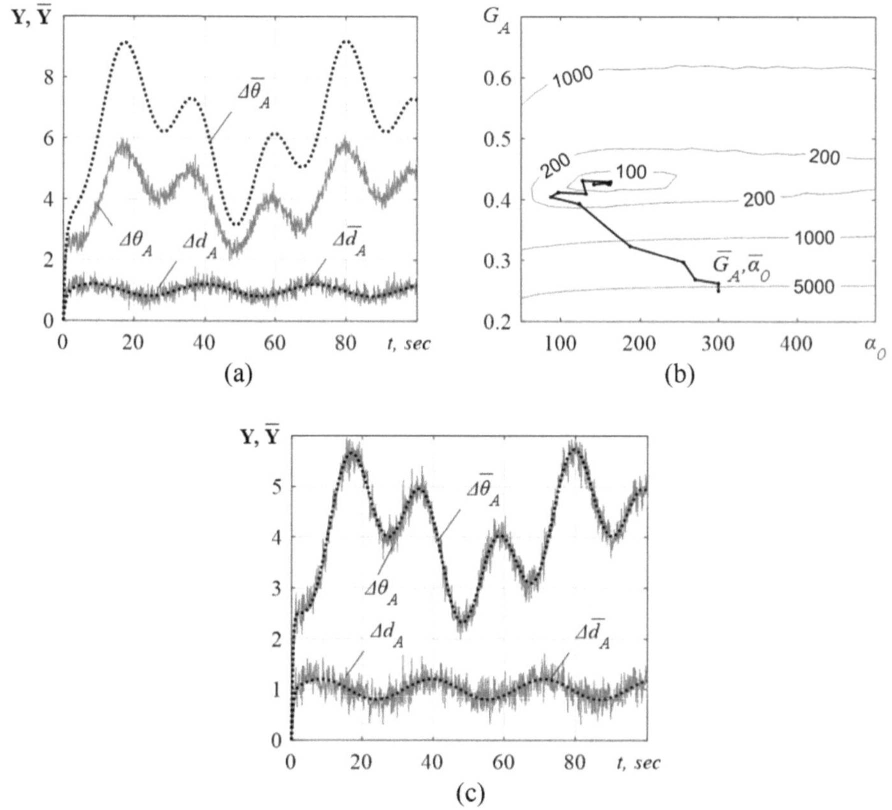

Fig. 8 The parametric identification of $\overline{\alpha}_0$ and $\overline{G}_A$ with harmonic influence $\mathbf{U}(t)$: **a** the simulation of transients before identification; the identification trajectory of parameters $\overline{\alpha}_0$ and $\overline{G}_A$ by criterion (7); the simulation of transients after identification parameters $\overline{\alpha}_0$ and $\overline{G}_A$

equations. Therefore, let us consider a discrete representation of continuous models (2) and (3).

The mathematical model (2) can be represented in a discrete form [22]

$$\overline{\mathbf{X}}_{k+1} = \overline{\mathbf{A}}_d\overline{\mathbf{X}}_k + \overline{\mathbf{B}}_d\mathbf{U}_k, \tag{10}$$

here $\overline{\mathbf{A}}_d = e^{\overline{\mathbf{A}}T_{KV}}$, $\overline{\mathbf{B}}_d = \int_0^{T_{KV}} e^{\overline{\mathbf{A}}(T_{KV}-t)}\overline{\mathbf{B}}dt$, T_{KV} is a sampling period.

To transform model (3) to a discrete form, it is sufficient to use the path of formal transition from the analog model in the Laplace domain to its discrete analogue in the z-domain [23]:

$$\mathbf{Y}(z) = \mathbf{W}(z)\mathbf{X}(z), \tag{11}$$

here $\mathbf{W}(z) = \dfrac{z-1}{z} Z\left\{\dfrac{\mathbf{W}(p)}{p}\right\}$ is the discrete transfer matrix of the multidimensional system, in which the multiplier $\dfrac{z-1}{z}$ takes into account the mathematical description of the zero-order extrapolator with the quantization period T_{KV}; or use a bilinear substitution, for example, $p \approx \dfrac{2}{T_{KV}}\left(\dfrac{z-1}{z+1}\right)$, which corresponds to the numerical approximation of the continuous model by the trapezium method. In this case it does not matter in principle, since these approaches have been studied and implemented as MatLAB functions.

Thus, the digital twin synthesis methodology for the electric heater consists of steps:

(1) the uncertain parameters identification ($\overline{\alpha}_0$ and $\overline{G}_A$) of mathematical model (2) or (3) by the considered algorithm;
(2) transition from the continuous model (2) or (3) to the discrete model (10) or (11), respectively, which is the digital twin;
(3) if during operation the digital twin accuracy has deteriorated (due to non-stationarity of the physical process) then go to step 1 to identify the parameters of the model.

5.1 Results of the Digital Twin Simulation

The proposed methodology was used to develop and simulate a digital twin model of an electric heater using the MatLAB software package. Let's consider the example of digital twin model simulation using model in state space (2). MatLAB was used to calculate the matrices $\overline{\mathbf{A}}_d$ and $\overline{\mathbf{B}}_d$ of the digital twin (10). Simulation results are shown in Fig. 9.

Figure 9a simulates the case with input influence $\mathbf{U}(t) = [1 + 0.5\sin(0.15t), 1 + 0.2\sin(0.15t), -0.2 + 0.1\sin(0.3t), 0.5 + 0.2\sin(0.05t)]^{\mathrm{T}}$, and initial conditions for the physical model $\mathbf{X}(0) = [2, 1, 15]^T$, $\alpha_0 = 161$, $G_A = 0.43$ and the identifiable model $\overline{\mathbf{X}}_k(0) = [0, 0, 0]^{\mathrm{T}}$, $\overline{\alpha}_0 = 250$, $\overline{G}_A = 0.8$. Figure 9b shows the surface isolines of criterion (6) and the minimization trajectory of its parameters $\overline{\alpha}_0$, $\overline{G}_A$, which resulted in finding $\overline{\alpha}_0 = 160.3$, $\overline{G}_A = 0.41$. After identification of model (2), using MatLAB function c2d(...) numerical values of digital twin model matrices (10) are calculated for sampling period $T_{KV} = 2$:

$$\overline{\mathbf{A}}_d = \begin{bmatrix} 0.0133 & 0 & 0.0829 \\ 0 & 0.0125 & 0 \\ 0.0552 & 0 & 0.7238 \end{bmatrix};$$
$$\overline{\mathbf{B}}_d = \begin{bmatrix} 0.9037 & 0 & -9.0374 & 0.547 \\ 0 & 0.9875 & 0 & 0 \\ 0.221 & 0 & -2.2097 & 6.1755 \end{bmatrix}.$$

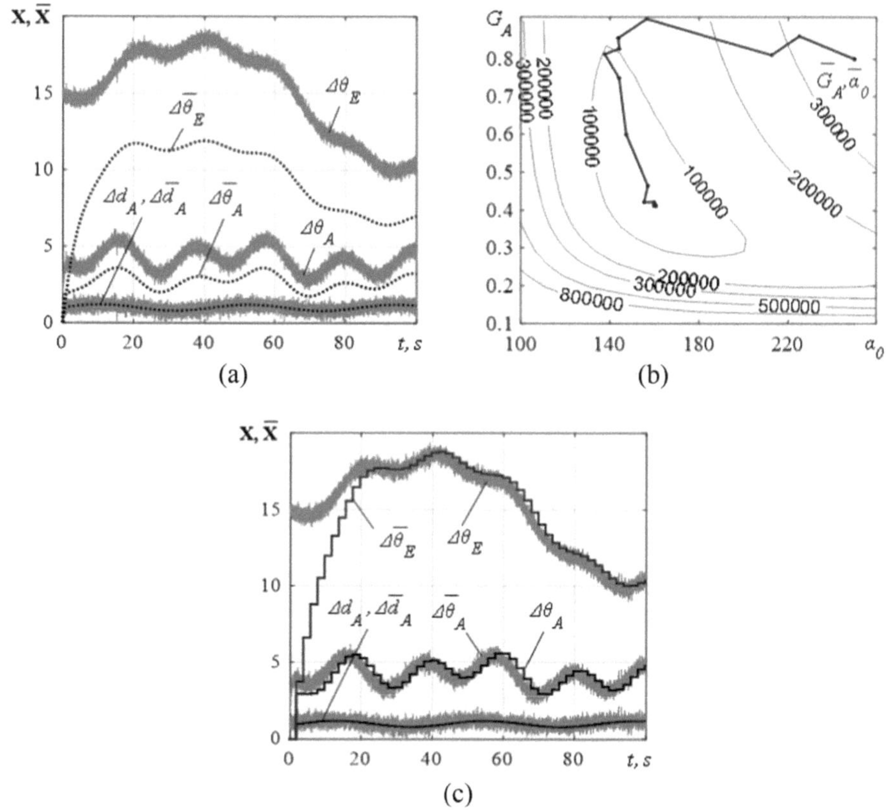

Fig. 9 The digital twin development for the electric heater HE 36/2: **a** the simulation of transients before identification; **b** the identification trajectory of parameters $\overline{\alpha}_0$ and $\overline{G}_A$ according to the proposed algorithm; **c** the simulation of transients for the physical model (2) and digital twin (10)

The following example demonstrates the digital twin development using the mathematical model (3). The results of the simulation are shown in Fig. 10. Figure 10a shows the case with input influence $\mathbf{U}(t) = [1 + 0.5\sin(0.15t), 1 + 0.2\sin(0.15t), -0.2 + 0.1\sin(0.3t), 0.2 + 0.8\sin(0.05t)]^{\mathrm{T}}$, and initial conditions for the reference model $\alpha_0 = 161$, $G_A = 0.43$ and the identifiable model $\overline{\alpha}_0 = 120$, $\overline{G}_A = 0.22$. Figure 10b shown the surface isolines of criterion (7) and the minimization trajectory of parameters $\overline{\alpha}_0, \overline{G}_A$, as a result of which $\overline{\alpha}_0 = 154$, $\overline{G}_A = 0.428$ are found.

Using these parameters, the transfer matrix of the identified model is calculated

$$\overline{\mathbf{W}}(p) = \begin{bmatrix} \dfrac{5.9p + 1}{2.5p^2 + 6.9p + 1} & 0 & -\dfrac{54.7p + 9.3}{2.5p^2 + 6.9p + 1} & \dfrac{0.0023}{2.5p^2 + 6.9p + 1} \\ 0 & \dfrac{1}{0.426p + 1} & 0 & 0 \end{bmatrix},$$

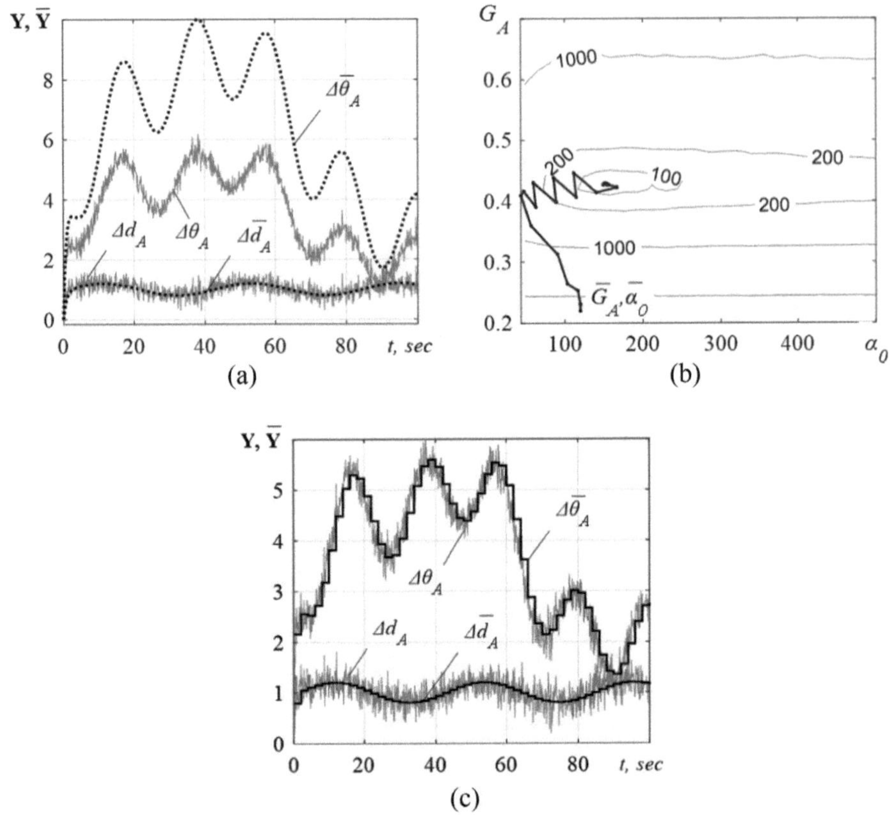

Fig. 10 The digital twin development for the electric heater HE 36/2: **a** the simulation of transients before identification; **b** the identification trajectory of parameters $\overline{\alpha}_0$ and $\overline{G}_A$ according to the proposed algorithm; **c** the simulation of transients for the physical model (3) and digital twin (11)

its numerical values are close to the reference transfer matrix (8). Further, using MatLAB function c2d(...), the discrete transfer matrix of the model (11) for the sampling period is $T_{KV} = 2$ found

$$\overline{\mathbf{W}}(z) = \begin{bmatrix} \dfrac{0.9127z - 0.6516}{z^2 - 0.742z + 0.004} & 0 & \dfrac{-8.539z + 6.09}{z^2 - 0.742z + 0.004} & \dfrac{0.0005z + 0.0001}{z^2 - 0.742z + 0.004} \\ 0 & \dfrac{0.991}{z - 0.009} & 0 & 0 \end{bmatrix}.$$

6 Conclusions

Digitalization of physical environs and processes involves the use of complex mathematical models with uncertain parameters, which need to be corrected in order to adapt them to specific application conditions. Using several examples, the methodology of developing mathematical support for a digital twin model is considered. On the basis of the system analysis methodology, the results obtained are summarized and the procedure for developing a digital twin model using the electric heater analytical model in conditions of conceptual uncertainties is proposed.

The peculiarity of the proposed procedure is the several key parameters identification of the analytical model, which are inaccurately defined and refined in the passive identification process. The use of an analytical model for a physical process makes it possible to abandon the choice of structure and model type, as well as the search for all its parameters. It is known that in modern system analysis methods, the choice of model structure and type plays an important role in further research and can require a lot of time and additional information to build an adequate model [2]. In the proposed procedure, the structure of the model is known, for which only the key uncertain parameters of the physical process are identified, rather than the model as a whole.

The identification examples of parameters $\overline{\alpha}_0$ and $\overline{G}_A$ for analytical models (2) and (3) are given. It is shown that identification of uncertain parameters $\overline{\alpha}_0$ and $\overline{G}_A$ for electric heater model belongs to one-extreme optimization problem. The procedure for the electric heater digital twin model synthesis was proposed and numerically investigated. Simulation results confirmed the proposed procedure effectiveness for developing the digital twin using the analytical model.

When developing digital twin using analytical model in the conditions of the conceptual uncertainty of its parameters, the qualifications and experience of the researcher play a huge role. It depends on the intuition and practical skills of the researcher to uncover the interrelated uncertainties of the model, which appear as perturbing factors of different nature at all stages of the physical process evolution, from design to operation. This creative task does not always have a positive result. Complex mathematical models in a number of cases give less significant results than expected, and simplifications do not allow an adequate "progress assessment" of a physical process. A researcher faces a task of determining a detailing limit a mathematical model for describe a physical process. On the one hand, the model must be simple for its study and use, and on the other hand, it must take into account the features of the physical process (significant nonlinearities and perturbations, interconnections of influence channels, etc.). The use of the digital twin toolkit allows us to find a compromise of contradiction for the mathematical model. The relationship between a real physical process and its digital copy allows the analytical model to be adapted to the real physical process.

In the future it is planned to develop a digital twin model for electric heater and other objects based on statistical models and to compare the obtained results with the existing ones. It is difficult to overestimate the use of digital twins concept for the production development. This toolkit allows identifying bottlenecks in the production life cycle, optimizing the structure of the real physical system, improving product quality, predicting and detecting equipment malfunctions and much more.

References

1. Digital Twins for Industrial Applications. Definition, Business Values, Design Aspects, Standards and Use Cases. An Industrial Internet Consortium White Paper, Version 1.0 (2020). https://www.iiconsortium.org/pdf/IIC_Digital_Twins_Industrial_Apps_White_Paper_2020-02-18.pdf
2. Bidyuk, P.I., Tymoshchuk, O.L., Kovalenko, A.E., Korshevniuk, L.O.: Decision Support Systems and Methods. Igor Sikorsky Kyiv Polytechnic Institute, Kyiv (2022)
3. Lee, E.A.: Fundamental limits of cyber-physical systems modeling. ACM Trans. Cyber-Phys. Syst. **1**(1), 1–26 (2016). https://doi.org/10.1145/2912149
4. Lee, E.A.: The past, present and future of cyber-physical systems: a focus on models. Sensors **15**(3), 4837–4869 (2015). https://doi.org/10.3390/s150304837
5. The Industrial Internet Reference Architecture. An Industry IoT Consortium Foundational Document (2022). https://www.iiconsortium.org/wp-content/uploads/sites/2/2022/11/IIRA-v1.10.pdf
6. Zgurovsky M.Z., Pankratova N.D.: System analysis: theory and applications, p. 475. Springer (2007). https://doi.org/10.1007/978-3-540-48880-4
7. Miskinis, C.: Technology. The history and creation of the digital twin concept (2019). https://www.challenge.org/insights/digital-twin-history
8. Grieves, M.: Origins of the digital twin concept. Florida Inst. Technol. (2016). https://doi.org/10.13140/RG.2.2.26367.61609
9. Parrott, A., Warshaw, L.: Industry 4.0 and the digital twin. Manufacturing meets its match (2017). https://www2.deloitte.com/us/en/insights/focus/industry-4-0/digital-twin-technology-smart-factory.html
10. Monsone, C.R., Mercier-Laurent, E., János, J.: The overview of digital twins in industry 4.0: managing the whole ecosystem. In: Proceedings of the 11th International Joint Conference on Knowledge Discovery, Knowledge Engineering and Knowledge Management, pp. 271–276 (2019). https://doi.org/10.5220/0008348202710276
11. What is digital twin technology and how does it work? TWI Ltd. (2023). https://www.twi-global.com/technical-knowledge/faqs/what-is-digital-twin
12. Jone, D., Snider, C., Nassehi, A., Yon, J., Hicks, B.: Characterising the digital twin: a systematic literature review. CIRP J. Manuf. Sci. Technol. **29**, 36–52 (2020). https://doi.org/10.1016/j.cirpj.2020.02.002
13. Wright, L., Davidson, S.: How to tell the difference between a model and a digital twin. Adv. Model. Simul. Eng. Sci. **7/13** (2020). https://doi.org/10.1186/s40323-020-00147-4
14. Rasheed, A., Omer, S., Kvamsdal, T.: Digital twin: values, challenges and enablers from a modeling perspective. IEEE Access **8**, 21980–22012 (2020). https://doi.org/10.1109/ACCESS.2020.2970143
15. Wang, Kung-Jeng., Lee, Tsung-Lun., Hsu, Yuling: Revolution on digital twin technology-a patent research approach. Int. J. Adv. Manuf. Technol. **107**, 4687–4704 (2020)
16. Ferguson, S.: Five reasons why Executable Digital Twins are set to dominate engineering in 2023 (2023). https://blogs.sw.siemens.com/simcenter/the-executable-digital-twin
17. Pankratova, N., Golinko, I.: Electric heater mathematical model for cyber-physical systems. Syst. Res. Inf. Technol. **2**, 7–17 (2021). https://doi.org/10.20535/SRIT.2308-8893.2021.2.01
18. Bidyuk, P.I., Trofymchuk, O.M., Fedorov, A.V.: Information system of decision-making support for prediction financial and economic processes based on structural-parametric adaptation

of models. In: Research Bulletin of the National Technical University of Ukraine. Kyiv Polytechnic Institute, No. 6, pp. 42–53 (2011)
19. Kornienko, Y.M., Lukach, Y.Y., Mikulonok, I.O., Rakytskyi, V.L., Ryabtsev, G.L.: Processes and equipment of chemical technology. NTUU "KPI". Kyiv (2012)
20. Kononyuk, A.E.: Fundamentals of optimization theory. Unconditional optimization. Kyiv: "Education of Ukraine", p. 544 (2011). ISBN 978-966-7599-50-8
21. Pankratova, N., Bidyuk, P., Golinko, I.: Decision support system for microclimate control at large industrial enterprises. In: CMIS-2020: Computer Modeling and Intelligent Systems, pp. 489–498 (2020). https://ceur-ws.org/Vol-2608/paper37.pdf
22. Dubovoi, V.M.: Identification and Modeling of Technological Objects and Control Systems. VNTU, Vinnytsia (2012)
23. Smith, S,: Digital Signal Processing: A Practical Guide for Engineers and Scientists. Newnes (2013). ISBN:9780080477329

Constructing Mathematical Models for Actuarial Processes

Petro Bidyuk, Yoshio Matsuki, and Vira Huskova

Abstract System analysis based methodology is proposed for constructing mathematical models for actuarial processes. An algorithm is also developed for parameter estimation of the models mentioned. As a basis for model structure estimation are used generalized linear models (GLM) that are convenient for formal description of linear and nonlinear processes using the family of exponential distributions. GLM represent expansion of linear regression to the cases when data distribution is distinguished from normal. Such approach to modeling provides a possibility for improving formal description of a process model structure and enhance its adequacy. The modeling methodology proposed is illustrated by actuarial process models constructed using actual statistical data from the sphere of car insurance. Among the candidate models constructed for the process mentioned the best one turned out to be the model based upon Poisson distribution of dependent variable and exponential link function.

Keywords System analysis · Generalized linear models · Modeling methodology · Bayesian approach · Monte Carlo estimation technique

1 Introduction

An environment of actuarial activities includes a set of processes related to insurance companies with the main purpose of compensating the consequences of some negative random events followed by definite material loss. However, some events like

---

P. Bidyuk · V. Huskova (✉)
Igor Sikorsky Kyiv Polytechnic Institute, Kyiv, Ukraine
e-mail: guskovavera2009@gmail.com

P. Bidyuk
e-mail: pbidyuke_00@ukr.net

Y. Matsuki
National University of Kyiv-Mohyla Academy, Kyiv, Ukraine
e-mail: matsuki@wdc.org.ua

M. Zgurovsky and N. Pankratova (eds.), *System Analysis and Artificial Intelligence*, Studies in Computational Intelligence 1107,
https://doi.org/10.1007/978-3-031-37450-0_2

games of chance or stock trading do not belong to the insurance sphere because in the case of game of chance its participants are consciously taking risk understanding that end of the game can be positive or negative regarding their material state. Similar situation takes place in the case of stock trading. The trading participants can lose but their risks are not insured because insurance companies are not liable to pay for any unsuccessful usage of capital being invested. Thus, in the insurance sphere there exist two types of risks: one of them is pure risk, and another one is speculative. The insurance companies are mostly interested in the pure risks only [1].

The insurance activities are directed towards re-distribution of capital assets available and their accumulation for carrying out insurance operations as well as for investing the capital into various other activities aiming future developments. Due to all these activities arise the problems of analysis and financial risks management as well as investment processes using mathematical modeling, estimation theory, forecasting, and decision support [2–5]. Thus, the sphere of insurance company activities requires development of new mathematical models for forecasting in conditions of uncertainty, availability of various risks for the participants of the processes.

The study is directed towards development of system analysis based modeling methodology for actuarial processes with the use of generalized linear models [6–10]. The models are helpful to expand the area of methodology application thanks to the possibility of using exponential family distributions. It is also proposed the method of GLM parameter estimation on the basis of Bayesian approach. Some examples of model constructing are provided with the use of actual data.

2 Problem Statement

Widely used in practice are auto-regression with moving average models (ARMA), and ARMA with extra exogenous variables (ARMAX) [11–13]. However, the methodologies available not always introduce clear definition of model structure, and nonlinearity is not paid necessary attention. Such approach may result in erroneous results regarding formal description of actuarial processes. The basic preliminaries on which historical forecasting models are grounded can be formulated as follows:–market can take into consideration all necessary influences; it means that selected variable value is a result and sufficient reflection of all substantial market forces;–the movement of prices is subject to some determined tendencies; generally it can be stated that market prices exhibit the periods of drops and growth in a way that within each period the basic tendency is preserved that is active until the movement is changed to opposite direction;–the history is repeating: a key to the future is hidden in the learning of the past.

Taking into consideration necessity of formal description and forecasting for actuarial processes the study is directed towards development of modified modeling methodology based on GLM applications that are widely used in analysis of insurance cases, for forecasting existing or new insurance agreements, development of new tariffs and purposeful marketing.

Thus, the purpose is stated as analysis of existing mathematical model structures aiming to their further application for forecasting in the insurance sphere. To reach the purpose stated the following problems are to be solved:

- to develop effective systemic methodology for constructing mathematical models of actuarial processes in the form of generalized linear models;
- to develop GLM parameters estimation algorithm to be used in the sphere of actuarial risks insurance;
- to propose illustrative example of the proposed methodology application aiming to actuarial processes modeling using actual statistical data.

3 Estimation of Mathematical Model Structure

Consider the methodology of mathematical model constructing for analysis and forecasting actuarial processes in the insurance sphere. As a basis for modeling generalized linear models will be used that includes linear and nonlinear regression, variance and covariance analysis, Log-linear models for analysis of random tables, nonlinear models of logit/probit type, Poisson regression and some others.

The basics of the time series models constructing were developed by Box and Junkins in the 1960s and further developed in [12, 13]. The modified modeling methodology proposed includes the following steps:

- systemic analysis of a process being analyzed on the basis of expert estimates, visual analysis of inputs and outputs, represented by time series observations, study of existing model structures, and other accessible information;
- preliminary data processing that includes application of appropriate normalizing and filtering techniques;
- analysis of data for stationarity and nonlinearity using appropriate set of statistical tests;
- estimation of candidate model structures taking into consideration structure of GLM: identification of stochastic and systemic components as well as link function; computing descriptive characteristics and correspondence of data to selected model (or class of models) using appropriate criteria (for example, maximum likelihood relation); determine statistical significance of predictors using, for example, Wald statistic; estimate characteristics of other structural elements of mathematical model (for example, residuals);
- selection and application of model parameter estimation method (or methods); it can be least squares (LS) method, generalized least squares (GLS), maximum likelihood (ML) or Monte Carlo based Bayesian approach;
- selection of the best model from the set of estimated candidates using appropriate statistical criteria.

Now consider some details of the stages mentioned above to clear out the possibilities of practical application of the methodology formulated.

4 Systemic Analysis of a Modeled Process

Generally mathematical model of some system includes description of a possible set of states and the transition law from one state to another. To construct such a model it is necessary to perform the system identification at the stage of system analysis. This approach supposes carrying out problem refinement and performing qualitative analysis of the system under study, i. e. determining the following:

- refinement of the basic modeling problem;
- initial conditions and constraints formulation;
- determine the system under study characteristics, its logical structure and the links between its elements;
- hypothesis formulation regarding behavior of system dynamics and its interaction with environment.

System and complex processes analysis is important stage of modeling, performing of which requires availability of practical experience of studying systems (objects) of various origin and it is directed towards solving the following problems [14–18]:

- determining the number of system inputs and outputs, i. e. identification of a system dimensionality;
- establishing logical relations between system variables and determining the possibilities for their correct formal mathematical description;
- identification of stochastic external disturbances, their distributions and parameters; in some cases disturbance may exhibit deterministic nature;
- determining the possibility of decomposition complex processes into separate elements that would be simpler from the point of view of their functioning and mathematical description; such approach to analysis is based on special mathematical methods and may require complicated computations;
- hierarchical structure of a system (process) should be decomposed into levels (say lower and upper) with clearly defined functions (like in the case of technology processes control) and logical/functional links between the levels;
- analysis and correct usage of models, knowledge, recommendations from previous publications and reports regarding special features of functioning of the process under study, its laws of development and available experience of research;
- the models discovered during retrospective analysis should be studied from the point of view of their advantages and disadvantages, and possibly used to solve the problem stated; such approach may result in substantial saving of time and other resources necessary for constructing and practical application of a new model.

The stage of systemic analysis should not be ignored because this may result in impossibility to construct a model exhibiting high adequacy and high quality of forecasting in practical application. All the information received on this stage can be used for preliminary estimating of a model structure or several candidate models, parameters of which will be estimated with statistical/experimental data. It is recommended to use information from alternative sources to avoid erroneous interpretations of available facts.

5 Preliminary Data Processing

The theory of time series analysis considers observations used for modeling as random variables containing deterministic component. It means that generally we have to process measurements influenced by random external disturbances and measurement errors. The purpose of preliminary data processing is in eliminating possible errors from measurements and improving conditions for determining the law of their distribution. When generalized linear models are hired for describing the processes under study the data available should correspond to one of known exponential distributions. The process of preliminary data processing in most cases includes the following operations:

- visual analysis, normalization and possible correcting of measurements; normalization means application of logarithmic operation or transforming the data to convenient interval for further processing;
- data correction means imputation of missing values, and processing extreme values that overcome some definite basic interval of their definition;
- computing (when necessary) first and higher order differences that are necessary for analysis of corresponding elements of a time series.

First and second order differences provide a possibility for modeling velocity and acceleration of dependent variable what can provide extra information regarding dynamic analysis of selected process. Sometimes sample mean is subtracted from the observations to get a possibility for processing deviations of measurements. This is necessary to perform in the case of constructing state space models widely used in creating control systems. In some cases, current sample mean is computed on short time intervals using appropriate recursive formula. It is clear that application of specific data preprocessing method depends on specific features of the data available.

6 Analysis of Data for Nonlinearity

To determine availability of nonlinearity visual analysis and formal statistical tests are applied. Visual analysis of observations can be helpful for detecting linear and nonlinear parts of trend, in some instants, availability of heteroscedasticity and extreme values that may create substantial influence on the resulted model [14]. The visual analysis should be supported by formal statistical tests that help to accept or reject the hypothesis formulated. There exist a number of formal tests for analyzing availability of nonlinearity, consider one of them. It supposes existence of a long enough time series that can be divided into several groups of data [15, 16]:

$$F = \frac{\frac{1}{m-2}\sum_{i=1}^{m} n_i \left(\bar{y}_i - \bar{y}_j\right)^2}{\frac{1}{n-m}\sum_{i=1}^{m}\sum_{j=1}^{n_i}\left(y_{ij} - \bar{y}_i\right)^2}, \tag{1}$$

where $\overline{y}_i$ is a mean for the $i - th$ data group; $\overset{|\Delta|}{y_i}$ is a mean for linear approximation of data; m is a number of data groups; n_i is a number of observations in $i - th$ group; n is total number of available observations. If $\overset{|\Delta|}{F}$ statistics with, $v_1 = m - 2$, and $v_2 = n - m$ degrees of freedom reaches or exceeds significance level then the hypothesis of linearity should be rejected.

The fact of nonlinearity existence with respect to some independent variable x can be established using the following sample nonlinear correlation function:

$$r_{yx^2}(s) = r_{y(k)x^2(k-s)} = \frac{1}{N}\frac{\sum_{k=s+1}^{N}\left\{[y(k) - \bar{y}][x(k-s) - \bar{x}]^2\right\}}{\sigma_y\sigma_x^2}, \quad s = 0, 1, 2, 3, \ldots \tag{2}$$

If at least some of the values computed with the function are distinct from zero in statistical sense, then the process $y(k)$ contains quadratic nonlinearity with respect to independent variable, $x(k)$.

Availability of nonlinear deterministic trend in data can be identified by estimating the following polynomial:

$$y(k) = a_0 + c_1k + c_2k^2 + \ldots + c_mk^m \tag{3}$$

where, $k = 0, 1, 2, \ldots$ is discrete time. If at least one of the coefficients c_i, $i = 2, \ldots, m$ is statistically significant, then the hypothesis on linearity of trend is rejected. In a case trend is quickly changing its direction of movement and its functional description is not adequate, then model of stochastic trend is constructed based upon combinations of random processes [13].

Model structure is automatically estimated by the Group Method of Data Handling (GMDH) that was successfully applied to wide class of nonlinear and nonstationary processes. The original method was further extended to fuzzy version with fuzzy representation of model parameters [14].

7 Estimation of Other Elements of GLM Model Structure

Taking into consideration possible distributions there are several GLM types presented in Table 1.

Regarding GLM, the notion of a model structure is considered from the point of view of its components and includes the following:

- stochastic component: independent variable that is characterized by distribution belonging to the exponential family with mean mu;
- systematic component including p independent variables that create so called "linear predictor" [13]: $\eta = X \cdot \beta$;
- establishing features of a link function taking into consideration their classification.

Table 1 GLM types considered in the study

Model type	Link function	Dependent variable distribution
Generalized linear model	$g(\mu) = \mu$	Normal
Log-linear model	$g(\mu) = \ln(\mu)$	Poisson
Logistic model	$g(\mu) = \ln\left(\frac{\mu}{1-\mu}\right)$	Binomial
Probit-analysis	$g(\mu) = \Phi^{-1}\mu$	Binomial
"Survival" analysis	$g(\mu) = \mu^{-1}$	Gamma-distribution, exponential

Besides the elements mentioned the notion of a model structure includes the following: model order (maximum order of equations comprising the model); model dimension (number of model equations); GLM residuals that are used for analyzing model adequacy, selection of a link function, variance function and elements of linear predictor; possible nonlinearities and their type; external disturbances and their type (deterministic or stochastic, additive or multiplicative).

It is supposed that combined influence of stochastic disturbances and other non-measurable factors can be represented (to some extent) by the random variable, $\varepsilon(k)$) . As far as it cannot be measured the only possible way to get its estimate after estimation of model parameters is as follows:

$$\hat{\varepsilon}(k) = e(k) = y(k) - \hat{y}(k) \tag{4}$$

where, $\hat{y}(k)$, is model estimate of $y(k)$; $y(k)$ is actual observation. To determine type of data distribution various statistics are applied, such as, χ^2, asymmetry, excess, Jark-Bera etc. Thus, this stage of modeling is finished with estimated structure for several candidate models with alternative distributions: normal, gamma, Poisson and corresponding type of a link function (logarithmic, identity). Usually, several candidates are estimated because it is impossible to identify single appropriate structure. Generally, reaching high model adequacy is iterative process that may take time and efforts. The next stage is devoted to model parameter estimation.

8 Model Parameter Estimation

After estimating model structure the next problem is to compute model parameters using available statistical/experimental data. Usually, the model structure estimated provides information on a number of parameters to be estimated and the estimation method to be applied in concrete case. In some cases, it is reasonable to use the saving principle that supposes the following: number of parameters to be estimated should not exceed their necessary quantity. Here "necessity" can be viewed as necessity to preserve in a model constructed basic statistical characteristics of the process being studied [19, 20].

Very often parameters of a generalized linear model can be estimated using ordinary LS (OLS) method. In a case of normal data distribution this is equivalent to application of maximum likelihood (ML) method. Generally, widely used methods for estimation of GLM parameters are the following: OLS, weighted LS (WLS), ML, method of moments and Monte Carlo techniques (iterative and non-iterative). As far as OLS in some cases results in biased parameter estimates due to its sensitivity to data outliers an alternative procedure for estimation is ML.

In a case of normal residuals logarithmic likelihood function, l, can be represented for n measurements as follows:

$$-2l = n\log\left(2\pi\sigma^2\right) + \sum_{i=1}^{n}\frac{(y_i-\mu_i)^2}{\sigma^2} \tag{5}$$

For a fixed σ^2 maximization of l is equivalent to minimization of quadratic deviations from mean: $\sum[y(k)-\mu]^2$; thus, for linear model we have:

$$\eta_i = \mu_i = \sum_{j=1}^{p} x_{ij}\beta_j \tag{6}$$

Very popular today method for GLM parameter estimation is Markov Chain Monte Carlo (MCMC) that can be applied to linear and nonlinear models [21]. The estimation algorithm based on this method is functioning as follows:

- sampling X_u^{i+1} from the distribution $P\left(X_u \mid X_0, \theta^i\right)$, where X_0 is experimental data;
- sampling parameter estimate, θ^{i+1}, from distribution $P\left(\theta \mid X_0, X_i^j\right)$.

The statistical series generated this way under weak regularity conditions has marginal stationary distribution, , that can be trivially transformed into . The MCMC techniques require rather high computational resources but they are very popular due to their universality, good scaling characteristics, capability to take into account non-observable variables, low estimation errors, and possibilities for parallel computing. Correctness of the conditions necessary for parameter estimation can be analyzed after computing the parameters. After computing the parameters, we get estimates of the random process as follows: $\hat{\varepsilon}(k) = e(k) = y(k) - \hat{y}(k)$, and analyze its statistical characteristics indicating correctness of the parameter values.

9 Model Diagnostics: Selection of the Best Model of the Candidates Constructed

At this stage model adequacy is analyzed that includes the steps given below.

1. Visual study of residual graph, $e(k) = y(k) - \hat{y}(k)$, where $\hat{y}(k)$ is dependent variable estimate computed with the model constructed. The graph should not exhibit outliers and long periods of time with large values of residuals (i.e., long periods of model inadequacy). In the case of constructing GLM such model characteristics can result from random selection of distribution law for dependent variable without substantial argumentation. It can result in substantial deviations of forecast estimates computed with the model from actual observations [19].
2. The estimated model residuals should not be auto-correlated. To analyze the autocorrelation level it is necessary to compute autocorrelation function (ACF) and partial ACF (PACF) for the series, $\{e(k)\}$, and analyze statistical significance of the functions with Q-statistics. Besides, the level of residuals autocorrelation can be additionally tested with Durbin-Watson statistic, $DW = 2 - 2\rho$, where, $\rho = E[e(k)e(k-1)]/\sigma_e^2$, is correlation between neighboring values of residuals; σ_e^2 is variance of the residual process; if $\rho = 0$, auto-correlation between residuals is absent, and ideal value for the DW statistic is 2.
3. Statistical significance of the model parameter estimates can be tested with Student t-statistic, $t = \left[\hat{a} - a^0\right]/SE_{\hat{a}}$, where $\hat{a}$ is parameter estimate; a_0 is zero-hypothesis regarding the estimate; $SE_{\hat{a}}$ is standard error for the estimate. Usually all the computer systems that include functionality for time series analysis provide all necessary information for a user regarding analysis of statistical significance of parameter estimates.
4. Determination coefficient, $R^2 = \mathrm{var}[\hat{y}]/\mathrm{var}[y]$, where, $\mathrm{var}(\hat{y})$ is variance of dependent variable estimated via the model constructed; $\mathrm{var}(y)$ is actual sample variance of dependent variable. The ideal value of the coefficient is, $R^2 = 1$, when the values of variance in nominator and denominator are the same. This statistical parameter can be interpreted as a measure of information contained in a sample and in a model. From this point of view R^2 compares volume of information represented by a model to the volume of information represented by data sample used for model constructing.
5. The sum of squared errors should be minimal for the best candidate model constructed: $\sum_{k=1}^{N} e^2(k) = \sum_{k=1}^{N} [\hat{y}(k) - y(k)]^2 \to \min_{\hat{\theta}}$.
6. For estimating constructed model adequacy Akaike information criterion (AIC) is used: $AIC = N \ln \left(\sum_{k=1}^{N} e^2(k)\right) + 2n$, and Bayes-Schwarz criterion: $BSC = N \ln \left(\sum_{k=1}^{N} e^2(k)\right) + n \ln(N)$, where, $n = p + q + 1$, is a number of estimated model parameters (p is a number of parameters for auto-regression part; q is a number of parameters for the moving average part; 1 is added in a case when the bias, a_0, is added). Both criteria exhibit minimum values for the best model among the candidates estimated.

7. Besides the statistics mentioned Fisher statistic $F \sim R^2/\left(1 - R^2\right)$ is applied that may show adequacy of a model as a whole after testing it for statistical significance. To formulate statistical inference regarding the model constructed Bayesian factor, $BF(i, j)$, is also used that is represented by the ratio of posterior probabilities to prior [21, 22]. The criteria of maximum marginal density of the distribution, $p\,(x \mid M_i)$, is used that corresponds to the following condition: $BF(i, j) > 1$.

Correct application of the methodology presented provides a possibility for constructing adequate mathematical model in the class of generalized linear models if the statistical/experimental data hired corresponds to the requirement of system representation and contain necessary information.

10 Example of Model Constructing

Structure of statistical data hired for model constructing is shown in Table 2. The data shows loss due to car insurance for three selected regions of Ukraine.

For the data described in Table 1, suppose that dependent variable is normally distributed, and accept as a link log-function. The histogram for dependent variable "loss" is presented in Fig. 1.

Results of log-normal model estimation are given in Fig. 2.

The mean value of dependent variable "Loss" is 1877.531. To select best model from the set of possible models constructed Akaike information criterion was used. Its minimum value corresponds to the best model. For log-normal model value of the criterion is 20.666. The drawback of the criterion is that its estimate asymptotically overestimates true value with non-zero probability. An alternative is Hannan-Quinn criterion that is based upon minimizing of corresponding sum instead of value itself.

Information Schwarz criterion usually selects the best model with a number of parameters that does not exceed number of parameters in the model selected by the Akaike criterion. The Schwarz criterion is asymptotically more reliable, and the Akaike criterion has a tendency to bias closer to selection of parametric models [23, 24]. However, for the example under consideration the values of Akaike, Hannan-Quinn, and Schwarz criteria are practically equal; that is why any of them is suitable for use. The standard deviation for dependent variable, "Loss", is 7492.245.

Table 2 Structure of statistical data

No	Characteristic of data	Value of the characteristic
1	Power of sample	9546
2	Dependent variable	Loss due to insurance
3	Region where police was sold	Kyiv, Crimea, Odesa
4	Year of a car production	Starting from 2006
5	Car brand	Mitsubishi, Toyota, VAZ, exponential

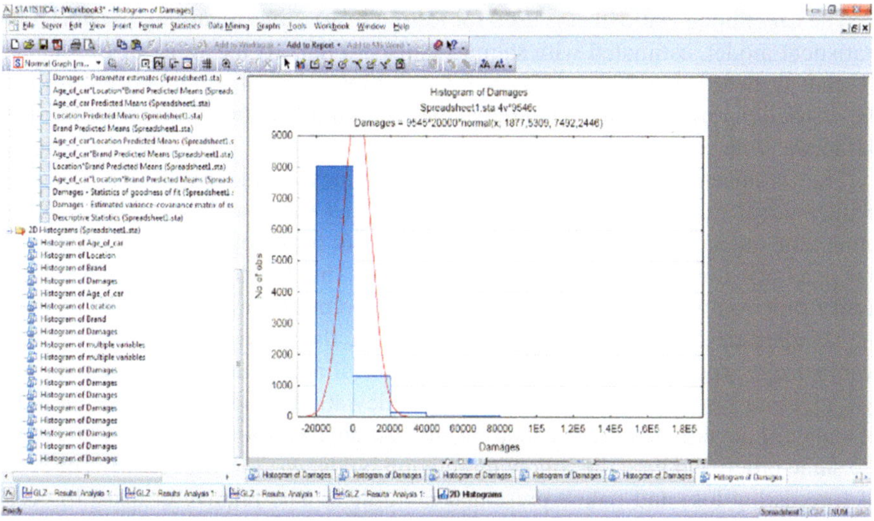

Fig. 1 The histogram for normally distributed dependent variable

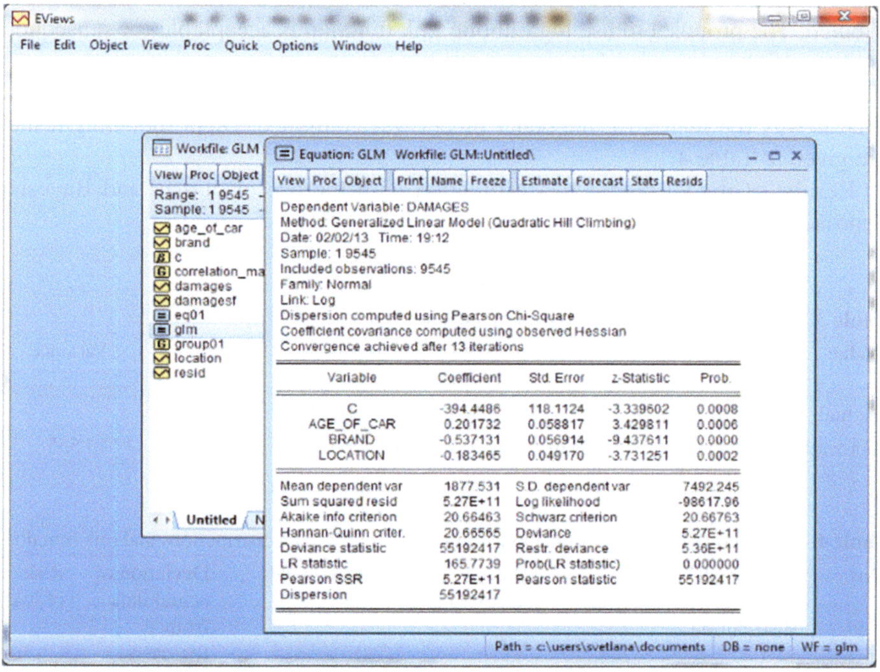

Fig. 2 Results of log-normal model estimation

The Likelihood Ratio Test is used for testing restrictions regarding parameters of statistical model, estimated with sample statistical data. If the value of LR-statistic exceeds critical value from χ-squared distribution at given significance level then the restrictions are neglected, and the model without restrictions is hired; otherwise, the model with restrictions is used.

The variance calculated shows how much the mean random value deviates from mathematical expectation; here it is important that this is quadratic value. At the same time the variance itself is not very convenient for practical analysis because it has quadratic dimensionality of a random value. That is why standard deviation is used further on as a measure of risk. From the point of view of financial analysis, the standard deviation is more important because the mean deviation of insurance results from expected return shows actual financial results and insurance risk. As a risk measure can also be used mean absolute deviation (mad). In practice the value of standard deviation is greater than mean absolute deviation but these values have the same order, and here the following relation holds: $mad = 0.7979 \cdot S$. The result of forecasting loss and risk is presented in Table 3. The relative error of forecasting with log-normal model amounted to 1.06%, what is high quality result for the model hired. The total forecasted loss amounts to 18111231.380, and actual loss was 17921032.581, what supports the fact that the variable "Loss" is normally distributed. The proposal to use logarithmic link function is acceptable for subsequent analysis of alternative models.

Thus, the log-normal model is acceptable but it is not the best for the dataset used. That is why the search for the better model was continued. Other modeling results are given in Table 4.

Results of model parameters estimation with the use of classic and Bayesian approach are presented in Table 5.

Table 3 Forecasting results with log-normal model

Value	Total loss	Mean	Std. deviance	Max	Min	Variance
Actual	17921032.58	18577.53	7492.24	151771.4	0.00	
Forecast	18111231.38	18976.45	9379.910	14010.97	634.0	49.535

Table 4 Other modeling results with application of alternative data distributions and link functions

No	Dependent variable distribution	Link function	Total forecasted loss	Actual total loss	Deviation of actual data from forecasts	Risk of loss
1	Gamma	LOG	102008320.905	17921032.581	84087288.32	1.003
2	Normal	LOG	18111231.380	17921032.581	190198.799	0.495
3	Poisson	LOG	17921032.574	17921032.581	0.007	0.547
4	Normal	Identity	17921032.589	17921032.581	0.009	0.532

Table 5 Results of model parameters estimation with the use of classic and Bayesian approach

No	Mean	Classic approach		Mean	Bayesian approach		
		Std. deviance	Variance, %		Std. deviance	Variance, %	R-squared
1	11805.69	15358.12	130.091	11804.346	15247.237	128.669	0.89735
2	1897.457	939.91	49.535	1897.294	939.94	49.4	0.99854
3	1877.531	1027.567	54.73	1877.301	1027.552	55.679	0.99887
4	1877.531	999.302	53.224	1876.909	999.751	53.809	1

Thus, after constructing models using various proposals regarding initial distributions of dependent variable and link function the following results were achieved:

1. the best data approximating model turned out to be the model with initial Poisson distribution of dependent variable and logarithmic link function what is proved by the forecasted total loss of 17921032.574 with practically zero errors of forecasting;
2. the value of risk for the models constructed was in the range 40–60%, what requires extra measures for minimizing this value;
3. the model based on normal distribution of data showed the risk about 49–50 %, but there were observed substantial deviations of forecasts from actual data;
4. comparing the models based upon normal data distribution with logarithmic and identity link functions it was established that Akaike criterion accepts the same value of about 20.66; that is why the model selection should be based upon forecasts of total loss;
5. the model based upon gamma-distribution with logarithmic link function showed maximum deviation from actual data (84087288.324) and maximum error: about 100%;
6. it is clear from results in Table 4 that the quality of model estimation with Bayesian approach are close to classic estimation with the maximum likelihood method but with better estimates of variance and standard deviation.

Thus, it can be concluded that the model based upon Poisson distribution and exponential link function is better for practical use from the point of view of forecasting, risk estimation, and parameter estimation.

11 Conclusions

The study was performed regarding the search of improved systemic methodology of constructing models for actuarial processes and financial processes of arbitrary origin. The multistep methodology was proposed and computational experiments proved its effectiveness with application to generalized linear models. An example of the methodology application is given that shows effectiveness of the methodology in application to actuarial processes. To estimate unknown GLM parameters it was proposed to apply Bayesian approach using prior and posterior parameter distributions as well as algorithms for selecting best models among possible candidates. The use of combined methods for parameter estimation and selection of the model using the best model based forecast of dependent variable provides the possibility for studying and improving existing modeling methodologies. The future studies should be touching upon analysis of the parameter estimation algorithms, their convergence analysis, and comparison with other methods, for example with Monte Carlo for Markov chains and others.

Application of the methodology proposed for modeling actuarial processes with the use of GLM and Bayesian parameter estimation guaranties high quality of risk estimation with minimum errors. It can also be stated that the insurance sphere under appropriate management, and with active use of mathematical methods, can create reliable source for stabilizing economy of a country.

References

1. Bowers, N.L., Gerber, H.U., Hickman, J.C., Jones, D.A., Nesbitt, C.J.: Actuarial Mathematics. Itasca, Illinois (1986)
2. Oleksyuk, O.S.: Financial Decision Support System on the Macro Level. Naukova Dymka, Kyiv (1998)
3. Rachev, S.T., Hsu, J.S.J., Bagasheva, B.S., Fabozzi, F.J.: Bayesian Methods in Finance. Wiley, Hoboken, NJ, USA (2008)
4. Johannes, M., Polson, N.: MCMC Methods for Continuous-Time Financial Econometrics. Columbia University, New York (2006)
5. Bidyuk, P., Gozhiy, O., Korshevnyuk, L.: Design of Computer Based Decision Support Systems. Petro Mohyla Black Sea State University (2011)
6. Gill, J.: Generalized Linear Models: A Unified Approach. New Delhi, USA (2001)
7. Dobson, A.J.: Introduction to Generalized Linear Models. Chapman & Hall/CRC, New York (2002)
8. Kozumi, H.: Bayesian Analysis of Generalized Linear Models with Gaussian Process Priors. Hokkaido University, Sapporo (2003)
9. Fritelli, M.R., Runggaldier, M.: Stochastic Methods in Finance. Springer, Berlin (2003)
10. Carlin, B.P., Louis, T.A.: Empirical Bayes Methods for Data Analysis. Chapman & Hall/CRC, New York (2000)
11. Bidyuk, P.: In: Bidyuk, P., Borusevich, A. (eds.) Estimation Parameters of Models Using Monte Carlo Markov Chain Technique, pp. 21–37. Petro Mohyla Black Sea State University (2008)
12. Tsay, S.: Financial Time Series Analysis. Wiley, Hoboken (New Jersey) (2010)
13. Enders, W.: Applied Econometric Time Series. Wiley and S, New York (1994)

14. Bidyuk, P., Romanenko, V., Tymoshchuk, O.:Time Series Analysis. NTUU 'KPI' (2013)
15. Sachs, L.: Statistische Auswertungsmethoden. Springer, Berlin (1968)
16. Himmelblau, D.M.: Process Analysis by Statistical Methods. John Wiley & Sons, Inc., New York (1970)
17. Zgurovsky, M.Z., Pankratova, N.D.: System Analysis: Problems, Methodology, Applications. Naukova Dumka, Kyiv (2011)
18. Katrenko, A.V.: System Analysis: System Analysis of Objects and Processes of Computerization. Novy Svit, Lviv (2003)
19. McCullagh, P., Nelder, J.A.: Generalized Linear Models. Chapman & Hall, New York (1990)
20. Trukhan, S.: In: Trukhan, S., Bidyuk, P. (eds.) Forecasting of Actuarial Processes Using Generalized Linear Models, pp. 14–20. Naukovi Visti NTUU 'KPI' (2014)
21. Bergman, N.: Recursive Bayesian Estimation: Navigation and Tracking Applications. Linkoping University (Sweden) (1999)
22. Besag, J.: In: Besag, J. (eds.) Markov Chain Monte Carlo for Statistical Inference, pp. 9–25. Working Paper, Center for Statistics and the Social Sciences (2001)
23. Kuznietsova, N., Bidyuk, P.: Theory and Practice of Financial Risk Analysis: Systemic Approach. Lira-K, Kyiv (2020)
24. Bidyuk, P.: In: Bidyuk, P., Belas A. (eds.) Selection of Quality Criteria for Estimation of Nonlinear Nonstationary Processes Forecasts, pp. 38–45. KPI Science News (2021)

Functional Planning Optimization of Exploiting Underground Space in Large Cities Using System Methodology

Hennadii Haiko, Illia Savchenko, and Yurii Haiko

Abstract The problems of managing rational exploitation of the cities' territorial resources, and the development of the corresponding scientific, system methodology-based guidelines for its implementation, are considered. The city building conditions and urban development factors, that form the structural and functional components of utilizing the benefits of underground construction, are analyzed. A network of morphological tables is developed for assessing the potential of underground construction territories, synthesizing the functional planning organization of city space, and the geology engineering factors of underground environment. The parameters of these morphological tables represent a wide spectrum of city building conditions, generalized for surface and underground urban development.

1 Introduction

Regulating urban development with the goal of increasing ecological standards and life safety of urbanized space is one of the most urgent, yet insufficiently studied and complex global problems. Urban underground construction is a prospective trend of sustainable development, being the integral constituent of modern metropolises that transcended the scope of local objects, becoming a system factor of the large cities' existence. For a long time the main factor for planning urban underground space was exclusively the geological environment that specified the advisability and cost of underground construction, whereas the structural and functional factors were only partially involved, so the choice of sites for underground objects (particularly

H. Haiko · I. Savchenko (✉)
Igor Sikorsky Kyiv Polytechnic Institute, Kyiv, Ukraine
e-mail: savil.ua@gmail.com

H. Haiko
e-mail: gayko.kpi@meta.ua

Y. Haiko
O.M. Beketov National University of Urban Economy, Kharkiv, Ukraine
e-mail: 33most33@i.ua

M. Zgurovsky and N. Pankratova (eds.), *System Analysis and Artificial Intelligence*, Studies in Computational Intelligence 1107,
https://doi.org/10.1007/978-3-031-37450-0_3

tunnel tracks) often had subjective nature. The models of urban development, and the respective condition groups, established in city-building and architectural activities, had minimal influence on the development of local underground facilities and complexes. Only the application of the system approach to planning underground space, and its envisioning as a basic component of urban environment, revised the existing situation. However, even in the most prominent works in this direction [1–4], the system is formed of mostly underground objects, and the ties to the surface buildings remain mostly auxiliary. To merge and scientifically fuse the planning decisions for surface and underground city components, we will consider the models and conditions for urban development as the constituents of system methodology from the reference points, common for surface and underground construction. Among other things, the relevance of this task is conditioned by the needs of efficient protection of civil population and critical infrastructure in underground city space, in case of military or terrorist threats, requiring a unified approach to urban space development (for both its surface and underground components).

Currently two models of urban development exist–the extensive one, when the city's territorial growth is achieved by claiming the adjacent free territories, and the sustainable development (intensive) model, when the efficiency of using urban territory is increased by condensation of construction, mastering the inconvenient construction places, valorization of construction objects, broad exploitation of underground space with the purpose of increasing ecological and safety standards of urbanized space. The extensive model mostly applies to the previous century, prioritizing the heavy infrastructure elements, and city growth on the periphery, while the intensive model corresponds to the modern socio-technical conditions of sustainable city development with the maximally efficient usage of space, energy and communications.

The city construction space, due to its diversity and complexity, may be viewed as a city building system consisting of the main fragments: the construction tissue (filled with various objects having spatially localized functions); the communication framework (a network of roads of different categories, and their intersections–junctions); the social activity nodes (the transport and communication nodes, saturated with social service functions); the engineering support network; the landscape and ecological framework (the open landscaped space related to the natural backbone of the city) [5].

The modern period of urbanization development is defined not only by the intensive growth of some cities, but also the emergence of large urbanized areas. Large cities grow rapidly, absorbing the adjacent territories, and they merge together, forming agglomerations, metropolises, as well as drastically different forms of spatial development–polycentric urbanized regions with a population of several million people (in some cases more than 20 million) [6].

Innovative approaches to urban development pay notable attention to the capacities of efficient underground space exploitation, that significantly shifts the past trend of constructing individual underground facilities as local objects, to large-scale projects of systemic exploitation of underground space as an integral part of the entire urbanized environment. They allow to complexly solve the urgent city building problems, including territorial, transport, power supply, ecological, social, safety concerns [1, 4, 7]. Among the most distinctive examples in Europe we can note the "underground twin" of Helsinki city (over 400 underground objects, interconnected and tied to the surface construction), which was designed using a single, previously approved master plan [3].

Planning the localization of underground facilities should involve the system approach, which factors the synergetic influence of three components:

1. the surface part of urban territory, including surface buildings, surface transport communications, engineering infrastructure, water environment, land resources;
2. the underground part of city (both existing and planned), including georesources, underground transport communications (metro tunnels and stations, car tunnels, pedestrian underground crossings, car parking lots etc.). the engineering underground objects (hydrotechnical and energy facilities, engineering networks), and the objects with social functions (trade and entertainment centers, sport and hotel complexes, museums etc.);
3. the geological environment that is characterized by specific geology engineering and hydrogeological properties, and which requires a choice of constructing geotechnologies for mastering.

These three constituents should be considered in the processes of planning, investment, design, construction and exploitation of objects located in the underground space. Forming a plan for complex mastering of underground space requires setting the priorities in constructing underground facilities of different purpose, which, in turn, requires the application of system methodology [4, 8].

2 City Building Conditions and Factors of Urban Development

Let us consider the conditions (Fig. 1) that influence the planning organization of the underground space of metropolises, and the reasoning for priority directions of its mastering in the framework of the system methodology:

- complex conditions of urban territorial development;
- natural conditions and processes;
- conditions of functional planning organization of city space;
- transport planning conditions, set by the configuration of the transport network.

The complex conditions for urban territorial development, defined by the resource categories and city building factors, that substantially influence the urban territorial and spatial development, are: territorial resources, water supply resources and drainage conditions, city ecological state conditions, sanitary and hygienic conditions on territory, city's transport links to the regions of raw material and labor resources, the labor resource itself, energy supply conditions [9].

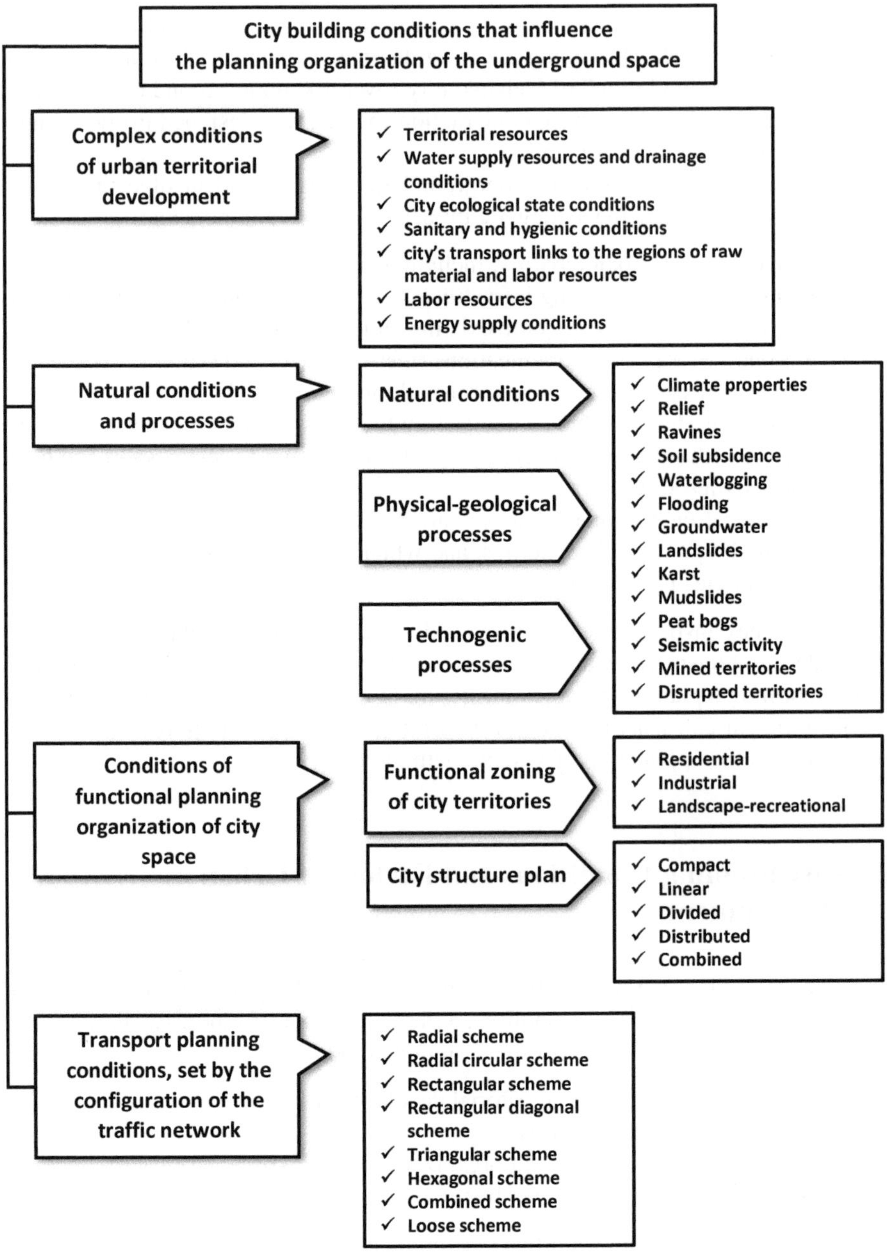

Fig. 1 City building conditions that influence the planning organization of the underground space

Territorial resources indicate the presence of reserve territories or areas that, according to their location relative to the city plan, their size, engineering constructive and city building requirements, could be used for city construction needs.

Water supply resources and drainage conditions also determine the city's development potential, its profile and prospects. The deficit of water resources causes the need to erect costly facilities for water supply (canals, water conduits etc.).

City ecological state conditions are profoundly analyzed to determine the necessary measures of their improvement, including reducing the level of smoke and fumes in air, contamination of water bodies and soil, industrial noise etc. These measures of improving the natural environment may impact the city development, particularly territorial development, as well as its functional zoning, the location of residential, industrial, social and other territories.

Sanitary and hygienic conditions on territory are studied to make a substantiated assessment of their state, and select the sites, which are favorable or hazardous from a sanitary standpoint, project the necessary sanitary requirements and health measures to be factored during design, determine the priorities of their implementation.

City's transport links to the regions of raw material and labor resources also essentially impact the functioning and development of the manufacturing complex, as well as the city's activity. Therefore, the throughput capacity of transport network, its ability to satisfy the current needs of different branches of national economy, the potential for growth and the respective investments should be analyzed.

Labor resources include the working age population, and the working people outside of the working age. The research of the city development conditions should detect and compare the amount of labor resource as a part of the city population, and their involvement in the social production, as well as the presence of unspent labor resource of the city and the adjacent settlements.

Energy supply conditions directly influence the city development and placement of the energy intensive industry branches. Determining optimal city development parameters, deciding the development strategy of its economy base, planning organization, and functional zoning of urban territories, the height and density of construction, the scale and methods of reconstruction cause significant fluctuation of energy consumption according to the city building situation.

Natural conditions and processes impact the functional zoning of territories, the height of construction, the tracing of the street network, the organization of transport connections, the placement of green areas, and other city building decisions, and also stipulate the favorability (or the lack of it) of mastering urban underground space.

Natural conditions include climatic, geomorphological, atmospheric (surface waters), geological, hydro-geological, hydrological [10].

Physical-geological processes include: flooding of urban areas by atmospheric waters or high river waters; flooding of urban areas by groundwater; ravine formations; landslides, landslips, screes; karst; mud flows; seismic phenomena.

Technogenic processes include: surface deformations during open and underground excavation or construction; territory disruption by waste from industrial and utility companies; flooding of territories by creating reservoirs, etc.

Urban territories after evaluation under a set of natural condition factors may be put into one of the three categories according to their city building suitability: favorable, insufficiently favorable, unfavorable.

The conditions of functional planning organization of city space are related to the zoning of urban territory that is conducted using the specifics of: functional utilization (functional zoning); maximum residents density per 1 hectare, and buildings' height (constructive zoning); ratios of built up to open areas, particularly landscaped areas (landscape zoning); location of construction objects regarding the city centre–central (according to the definition of the city center core limits for the largest, large and big cities), intermediate and periphery segments [11].

Broad multifunctional exploitation of the underground space by metropolises is the most relevant for central, most attended areas, as well as in various specialized centers, transport junction zones, and other social and transport complexes. However, the underground facilities could be situated almost anywhere, in different functional and territorial zones.

Zoning the territories of settlements is conducted according to the environmental protection, ecological, historical and cultural, and other planning restrictions. The bounds of the zones are determined considering the specifics of natural factors, historical evolution of city planning, peculiarities of transport and engineering infrastructure networks.

Functional zoning of urban territories entails the selection of the following city territories: residential, industrial, landscape-recreational, which should be taken into account during justification of the underground object design.

Residential territories are intended for organizing a welcome living environment, adhering to social, ecological and city building conditions, supporting the population's life activity processes, related to its demographic and social reproduction. Residential areas are formed mostly as the zones of household and social buildings, landscaped territories of common use, and other functional elements.

Household building zone comprises land plots and territories: multi-apartment buildings, manor residential buildings, residential buildings with premises of social purpose. Within the household building zone, aside from the residential buildings, other types of objects are allowed: social purpose objects; industrial objects, provided the lack of harmful emissions requiring the organization of sanitary protective zones; objects of recreational and wellness purpose; landscaped territories of common and restricted use; objects and networks of transport and engineering infrastructure.

The underground space of a residential zone is most rationally exploited by complexly allocating in it the underground transport infrastructure, including parking lots and garages, trading enterprises, catering and utility service facilities, civil defense shelters, all types of subsidiary objects of engineering infrastructure.

Social building zone is intended for concentrated allocation of institutions and establishments for service of urban and suburban population.

Social building zone is formed at the places of largest concentration of population that at daytime is located within the city center, and along the main roads and squares. The territories of multifunctional social centers in cities depend, aside from their size and place within the administrative territorial structure and settling system, on the specifics of functional planning of the city structure; historical, natural and landscape factors.

In the largest cities the social building zones should be formed as a concept of a general city center that, beside the central core, consists of social centers of planning zones, and centers of residential (industrial, recreational) districts and microdistricts.

Aside from the service institutions and establishments, the social centers of planning zones contain administrative, business centers, transport infrastructure objects (transport junctions, pedestrian areas, vehicle parking territories etc.), as well as land plots for residential buildings.

The main zone for complex exploitation of underground space is the system of a social city center, including: the central city core; main arterial roads; planned district and city zone centers; key social and transport nodes. All of these can be duplicated in the underground space, significantly unloading the city centers.

Manufacturing territories take second place by significance in the city planning structure and its territorial balance. They include industrial territories (industrial zones, industrial districts, clusters of enterprises, factories), innovative development territories (technology parks, industrial parks), utility companies, transport infrastructure, warehousing areas.

The *utility zone* is intended for allocating enterprises that provide services to objects and systems of social, transport, engineering infrastructure, as well as municipal engineering, and provide utility services to the city dwellers.

The *transport and warehousing zone* (logistics and warehouse centers) is commonly located at the peripheral zone of a settlement, or beyond its limits near the appropriate transport communications. These transport and warehousing zones in the largest cities sometimes become scattered. The warehousing complexes that aren't directly involved in the service to population, are located beyond the city limits, closer to the external transport nodes.

The enterprises and warehousing complexes can be situated in the underground space, if it is conditioned by the safety and energy saving requirements.

Landscape and recreational territories represent a network of landscaped sites and other open spaces of varying purposes, located both in urban or suburban areas, and in the inter-city areas, and include landscape complexes, recreational zones, health resorts and areas, objects of cultural heritage and tourist areas, territories of natural reserve and aquatic funds, windbreak and buffer strips, transport-separating green lanes and other objects of greenery management. Connecting these territories to other urban territories by underground transport is advisable, as it would favorably impact their ecological sustainability.

3 Functional and Spatial Differentiation of Urban Territories

The attempts to select a dominating city function, and introduce certain classifications of other functions, were done by various authors. However, the complexity of this task is related to the fact that in most of the existing cities the functions are mutually superimposed, creating an extremely eclectic canvas of functional exploitation of territories, making it exceedingly hard to define the dominating function. This causes city building problems related to the optimization of functional exploitation of urban territories, as a part of the general problem of spatial planning in city organization [12]. The modern city building practice proves that the actual life activity processes in a city do not constrain to the "traditional" scheme of the urban functional zoning "work–living–rest". Therefore, the modern trend of managing the city's life space potential rejects the rigid functional zoning of a city, and moves to creating multifunctional territories, where all main functions of a modern city can be implemented.

The city planning structure characterizes the urban entity as a unity of its interacting parts (elements), i.e. the totality of functional zones and transport arteries. We can say that the functional zoning scheme, and the territory differentiation by the intensity of its development (selecting a scheme of planning structure) are the two complementary models of urban planning, reflecting also the mastering of the underground space.

The city planning structure conditions take into account the *city structure plan*: *compact*, *linear*, *divided*, *distributed*, or *combined*.

The *compact (centric) structure plan* is characterized by allocating all of the city's functional zones within the same perimeter. The compact form of the plan, and the center accessibility can only be fully implemented for limited settlement size. Growing city territory raises the functional load on the city center, which at some point fails to cope with the role of the main node of the communication network.

Linear structures are less compact, and tend to condense along the main transport artery, which is the compositional axis of the city plan. The important advantages of the concentrated construction along the main arterial road are the economy of transit time, the capability to evolve without drastic reconstruction of the already formed districts, proximity of a settlement to the natural environment. However, the extensive length of longitudinal communications requires significant expenses on engineering, technical and transport facilitation of the city, whereas the continuous strips of buildings create the danger of artificial disruption of a natural landscape.

Divided type emerges when the city territory is sliced by rivers, ravines or transit railroads, significantly increasing the need to connect the divided territories by transport communications.

Distributed type relates to several city planning formations, tied by transport communications. The emergence of distributed type is often caused by the nature of a city-forming group of enterprises of this city (e.g. the mining industry), or the natural and climate conditions.

Combined structure plans have the traits of several types simultaneously. A *divided linear* system can appear when the city is situated along the bank of a wide river; the city usually does not reach too far in the transversal direction, instead extending along the river to large distances. In these cases the longitudinal connections acquire special significance, requiring high-speed transport.

The considered types of structure plans stipulate one of the most important characteristics to take into account when determining the most efficient place for functional placement (in our case, placement of the underground object function), that is–the evaluation of a location for an envisioned functional object regarding the different centers of gravity so that the effort of interacting with these planning structure centers will be minimal.

Transport planning conditions, set by the configuration of the transport network. The basis of the city structure plan is formed by the main arterial roads, which, along with other city roads and lanes, constitute the traffic network providing the interconnection between all parts of the city. The city traffic structure not only cements the planning structure, but also in many ways defines its further evolution, since the traffic management objects–arterial roads, overpasses, public transport lines (especially metropolitan in largest cities)–are incredibly costly, and so become the most stable elements of city planning.

Highlighting the arterial roads in a traffic network of the city, and viewing them as a basis of a city plan clearly demonstrates the geometric scheme of city planning. The configuration of the transport infrastructure is anchored in the city plan by the areas with the most intensive urban space mastering gravitating towards it, forming the relatively immutable, stable in time basis of the spatial planning city organization–its transport planning framework. Its duplication by ecologized car tunnels and metro lines significantly solves the transport and ecology problems of the cities.

Let us consider the varieties of traffic network schemes that define the conditions of transport ties planning organization for objects of the urban underground space: radial, radial circular, rectangular, rectangular diagonal, triangular scheme, hexagonal, combined, and loose.

The *radial* planning scheme is characterized by overloading the central city district, and complicated connections with the peripheral districts.

The *radial circular* planning scheme is essentially the radial scheme, complemented by the circular arterial roads that distribute the excessive load from the central city district by introducing convenient connections between districts, bypassing the central city core.

The *rectangular* planning scheme is a system of parallel and perpendicular transport arteries, characterized by high throughput and distribution of transport among parallel streets, and also by significant lengthening of routes that connect diagonally opposite districts and regions of the city.

The *rectangular diagonal* planning scheme simplifies the connections between different peripheral regions, or peripheral regions and center, due to the diagonal arterial roads. The drawback is the presence of multi-road junctions that complicate the traffic organization.

The *triangular* planning scheme was formed as an evolution of a rectangular diagonal or a radial scheme. However, it complicates the organization of traffic because of the large number of emerging junctions where multiple roads intersect in acute angles, and poor visibility on crossroads.

The *hexagonal* scheme is formed as a combination of hexagons, and allows to avoid complex intersections by switching to Y-shaped ones [13]. The road network that adheres to this principle is proposed for residential areas, where one-way, "quiet" traffic is necessary.

Combined transport planning scheme is a blend of schemes described above, and is typical for large cities where the old districts have the radial circular scheme, and newer districts have the rectangular scheme.

The *loose* scheme is usually conditioned by specific natural or artificial circumstances, e.g. relief, water bodies, railroad and other external factors. Sometimes a loose scheme becomes the result of a certain architectural intent.

The city building evaluation of all schemes may be conducted using the average weighted *nonlinearity coefficient* that reflects the generalized ratio of the actual distance between the destinations in a network, to the shortest route length between them [13]. The calculation of the traffic network functioning characteristics is made in the following order: developing the topological scheme of a traffic network, determining the volumes of traffic stream generation and absorption, calculating the movement time between the nodes of the network, calculating correspondences between the nodes of the network, distributing the traffic streams over the network. The efficiency assessment for measures of improving the city's traffic network is made by comparing the main expenses related to the functioning of traffic network for the basic and the suggested options, as well as the measure's payback period.

Strategic traffic planning uses the traffic model of a city, which is an efficient tool for quantitative assessment of the suggested variants for the traffic network development, their subsequent comparison and justified conclusions regarding the expediency of investments into the traffic infrastructure development projects. Transport models require extremely powerful software complexes that, basing on the functional and spatial characteristics of a city, combined with the present data on traffic supply and demand, calculate the most likely distribution of transport and passenger streams in a traffic system [14].

4 Application of System Methodology in Structural and Functional Planning of Underground Space

Structural and functional planning of underground space is attributed to the class of weakly structured problems, where the goals, structure and conditions are only partially known, and are characterized by a number of un-factors: uncertainty, imprecision, incompleteness, indistinctness of data describing the objects. Unlike the decision-making problems with quantifiable variables and dependencies between them, that can be solved using methods and techniques of operation research, econometrics and other similar methods, the weakly structured problems require specific decision-making methods. The description of weakly structured problems is usually performed by experts, specialists in the given field. The assessments they give are subjective, and generally are verbal in nature.

Methods of system analysis, commonly involved in tasks like these, include Delphi method [15], cross-impact analysis [16], morphological analysis [17, 18], analytic hierarchy or network processes [19–21], and other expert estimation-based, qualitative analysis methods. The topmost advantage of these methods is the capability to use the experience, knowledge, intuition of the specialists in the given field to study the problem, and make decisions regarding objects that are difficult to formalize using other methods. Delphi method is usually applied for high-level strategic decision-making support, its main strength lies in the coordinated work of a group of experts regarding the solution of a stated problem. Cross-impact analysis is important for constructing and evaluating scenarios accompanied by the emergence of interdependent varying events. Analytical hierarchy process and morphological analysis method are good at estimating decision priorities from different standpoints–according to an hierarchy of competing goals (analytical hierarchy process), or the potential of influencing the object having a number of uncertain parameters (morphological analysis method).

The authors already conducted several studies devoted to assessing potential underground construction sites, considering the geology engineering, structural and functional, and risk factors, using qualitative analysis methods, particularly a combination of analytical hierarchy process and BOCR [22], and modified morphological analysis [23–26]. The morphological analysis method should be paid special attention, as its essential procedure includes the multi-aspect classification of an object, whereas previous sections showed that multiple classifications of territories are critical in the process of structural and functional planning of city space. Additionally, as stated above, the classification should not be rigid, as some parameters may assume the attribution of an urban territory to several classes simultaneously, with different belonging measures–which is exactly the operating principle of the modified morphological analysis method [17].

In the authors' previous papers the functional planning optimization of underground space was studied separately in the context of geology engineering factors [23, 24], or, in the other cases, considering also structural and functional factors, but only for specific types of underground objects (parking lots [25], tunnels [26]). Complex factoring of the city building conditions that influence the underground space organization, will allow to consider a much general problem of planning the underground urban territories. This extended problem requires an improved morphological table network, comparatively to the previous studies. The following network of morphological tables is proposed (Fig. 2).

Let us describe in detail the potential parameter composition of each of the tables that are included in the network on Fig. 2.

1. *Functional planning organization of urban space*. The table describes the direct characteristics of the urban territory itself, as well as its place in the city's functional

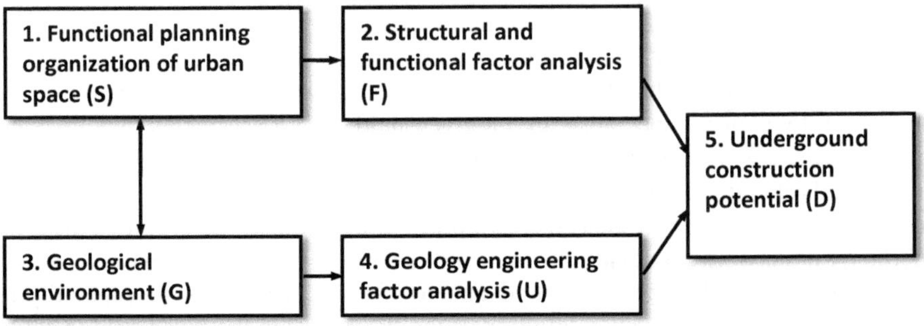

Fig. 2 A network of morphological tables for evaluating the underground construction potential of a territory

planning organization. The choice of parameters for this table is based on the city building conditions (Fig. 1), and may contain the following parameters, that are directly assessed or determined:

- functional planning organization of the city space (compact, linear, divided, distributed);
- downtown factor (distances to the city center, city limits);
- landscape conditions (relief, water bodies, park zones);
- traffic planning conditions scheme (radial, radial circular, rectangular, rectangular diagonal, triangular, hexagonal, loose);
- average traffic speed;
- passenger throughput of public transport (metropolitan, bus, trolleybus, tram);
- building types (residential, public, trade and entertainment centers, industrial, warehousing and transport, natural recreation areas, undeveloped areas);
- architectural and historical value of the objects (presence and number of showplaces, architectural monuments);
- existing underground objects and facilities;
- building density;
- population density (number of permanent dwellers);
- amount of workplaces;
- level of harmful industrial and transport emissions;
- etc.

Thus, a possible presentation of this morphological table is given in Table 1.

It should be noted that, according to the modified morphological analysis method, for every parameter, each of its alternatives obtains a separate estimate, allowing to form combined descriptions of conditions–for example, mixed types of functional planning organizations of urban space. The similar principle applies to the parameters which are hard to precisely assess (e.g., the population density, amount of workplaces); the method supports the use of indistinct parameter assessment.

Table 1 The morphological table for functional planning organization of urban space

S1. Urban space organization	S2. Distance to city center	S3. Distance to city limit
S1.1 Compact S1.2 Linear S1.3. Divided S1.4. Distributed	S2.1. Center S2.2. Territories close to center (downtown zone) S2.3. Intermediate territories close to peripheral zone S2.4. Peripheral zone	S3.1. Boundary zone (0–0,1L[a]) S3.2. Small distance (0,1–0,3L) S3.3. Average distance (0,3–0,6L) S3.4. Large distance (0,6–1L)
S4. Water bodies	S5. Forest park zones	S6. Traffic planning conditions scheme
S4.1. No impact in neighborhood S4.2. Small rivers S4.3. Small lakes, ponds S4.4. Coastal territory of large rivers, lakes, seas	S5.1. Absent S5.2. Small organized parks, squares S5.3. Park zones S5.4. Forest zones	S6.1. Radial S6.2. Radial circular S6.3. Rectangular S6.4. Rectangular diagonal S6.5. Triangular S6.6. Hexagonal S6.7. Loose
S7. Average traffic speed	S8. Public transport type	S9. Building type
S7.1. High (over 50 km/h) S7.2. Average (30–50 km/h) S7.3. Slowed (15–30 km/h) S7.4. Low (below 15 km/h)	S8.1. None in neighborhood S8.2. Metro S8.3. Bus S8.4. Trolleybus S8.5. Tram	S9.1. Residential S9.2. Social S9.3. Trade and entertainment centers S9.4. Industrial S9.5. Warehousing and transport S9.6. Recreation zones S9.7. Undeveloped areas
S10. Architectural and historical value	S11. Existing underground objects and facilities	S12. Building density
S10.1. Insignificant S10.2. Solitary objects S10.3. Large number of objects S10.4. Most of the neighborhood	S11.1. Absent S11.2. Tunnel communications S11.3. Metro stations, parking lots and affiliated facilities S11.4. Underground trade and entertainment malls	S12.1. Very low S12.2. Low S12.3. Average S12.4. High
S13. Population density	S14. Workplaces	S15. Harmful industrial and transport emissions
S13.1. Up to 1000 S13.2. 1000–3000 S13.3. 3000–6000 S13.4. 6000–10000 S13.5. Over 10000	S14.1. Up to 500 S14.2. 500–1000 S14.3. 1000–3000 S14.4. 3000–5000 S14.5. Over 5000	S15.1. Very low S15.2. Low S15.3. Average S15.4. High S15.5. Very high

[a]L is the distance from the city limits to the center of the city (main post office building)

2. *Analysis of structural and functional factors*. The table describes a site according to the factors, which are derivative from the first table, i.e. they are calculated from the basic characteristics of Table 1, using the two-stage modified morphological analysis procedure. The table allows to obtain an extended presentation of the area, the potential for its exploitation. This table may contain the following parameters:

- zone type: residential, industrial, landscape-recreational, commercial;
- proximity to the city center (center itself, downtown area, intermediate zone, periphery), or the relative distance to the center (the ratio of distance to center, to the distance between the center and the city limit in that direction);
- territory potential (the distribution of attraction factors for the territory) for underground construction;
  - proximity to large shopping malls;
  - proximity to transport nodes (transport communications);
  - architectural and historical significance;
  - ecological factor;
  - social factor (based on population density, construction quality, workplaces, social support, engineering and technical support);
  - safety factor (protection of population and critical infrastructure from military and terrorist threats).

The possible view of the morphological table is given in Table 2.

3. *Geological environment*. The table considers the direct characteristics of the geological environment that are assessed using expert estimation, if the analysis is conducted on the pre-project construction stage. The morphological table from the previous research [23] is proposed, with the parameters given in Table 3:

- level of dynamic load;
- static load from surface buildings;
- static load from soil;

Table 2 The morphological table for structural and functional factor analysis

F1. Zone type	F2. Location	F3. Territory potential
F1.1. Residential	F2.1. Center	F3.1. Proximity to center
F1.2. Industrial	F2.2. Downtown	F3.2. Proximity to shopping malls
F1.3. Landscape-recreational	F2.3. Intermediate zone	F3.3. Proximity to transport nodes
F1.4. Commercial	F2.4. Peripheral zone	F3.4. Architectural and historical significance
		F3.5. Ecological factor
		F3.6. Social factor
		F3.7. Safety factor

Table 3 The morphological table for geology engineering factors

Parameter	Alternatives
G1. Level of dynamic load	G1.1. Low (46–53 dB)
	G1.2. Medium (53–73 dB)
	G1.3. Increased (73–96 dB)
	G1.4. High (over 96 dB)
G2. Static load from surface buildings	G2.1. Insignificant ($Ksl < 1$)
	G2.2. Medium ($1 < Ksl < 2$)
	G2.3. Increased ($2 < Ksl < 3, 5$)
	G2.4. High ($Ksl > 3, 5$)
G3. Static load from soil	G3.1. Insignificant ($Kmas < 0, 05$ MPa)
	G3.2. Medium ($0, 05 < Kmas < 0, 3$ MPa)
	G3.3. High ($0, 3 < Kmas < 0, 5$ MPa)
	G3.4. Very high ($Kmas > 5$ MPa)
G4. Influence of existing underground objects	G4.1. Absent (distance over 50 m)
	G4.2. Slight (distance 20–50 m)
	G4.3. Significant (distance 10–20 m)
	G4.4 Hazardous (distance less than 10 m)
G5. Genetic type and lithologic composure of soil	G5.1. Unweathered clays and average density sands
	G5.2. Technogenic deposits (alluvial and bulk types)
	G5.3. Deluvial clay soils (water-saturated), water-saturated overfloodplain sands
	G5.4. Sedentary soils, soils with special properties (loess, peat, silt)
G6. Effective soil strength	G6.1. Very strong soils > 300 kPa
	G6.2. Strong soils 200−300 kPa
	G6.3. Average strength soils 150−200 kPa
	G6.4. Relatively strong soils < 150 kPa
G7. Influence of aquifers and perched groundwater	G7.1. Water-bearing horizons at P-N1np
	G7.2. Groundwater depth > 3 m, pressurized groundwater > 10 m
	G7.3. Groundwater depth < 3 m, pressurized groundwater < 10 m
	G7.4. Flooded areas/quicksand are present
G8. Landscape type and morphometrics	G8.1. Flat areas of overfloodplain terraces, morainic-glacial plains
	G8.2. Slightly tilted overfloodplain terraces, watershed ares
	G8.3. Small river valleys, slightly irregular slopes, high floodplain
	G8.4. Slope areas with ravines and steep banks, low floodplain
G9. Geological engineering processes	G9.1. Absent
	G9.2. Stabilized
	G9.3. Low displacement processes
	G9.4. Active manifestations of subsidence, underflooding, gravitational processes
G10. Geotechnologies of underground construction	G10.1. Open
	G10.2. Underground

- influence of existing underground objects;
- genetic type and lithologic composure of soil;
- effective soil strength;
- influence of aquifers and perched groundwater;
- landscape type and morphometrics;
- geological engineering processes;
- geotechnologies of underground construction.

4. *Analysis of geology engineering factors*. The table contains the conclusion regarding the suitability of the site from the geology engineering and related risks point of view, obtained using modified morphological analysis method. Here also a slightly modified table from the previous studies [23, 24] is suggested, with the following parameters (Table 4):

- site suitability (considering geology engineering factors);
- advisable object scale;
- advisable construction depth;
- geology engineering-related risk factors (the distribution between different groups);
- degree and level of risk in underground construction.

5. *Underground construction potential*. The table contains the generalized decisions on territory in the framework of the structural and functional planning of the underground space, assessed on the base of analysis of the structural and functional, and the geology engineering factors. As decision elements, the following parameters may be taken:

- social purpose objects–underground malls, sports, cultural, recreational objects;
- transport purpose objects–metropolitan, car tunnels, car parkings/garages, hydrotechnical and engineering communications;
- double purpose objects (civil defense shelters as the other function);
- construction scope (size, spatial organization: single-level, multi-level);
- potential integration with surface buildings and transport system;
- potential integration with existing underground objects.

A sample structure of this morphological table is given in Table 5.

Thus, the application of the modified morphological analysis procedure for the network from Fig. 2 allows to conduct analysis and fuzzy classification of territories by their structural and functional, and the geology engineering factors, as well as to obtain the decision alternative priorities regarding the different purpose exploitation of the site in underground construction, on the base of the observed or estimated input characteristics.

We would also like to note the potential of integrating the technique with geoinformational systems (GIS). GIS-based automated or semi-automated data input for Tables 1 and 3 allows to build a city-wide "temperature map" of construction advisability regarding the underground objects of different type, highlighting the most prospective areas.

Table 4 The morphological table for analysis of geology engineering factors

Parameter	Alternatives
U1. Site suitability	U1.1. Suitable
	U1.2. Not suitable
U2. Object scale (cross-section)	U2.1. Cross-section up to 10 m^2
	U2.2. Cross-section up to 25 m^2
	U2.3. Cross-section up to 40 m^2
	U2.4. Cross-section up to and over 40 m^2
U3. Underground object length	U3.1. Up to 50 m
	U3.2. 50–200 m
	U3.3. 200–1000 m
	U3.4. Over 1000 m
U4. Construction depth	U4.1. 0–10 m
	U4.2. 10–20 m
	U4.3. 20–60 m
	U4.4. below 60 m
U5. Risk factor	U5.1. Construction failure, malfunction
	U5.2. Increasing construction and operation cost
	U5.3. Dangerous influence on surface or neighboring underground objects
	U5.4. Initiating displacements and other unwanted geological processes
U6. Risk degree	U6.1. < 3%
	U6.2. 3–10%
	U6.3. 10–20%
	U6.4. 20–50%
	U6.5. > 50%
U7. Risk level	U7.1. 0,1–5% Q[a]
	U7.2. 5–20% Q
	U7.3. 20–50% Q
	U7.4. > 50% Q

[a] Q is the underground object construction costs (including materials and other expenses)

5 Conclusion

Guaranteeing sustainable and balanced development of large cities requires intensive and large-scale mastering of their underground space, with urban environment planning that synthesizes the simultaneous development of surface and underground facilities. Contrary to the surface construction, where the density is extremely high, and the amount of free construction sites is nearing exhaustion (at least, in downtown areas), the underground space remains the primary territorial resource (in most metropolises the underground objects take no more than 5–10% of the city territory),

Table 5 The morphological table for underground construction potential

D1. Social purpose objects	D2. Transport purpose objects	D3. Industrial and power supply objects	
D1.1. Inadvisable D1.2. Underground malls D1.3. Sports, cultural objects D1.4. Recreational objects	D2.1. Metropolitan D2.2. Car tunnels D2.3. Car parkings/garages, railway stations D2.4. Hydrotechnical and engineering communications	D3.1. Inadvisable D3.2. Underground factories D3.3. Underground power plants D3.4. Separate underground facilities of surface objects	
D4. Double purpose object potential	D5. Spatial organization	D6. Potential integration with surface buildings	D7. Potential integration with existing underground objects
D4.1. Minimal D4.2. Limited D4.3. Significant	D5.1. Single-level D5.2. Multi-level	D6.1. Minimal D6.2. Limited D6.3. Significant	D7.1. Minimal D7.2. Limited D7.3. Significant

enabling to search among a broad set of alternatives for the optimal territory to place the underground complexes. This choice should take into account as large number of city building conditions, urban development and geological environment factors as possible, constituting a weakly structured task, that could be most efficiently implemented using system methodology, particularly the modified morphological analysis method. A pioneer network of morphological tables is proposed, that evaluates the territory's underground construction potential by synthesizing the functional planning organization of urban space, and the geology engineering factors of the geological environment on the studied areas. The combination of this approach with the capacities of geoinformational systems opens up new possibilities for functional planning optimization of the underground space exploitation.

References

1. Gilbert, P., et al.: Underground Engineering for Sustainable Urban Development. The National Academies Press, Washington (2013)
2. Sterling, R., Admiraal, H., Bobylev, N., Parker, H., Godard, J., Vähäaho, I., Shi, X., Hanamura, T.: Sustainability issues for underground spaces in urban areas. In: Proceedings of ICE. Urban Design and Planning, vol. 165, no. 4, pp. 241–254 (2012)
3. Vähäaho, I.: Underground space planning in Helsinki. J. Rock Mech. Geotech. Eng. **6**, 387–398 (2014). https://doi.org/10.1016/j.jrmge.2014.05.005
4. Pankratova, N.D., Haiko, H.I., Savchenko, I.O.: Development of Urban Underground Studies as a System of Alternative Project Configurations. Naukova Dumka, Kyiv (In Ukrainian) (2020)
5. Gutnov, A.Y.: Evolution of City Building. Stroyizdat, Moscow (In Russian) (1984)
6. Bezliubchenko, O.S., Hordiienko, S.M., Zavalnyi, O.V.: City Planning and Transport: Reference Book. O.M, Beketov KhNUMH, Kharkiv (In Ukrainian) (2021)
7. Admiraal, H., Cornaro, A.: Underground Spaces Unveiled: Planning and Creating the Cities of the Future. ICE Publishing, London (2018)

8. Resin, V.I., Popkov, Y.S.: Development of Large Cities in Conditions of Transitional Economics. System Approach. Bookhouse “LIBROCOM”, Moscow (In Russian) (2013)
9. Kliushnichenko, Y.Y.: Technical and Economic Foundations in City Building: Textbook. Budivelnyk, Kyiv (In Ukrainian) (1999). https://core.ac.uk/download/pdf/187725644.pdf. Cited 7 Apr 2023
10. Lynnyk, I.E., Zavalnyi, O.V. (eds.): Urban Territory Design: Textbook. Part 2. O.M. Beketov KhNUMH, Kharkiv (In Ukrainian) (2019)
11. DBN N.2.2-12:2019: Urban construction and territory building (2019) State inspection of architecture and city building of Ukraine, Kyiv (In Ukrainian) (2019). https://dbn.co.ua/load/normativy/dbn/b_2_2_12/1-1-0-1802. Cited 7 Apr 2023
12. Pleshkanovska, A.M.: Functional Planning Optimization of Urban Territory Exploitation. Logos, Kyiv (In Ukrainian) (2005). https://www.researchgate.net/publication/351010233_Funkcionalno-planuvalna_optimizacia_vikoristanna_miskih_teritorij. Cited 7 Apr 2023
13. Lynnyk, I.E., Zavalnyi, O.V. (eds.): Urban Territory Design: Textbook. Part 1. O.M. Beketov KhNUMH, Kharkiv (In Ukrainian) (2019)
14. Osietrin, M.M., Bespalov, D.O., Dorosh, M.I.: Main Principles of Creating a Traffic Model of a City. City Building and Territorial Planning: Scientific Technical Digest, KNUBA, vol. 57, pp. 309–320 (In Ukrainian) (2015). https://repositary.knuba.edu.ua/handle/987654321/7449?locale-attribute=en. Cited 7 Apr 2023
15. Pankratova, N., Malafieieva, L.: Delphi Method. Theory and Applications. Reference Book. Naukova Dumka, Kyiv (In Ukrainian) (2017)
16. Weimer-Jehle, W.: Cross-impact balances: a system-theoretical approach to cross-impact. Technol. Forecast. Soc. Chang. **73**, 334–361 (2006). https://doi.org/10.1016/j.techfore.2005.06.005
17. Pankratova, N., Savchenko, I.: Morphological Analysis. Problems, Theory, Application. Naukova Dumka, Kyiv (In Ukrainian) (2015)
18. Ritchey, T.: Problem structuring using computer-aided morphological analysis. J. Oper. Res. Soc. **57**, 792–801 (2006). https://doi.org/10.1057/palgrave.jors.2602177
19. Saaty, T.L.: How to make a decision: the analytic hierarchy process. Eur. J. Oper. Res. **48**(1), 9–26 (1990). https://doi.org/10.1016/0377-2217(90)90057-I
20. Pankratova, N., Nedashkivska, N.: Models and Methods of Hierarchy Analysis. Theory. Applications: Reference Book. “Polytechnica” Publishing, Kyiv (In Ukrainian) (2010)
21. Nedashkivs’ka, N.: A systematic approach to decision support based on hierarchical and network models. Syst. Res. Inf. Technol. **1**, 7–18 (2018). (In Ukrainian)
22. Pankratova, N., Nedashkovskaya, N., Haiko, H., Biletskyi, V.: Assessment of environmental risks of underground transport infrastructure development by BOCR method. V.N. Karazin Kharkiv National University reports, “Geology. Geography. Ecology” series, vol. 55, pp. 130–143 (2021). https://doi.org/10.26565/2410-7360-2021-55-21
23. Pankratova, N., Savchenko, I., Gayko, G., Kravets, V.: Evaluating perspectives of urban underground construction using modified morphological analysis method. J. Autom. Inf. Sci. **50**(10), 34–46 (2018). https://doi.org/10.1615/JAutomatInfScien.v50.i10.30
24. Haiko, H., Savchenko, I., Matviichuk, I.: Development of a morphological model for territorial development of underground city space. Naukovyi Visnyk Natsionalnoho Hirnychoho Universytetu **3**, 92–98 (2019). https://doi.org/10.29202/nvngu/2019-3/14
25. Haiko, H., Savchenko, I., Matviichuk, I.: A morphological analysis method-based model of assessing territories for underground parking lots. In:2020 IEEE 2nd International Conference on System Analysis and Intelligent Computing, SAIC 2020 (2020)
26. Haiko, H., Savchenko, I.: Assessing territories for urban underground objects using morphological analysis-based model. In: Zgurovsky, M., Pankratova, N. (eds.) System Analysis & Intelligent Computing. Studies in Computational Intelligence, vol. 1022. Springer, Cham (2022). https://doi.org/10.1007/978-3-030-94910-5_6

Analysis and Modelling of the Underground Tunnel Planning in Uncertainty Conditions

Nataliya Pankratova, Vladymyr Pankratov, and Danylo Musiienko

Abstract Analysis and modeling of underground tunnel planning under uncertainty for megacities based on the use of foresight and cognitive modeling methodologies are proposed. The use of foresight methodology makes it possible to construct alternative scenarios with quantitative characteristics, which are input data for cognitive modeling. An algorithm is created to build a numerically stable cognitive map, the vertices of which contain reliable data from the foresight stage. The application of cognitive impulse modeling makes it possible to build scenarios for underground tunnels, taking into account uncertainties of various kinds, to analyze the impact of any threats types on them, to select the more reliable ones and justify of their creation priority.

Keywords Foresight · Cognitive · Impulse modelling · Planning · Scenarios · Underground tunnel

1 Introduction

The post-war restoration of Ukraine's large cities requires making systemic functional and planning decisions that address the new scale and increased pace of urban growth. Regulating urban development with the goal of increasing comfort, ecological standards and life safety in the constantly growing metropolises is one of the most urgent though insufficiently studied and complex world problems [1]. The most characteristic threats to the urban environment are air pollution by motor vehicle emissions, dynamic and noise manifestations, the negative effects of industrial zones, waste and garbage storage facilities, breakdowns in sewage, water and energy supply systems, overloaded sewage treatment plants, traffic accidents and traffic jams, landslides, other dangerous processes of the geological environment, which attract the

N. Pankratova (✉) · V. Pankratov · D. Musiienko
Institute for Applied System Analysis, Igor Sikorsky Kyiv Polytechnic Institute,
Peremogy avenue 37, Kyiv 03056, Ukraine
e-mail: natalidmp@gmail.com

M. Zgurovsky and N. Pankratova (eds.), *System Analysis and Artificial Intelligence*,
Studies in Computational Intelligence 1107,
https://doi.org/10.1007/978-3-031-37450-0_4

constant attention of researchers and environmental organizations [2]. World concepts of safe development and ecologization of urban environment pay significant attention to the capacity of underground space to assume the functions of the most critical and vulnerable surface objects and communications, providing the minimization of technogenic and ecological risks, military and terrorist threats. Underground urban planning is a complex system in many aspects. Firstly, this system consists of many interconnected subsystems and objects. Secondly, the processes occurring in this system during construction and during operation are also complex and in some cases poorly predictable, because they are largely associated with various geological processes an different types of uncertainty. The problems that accompany underground urban development can be attributed to poorly structured problems. Each major city has objects of critical environmental and technogenic threats. For example, for the city of Kiev the traffic jams of the downtown arteries and bridges over the Dnipro give rise to a number of transport and environmental problems, the system solution of which is possible only through the construction of underground road tunnels. In accordance with the draft master plan of Kyiv development until 2025, long road tunnels will be a new and very important direction of the development of the Ukrainian capital's transport infrastructure, which should significantly increase the capacity of transport routes, help to preserve the historic buildings in the city center, unload it from the transit and significantly improve the environmental performance of the center (Fig. 1) [3]. In contrast to the relatively small construction sites of parking lots, which are localized mostly in a single-type geological environment, tunnels have a considerable length (thousands of meters) and can be in different engineering-geological and morphological conditions (to have variable alternatives of impact parameters). To display this situation correctly, it is necessary to take into account the variability of parameter alternatives along the tunnel route by having an expert enter such an estimate for each alternative in the survey forms, which corresponds to the length of the area with the alternative parameter referred to the entire length of the tunnel. It is also quite important to take into account the structural and functional characteristics of the area around which the tunnel passes. The number of road tunnels according to the Master Plan is 8, three of which will go under the Dnipro, and five will connect the transport arteries of the city within the right bank.

According to the "optimistic scenario" over 20 km of transport tunnels can be built in Kiev in the next ten years. At the same time it is necessary to justify the expediency and reliability of the tunnel construction taking into account the development characteristic to the territory in question, the road network, and the characteristics of traffic in the area of the potential tunnel.

All of the above allows to propose a methodology of foresight and cognitive modeling of complex systems [4, 5] for analysis and modeling of the tunnels development planning in the capital under conditions of environmental, man-made and terrorist threats.

Analysis and modelling of the underground tunnel planning in uncertainty conditions

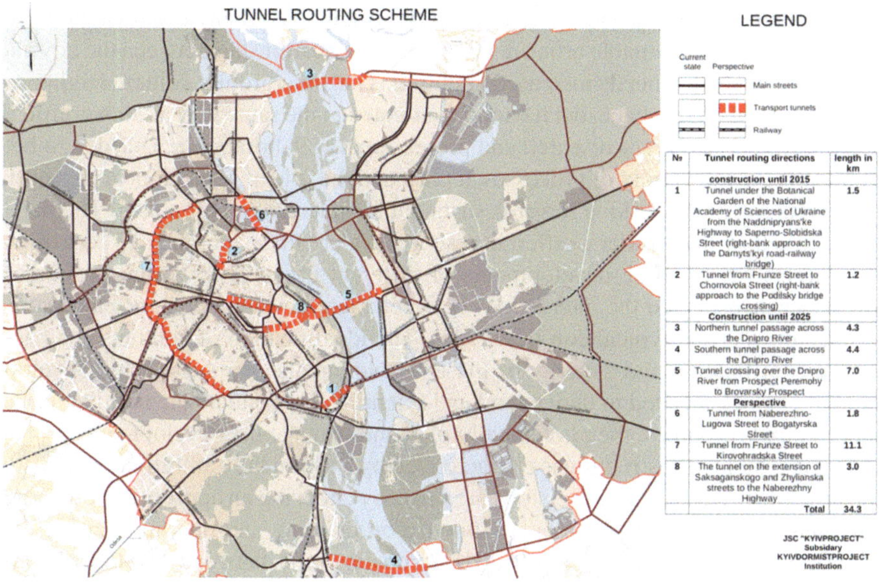

Fig. 1 Scheme of the road tunnels tracing (Master Plan for the Development of Kiev until 2025)

2 Related Papers

The modern approach to the planning of megacities is based on the systemic concept of sustainable development of the city, within which the modern needs of society must be satisfied without loss and damage to future generations. An important aspect of the sustainable development of cities is the possibility of timely response to changes in the structural, functional and natural environment and minimization of technogenic and environmental risks. This concept affects the scale and strategy of many engineering projects and involves changing the traditional vision of local problems in the direction of considering projects from the standpoint of large natural-technical and social systems. Article [6] focuses on sustainability issues related to the use of urban underground space, including the contribution to an environmentally sustainable and aesthetically acceptable landscape, the expected longevity of the structure, and the preservation of opportunities for urban development by future generations. However, due to their initial cost and the constant change of the underground environment, when working with underground structures, special attention is paid to long-term planning, which takes into account life cycle costs and benefits and the selection of projects that contribute most to the sustainability of the city, rather than a fixed term for individual need. An analysis of the use of urban underground space and its impact on the sustainability of cities is presented in [7]. Sustainable underground urbanism develops around two main concepts or spaces: first, the urban underground space as opposed to above ground; secondly, the sustainability and role or potential role of underground space in the work towards sustainable urban development. In [8] is argued that for sustainable urban development the relationship between the city and the ground beneath it needs increased attention. The underground volume might provide additional urban space, but it cannot be treated in the same way as above-ground space. Cross-disciplinary research and professional collaboration are needed to better understand the variety of processes at play, and the role of geotechnical engineers and geoscientists in working towards sustainable underground urbanism. The paper [9] discuss how to consider a different resources in urban development. There might be conflicts between the developments of different urban underground resources. The paper investigated the interactions between these developments, revealed some serious impacts and typical conflict modes. The identification of conflicts is a basis for the coordination and synergy of these developments. For sustainable development of city, it is needed to understand and scientifically evaluate the multiple urban underground resources, then holistically plan and manage the developments.

Underground infrastructure is characterized by a technically ambitious and costly construction process. Basic decisions are made by choosing an appropriate construction method. These decisions have influence on building and operating costs. The specific nature of such decisions needs a consideration between ecological, economical, technical and social aspects. The method of Analytic Hierarchy Process (AHP) is implemented for decision problems [10]. At the stage of designing of underground structures it is necessary to consider and justify the socio-economic feasibility and technical feasibility of construction of underground structures in the mining-geological conditions and under the influence of technology of construction works, functional purpose of underground building facilities. This peculiarity is pointed out in the article [11]. When planning a tunnel construction project, finding the best possible alignment is a very important and substantial task. The main goal is to carry out a solution which is realizable and which minimizes risks and costs at the same time. Not only the general feasibility and the result of risk and cost assessments have to be considered, but also many other requirements. These include, for example, the consideration of existing structures, specific driving dynamics and safety requirements depending on the function of the tunnel and mode of transportation, or the tunnel design itself, such as the overall diameter or number of tubes. In [11]. presenting an approach which enables the integration of all planning data based on open standards and automatically checks all necessary requirements. In the operating phase of a road tunnel, not only maintaining or increasing the availability in the network but the economic optimization regarding the life cycle costs of the structure are also important priorities. A consistent application of the Building Information Modelling (BIM) methodology can theoretically make a useful and targeted contribution, as it provides a complete digital model of the structure with all installed elements and the information required for the operator tasks. In the article [12] the results of the research are presented on the basis of specific use cases of a BIM-based operation and maintenance management. Much attention when planning underground facilities is paid to environmental issues, reducing ground noise sources, minimizing vibrations in the environment, both caused by the railway and created by the explosion of rocks [13–17]. When laying underground tunnels in megacity complex systems, it is necessary to take into account the latent indicators of practical necessity, technological possibility and economic effectiveness of planed tunnel [18].

The issues of creating tunnels and the rationale for their necessity and expediency are being discussed in many countries. In particular, significant work is underway in India in the construction of tunnels for hydroelectric power plants, irrigation, construction of roads and railways. Tunneling technologies for planning, design and construction are presented in the article [19]. In Shanghai, during the construction of a tunnel, some surface deformation characteristics were analyzed to assess the current level of construction technology in soft ground [20]. A simple classification and influencing factors associated with unfavorable geological conditions is presented in [21]. It discusses the main associated problems and the corresponding mitigation measures for tunneling in adverse geological conditions. In 1998, the Finnish Ministry of the Environment appointed a committee to review existing underground construction planning systems [20]. The correlations between urban indicators and underground space development, as well as between each underground indicator and the other, are considered in [23]. Based on summarizing underground space evaluation indicators and the city indicators that affect underground space development, 12 indicators from categories of economy, society and underground are chosen to assess urban underground space (UUS) development.The 7 urban indicators that may affect the development of UUS. Based on sample analysis and other detailed works, four main conclusions of UUS developments in Chinese cities are put forward. The paper [24] presents the current status of spatial planning regulations in Germany concerning underground space, and investigates possible planning solutions for its coordinated use. Findings of this study indicate that dedicated underground spatial planning should replace the common procedures towards designating exclusive areas for certain purposes in order to ensure the sustainable use of underground space.

It follows from the above review that one of the complex problems is underground tunnels, which ensure the vital activity of both surface and underground urban planning. This paper examines the issues of the analysis and modeling of the underground tunnels planning in uncertainty conditions and the rationale for the priority of their creation.

3 Models and Methods

The analysis and modeling of underground tunnel planning under conditions of uncertainty for megacities is proposed based on the simultaneous use of foresight and cognitive modeling methodologies. The use of foresight methodology allows using the procedures of expert evaluations to identify critical technologies and build alternative scenarios with quantitative characteristics. Cognitive modeling is used to justify the implementation of a particular scenario, which allows to build causal relationships based on knowledge and experience, to understand and analyze the behavior of a complex system for a strategic perspective with a large number of interconnections and interdependencies. The analysis of the properties of underground infrastructure and consideration of the set of problems arising from their study allows us to point out the uncertainties and peculiarities of their study:

- underground infrastructure contains a large number of heterogeneous subsystems with a large number of interconnections–carriers of information of different nature;
- to make decisions to purposefully change the behavior of underground infrastructure, it is necessary to solve a large number of interrelated problems using various methods of both quantitative and qualitative analysis, human knowledge and experience;
- in the study of underground infrastructure there are factors of uncertainty and multifactorial risks of different nature.

These features determine the focus of the use of methods and tools of systems analysis methodology in the study of underground construction. A set of solutions in the form of a specific scenario, providing targeted research of underground infrastructure on the basis of cognitive modeling, is taken on the basis of the following source material:

- information about the current state of all technological and organizational subsystems;
- monitoring data;
- the recommendations of the expert system when involving methods of qualitative analysis;
- a set of given or formulated by the decision-maker criteria and conditions of anthropogenic environmental factors;
- heuristic knowledge and assumptions of the expert group.

As a result of the comparative analysis of cognitive modeling systems and system dynamics, it should be noted that the main functions of such systems are:

- description of the situation;
- determination of target factors;
- determination of controlling factors;
- determination of measures to influence the situation;
- definition of functional relationships to build a cognitive model in the form of patterns of statistical information about the situation under study, presentation of values in the form of a fuzzy set, qualitative evaluations (expert assessment), attributing values on a scale of strength of relationships, laws of functioning;
- determination of trends in the development of situations through simulation modeling;
- developing strategies and analyzing their perspective in the context of modeling goals.

Almost all systems perform the same functions in one way or another, but the objective reason for their existence is the unresolved problems:

- selection and ranking (selection of basic and secondary) factors at the stage of constructing a cognitive map;

- determination of the degree of mutual influence of factors for assigning weight coefficients to elements of the cognitive model.

A systematic approach to the modeling and scenario analysis of underground tunnel planning under environmental, man-made and terrorist threats, based on the joint application of the methodologies of foresight and cognitive modeling is used [25, 26]. It is proposed at the first stage to apply the methodology of foresight using the method of morphological analysis. The results obtained are used as background information to find ways to build an alternative of this or that scenario in the form of a cognitive map. An algorithm is created to build a numerically stable cognitive map, the vertices of which contain reliable data obtained at the stage of the method of morphological analysis applying [27]. The application of cognitive impulse modeling makes it possible to build scenarios for underground tunnels, taking into account uncertainties of various kinds, to analyze the impact of any threats types on them, to select the more reliable ones and justify of their priority creation scenario.

To justify the implementation of this or that scenario alternative, cognitive modeling methodology is involved [28]. In the study of the underground tunnel planning problem at the first stage were used cognitive models such as a cognitive map - sign oriented graph (1) and a functional graph in the form of a weighted sign digraph

$$V = \langle V, E \rangle \,, \tag{1}$$

where: V–set of nodes $V_i \in V, i = 1, 2, \ldots, k$, which are elements of the system under study; E–set of arcs $e_{ij} \in E$; $i, j = 1, 2, \ldots, N$, reflecting the relationship between the nodes V_i and V_j; the impact can be positive (sign "+" above the arc), when the increase (decrease) of one factor leads to an increase (decrease) of another, negative (sign "–" above the arc), when the increase (decrease) of one factor leads to a decrease (increase) of another, or absent (0).

Vector graph

$$\Phi = \langle G, X, F(X, E), \Theta \rangle \,, \tag{2}$$

where G is a cognitive map; X is a set of node parameters, Θ–vertex parameter space; $F(X, E)$ - arc transformation functional.

At the next stage of cognitive modeling, a pulse process model (cognitive modeling of perturbation propagation) was used to determine the possible development of processes in a complex system and develop development scenarios [29]. Involving the impulse cognitive modelling with disturbing factors in the form of the potential unfavorable events that may include both natural emergencies and catastrophes, and events of technogenic or anthropogenic origin (including those with malicious intent–military action, terrorism acts), potentially influencing the urban underground object or a class of objects, provides the opportunity to construct the justified scenarios for further decision making

$$x_{v_i}(n+1) = x_{v_i}(n) + \sum_{j=1}^{N} f(x_i, x_j, e_{ij}) P_j(n) + Q_{v_i}(n). \tag{3}$$

Where $x(n)$, $x(n+1)$ are the values of the index in the vertex V_i at the simulation steps at the moment $t = n$ and following it $t = n + 1$; $P_j(n)$ is the momentum in the vertex V_j at the moment $t = n$; $Q_{v_i}(n) = (q_1, q_2, ..., q_k)$ is the vector of external momentum (disturbing or controlling actions) introduced in the vertices V_i at time n.

Simulation cognitive modeling, especially at the design stage of underground space development, is extremely necessary. A serious reason for this may be the fact that it is necessary to anticipate and exclude or reduce the risks, which are inevitably inherent in the underground urban development. Therefore, the basic principle of research on improvement of methods of design and construction of underground facilities is the principle of minimization of damage from consequences of negative manifestations of risks taking into account interaction of all natural, technical, technological and other factors. A possibility of catastrophes caused not only by unpredictable natural situations but also by design errors, imperfection of existing technologies requires formulation and solving problems of object viability in extreme and emergency situations. One of the complex problems is the underground tunnels providing life for both surface and underground urban development. This paper explores the construction of underground tunnels and the justification of the priority and reliability of their construction.

4 Algorithm for Converting to a Numerically Stable Matrix

Before starting to convert the original matrix $\hat{A} \in M_{n\times n}([-1; 1])$ with elements $\hat{a}_{ij}$ to a numerically stable matrix $A \in M_{n\times n}([-1; 1])$ with elements a_{ij} such that $\rho(A) < 1$, let's define a set of possible solutions as a set of matrices $S_{\hat{A}} \subset M_{n\times n}([-1; 1])$, such that: $\forall A \in S_{\hat{A}} \forall \hat{a}_{ij} : a_{ij} = 0 \wedge \forall A \in S_{\hat{A}} \forall \hat{a}_{ij} \neq 0 : a_{ij}\hat{a}_{ij} > 0$, that means that the elements with indices (i, j) which are equal to zero in the initial matrix $\hat{A}$ should be equal to zero in the matrix A, and also the elements should not change their sign in the process of finding a numerically stable matrix.

In the general case, exists $p > 1$, dividing by which all elements of the matrix $\hat{A}$, we obtain a numerically stable matrix, such that $\exists d > 0 : \rho\left(\frac{1}{p}\hat{A}\right) < 1$.

Proceeding from the Gershgorin theorem [30], the constant p can be chosen from the set of solutions of the inequality $\hat{p} > \max_i \sum_{j\neq i} \left|\hat{a}_{ij}\right|$, where $\hat{a}_{ij}$. Note that the constant p is positive because of the fact that with negative p, the matrix $\frac{1}{p}\hat{A} \notin S_{\hat{A}}$. In practice, using this approach, the obtained matrix A does not guarantee reliable results, since this approach does not take into account the values obtained from the morphological analysis, which are used in the original matrix $\hat{A}$. The approach to finding a numerically stable matrix A from the original matrix $\hat{A}$ which compensates this drawback is described below.

Let us define two two sets $C^+, C^- \subset I^2$, consisting of such pairs of vertex indices, that belonging of pair of indices (i, j) to set C^+ means that increasing of value in vertex V_i at simulation step k leads to increasing of value in vertex V_j at all simulation steps starting from $k + 1$, provided that value in all other vertices at simulation step k equals 0. The belonging of the pair of indices (i, j) to the set C^- means that the increase of value in vertex V_i at simulation step k leads to the decrease of value in vertex V_j at all simulation steps starting from k+1, provided that the value in all other vertices at simulation step k equals 0. Note that $C^+ \cap C^- = \emptyset$, because belonging of the pair of vertex indices (i, j) to two sets C^+ and C^- means that the increase of value in vertex V_i at simulation step k will simultaneously lead to increase and decrease of value in vertex V_j at any simulation step starting from k+1, which is impossible.

Let us define a function $w(A, i, j, k)$ that will obtain the value 0 if the condition of increasing or decreasing the value in vertex V_j is satisfied when $(i, j) \in C^+$ or $(i, j) \in C^-$ for matrix A and simulation step k, respectively, and will obtain the value 1 otherwise (4).

$$w(A, i, j, k) = \begin{cases} 1, & x_{k_j} - x_{(k-1)_j} \leq 0, (i, j) \in C^+ \\ 1, & x_{k_j} - x_{(k-1)_j} \geq 0, (i, j) \in C^- \\ 0, & x_{k_j} - x_{(k-1)_j} > 0, (i, j) \in C^+ \\ 0, & x_{k_j} - x_{(k-1)_j} < 0, (i, j) \in C^- \end{cases} \tag{4}$$

where $\mathbf{x}_k$ is a vector composed of values of vertices from the set of vertices V at simulation step k and:

- $\mathbf{x}_0 = \mathbf{0}$,
- $\mathbf{x}_1 = (0, \ldots, x_{1_j} = 1, \ldots, 0)$,
- $\mathbf{x}_{k+1} = \mathbf{x}_k + (\mathbf{x}_k - \mathbf{x}_{k-1})A$.

Let's state that number of simulation steps $N > 2$, then the sum of values of function w for all pairs of indexes of vertices (i, j) from sets C^+ and C^- and all simulation steps $k = \overrightarrow{2, N}$ is shown in (4).

$$\sum_{(i,j)\in C^+\cup C^-} \sum_{k=2}^{N} w(A, i, j, k)$$

Then the minimum value of the sum (4) equals to 0 if the function w becomes 0 for all pairs of vertex indices $(i, j) \in C^+ \cup C^-$ at each simulation step $k = \overrightarrow{2, N}$, and a maximum value of $(N - 1) * \left|C^+ \cup C^-\right|$ if the function w becomes 1 for all pairs of vertex indices $(i, j) \in C^+ \cup C^-$ at each simulation step $k = \overrightarrow{2, N}$. Let us define the function m as

$$\frac{\sum_{(i,j)\in C^+\cup C^-} \sum_{k=2}^{N} w(A, i, j, k)}{(N - 1)\left|C^+ \cup C^-\right|}$$

Thus, the function m takes values ranging from 0 to 1. To estimate the spectral radius of matrix A of dimension $n \times n$, let us define the function s as

$$s(A) = \begin{cases} 0, & \rho(A) < 1 \\ \dfrac{\rho(A)}{\max_i \sum_j \psi\left(a_{ij}\right)} & \rho(A) \geq 1, \end{cases} \tag{5}$$

Where $\psi\left(a_{ij}\right) = \begin{cases} 0, & a_{ij} = 0 \\ 1, & otherwise \end{cases}$

Note that since all the elements of the matrix A ranging from -1 to 1, and all the diagonal elements of the matrix are equal to zero, and the off-diagonal elements of the matrix which are equal to zero are not subject to change, based on the Gershgorin theorem [30], the eigenvalues of the matrix will be in the Gershgorin circle $D\left(0, \max_i \sum_j \psi(a_{ij})\right)$. In this case the maximal value of eigenvalues modulo, of matrix A will not exceed the value of $\max_i \sum_j \psi(a_{ij})$, which implies that $\max_i \sum_j \psi(a_{ij})$ is the maximal possible value of the spectral radius of matrix A of dimension n×n with the elements $a_{ij} \in [-1; 1]$. In this case, since $\rho(A) \leq \max_i \sum_j \psi(a_{ij})$, function s ranges from 0 to 1.

To estimate the spectral radius of matrix A, let us define the function d as

$$d(A, \hat{A}) = \frac{\|A - \hat{A}\|}{max_{\tilde{A} \in S_{\hat{A}}} \|\tilde{A} - \hat{A}\|}, \tag{6}$$

where the value of $max_{\tilde{A} \in S_{\hat{A}}} \|\tilde{A} - \hat{A}\|$ depends on the chosen matrix norm. If the Euclidean norm is used, the value of $\sqrt{\sum_{\hat{a}_{ij} \neq 0} \left(\max\left(|\hat{a}_{ij}|, 1 - |\hat{a}_{ij}|\right)\right)^2}$ can be taken. Thus, the function d ranges from 0 to 1.

Using expressions (4), (5), (6), let us define the function F as

$$F(A, \alpha, \beta, \gamma) = \alpha d\left(A, \hat{A}\right) + \beta s(A) + \gamma m(A, N), \tag{7}$$

where

1. N is the number of modelling steps used in the function m,
2. $\alpha, \gamma \in [0, 1]$, $\beta \in \mathbb{R}^+$ are weight coefficients for the functions d, s, m.

Reduce the problem of finding a numerically stable matrix A to the problem of constrained optimization (8):

$$\begin{cases} F(A, \alpha, \beta, \gamma) \rightarrow \min \\ A \in S_{\hat{A}} \end{cases} \tag{8}$$

It is necessary to note that the weight coefficients α and γ are freely set by the researcher. If the approximation of the resulting matrix A to the initial matrix $\hat{A}$ is important (e.g. when using a morphological table for the matrix A formation), recommended to increase the value of the coefficient α. If the model should maximally meet the conditions of increasing/decreasing values in the corresponding vertices $(i, j) \in C^+ \cup C^-$, recommended to increase the value of coefficient γ.

Once the coefficients α, γ are given, it is necessary to find $\beta > 0$, such that $\rho(A) < 1$. To solve this problem we can use the method of half division, i.e. apply the following algorithm:

1. Define the search interval $\hat{\beta} = [h_a; h_b]$, $h_a = 0, h_b > 0$, such that by solving the problem (8) for $\beta = h_b$ we have $\rho(A) < 1$.
2. Define the search accuracy $\varepsilon > 0$.
3. Solve problem (8) for $\beta = \frac{h_a + h_b}{2}$.
4. If $h_b - h_a < \varepsilon$, terminate the search.
5. If $\rho(A) < 1$, put $h_b = \frac{h_a + h_b}{2}$ and return to step 2.
6. If $\rho(A) \geq 1$, put $h_a = \frac{h_a + h_b}{2}$ and return to step 2.

5 Case Study

Let us carry out (in accordance with the General Development Plan of Kiev until 2025, Fig. 1a study of the reliability planning of two types of underground tunnels: Tunnel 1 through the built-up part of the city and Tunnel 5 through the Dnipro River. Table 1 presents the data of the vertices (concepts) of the cognitive model G_5 of Tunnel 5.

The graphical view of the cognitive map G_5 of the Tunnel 5 is shown in the Fig. 2. The solid and dashed lines show the increasing and decreasing impact respectively.

The cognitive map G_5 of Tunnel 5 contains 36 cycles, 13 of them are positive and 23 are negative. An analysis of the ratio of the number of stabilizing cycles (23 negative feedbacks) and process accelerator cycles (13 positive feedbacks) indicates the structural stability of such a system [5].

The cognitive map G_1 of Tunnel 1 has no vertex V31 and has no connection from V4.1 to V24 because underground shifts cannot lead to the flooding in the case of Tunnel 1. It contains 18 cycles, 7 of them are positive and 11 are negative. Since the number of the negative cycles is more, the cognitive map G_1 is structurally stable.

Before using the cognitive model to determine its possible behavior of modeling analyzes the various properties of the model is fulfilled. In this case, the stability properties of the model must be analyzed. The initial cognitive maps were numerically unstable. In order to bring to numerically stable matrices, is proposed to use the developed algorithm (Sect. 4), which allows us to bring the original matrices to the numerically stable using the results of the consistency matrix of the modified morphological analysis method [27]. After applying the algorithm from Sect. 4 to original matrices, the spectral radius of cognitive map G_1 is equal to 0.88098 and the spectral radius of cognitive map G_5 is equal to 0.88867, which means that both of the maps are numerically stable. Since both of the maps are structurally and numerically stable, we have an ability to carry out the impulse modelling of various scenarios by acting on the control vertices of the graph.

Table 1 The vertices of the hierarchical cognitive map G_5 of Tunnel 5

Code	Name of the vertex	Assignment of the vertex
V1	State of the tunnel	Indicative
V2	Anthropogenic activities	Perturbing
V2.1	Terrorism or sabotage	Perturbing
V2.2	Human error	Perturbing
V3	Technogenic events	Perturbing
V3.1	Technical	Perturbing
V3.2	Technological	Perturbing
V4	Natural disasters, weather catastrophes	Perturbing
V4.1	Shifts	Perturbing
V4.2	Storms	Perturbing
V4.3	Temperature fluctuations	Perturbing
V5	Security	Basic
V6	Impact of an undesirable event	Disturbing
V7	Ability to function	Disturbing
V8	Time to restore functioning	Disturbing
V9	Environmental consequences	Disturbing
V10	Economic consequences	Disturbing
V11	Consequences for life	Disturbing
V12	The number of injured	Disturbing
V13	Organizational, technical, etc. capabilities	Basic
V14	Investment	Basic
V15	Level of damage	Disturbing
V15.1	Integrity of the ruin system	Basic
V16	Material damage	Disturbing
V17	Geotechnology of construction	Basic
V18	Capacity of land routes	Basic
V19	Population	Basic
V20	Intensity of movement	Disturbing
V21	Average speed	Disturbing
V22	Underground vibrations	Disturbing
V23	Atmospheric pollution	Disturbing
V24	Flooding	Disturbing
V25.1	Fire	Perturbing
V25.2	Explosions	Perturbing
V26	Tunnel corrosion	Disturbing
V27	Ventilation involvement	Disturbing
V28	Training of personnel	Basic
V29	Military threat	Perturbing
V29.1	Rocket attack	Perturbing
V30	Tunnel location depth	Basic
V31	Water depth	Basic

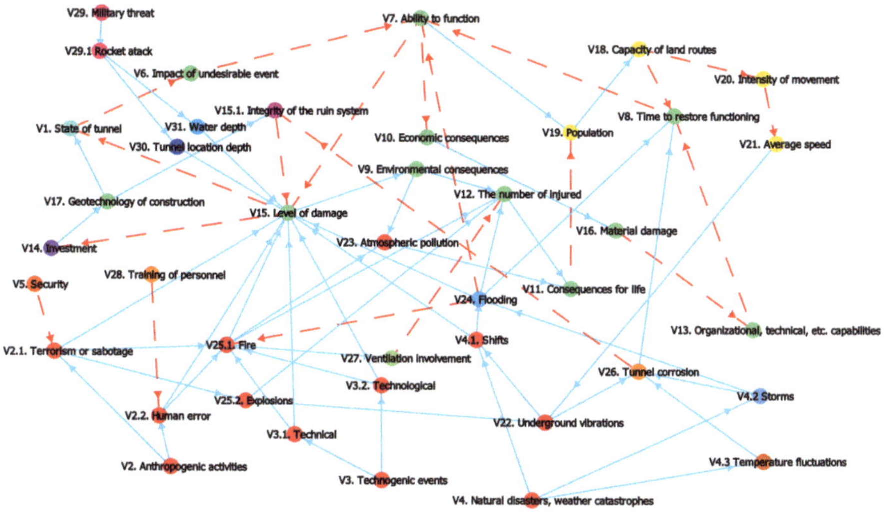

Fig. 2 Sustainable cognitive map G_5 of the Tunnel 5

6 Computational Experiment

Applying pulse modeling, numerous computational experiment was carried out, with the setting of disturbing factors in the vertices in the form of potential adverse events, which can include both natural emergencies and disasters (vibration, shifts, storms and others) and events of man-made or anthropogenic origin (including with malicious intent–military actions, terrorist acts). Taking into account military actions in Ukraine, scenarios regarding malicious intent related to terrorism (or sabotage) are analyzed.

Scenario №1 Terrorism or sabotage attacks
This scenario implies the use of explosive devices, or the use of incendiary mixtures inside the tunnel. To obtain the model, is applied the impulse +1 to the vertex of $V_{2.1}$, so the disturbing vector $Q = (q_{V_1} = 0, \ldots, q_{V_{2.1}} = +1, \ldots)$. The results of modelling for tunnels 1 and 5 are shown on Figs. 3 and 4 respectively.

The comparison results for both tunnels are shown in Table 2.

The results show that a terrorist or sabotage attack greatly reduces the condition of the tunnels and their ability to function as well as increases economic consequences and consequences for life in the town. Particular attention should be paid to the number of injured, which is much higher in the tunnel number 5, because in case of explosions, the tunnel can flood, and the evacuation of the people is difficult. Based on this fact and all other results obtained, we can conclude that Tunnel 1 is more resistant to terrorist attacks and sabotage than Tunnel 5.

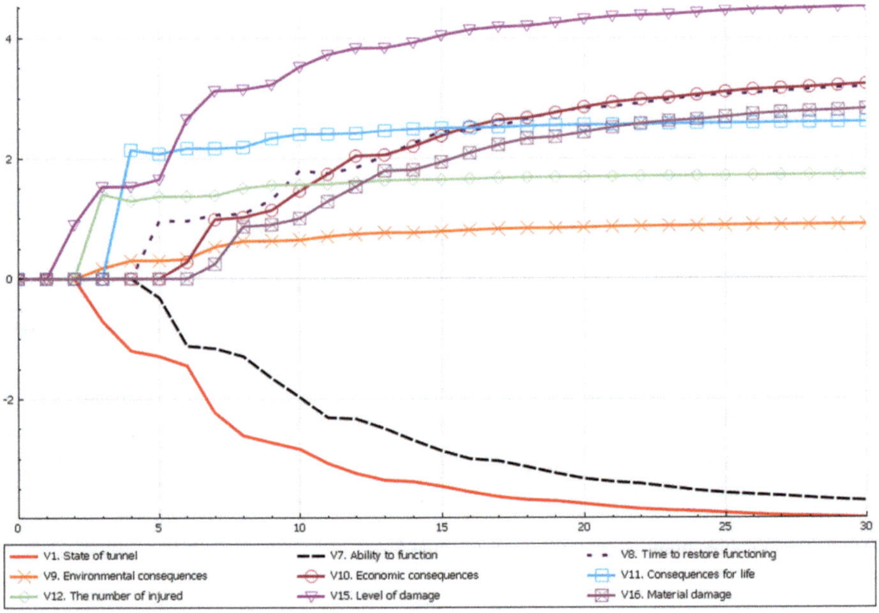

Fig. 3 Model of scenario $\mathcal{N}_{\underline{o}}1$ for Tunnel 1

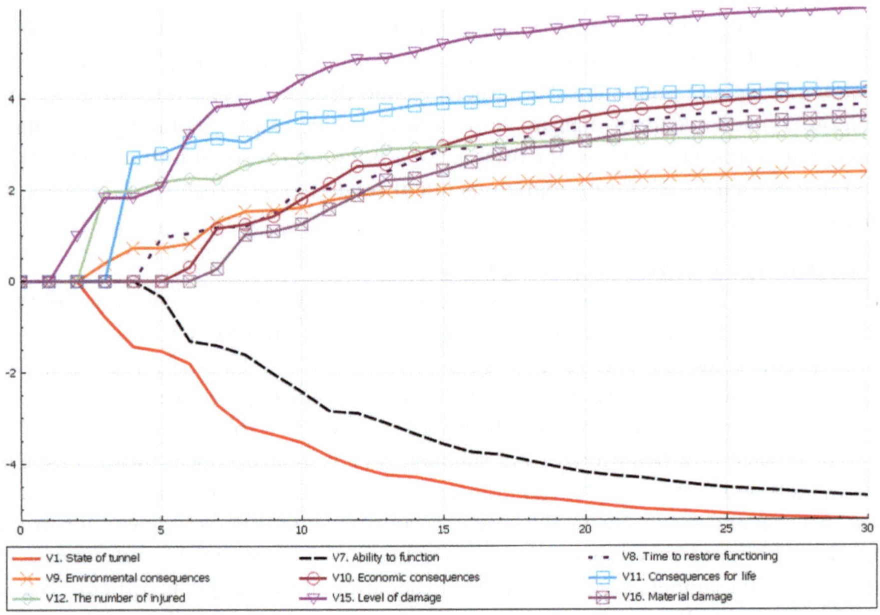

Fig. 4 Model of scenario $\mathcal{N}_{\underline{o}}1$ for Tunnel 5

Table 2 Comparison results of scenario $\mathcal{N}_{\underline{o}}1$

Name of vertex (V_i)	Value of V_i in Tunnel 1	Value of V_i in Tunnel 5	Difference (%)
V_1. State of tunnel	–3.97441	–5.22577	23.95
V_7. Ability to function	–3.69267	–4.70607	21.53
V_8. Time to restore functioning	3.18213	3.84968	17.34
V_9. Environmental consequences	0.900955	2.36986	61.98
V_{10}. Economic consequences	3.23239	4.11704	21.49
V_{11}. Consequences for life	2.60032	4.20084	38.1
V_{12}. The number of injured	1.72253	3.15836	45.46
V_{15}. Level of damage	4.51939	5.94824	24.02
V_{16}. Material damage	2.82323	3.59316	21.43

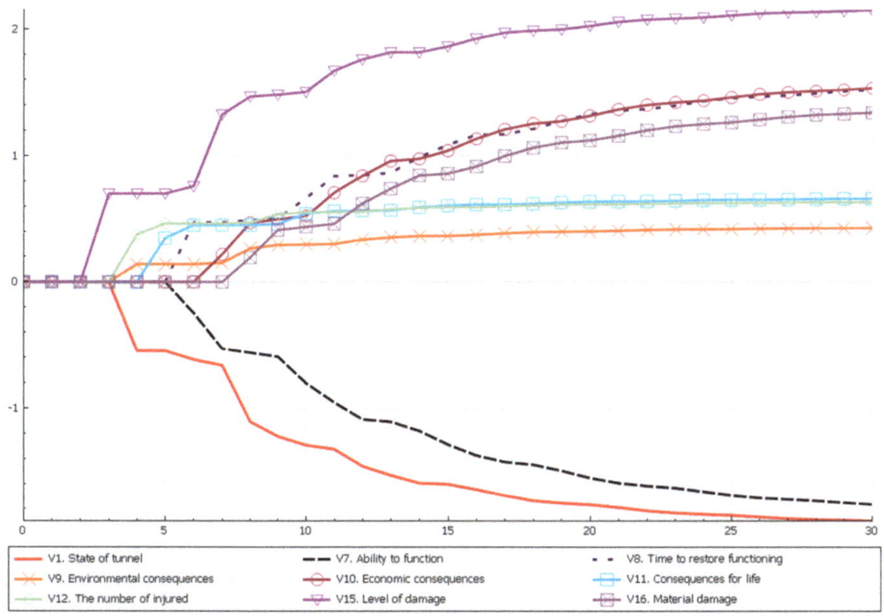

Fig. 5 Model of scenario $\mathcal{N}_{\underline{o}}2$ for Tunnel 1

Scenario $\mathcal{N}_{\underline{o}}$2 Rocket attacks
This scenario implies a direct impact of the rocket on the surface above the tunnels. To obtain the model, is applied the impulse +1 to the vertex of $V_{29.1}$, so the disturbing vector $Q = (q_{V_1} = 0, \ldots, q_{V_{29.1}} = +1, \ldots)$. The results of modelling for tunnels 1 and 5 are shown on Figs. 5 and 6 respectively.

The comparison results for both tunnels are shown in Table 3.

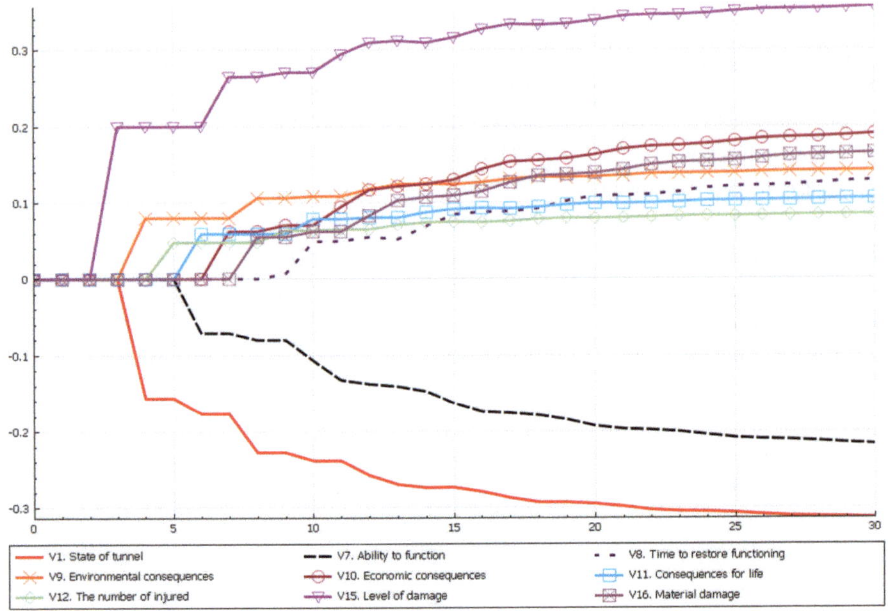

Fig. 6 Model of scenario $\mathcal{N}_{\underline{o}}2$ for Tunnel 5

Table 3 Comparison results of scenario $\mathcal{N}_{\underline{o}}2$

Name of vertex (V_i)	Value of V_i in Tunnel 1	Value of V_i in Tunnel 5	Difference (%)
V_1. State of tunnel	–1.89418	–0.313217	83.46
V_7. Ability to function	–1.76183	–0.216902	87.69
V_8. Time to restore functioning	1.52343	0.128594	91.56
V_9. Environmental consequences	0.429321	0.141977	66.93
V_{10}. Economic consequences	1.53843	0.189361	87.69
V_{11}. Consequences for life	0.66198	0.105215	84.11
V_{12}. The number of injured	0.634147	0.0845262	86.67
V_{15}. Level of damage	2.15542	0.356507	83.46
V_{16}. Material damage	1.34163	0.164859	87.71

Analyzing the results, we can conclude that underground tunnels are much less susceptible to missile attacks than to terrorist attacks. Especially, this can be said about Tunnel 5, because the water layer above this tunnel significantly (approximately 80%) reduces the missile's combat effectiveness, so Tunnel 5 is more resistant to such attacks.

7 Conclusion

Analysis and modeling of underground tunnel planning for megacities is carried out for the purpose of reasonable choice of tunnel type, its reliability, location, ensuring minimization of man-made and ecological risks, military and terrorist threats. The research is based on the system approach to the modeling and scenario analysis of underground tunnel planning under environmental, man-made and terrorist threats, based on the joint application of the methodologies of foresight and cognitive modeling. Using the results of the consistency matrices of the modified morphological analysis method with application of the proposed optimization algorithm, a numerically stable cognitive maps are formed, the vertices of which contain the reliable data of the problem in question. The use of impulse cognitive modeling with specifying perturbing factors in the vertices in the form of potential adverse events, which may include both natural emergencies and disasters and events of man-made or anthropogenic origin (including with malicious intent - military actions, terrorist acts) potentially affecting underground tunnels, makes it possible to build justified scenarios for further decision-making. The analysis of different types underground tunnels modeling has shown the possibility of analyzing the impact of various types of threats on them at the planning stage, choosing the more reliable objects and justifying the priority of their creation.

The peculiarity of foresight methodology and cognitive modeling joint application is the possibility to plan minimization of military and man-made threats through system management of critical infrastructure development and transfer of functions from the most dangerous and vulnerable surface objects and communications to underground facilities, ensuring reliable protection of critical infrastructure and logistics of underground space use for safe and sustainable development of urban environment.

References

1. World Urbanization Prospects 2018: Highlights. United Nations. New York (2019). https://population.un.org/wup/Publications/Files/WUP2018-Highlights.pdf
2. Regional report on the environmental state in Kyiv for 2017. Kyiv City Administration, Kyiv, p. 128 (2018). Access: https://ecodep.kyivcity.gov.ua/files/2019/1/22/REG_DOP_2017.pdf
3. Master plan for the development of Kyiv and its suburban area until 2025 (draft). Access: https://kga.gov.ua/generalnij-plan/genplan2025
4. Pankratova, N.D., Savchenko, I.A., Gayko, G.I. Development of underground urbanism as a system of alternative design configurations. Scientific thought, Kyiv (2019)
5. Gorelova G.V., Pankratova N.D. (eds.): Innovative Development of Socio-economic Systems Based on Methodologies of Foresight and Cognitive Modeling. Collective monograph, p. 464. Naukova Dumka (2015)
6. Sterling, R., Admiraal, H., Bobylev, N., Parker, H., Godard, J.-P., Vähäaho, I., Hanamura, T.: Sustainability issues for underground space in urban areas. In: Proceedings of the Institution of Civil Engineers-Urban Design and Planning, vol. 165, no. 4, pp. 241–254 (2012). https://doi.org/10.1680/udap.10.00020

7. Bobylev, N.: Mainstreaming sustainable development into a city's Master plan: a case of Urban Underground Space use. Land Use Policy **26**(4), 1128–1137 (2009). https://doi.org/10.1016/j.landusepol.2009
8. von der Tann, L., Ritter, S., Hale, S., Langford, J., Salazar, S.: From urban underground space (UUS) to sustainable underground urbanism (SUU): shifting the focus in urban underground scholarship. Land Use Policy **109**(C) (2021). https://doi.org/10.1016/j.landusepol.2021.105650
9. Li, X.Z., Li, C., Parriaux, A., Wenbo, W., Li, H.Q., Sun, L., Liu, C.: Multiple resources and their sustainable development in Urban Underground Space. Tunn. Undergr. Space Technol. **55**, pp. 59–66 (2016). https://doi.org/10.1016/j.tust.2016.02.003
10. Thewes, M., Kamarianakis, S.: Decision analysis for underground infrastructure using uncertain data and fuzzy scales. In: Underground-The Way to the Future: Proceedings of the World Tunnel Congress (Geneva, 2013). CRC Press, London (2013). https://doi.org/10.1201/b14769
11. Stepien, M., Jodehl, A., Vonthron, A., König, M., Thewes M.: An approach for cross-data querying and spatial reasoning of tunnel alignments. Adv. Eng. Inf. **54**(C01) (2022). https://doi.org/10.1016/j.aei.2022.101728
12. Vollmann, G., Stepien, M., Riepe, W., König, M., Lehan, A., Thewes, M., Wahl, H.: Use of BIM for the optimized operation of road tunnels: modelling approach, information requirements, and exemplary implementation(Article). Geomechanik und Tunnelbau **15**(2), pp. 167–174 (2022). https://doi.org/10.1002/geot.202100074.
13. Andersen, L., Nielsen, S.R.K.: Reduction of ground vibration by means of barriers or soil improvement along a railway track. Soil Dyn. Earthq. Eng. **25**(7–10), 701–716 (2005). https://doi.org/10.1016/j.soildyn.2005.04
14. Auersch, L.: Mitigation of railway induced vibration at the track, in the transmission path through the soil and at the building. Proc. Eng. **199**, 2312–2317 (2017). https://doi.org/10.1016/j.proeng.2017.09.1
15. Berta, G.: Blasting-induced vibration in tunnelling. Tunn. Undergr. Space Technol. **9**(2), 175–187 (1994). https://doi.org/10.1016/0886-7798(94)90029-9
16. Yang, W., Hussein, M.F.M., Marshall, A.M.: Centrifuge and numerical modelling of ground-borne vibration from an underground tunnel. Soil Dyn. Earthq. Eng. **51**, 23–34 (2013). https://doi.org/10.1016/j.soildyn.2013.04
17. Yuan, C., Wang, X., Wang, N., Zhao, Q.: Study on the effect of tunnel excavation on surface subsidence based on GIS data management. Proc. Environ. Sci. **12**, 1387–1392 (2012). https://doi.org/10.1016/j.proenv.2012.01.4
18. Pankratova, N.D., Gorelova, G.V., Pankratov, V.A.: Strategy for simulation complex hierarchical systems based on the methodologies of foresight and cognitive modelling. In: Advanced Control Systems: Theory and Applications. River Publishers Series in Automation, Control and Robotics. Chapter 9, pp. 257–288 (2021)
19. Goel, R.K.: Status of tunnelling and underground construction activities and technologies in India. Tunn. Undergr. Space Technol. **16**(2), 63–75 (2001). https://doi.org/10.1016/s0886-7798(01)00035-9
20. Lu, Z.P., Liu, G.B.: Analysis of surface settlement due to the construction of a shield tunnel in soft clay in Shanghai, Geotechnical Aspects of Underground Construction in Soft Ground-Ng, Huang & Liu (eds.) 2009, pp. 805–810. Taylor & Francis Group, London (2009). ISBN 978-0-415-48475-6
21. Gong, Q., Yin, L., Ma, H., Zhao, J.: TBM tunnelling under adverse geological conditions: an overview. Tunn. Undergr. Space Technol. **57**, 4–17 (2016). https://doi.org/10.1016/j.tust.2016.04.002
22. Rönkä, K., Ritola, J., Rauhala, K.: Underground space in land-use planning. Tunn. Undergr. Space Technol. **13**(1), 39–49 (1998). https://doi.org/10.1016/s0886-7798(98)00029-7
23. Chen, Z.-L., Chen, J.-Y., Liu, H., Zhang, Z.-F.: Present status and development trends of underground space in Chinese cities: evaluation and analysis. Tunn. Undergr. Space Technol. **71**, pp. 253–270 (2018). https://doi.org/10.1016/j.tust.2017.08.027

24. Bartel, S., Janssen, G.: Underground spatial planning-Perspectives and current research in Germany. Tunn. Undergr. Space Technol. **55**, 112–117 (2016). https://doi.org/10.1016/j.tust.2015.11.023
25. Pankratova, N., Savchenko, I., Haiko, H., Kravets, V.: System approach to planning urban underground development. Int. J. Inf. Content Process. **6**(1), 3–17 (2019)
26. Pankratova, N.D., Pankratov, V.D.: System approach to the underground construction objects planning based on foresight and cognitive modelling methodologies. Syst. Res. Inf. Technol. **2**, 7–25 (2022). https://doi.org/10.20535/SRIT.2308-8893.2022.1.01
27. Pankratova, N.D., Haiko, H.I., Savchenko, I.O.: Morphological model for underground crossings of water objects. Syst. Res. Inf. Technol. **4**, 53–67 (2021). https://doi.org/10.20535/SRIT.2308-8893.2021.4.04
28. Pankratova, N.D., Gorelova, G.V., Pankratov, V.A.: Study of the plot suitability for underground construction: cognitive modelling. In: ISDMCI 2020: Lecture Notes in Computational Intelligence and Decision Making, pp. 246–264 (2020). https://doi.org/10.1007/978-3-030-54215-3_16
29. Roberts, F.: Graph Theory and its Applications to Problems of Society. Society for Industrial and Applied Mathematics, Philadelphia (1978)
30. Gershgorin, S.A.: Über die Abgrenzung der Eigenwerte einer Matrix, pp. 749–754 (1931)

Stabilization of Impulse Processes of the Cognitive Map of Cryptocurrency Usage with Multirate Sampling and Coordination Between Some Nodes Parameters

Viktor Romanenko, Yurii Miliavskyi, and Heorhii Kantsedal

Abstract In this paper, based on the cognitive map (CM) of the cryptocurrency usage in financial markets developed in Romanenko et al. [1], models of impulse processes with multirate sampling are applied to describe the dynamics of fast and slow measurable coordinates of the CM nodes in the format of systems of difference equations (Roberts equations [2]). Control subsystems with multirate sampling for fast and slow measurable nodes coordinates to stabilize them at the given levels under the effect of random disturbances were developed. External disturbances include various informational influences that affect the cryptocurrency application system. Mutual influences of related subsystems of CM impulse processes are classified as internal disturbances. External control vectors for closed-loop subsystems with fast and slow measurable coordinates of the CM nodes are implemented by varying some coordinates that can be changed by the decision-maker in closed-loop control systems. Additionally, in the closed-loop control subsystem with fast measured coordinates, the subsystem of dynamic coordination of a given ratio of the coordinates of the CM nodes describing the volume of supply and demand for cryptocurrency on financial exchanges is implemented. The designed discrete controllers were studied using digital simulation.

Keywords Cognitive map · Multirate sampling · Cryptocurrency · Control subsystem · Optimality criteria

1 Introduction

In [1] a cognitive map of the CM of the cryptocurrency usage in financial markets, shown on Fig. 1, was developed. The CM includes the following nodes:

1. Cryptocurrency rate.
2. The volume of cryptocurrency trading (directly on the spot market).

V. Romanenko · Y. Miliavskyi (✉) · H. Kantsedal
Igor Sikorsky Kyiv Polytechnic Institute, Kyiv, Ukraine
e-mail: romanenko.viktorroman@gmail.com

M. Zgurovsky and N. Pankratova (eds.), *System Analysis and Artificial Intelligence*,
Studies in Computational Intelligence 1107,
https://doi.org/10.1007/978-3-031-37450-0_5

3. Supply of cryptocurrency.
4. Demand for cryptocurrency.
5. The amount of speculation in various cryptocurrency contracts (derivatives market, futures contracts).
6. Risks of a cryptocurrency exchange rate collapse.
7. Number of cryptocurrency users.
8. Volume of investments (interest in cryptocurrency from institutional investors, they include cryptocurrency holders and developers of related infrastructure such as exchanges, wallets, for relationships).
9. Amount of capitalization.
10. Indirect income.
11. The level of trust in cryptocurrency as a means of payment and accumulation.
12. The risk of losing the number of users (without users, the value of cryptocurrencies goes to zero).
13. Information disturbances of various natures, representing external disturbances.

Control actions are implemented externally by varying the resources of the coordinates of the CM nodes. The following nodes can be used:

1. Variation in the volume of trades using cryptocurrency $\Delta u_2(k)$
2. Variation in the amount of cryptocurrency supply $\Delta u_3(k)$
3. Variation in the amount of demand for cryptocurrency $\Delta u_4(k)$
4. Variation in the amount of cryptocurrency speculation $\Delta u_5(k)$

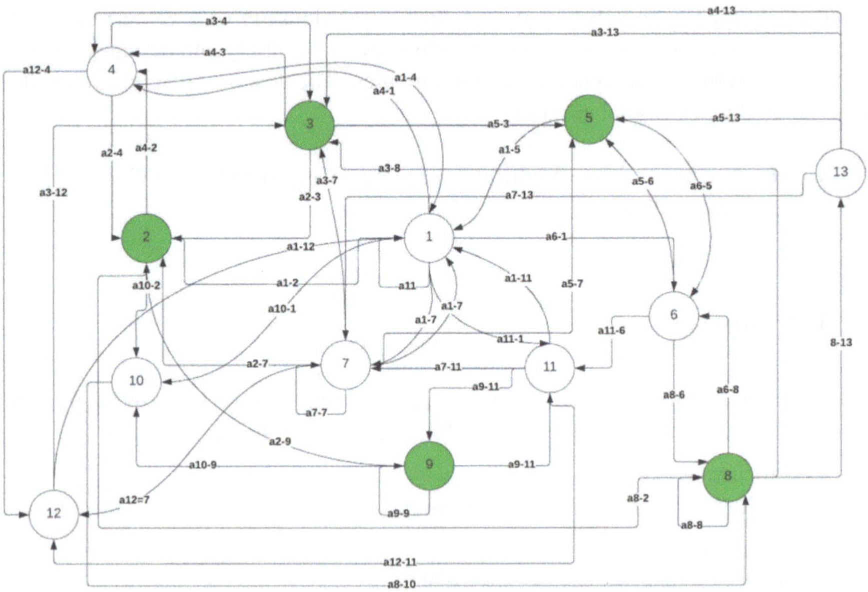

Fig. 1 The cognitive map of the use of cryptocurrency used in the article. Controlled nodes are marked in green

5. Variation in the volume of investments $\Delta u_8(k)$
6. Variation in the amount of capitalization $\Delta u_9(k)$

All the coordinates of the CM nodes, except for node 13, are measured or calculated from indirect indicators. Value-at-risk (VaR) financial risk assessment methods are used to calculate the risks of a cryptocurrency exchange rate collapse (node 6), which is based on an analysis of the statistical nature of the market and offers a universal method for assessing various types of risks: price, currency, credit, and liquidity risks. The VaR technique is currently used as a standard for risk assessment [3, 4]; parametric VaR is calculated as follows [5]:

$$VaR = \alpha\delta OP\sqrt{P}, \tag{1}$$

where α–is the quantile of the confidence interval, δ–volatility (displacement norm), OP–is the value of the overriding position, P–is the forecasting period. Volatility δ is determined based on a preliminary calculation of the dispersion of the cryptocurrency rate (node 1), which is determined on the NT_0 interval, where T_0 is the sampling period, as follows:

$$\delta_{y_1}^2(kT_0) = \frac{1}{N}\sum_{i=1}^{N}\left\{y_1[(k-i)T_0] - \frac{1}{N}\sum_{i=1}^{N}y_1[(k-i)T_0]\right\}^2, \tag{2}$$

where the sampling interval NT_0 is chosen experimentally.

We also use the VaR method (1) to determine the risk of spending the number of users (node 12), in which the variance of the coordinate y_7 (number of users) is calculated according to the recurrent procedure:

$$\delta_{y_7}^2\left(\left\lfloor\frac{k}{m}\right\rfloor h\right) = \frac{1}{N_1}\sum_{i=1}^{N_1}\left\{y_7\left[\left(\left\lfloor\frac{k}{m}\right\rfloor - i\right)h\right] - \frac{1}{N_1}\sum_{i=1}^{N_1}y_7\left[\left(\left\lfloor\frac{k}{m}\right\rfloor - i\right)h\right]\right\}^2 \tag{3}$$

The sampling period is $h = mT_0$, and $\left\lfloor\frac{k}{m}\right\rfloor$–is an integer from the division of k by m. The nature of the main external disturbances (node 13) is described in [1].

Dynamic processes between the nodes of the CM are described in [2] in the form of differential equations of impulse processes:

$$\Delta y_i(k+1) = \sum_{j=1}^{n} a_{ij}\Delta y_j(k), \tag{4}$$

where $\Delta y_i(k+1) = y_i(k) - y_i(k-1), i = 1, ..., n,$ where $n = 12$; a_{ij}–is the weight of the edge that connects the j-th node of the CM with the i-th one. Equation (4), which describes the free movement of the coordinate vector $\bar{y}$ of the nodes of CM, can be written in the vector-matrix form:

$$\Delta \bar{y}_i (k + 1) = A\Delta \bar{y} (k), \tag{5}$$

where A–is the weighted adjacency matrix of the CM $(n \times n)$. In order to implement the control impulse process of CM based on modern control theory, the model of the controlled impulse process is applied in [1]:

$$\Delta \bar{y}_i (k + 1) = A\Delta \bar{y} (k) + B\Delta \bar{u} (k), \tag{6}$$

where $\Delta \bar{u} (k + 1) = \bar{u} (k) - \bar{u} (k - 1)$ is the vector of increments of control actions. For the CM vector increments of control actions are equal to $\Delta \bar{u} (k) = [\Delta u_2 (k), \Delta u_3 (k), \Delta u_4 (k), \Delta u_5 (k), \Delta u_8 (k), \Delta u_9 (k)]^T$, the size is $r = 6$. The control matrix $B(n \times r)$, which is usually developed by the operator-programmer, consists of ones and zeros.

The cognitive map (Fig. 1) contains fast coordinates of the nodes of $\bar{y}_f$, which can be measured (or calculated) at discrete moments with a small sampling period T_0. Such nodes include the concepts of the above-described nodes 1, 2, 3, 4, 5, 6. At the same time, nodes 7, 8, 9, 10, 11, 12 can be measured only at a slow rate at discrete moments with a sampling period $h = mT_0$, where m is an integer greater than one. Then the dynamics of the fast measurable coordinates $\bar{y}_f$ in [1] is described by the differential equation of the fast subsystem:

$$\Delta \bar{y}_f [rh + (l + 1)T_0] = A_{11}\Delta \bar{y}_f (rh + lT) + A_{12}\Delta \bar{y}_s (rh), \tag{7}$$

where $r = \left\lfloor \frac{k}{m} \right\rfloor$–is an integer from the division of k by m; $l = 0, 1, ..., (m - 1)$, $\Delta \bar{y}_s (rh)$ is added only when $l = 0$. Matrices A_{11}, A_{12} have sizes $(r \times r)$ and $(r \times (n - r))$ respectively.

With the control vector $\Delta \bar{u}_f (k)$and external disturbances (node 13) model (7) will have the following form:

$$\begin{aligned} &\Delta \bar{y}_f [rh + (l + 1)T_0] = A_{11}\Delta \bar{y}_f (rh + lT_0) + B_f \Delta \bar{u}_f (rh + lT_0) + \\ &A_{12}\Delta \bar{y}_s (rh) + \bar{\psi}_f \Delta \bar{\xi}_f (rh) (rh + lT_0) \end{aligned} \tag{8}$$

where $\Delta \bar{\xi}_f$ is the increment in informational spurts that affect the vector $\bar{y}_f$. The matrix $B_f(p \times 4)$ is formed from ones and zeros. For this CM values $n = 12$ and$p = 6$. The vector of coefficients $\bar{\psi}_f$ has size (6×1).

The dynamics of the slowly measured CM coordinates $\bar{y}_s = [y_7, y_8, y_9, y_{10}, y_{11}, y_{12}]^T$ in work [1] is described by the following differential equation of the slow subsystem:

$$\begin{aligned} &\Delta \bar{y}_s [rh + (i + 1)T_0] = A_{22}\Delta \bar{y}_s (rh + iT_0) + A_{21}\Delta \bar{y}_f (rh + iT_0) + \\ &+ \bar{\psi}_s \Delta \bar{\xi}_s (rh) (rh + iT_0) \end{aligned}$$

When the control vector $\Delta \bar{u}_s (rh + iT_0) = \left[\Delta u_{s_8} (rh + iT_0), \Delta u_{s_9} (rh + iT_0)\right]^T$ exists, the equation of the controlled impulse process will have the form:

$$\begin{aligned}\Delta \bar{y}_s\left[rh + (i + 1)T_0\right] = A_{22}\Delta \bar{y}_s\left(rh + iT_0\right) + A_{21}\Delta \bar{y}_f\left(rh + iT_0\right) + \\ + B_s\Delta \bar{u}_s\left(rh + iT_0\right) + \bar{\psi}_s\Delta \bar{\xi}_s\left(rh + iT_0\right), i = 0, 1, ..., (m - 1)\end{aligned} \tag{9}$$

where matrices A_{22}, B_s, A_{21} respectively have sizes $((n - p) \times (n - p))$, $((n - p) \times 2)$, $((n - p) \times p)$, the vector of coefficients $\bar{\psi}_s$ has size $((n - p) \times 1)$.

Since for this CM the coordinates of the nodes $\bar{y}_s$ are measured at discrete moments with the sampling period $h = mT_0$, and the control vector is $\Delta \bar{u}_s\left(rh + iT_0\right) = \Delta \bar{u}_s\left(rh\right), i = 0, 1, ..., (m - 1)$, the previous model could be presented with a different time sampling using the iterative procedure [1]:

$$\begin{aligned}\Delta \bar{y}_s\left[rh + (i + 1)T_0\right] = A_{22}^m\Delta \bar{y}_s\left(rh + iT_0\right) + \sum_{i=0}^{m-1} A_{22}^i B_s\Delta \bar{u}_s\left(rh\right) + \\ + \sum_{j=0}^{m-1} A_{22}^j A_{21}\Delta \bar{y}_f\left(rh + (m - 1 - j)T_0\right) + \sum_{\mu=0}^{m-1} A_{22}^{\mu}\bar{\psi}_s\Delta \bar{\xi}_f\left(rh + (m - 1 - \mu)T_0\right),\end{aligned} \tag{10}$$

where

$$\Delta \hat{Y}_f\left[rh + (m - 1 - i)\,T_0\right] = \hat{Y}_f\left[rh + (m - 1 - i)\,T_0\right] - \hat{Y}_f\left[(r - 1)\,h + (m - 1 - i)\,T_0\right]$$

$$\Delta \overline{\xi}_f\left[rh + (m - 1 - i)\,T_0\right] = \overline{\xi}_f\left[rh + (m - 1 - i)\,T_0\right] - \overline{\xi}_f\left[(r - 1)h + (m - 1 - i)\,T_0\right]$$

Multirate subsystem models (8) and (10) are stable, confirmed by the eigenvalues of the matrices A_{11} and A_{22}^m, which are less than one in modulus.

Also, models (8) and (10) are controlled, where the control matrices have the following form:

$$\begin{aligned}P_f &= \left[B_f, A_{11}B_f, A_{11}^2B_f, ..., A_{11}^5B_f\right], \\ P_s &= \left[\sum_{i=0}^{m-1} A_{22}^i B_s, A_{22}^m\sum_{i=0}^{m-1} A_{22}^i B_s, \left(A_{22}^m\right)^2\sum_{i=0}^{m-1} A_{22}^i B_s, ..., \left(A_{22}^m\right)^5\sum_{i=0}^{m-1} A_{22}^i B_s\right]\end{aligned}$$

where the ranks of the control matrix $p = 6$ and $(n - p) = 6$.

1.1 Problem Statement

The first problem, which is solved in this paper, is the development of interconnected closed-loop control subsystems with multirate sampling, for the fast-acting and slow-acting subsystems (8) and (9) of the cognitive map to stabilize the coordinates of the nodes $\bar{y}_f$ and $\bar{y}_s$ at the given levels ($\bar{G}_f$ and $\bar{G}_s$ respectively). Additionally, in the subsystem with fast measured coordinates (8), it is necessary to implement dynamic coordination of a given ratio of the coordinates of the CM nodes, which describe the supply and demand of cryptocurrency, as well as the volume of trading directly in cryptocurrency and trading in the form of cryptocurrency-related products (smart contracts, derivatives etc.).

2 Materials and Methods

2.1 Coordinating Control of the Impulse Process of a Fast CM Subsystem in a Stochastic Environment

Consider the controlled impulse process (8) in the full nodes coordinates $\bar{y}_f$ with a reverse shift for one sampling period T_0:

$$\begin{aligned}&\left[I_f-(I_f+A_{11})q_1^{-1}+A_{11}q_1^{-2}\right]\bar{y}_f\left[rh+lT_0\right]+B_fq_1^{-1}\Delta\bar{u}_f\left(rh+lT_0\right)+\\&+A_{12}\Delta\bar{y}_s\left(rh-T_0\right)+\bar{\psi}_f\Delta\bar{\xi}_f\left(rh+(l-1)\,T_0\right),l=0,1,...,(m-1)\end{aligned} \tag{11}$$

where q_1 is the shift operator for one sampling period T_0, $\bar{y}_f\left[rh+lT_0\right]$is the vector of fast measurable coordinates of the CM nodes with size (6×1), $\Delta\bar{u}_f\left(rh+lT_0\right)$ is the vector of the increment of control actions (4×1), $\Delta\bar{y}_s\left(rh-T_0\right)$ is the vector of slow measurable coordinates of the node CM, $\Delta\bar{\xi}_f$– stochastic perturbation from information influences (node 13). For the convenience of further explanations, we will consider the following vector of coordinates:

$$\begin{aligned}&\bar{\bar{y}}_f\left[rh+(l-1)T_0\right]=(I_f+A_{11})\bar{y}_f\left[rh+lT_0\right]-A_{11}\bar{y}_f\left[rh+(l-1)T_0\right]+\\&+A_{12}\Delta\bar{y}_s\left(rh-T_0\right)\end{aligned}, \tag{12}$$

which is known part of the vector $\bar{y}_f\left[rh+(l-1)T_0\right]$, and as usual $A_{12}\Delta\bar{y}_s\left(rh-T_0\right)$ is added only once during long period h.

To synthesize the fast discrete controller [6, 7], we first apply the quadratic criteria of optimality in the form of minimizing the generalized variance–residual between $\bar{y}_f\left[rh+(l-1)T_0\right]$ and vector of desired values $\bar{G}_f\left[rh+lT_0\right]$ and increments of control actions $\Delta\bar{u}_f\left(rh+lT_0\right)$:

$$J_{f_{\bar{G}}}\left[rh+(l+1)T_0\right]=E\left\{\begin{array}{l}\left(\bar{y}_f\left[rh+(l+1)T_0\right]-\bar{G}_f\left[rh+lT_0\right]\right)^T*\\ *\left(\bar{y}_f\left[rh+(l+1)T_0\right]-\bar{G}_f\left[rh+lT_0\right]\right)+\\ +\Delta\bar{u}_f^T\left(rh+lT_0\right)R_f\Delta\bar{u}_f\left(rh+lT_0\right)\end{array}\right\}\to\min,$$
$$l=0,..,(m-1) \tag{13}$$

where E is the operator of expectation; the matrix R_f is chosen to be positive definite and symmetric so that $\left(B_f^TB_f+R_f\right)$ is not degenerated.

Taking into account (12), we write model (11) in the form:

$$\begin{aligned}&\bar{y}_f\left[rh+(l+1)T_0\right]=\bar{\bar{y}}_f\left[rh+(l+1)\,T_0\right]+B_f\Delta\bar{u}_f\left(rh+lT_0\right)+\\&+\bar{\psi}_f\Delta\bar{\xi}_f\left(rh+lT_0\right)\end{aligned} \tag{14}$$

Taking into account this equality, criterion (13) can be written as follows:

$$J_{f_{\bar{G}}}\left[rh+(l+1)T_0\right]=E\left\{\begin{array}{l}\left(\begin{array}{l}\tilde{\bar{y}}_f\left[rh+(l+1)\,T_0\right]+B_f\Delta\bar{u}_f\left(rh+lT_0\right)+\\+\bar{\psi}_f\Delta\bar{\xi}_f\left(rh+lT_0\right)-\bar{G}_f\left[rh+lT_0\right]\end{array}\right)^T*\\*\left(\begin{array}{l}\tilde{\bar{y}}_f\left[rh+(l+1)\,T_0\right]+B_f\Delta\bar{u}_f\left(rh+lT_0\right)+\\+\bar{\psi}_f\Delta\bar{\xi}_f\left(rh+lT_0\right)-\bar{G}_f\left[rh+lT_0\right]\end{array}\right)+\\+\Delta\bar{u}_f^T\left(rh+lT_0\right)R_f\Delta\bar{u}_f\left(rh+lT_0\right)\end{array}\right\} \tag{15}$$

For known value of $\tilde{\bar{y}}_f\left[rh+(l+1)\,T_0\right]+B_f\Delta\bar{u}_f\left(rh+lT_0\right)-\bar{G}_f\left[rh+lT_0\right]$, the equality will be true:

$$E\left\{\begin{array}{l}\left(\tilde{\bar{y}}_f\left[rh+(l+1)\,T_0\right]+B_f\Delta\bar{u}_f\left(rh+lT_0\right)-\bar{G}_f\left[rh+lT_0\right]\right)^T*\\*\left(\tilde{\bar{y}}_f\left[rh+(l+1)\,T_0\right]+B_f\Delta\bar{u}_f\left(rh+lT_0\right)-\bar{G}_f\left[rh+lT_0\right]\right)\end{array}\right\}=$$
$$=\left(\tilde{\bar{y}}_f\left[rh+(l+1)\,T_0\right]+B_f\Delta\bar{u}_f\left(rh+lT_0\right)-\bar{G}_f\left[rh+lT_0\right]\right)^T*$$
$$*\left(\tilde{\bar{y}}_f\left[rh+(l+1)\,T_0\right]+B_f\Delta\bar{u}_f\left(rh+lT_0\right)-\bar{G}_f\left[rh+lT_0\right]\right)$$

Then the criterion (15) after the transformations will have the next form:

$$J_{f_{\bar{G}}}\left[rh+(l+1)T_0\right]=\left(\begin{array}{l}\tilde{\bar{y}}_f\left[rh+(l+1)\,T_0\right]+B_f\Delta\bar{u}_f\left(rh+lT_0\right)-\\-\bar{G}_f\left[rh+lT_0\right]\end{array}\right)^T*$$
$$*\left(\tilde{\bar{y}}_f\left[rh+(l+1)\,T_0\right]+B_f\Delta\bar{u}_f\left(rh+lT_0\right)-\bar{G}_f\left[rh+lT_0\right]\right)+$$
$$+2E\left\{\begin{array}{l}\left(\tilde{\bar{y}}_f\left[rh+(l+1)\,T_0\right]+B_f\Delta\bar{u}_f\left(rh+lT_0\right)-\bar{G}_f\left[rh+lT_0\right]\right)^T*\\*\left(\bar{\psi}_f\Delta\bar{\xi}_f\left(rh+lT_0\right)\right)\end{array}\right\}+$$
$$+E\left\{\left(\bar{\psi}_f\Delta\bar{\xi}_f\left(rh+lT_0\right)\right)^T*\left(\bar{\psi}_f\Delta\bar{\xi}_f\left(rh+lT_0\right)\right)\right\}+$$
$$+\Delta\bar{u}_f^T\left(rh+lT_0\right)R_f\Delta\bar{u}_f\left(rh+lT_0\right) \tag{16}$$

In Eq. (16), the term before the last does not depend on $\Delta\bar{u}_f^T\left(rh+lT_0\right)$. Therefore, it is unnecessary to consider it when minimizing the criterion. Consider the second addendum:

$$E\left\{\begin{array}{l}\left(\tilde{\bar{y}}_f\left[rh+(l+1)\,T_0\right]+B_f\Delta\bar{u}_f\left(rh+lT_0\right)-\bar{G}_f\left[rh+lT_0\right]\right)^T*\\*\left(\bar{\psi}_f\Delta\bar{\xi}_f\left(rh+lT_0\right)\right)\end{array}\right\}\approx 0$$

since, by assumption, the perturbations $\Delta\bar{\xi}_f\left(rh+lT_0\right)$ have a zero mean and are uncorrelated with other coordinates.

Criterion (16) is minimized based on $\frac{\partial J_{f_{\bar{G}}}[rh+(l+1)T_0]}{\partial\Delta\bar{u}_f(rh+lT_0)}=0$. Then, using the method of taking the derivative of the quadratic form described in [8], we obtain:

$$\frac{\partial J_{f_{\bar{G}}}[rh+(l+1)T_0]}{\partial\Delta\bar{u}_f(rh+lT_0)}=2B_f^T\left(\tilde{\bar{y}}_f\left[rh+(l+1)\,T_0\right]+B_f\Delta\bar{u}_f\left(rh+lT_0\right)-\bar{G}_f\left[rh+lT_0\right]\right)+$$
$$+2R_f\Delta\bar{u}_f\left(rh+lT_0\right)=0$$

This is the optimal discrete controller equation for stabilizing the vector of fast-measured coordinates $\bar{y}_f$ at a given level $\bar{G}_f$. The control law will have the next form:

$$\Delta\bar{u}_{f_G}(rh + lT_0) = -\left(B_f^T B_f + R_f\right)^{-1} B_f^T \left(\tilde{\bar{y}}_f\left[rh + (l+1)\,T_0\right] - \bar{G}_f\left[rh + lT_0\right]\right) \tag{17}$$

where $\tilde{\bar{y}}_f\left[rh + (l+1)\,T_0\right]$ is calculated on the basis of (12).

Let us go directly to the problem of coordinating control. Assume that the set of coordinate ratios of the nodes $\bar{y}_f$ of the CM is given in the form:

$$S\bar{y}_f\left[rh + lT_0\right] = \bar{b},\, l = 0, 1, \ldots, (m-1) \tag{18}$$

where $\bar{b}$ is the given vector of size m, and S is the given matrix of size $(m \times n)$, and $rank\left(SB_f\right) = M$.

Coordinating control requires that relation (18) be fulfilled in dynamics as accurately as possible at each sampling period T_0. To do this, we introduce the minimization of the ratio discrepancy variance:

$$J_{f_{\bar{b}}}\left[rh + (l+1)T_0\right] = E\left\{\begin{array}{l}\left(S\bar{y}_f\left[rh + (l+1)\,T_0\right] - \bar{b}\right)^T * \\ *\left(S\bar{y}_f\left[rh + (l+1)\,T_0\right] - \bar{b}\right)\end{array}\right\} \to \min \tag{19}$$

Let us consider the option when the vector of specified actions $\bar{G}_f$ is determined in advance, but simultaneously the relation (18) must be satisfied based on $S\bar{G}_f(k) = \bar{b}$. We will solve the problem of minimizing criterion (19) by $\Delta\bar{u}_f(rh + lT_0)$ applying the previous method (14–16):

$$\begin{array}{l}J_{f_{\bar{b}}}\left[rh + (l+1)T_0\right] = \left(S\tilde{\bar{y}}_f\left[rh + (l+1)\,T_0\right] + SB_f\Delta\bar{u}_f(rh + lT_0) - \bar{b}\right)^T * \\ *\left(S\tilde{\bar{y}}_f\left[rh + (l+1)\,T_0\right] + SB_f\Delta\bar{u}_f(rh + lT_0) - \bar{b}\right) + \\ +2E\left\{\begin{array}{l}\left(S\tilde{\bar{y}}_f\left[rh + (l+1)\,T_0\right] + SB_f\Delta\bar{u}_f(rh + lT_0) - \bar{b}\right)^T * \\ *S\left(\bar{\psi}_f\Delta\bar{\xi}_f(rh + lT_0)\right)\end{array}\right\} + \\ +E\left\{\begin{array}{l}\left(\bar{\psi}_f\Delta\bar{\xi}_f(rh + lT_0)\right)^T S^T * \\ *S\left(\bar{\psi}_f\Delta\bar{\xi}_f(rh + lT_0)\right)\end{array}\right\}\end{array} \tag{20}$$

The last term does not depend on $\Delta\bar{u}_f(rh + lT_0)$; therefore, it can be ignored when minimizing the criterion. The penultimate term in criterion (20) is equal to the square of the Euclidean norm of the vector $S\tilde{\bar{y}}_f\left[rh + (l+1)\,T_0\right] + SB_f\Delta\bar{u}_f(rh + lT_0) - \bar{b}$. It is known that the norms of a vector are non-negative and equal to zero if and only if the vector is zero. Therefore, if the equation $S\tilde{\bar{y}}_f\left[rh + (l+1)\,T_0\right] = -\left[SB_f\Delta\bar{u}_f(rh + lT_0) - \bar{b}\right]$ has a solution, then this solution is the point of the global minimum of the criterion. According to the condition in the given CM, the vector $\Delta\bar{u}_f(rh + lT_0)$ has size $M < 4$, and $rank\left(SB_f\right) = M$. Therefore, according to the Kronecker-Capelli theorem, this equation has a set of solutions on which the minimum of criterion (19) is reached:

$$S\tilde{\bar{y}}_f\,[rh + (l+1)\,T_0] = -\left[SB_f\Delta\bar{u}_f\,(rh + lT_0) - \bar{b}\right] \tag{21}$$

In general, control (17) does not satisfy equality (21). Thus, we have a multi-criteria optimization problem when forming $\Delta\bar{u}_f\,(rh + lT_0)$ according to criteria (13) and (19). At the same time, the fulfillment of criterion (19) is a higher priority condition, but it has an ambiguous solution.

2.2 The Conditional Minimization of the Ratio Discrepancies Variance and the Generalized Variance of the Fast-Measured CM Nodes

When solving the given problem, we will consider equality (21), a constraint that must necessarily be fulfilled when minimizing the optimality criterion (13). We will reduce the problem of unconditional multi-criteria optimization to the problem of conditional single-criteria optimization.

The conditional optimization problem is solved by the Lagrange multipliers [7, 8], and then the Frobenius theorem is applied to invert the block matrix. The following theorem determines the optimal value of $\Delta\bar{u}_f\,(rh + lT_0)$.

Theorem 1 *Let the CM impulse process be given in the form* (11), *it is necessary to form such a vector of control actions at each sampling period* T_0 *that will minimize the optimality criteria* (19) *and* (13) *and the criterion* (19) *is given a higher priority. Then the optimal value of* $\Delta\bar{u}_f\,(rh + lT_0)$ *will be determined as follows:*

$$\Delta\bar{u}_f\,(rh + lT_0) = -\left(B_f^T B_f + R_f\right)^{-1}\left\{\begin{array}{l}\left[I_f - B_f^T S^T L^{-1} S B_f\left(B_f^T B_f + R_f\right)\right] * \\ * \left[B_f^T\left(\begin{array}{l}\tilde{\bar{y}}_f\,[rh + (l+1)T_0] - \\ -\bar{G}_f\,[rh + lT_0]\end{array}\right)\right] + \\ + B_f^T S^T L^{-1}\left(S\tilde{\bar{y}}_f\,[rh + (l+1)T_0] - \bar{b}\right)\end{array}\right\} \tag{22}$$

if $\left(B_f^T B_f + R_f\right)$ *and* $L = SB_f\left(B_f^T B_f + R_f\right)^{-1} B_f^T S^T$ *are not degenerate.*

The proof of the theorem for the CM impulse process model with unirate sampling is given in [9].

2.3 Stabilization of Impulse Processes of the CM Coordinate Subsystem with Slowly-Measurable Node Coordinates

Using the first differences $\Delta\bar{y}_s[(r+1)h] = \bar{y}_s[(r+1)h] - \bar{y}_s(rh)$, $\Delta\bar{y}_s(rh) = \bar{y}_s(rh) - \bar{y}_s[(r-1)h]$, we will get the equation of the slow subsystem (10) of the CM in the full nodes coordinates $\bar{y}_s(rh)$:

$$\begin{aligned}&\bar{y}_s[(r+1)h] = \left(I_s + A_{22}^m\right)\bar{y}_s(rh) - A_{22}^m\bar{y}_s[(r-1)h] +\\&+\textstyle\sum_{i=0}^{m-1} A_{22}^i B_s\Delta\bar{u}_s(rh) + \sum_{j=0}^{m-1} A_{22}^j A_{21}\Delta\bar{y}_f(rh + [m-1-j]T_0) +\\&+\textstyle\sum_{j=0}^{m-1} A_{22}^j A_{21}\Delta\bar{\xi}_f(rh + [m-1-j]T_0)\end{aligned} \tag{23}$$

where $\Delta\bar{y}_f(rh + [m-1-j]T_0)$ and $\Delta\bar{\xi}_f(rh + [m-1-j]T_0)$ are defined below formula (10).

To design control actions $\Delta\bar{u}_s(rh)$, we apply the quadratic criterion of optimality in the form of the generalized variance of the differences between the vector of the slow coordinates $\bar{y}_s[(r-1)h]$ and the vector of desired values $\bar{G}_s[rh]$ and increments of control actions $\Delta\bar{u}_s(rh)$:

$$J_s[(r+1)h] = E\left\{\begin{array}{l}\left(\bar{y}_s([r+1]h) - \bar{G}_s[rh]\right)^T * \\ * \left(\bar{y}_s([r+1]h) - \bar{G}_s[rh]\right) + \\ +\Delta\bar{u}_s^T(rh) R_s \Delta\bar{u}_s(rh)\end{array}\right\} \tag{24}$$

where E is the operator of mathematical expectation, R_s is a positive definite symmetric matrix. For the convenience of further explanations in the model (23) for $\bar{y}_s([r+1]h)$, we will first take its known part into consideration:

$$\begin{aligned}&\tilde{\bar{y}}_s([r+1]h) = \left(I_s + A_{22}^m\right)\bar{y}_s(rh) - A_{22}^m\bar{y}_s([r-1]h) +\\&+\textstyle\sum_{j=0}^{m-1} A_{22}^j A_{21}\Delta\bar{y}_f(rh + [m-1-j]T_0)\end{aligned} \tag{25}$$

Then model (23) will take the form:

$$\begin{aligned}&\bar{y}_s([r+1]h) = \tilde{\bar{y}}_s([r+1]h) + \textstyle\sum_{i=0}^{m-1} A_{22}^i B_s\Delta\bar{u}_s(rh) +\\&+\textstyle\sum_{j=0}^{m-1} A_{22}^j A_{21}\Delta\bar{\xi}_f(rh + [m-1-j]T_0)\end{aligned} \tag{26}$$

After transformations taking into account (26), the optimality criterion (24) can be written as follows:

$$
\begin{aligned}
&J_s[(r+1)h] = \left[\bar{\tilde{y}}_s([r+1]h) + \sum_{i=0}^{m-1} A_{22}^i B_s \Delta\bar{u}_s(rh) - \bar{G}_s[rh]\right]^T * \\
&* \left[\bar{\tilde{y}}_s([r+1]h) + \sum_{i=0}^{m-1} A_{22}^i B_s \Delta\bar{u}_s(rh) - \bar{G}_s[rh]\right] + \Delta\bar{u}_s(rh) R_s \Delta\bar{u}_s^T(rh) + \\
&+2E\left\{\begin{array}{l} \left[\bar{\tilde{y}}_s([r+1]h) + \sum_{i=0}^{m-1} A_{22}^i B_s \Delta\bar{u}_s(rh) - \bar{G}_s[rh]\right]^T * \\ * \left[\sum_{j=0}^{m-1} A_{22}^j A_{21} \Delta\bar{\xi}_f(rh + [m-1-j]T_0)\right] + \end{array}\right\} + \\
&+E\left\{\begin{array}{l} \left[\sum_{j=0}^{m-1} A_{22}^j A_{21} \Delta\bar{\xi}_f(rh + [m-1-j]T_0)\right]^T * \\ * \left[\sum_{j=0}^{m-1} A_{22}^j A_{21} \Delta\bar{\xi}_f(rh + [m-1-j]T_0)\right] \end{array}\right\}
\end{aligned} \tag{27}
$$

At the same time, it was taken into account that for a known value $\bar{\tilde{y}}_s([r+1]h) + \sum_{i=0}^{m-1} A_{22}^i B_s \Delta\bar{u}_s(rh) - \bar{G}_s[rh]$, equality will be fulfilled:

$$
\begin{aligned}
&E\left\{\left[\bar{\tilde{y}}_s([r+1]h) + \sum_{i=0}^{m-1} A_{22}^i B_s \Delta\bar{u}_s(rh) - \bar{G}_s[rh]\right]^T \left[\bar{\tilde{y}}_s([r+1]h) + \sum_{i=0}^{m-1} A_{22}^i B_s \Delta\bar{u}_s(rh) - \bar{G}_s[rh]\right]\right\} = \\
&= \left[\bar{\tilde{y}}_s([r+1]h) + \sum_{i=0}^{m-1} A_{22}^i B_s \Delta\bar{u}_s(rh) - \bar{G}_s[rh]\right]^T \left[\bar{\tilde{y}}_s([r+1]h) + \sum_{i=0}^{m-1} A_{22}^i B_s \Delta\bar{u}_s(rh) - \bar{G}_s[rh]\right]
\end{aligned}
$$

The last term (27) does not depend on $\Delta\bar{u}_s(rh)$, so it is not considered when minimizing the criterion. The penultimate term is equal to zero because of the mathematical expectation of disturbances, according to the definition, equal to zero:

$$
E\left\{\left[\sum_{j=0}^{m-1} A_{22}^j A_{21} \Delta\bar{\xi}_f(rh + [m-1-j]T_0)\right]\right\} = 0
$$

Then, using the method of calculating the derivative of the quadratic form to minimize criterion (27), we make differentiation:

$$
\begin{aligned}
\frac{\partial J_s[(r+1)h]}{\partial \Delta\bar{u}_s(rh)} &= 2\left(\sum_{i=0}^{m-1} A_{22}^i B_s\right)^T \left[\bar{\tilde{y}}_s([r+1]h) + \sum_{i=0}^{m-1} A_{22}^i B_s \Delta\bar{u}_s(rh) - \bar{G}_s[rh]\right] \\
&+ 2R_s \Delta\bar{u}_s(rh) = 0
\end{aligned}
$$

From this equation, taking into account (25), we determine the control law of the discrete controller for stabilizing the slowly measurable CM coordinates at the given levels $\bar{G}_s$:

$$
\begin{aligned}
&\Delta\bar{u}_s(rh) = \left[\left(\sum_{i=0}^{m-1} A_{22}^i B_s\right)^T \left(\sum_{i=0}^{m-1} A_{22}^i B_s\right) + R_s\right]^{-1} \left(\sum_{i=0}^{m-1} A_{22}^i B_s\right)^T * \\
&* \left[\left(I_s + A_{22}^m\right)\bar{y}_s(rh) - A_{22}^m \bar{y}_s([r-1]h) + \sum_{j=0}^{m-1} A_{22}^j A_{21} \Delta\bar{y}_f(rh + [m-1-j]T_0) - \bar{G}_s(rh)\right]
\end{aligned} \tag{28}
$$

We will show that the value $\Delta\bar{u}_s(rh)$ from this equation is the minimum point of criterion (24):

$$\frac{\partial J_s\left[(r+1)h\right]}{\partial \Delta \bar{u}_s(rh)} = 2\left[\left(\sum_{i=0}^{m-1} A_{22}^i B_s\right)^T \left(\sum_{i=0}^{m-1} A_{22}^i B_s\right) + R_s\right],$$

$$\left(\sum_{i=0}^{m-1} A_{22}^i B_s\right)^T \left(\sum_{i=0}^{m-1} A_{22}^i B_s\right) \geq 0,\ R_s > 0$$

Then $\left(\sum_{i=0}^{m-1} A_{22}^i B_s\right)^T \left(\sum_{i=0}^{m-1} A_{22}^i B_s\right) + R_s \geq 0$, and by condition this matrix is not degenerate.

As a result, the second derivative $\frac{\partial^2 J_s[(r+1)h]}{\partial \Delta \bar{u}_s^2(rh)} > 0$, therefore, the determination of the value $\Delta \bar{u}_s(rh)$ based on the control law (28) is the minimum point of the criterion (24).

3 Results

3.1 Experimental Study of the Stabilization System of Impulse Processes of the Fast-Measurable Subsystem of the CM with Coordination of the Coordinates Ratios

The impulse process model (11) is used for the experimental study of a closed control subsystem with fast-measurable coordinates of the CM nodes, which is written in the form:

$$\begin{aligned}
&\bar{y}_f(rh + lT_0) = (I_f + A_{11})\bar{y}_f\left[rh + (l-1)T_0\right] - A_{11}\bar{y}_f\left[rh + (l-2)T_0\right] + \\
&+ B_f \Delta \bar{u}_f\left[rh + (l-1)T_0\right] + A_{12}\Delta \bar{y}_s(rh) + \bar{\psi}_f \Delta \xi_f(rh + lT_0), \\
&l = 0, 1, \ldots, (m-1)
\end{aligned} \tag{29}$$

The fast-measurable coordinates vector $\bar{y}_f$ of the nodes includes:

1. $\bar{y}_{f_1}$–cryptocurrency exchange rate
2. $\bar{y}_{f_2}$–the volume of cryptocurrency trading
3. $\bar{y}_{f_3}$–offer of cryptocurrency
4. $\bar{y}_{f_4}$–demand for cryptocurrency
5. $\bar{y}_{f_5}$–the amount of speculations in the form of cryptocurrency related exchange products
6. $\bar{y}_{f_6}$–the risk of a collapse of the cryptocurrency

A small sampling period T_0 is selected for these coordinates, which is determined in relation to the most rapidly changing coordinate $\bar{y}_f$.

The slowly-measured coordinates of the nodes $\Delta\bar{y}_s$, which in model (29) represent the internal perturbations of the CM, include the increments of the following coordinates:

1. $\Delta\bar{y}_{s_7}$–number of users
2. $\Delta\bar{y}_{s_8}$–volume of investments
3. $\Delta\bar{y}_{s_9}$–amount of capitalization
4. $\Delta\bar{y}_{s_{10}}$–indirect income
5. $\Delta\bar{y}_{s_{11}}$–the level of trust in cryptocurrency
6. $\Delta\bar{y}_{s_{12}}$–the risk of losing the number of users

For these coordinates, the sampling period $h = mT_0$ is used, where $m > 1$. The control vector includes control increments $\Delta\bar{u}_f$ of fast-measurable coordinates:

1. $\Delta\bar{u}_{f_2}$–the volume of trading
2. $\Delta\bar{u}_{f_3}$–offer of cryptocurrency
3. $\Delta\bar{u}_{f_4}$–demand for cryptocurrency
4. $\Delta\bar{u}_{f_5}$–the amount of speculations

These actions are calculated by a discrete controller with the control law (22). They are provided to the corresponding nodes of the CM by varying them by a decision-maker on financial exchanges.

Increments of informational disturbances $\Delta\xi_f$ (node 13) are used as external disturbances in the model (29). According to Fig. 1 matrices A_{11}, A_{12}, B_f and vector $\bar{\psi}_f$ are equal to:

$$A_{11} = \begin{bmatrix} 0.1 & 0.4 & 0 & 0.1 & 0.4 & 0 \\ 0 & 0 & 0.2 & 0.1 & 0 & -0.6 \\ 0 & 0 & 0 & 0.7 & 0 & 0 \\ -0.2 & 0.8 & -0.5 & 0 & 0.7 & 0 \\ 0 & 0.5 & 0.6 & 0 & 0 & -0.5 \\ 1 & 0 & 0 & 0 & 0-0.15 & 0 \end{bmatrix},$$

$$A_{12} = \begin{bmatrix} 0.4 & 0.3 & 0.15 & 0 & 0.3 & -0.3 \\ 0.3 & 0.5 & 0.1 & 0 & 0.2 & 0 \\ 0.8 & 0.3 & 0 & 0.1 & 0 & -0.5 \\ 0 & 0 & 0 & 0 & 0.7 & 0 \\ 0.1 & 0 & 0 & 0 & 0 & 0 \\ 0 & -0.2 & 0 & 0 & 0 & 0 \end{bmatrix},$$

$$B_f = \begin{bmatrix} 0 & 0 & 0 & 0 \\ 1 & 0 & 0 & 0 \\ 0 & 1 & 0 & 0 \\ 0 & 0 & 1 & 0 \\ 0 & 0 & 0 & 1 \\ 0 & 0 & 0 & 0 \end{bmatrix}, \bar{\psi}_f = \begin{bmatrix} 0 \\ 0 \\ -0.1 \\ -0.2 \\ -0.5 \\ 0 \end{bmatrix}$$

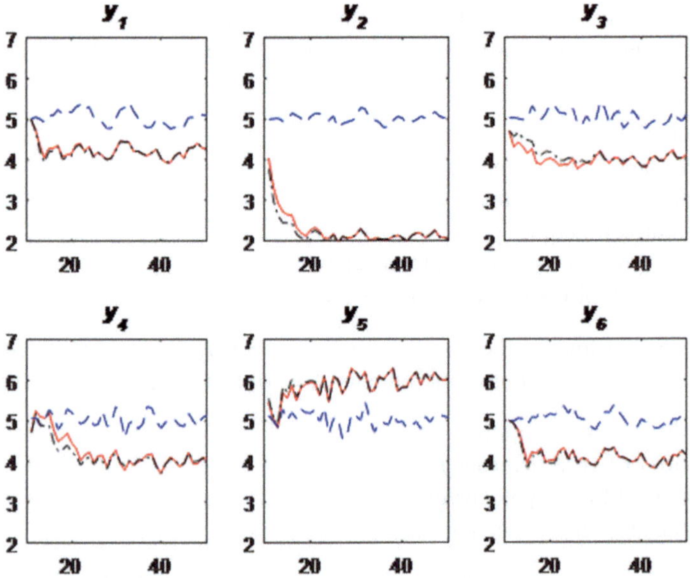

Fig. 2 Transient processes of node coordinates

The problem of coordinating the ratio is to maintain the equality of supply and demand for cryptocurrency, i.e. $\bar{y}_{f_3} = \bar{y}_{f_4}$, and the amount of speculation in the process of trading, which takes place in the form of cryptocurrency contracts, must exceed 65% of the volume of trading directly in cryptocurrency, i.e. $\bar{y}_{f_2} = 0.35\bar{y}_{f_5}$. Then the matrix S in (18) will have the form:

$$S = \begin{bmatrix} 0\;1\;0\;0\;-0.35\;0 \\ 0\;0\;1\;-1\;0\;0 \end{bmatrix}$$

and vector $\bar{b} = 0$. Then the control vector $\Delta\bar{u}_f\ (rh + lT_0)$ will be calculated according to the control law (22).

For model the transient processes of the coordinates of the CM nodes, we assume that the normalized coordinates change in intervals from 0 to 10. Let the initial value of the coordinates be equal to $\bar{y}_f\ (0) = \begin{bmatrix} 5\;5\;5\;\;5\;5\;5 \end{bmatrix}^T$ and the desired values are constant and equal to $\bar{G}_f = \begin{bmatrix} 6\;2.1\;\;4\;\;4\;6\;5 \end{bmatrix}^T$(satisfy the ratios introduced above). In the optimality criterion (13), we put $R_f = 2I$. A normal signal with zero mean and variance 0.1 is applied to node 13 as an external disturbance ξ_f during modeling. It should be noted that the simulation of fast-measurable (29) and slow-measurable (23) subsystems is done simultaneously.

In Fig. 2 the transient processes of node coordinates $\bar{y}_f\ (rh + lT_0)$ are shown: with stabilizing only control (17)–solid line, with coordinating control (22)–dashed-dot line and in the absence of control, that is, in free movement,–dashed line.

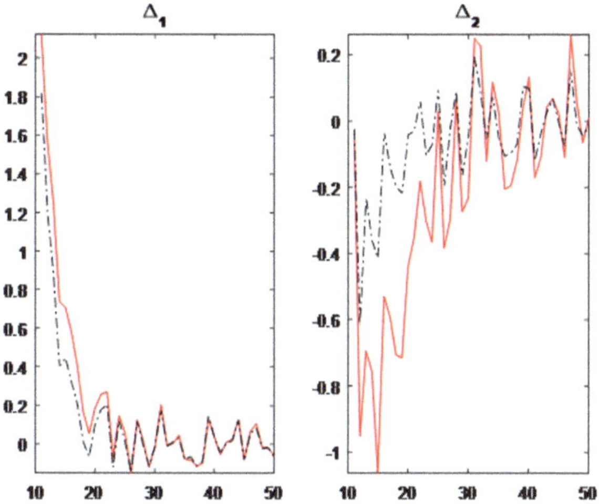

Fig. 3 Charts of ratios errors (discrepancies)

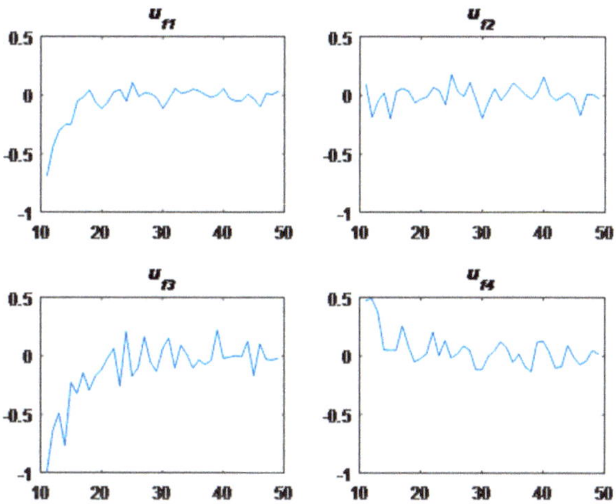

Fig. 4 Coordinating control actions

In Fig. 3 charts of ratios errors (discrepancies) $\Delta_1 = y_{f_3}(k) - y_{f_4}(k)$, $\Delta_2 = y_{f_2}(k) - 0.35 y_{f_5}(k)$ are shown for algorithms based on the control law with coordination (22) (dashed line) and without coordination (17) (solid line). Figure 4 shows coordinating control actions $\Delta \bar{u}_f$ according to (22).

3.2 Experimental Study of the System of Stabilization of Impulse Processes of Slow-Measurable Coordinates of CM

The impulse process model (23) is used for simulation, where the matrices A_{22}, A_{21}, B_s and the vector $\bar{\psi}_s$ are equal to:

$$A_{22} = \begin{bmatrix} 0.1 & 0 & 0 & 0 & 0.5 & 0 \\ 0 & 0.1 & 0 & 0.2 & 0 & -0.2 \\ 0 & 0.2 & 0.1 & 0 & 0.15 & 0 \\ 0 & 0 & 0.3 & 0 & 0 & 0 \\ 0 & 0.2 & 0.1 & 0 & 0 & 0 \\ 1 & 0 & 0 & 0 & -0.2 & 0 \end{bmatrix},$$

$$A_{21} = \begin{bmatrix} 0.2 & 0 & 0 & 0 & 0 & -0.6 \\ 0 & 0.3 & 0 & 0 & 0 & -0.4 \\ 0.1 & 0 & 0 & 0 & 0 & 0 \\ 0.15 & 0 & 0 & 0 & 0 & 0 \\ 0.4 & 0 & 0 & 0 & 0 & -0.3 \\ 0 & 0 & 0 & 0 & 0 & 0 \end{bmatrix},$$

$$B_s = \begin{bmatrix} 0 & 0 \\ 1 & 0 \\ 0 & 1 \\ 0 & 0 \\ 0 & 0 \\ 0 & 0 \end{bmatrix}, \bar{\psi}_s = \begin{bmatrix} -0.2 \\ -0.3 \\ 0 \\ 0 \\ -0.4 \\ 0 \end{bmatrix}$$

As controls for the slow-measurable subsystem, we used: Δu_{s_8}- variation in the amount of investments and Δu_{s_9}- variation in the amount of capitalization. These control actions are used to set the slow nodes coordinates at the desired values while minimizing the generalized variance (24). The sampling period $h = 5T_0$ is used to sample coordinates $\bar{y}_s$ and control actions Δu_{s_8}, Δu_{s_9}. The vector of control actions $\Delta \bar{u}_s\ (rh)$ is formed based on the control law (28), where the matrix R_s (2×2) is equal to $2I$. The initial values of the coordinates are equal to $\bar{y}_s\ (0) = \begin{bmatrix} 5\ 5\ 5\ 5\ 5\ 5 \end{bmatrix}^T$ taking into account the constant vector of desired values $\bar{G}_s = \begin{bmatrix} 5\ 5\ 6\ 5\ 5\ 5 \end{bmatrix}^T$.

Figure 5 shows the transient processes of coordinates $\bar{y}_s\ (rh)$ with control actions (28) (solid line) and without them (dashed line). Figure 6 shows corresponding controls.

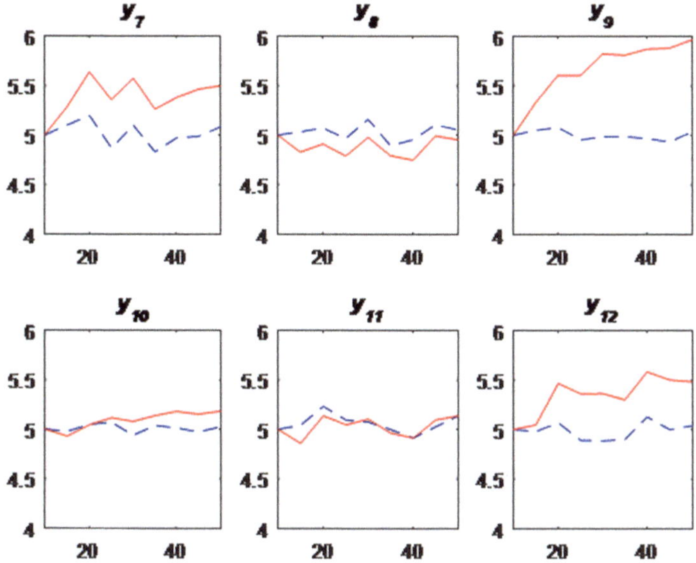

Fig. 5 Transient processes of coordinates

Fig. 6 Corresponding controls

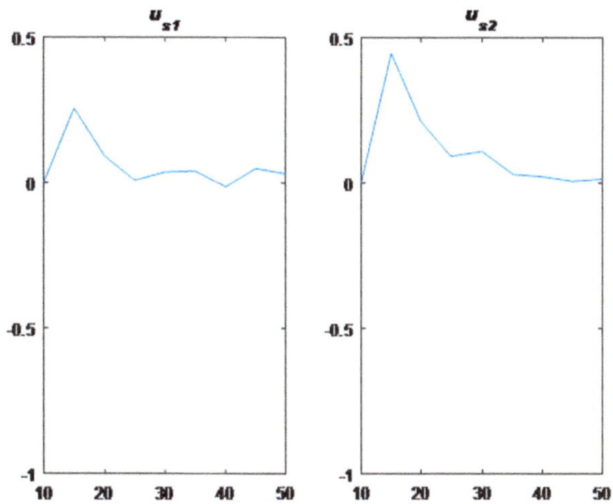

4 Conclusion

As expected, impulse processes with stabilizing and coordinating controls look similarly and both help to set nodes coordinates values at the desired levels (while uncontrolled motion is oscillating around initial values). But the discrepancy of the desired ratios stabilizes at zero much quicker under the coordinating control compared to stabilizing only control. This confirms the usefulness of the suggested coordinating control in fast subsystem. The slow subsystem is successfully stabilized at the desired levels. Control actions have quite small amplitude, so they can be physically implemented.

In general, the research and simulation provided demonstrate the possibility and effectiveness of coordinating control of the cryptocurrency usage dynamic system described by the multirate CM impulse process.

References

1. Romanenko, V., Miliavskyi, Y., Kantsedal, H.: In: Zgurovsky, M., Pankratova, N. (eds.) System Analysis & Intelligent Computing: Theory and Applications, pp. 115–137. Springer International Publishing, Cham (2022). https://doi.org/10.1007/978-3-030-94910-5_7
2. Roberts, F.: Discrete Mathematical Models, with Applications to Social, Biological, and Environmental Problems. Prentice-Hall, Englewood Cliffs (1976)
3. Perepelytsya, V.: Mathematical Models and Methods of Risk Assessment of Economic, Social and Agrarian Processes. Center for Educational Literature, Kyiv (2013)
4. Bashkirov, O.: Comparative VaR method of assessing the risk of bank assets. Problems and Prospects of Development of the Banking System of Ukraine. UABS NBU, Sumy (2005)
5. Kuznetsova, N., Bidyuk, P.: Comparative VaR method of assessing the risk of bank assets. Problems and Prospects of Development of the Banking System of Ukraine. Lira Publishing House, Kyiv (2020)
6. Zhaldak, M.: Basics of Optimization Theory and Methods. Brama, Cherkasy (2005)
7. Sukharev, A.: Optimization Methods Course. Nauka, Moscow (1986)
8. Magnus, Y.: Matrix Differential Calculus with Applications to Statistics and Econometrics. Fizmat, Moscow (2002)
9. Romanenko, V.: Problems of control and informatics, no. 4, pp. 49–58 (2022)

Wireless Sensor Networks for Healthcare on SoA

Anatolii Petrenko and Oleksii Petrenko

Abstract This article describes how to eliminate the two significant mHealth barriers: the lack of standardization of interoperable services and their absence in general. To minimize these barriers, the service-oriented approach is used—to develop the Repository of services, which can be the service source for any necessary personal healthcare platform for chronic diseases. Monitoring patients' vital signs parameters (measured at home) is achieved using modern Internet of Things technology and the Body Area Network (BAN). It provides networkable connections between portable diagnostic sensors, patients' cell phones, cloud data storage with patients' Personal Health Records, and professional health providers.

Keywords Wireless sensor networks · Health monitoring · Personal health systems · Cloud services · Decision support systems · Web-services · Service-Oriented Architecture (SOA) · Respiratory diseases · Deep learning · Cloud · Edge computing

1 Introduction

The world economy focuses on the service industry, which leads to the emergence of the interdisciplinary science of services and the spread of service approaches to technical systems (including software structures). This, in turn, motivates economists, sociologists, mathematicians, programmers, legislators, etc., to work together to achieve fundamental goals: analysis, construction, management, and development of complex service systems as a basis for *the digitalization of society*. Digitalization

A. Petrenko (✉)
The National Technical University of Ukraine "Igor Sikorsky Kyiv Polytechnic Institute", Kyiv, Ukraine
e-mail: tolja.petrenko@gmail.com

O. Petrenko
Miratech, Kyiv, Ukraine
e-mail: petrenko.chiffa@gmail.com

M. Zgurovsky and N. Pankratova (eds.), *System Analysis and Artificial Intelligence*, Studies in Computational Intelligence 1107,
https://doi.org/10.1007/978-3-031-37450-0_6

in the country determines the degree of its development and security. So, there is an important task to increase the number of citizens involved in society's digitalization. In the current modern conditions, it is necessary to widely and comprehensively use advances of the IoT and IT, in general, in various areas. As a result, here we focused on the qualitatively new level of service to the country's health needs, namely: service-oriented architectures (SOA), cloud computing, web services, semantic technologies, the Internet of Things, and knowledge bases [1–5].

Many people receive detailed information about their medical measurements and health state, the environment around them, and risk factors in many countries already today from their mobile devices. And now, still more than ever, there is an urgent need for quality medical care in the current pandemic environment, which allows remote monitoring of patients, which minimizes the need for a doctor to visit patients or hospital visits. It is facilitated by mobile medicine (m-Health), which involves the use of mobile devices to collect aggregated data on the patient's health, providing this information to doctors, researchers, and patients themselves, as well as real-time monitoring of vital patient organs and direct medical care (via mobile telemedicine).

The platform for remote monitoring and diagnosis of patients, *which binds together and unifies the development of applications for patients and doctors, and the central server* by orchestrating and composing web services from a shared cloud repository, is considered. It allows, on the one hand, doctors to monitor patients' health in real-time, regardless of their location, without directly visiting the medical institution, detecting diseases at an early stage and thus preventing their development and reducing costs for further treatment, etc., and on the other hand, helps patients to contact the necessary experts to receive faster and accurately an initial consultation, which means saving not only doctors work time, but also expenses of the relocations to medical institutions. Due to this approach, *the created applications can be adapted to the particular patient, his disease, and his treatment plan at home.*

Many specialists in the world believe that patient monitoring can transform medical care. Wireless sensors, smartphones, modernized clinical trials, internet connectivity, advanced diagnostics, targeted therapies, and cloud computing, together with other science, enable the individualization of medicine and force overdue radical change in how the treatment is delivered, regulated, and reimbursed. In any case of emergency, mobile devices are being used to capture data at that specific moment when care is hugely needed; and at the same very moment, it keeps the line of communication open no matter where the doctor is. Together with that, mobile devices are used at home during regular life routines to record and send vital health data to the caregiver, which turns back to the patient as healthcare management information as rules or advice.

There was proposed to build a smart healthcare ecosystem where patients' health parameters can be permanently monitored by the networked medical heterogeneous devices and then used for decision-making together with structured EHR data, unstructured clinical notes, medical imaging data, etc. [7–11]. Let's imagine that every medical sensor (or another data resource) of that ecosystem has its own URI allowing doctors and patients to interact with it via the web browser or application, and, at the same time, each sensor can have the software interface – a set of web services allowing intelligent software agents to interact with it (collect and analyze the data, etc.) on behalf of doctors and patients.

Indeed, integrating these innovative capabilities with classical medical records is vital. Daily, the e-Health systems (which includes m-Health) generate an enormous amount of data, which is used for making a more accurate diagnosis as well as in prescribing treatment and assisting in the production and testing of new medicines, in the prevention of disease onset, and in enhancing care quality. E-Health becomes one of the Big Data consumers because data collections are so extensive and complex that it becomes difficult to process using the traditional on-hand database management tools or data processing applications.

Healthcare was once about trying to heal the sick patient. Healthcare organizations worldwide now have an opportunity to shift this focus to one keeping the public healthy and anticipating health issues before they become a problem. Physicians and other providers seek to monitor patients remotely through new technologies, aiming to identify problems early and cut costs and inefficiencies in the healthcare system. Identifying people at risk of becoming ill or developing a severe condition and providing the foresight to prescribe preventive measures is a genuine possibility.

This paper is structured as follows. We start by presenting the necessary objectives for accelerating the implementation of inter-operable Health systems. It highlights requirements and recommendations, security as well as appropriate technologies. The following section will describe existing cloud-based technology together with health IoT. Standardization activities used SOA, and the proposed services repository for BSN concluded in the last part.

2 Activity Overview

Five objectives are necessary to accelerate the implementation of Big-Data capabilities in patient-generated inter-operable Health systems:

Objective 1: Ensure the continued contribution in the cross-border European Big Data activities for e-Health by investigating ways of realization of upper-level of health analytic technology "Personalized Medicine & Prescriptive Analytics" proposed in "Healthcare Analytics Adoption Model: A Framework and Roadmap" [1].

At this level, the analytic motive expands to wellness management, physical and behavioral-functional health, and mass customization of accurate, patient-tailored care under direct contracts with patients and doctors. Healthcare organizations are entirely engaged in a data-driven culture and a shift from a fixation with care delivery to an obsession with risk intervention, health improvement, and preventive medicine. They will only benefit from big data if they take a more holistic, patient-centered approach to value that focuses equally on healthcare spending and treatment outcomes.

Improved qualities of care and better clinical outcomes are the observable benefits of this future medical care revolution, as was declared in the GREEN PAPER on m-Health [2], the Action Plan 2012–2020 [3], and recommendations of the eHealth Stakeholder Group [4], the EU eHealth Governance Initiative and the eHealth Task Force [5].

Objective 2: Creation of a one-stop shop for health information by building a smart healthcare ecosystem where patients' health parameters can be permanently monitored by the networked medical heterogeneous devices and solutions and then used for decision-making together with the structured EHR data, unstructured clinical notes, medical imaging data, etc. Effectively integrating and efficiently analyzing various forms of healthcare data over a while can answer many of the impending healthcare problems.

Current health data is diverse, comprising structured and unstructured information in various formats. It is vitally essential for healthcare organizations to get the tools, infrastructure, and techniques in place now to deal with big data effectively. Of the diversity of the participating devices, we should provide the integration framework for Patient Monitoring (both healthy and sick patients) and personalizing healthcare for every patient.

EHRs and automation tools already exist. They are used to identify and stratify patients needing special attention or care. But there is a lack of semantic interoperability. No standards exist on healthcare data interoperability, and there are only syntactic standards (e.g., HL7 from the US). Although there are ongoing efforts to develop OWL-DL ontology to solve HL7 v.2 and v.3 interoperability problems, there are still no adopted solutions. A draft standard FHIR (Fast Health Interoperability Resources), the younger sibling of interoperability standard HL7, is already being used by developers. However, developing an ontology for FHIR is still urgent [6].

Objective 3: Using a semantically enhanced event-driven service-oriented architectural model (SEMSOA) to develop an open ecosystem (based on the Open Science Commons) for providing more comprehensive interoperability of healthcare services for patients groups with different illnesses and for supporting a policy of standardization of these healthcare services [39].

Patients, doctors, and device-trigger outcomes can initiate events in this ecosystem. Due to such an ecosystem, healthcare migrates from episodic and fragmented illness responses to a patient-centric care delivery model.

We propose to develop a repository of services related to data collection and storing processes from all data sources (and develop the ontology of these services). This Repository will include the following:

- Ontology-based services for gathering the various portable wearable devices' output data.
- Ontology-based services for managing the EHR data, clinical notes, claims, medical imaging data, etc.
- Ontology-based services for capturing information on patients' behavior.
- Services for identifying patients at risk of becoming ill or developing a severe condition by constant patient monitoring and personalizing healthcare.
- Services for insights synthesis and information of patients, health professionals, health payers and insurers, and life science companies' actions.

- Services for aggregating individual patients' data across a community into a broader, meaningful view of health and healthcare in a particular region to support healthcare migrating from episodic and fragmented illness response to a patient-centric public healthcare model.

The customer-driven medical applied software development mechanism will be provided by compositing and orchestrating dynamically discovered services from the developed Repository to form the individual patient pathway (patient-specific work-flows) of monitoring and treatment, considering different existing rules, regulations, and standards.

Objective 4: Establishing connections between data sources, patients, doctors, and healthcare organizations by developing a healthcare ecosystem in the **European Open Science Cloud for Research** with the aim of data sharing across the entire ecosystem, using SaaS, PaaS, and IaaS technologies to host eHealth community-support services, services for analytic processing of data across the healthcare ecosystem.

Expanding solutions across the health continuum, from healthy living, prevention, diagnosis, treatment, recovery, and home care, to truly impact patients' health at the individual and population levels and to the most clinically and cost-effective treatments will be applied. The solutions will be offered to large and medium size healthcare organizations, small research communities, and the long-tail of science, education, industry, and SMEs. Consequently, these customers will be able to manage patient populations to improve health, improve outcomes and reduce costs by identification of patients with the same patient's characteristics. Even more, it is possible to decrease treatment error rates and liability claims by analyzing doctors' prescriptions and comparing them against medical guidelines, especially those arising from clinical mistakes, and, perhaps, to find more effective approaches to treating different conditions. Cloud computing security technologies had to be applied, which allows practitioners access to patient data with ever storing protected health information on mobile or remote devices.

Objective 5: Promote the integration of the developed Repository of services with the EGI generic Environment Supporting services and extend them with new capabilities through user co-development, in particular, by collaborating with Competence Centers BBMRI (Biobanking and Bimolecular Research Infrastructure), ELIXIR (Life-sciences Infrastructure for biological information) and Life Watch (E-Science Infrastructure for Biodiversity and Ecosystem Research), ECRIN (European Clinical Research Infrastructure Network) and Open AIRE (Open Access Infrastructure for Research in Europe) for adopting their possibilities to support also Public Health needs.

As a result, a typical service marketplace may be formed to promote the commercial adoption of the state-of-the-art customer-driven technology of eHealth applications development to increase their potential for innovation in eHealth. The EGI e-Infrastructure Commons solution will offer services for identity management in a federated environment, authentication and authorization, accounting and monitoring, identification of permanent digital objects, etc. [17]. The integrative solution for practitioners and patients will demonstrate Big Data solutions for eHealth based on services composition and orchestration in the EGI e-Infrastructure Commons, aiming at ill citizens. It should benefit end users from prior research and development on data aggregation and processing. The practitioner, on another side, should be able to compare data from the respective patient treatment to gold standard care paths to optimize the treatment outcome. In parallel, the patient should profit at the maximum level from the data collection to guarantee acceptance and usage of the solution.

Processing and comparing the collected data with the existing anonymous data will enable the practitioner to identify warning signals about the patient's status and to adapt the care path of the patient in accordance to maximize their treatment outcome. Furthermore, the data can be anonymized and contributed to the existing data, enhancing the capabilities of the foresight and care path management tools and generating even better output.

The proposed objectives address the challenges in the recent Horizon Europe Cluster 1 "HEALTH" [18]. The proposed healthcare ecosystem may be used for expert functions and modeling developing health systems in different countries. As a result, some dynamic, innovative healthcare model can be established, considering both the current technical capabilities and emerging technologies. This will allow quick response to the market dynamics and offer the developers a system of innovation implementation to provide consulting services for the performance of new healthcare systems in various markets and hold the bulk of the information in this area. Having a centralized repository of services and data makes it possible to develop a universal adapting software product to embedded software solutions from third parties that is compatible with various sensors and provides the ability to get medical help anywhere in Europe.

3 Related Work

The most accessible tool for maintaining and improving the above factors is the implementation of e-Health systems, particularly telemedicine and mobile medicine, which constantly monitor the patient's condition. Mobile health care (m-Health) involves the use of mobile devices to collect, aggregate, and transmit patient data, sending it to doctors with instant availability of feedback (for mobile telemedicine). Provides real-time health monitoring outside clinics using Body Sensor Networks (BSN) and Medical Internet of Things (m-IoT) systems and healthcare knowledge management, providing physicians and researchers access to the most modern methods and practices.

The BSN architecture shown in Fig. 1 is a general architecture that integrates body sensor networks, environmental sensors, server data management, data analysis system, and local or remote user interfaces. Wearable sensors are used to monitor a patient's vital parameters (for example, a person's blood glucose level, blood pressure, ECG, or to detect a fall). Environmental sensors are used to monitor temperature,

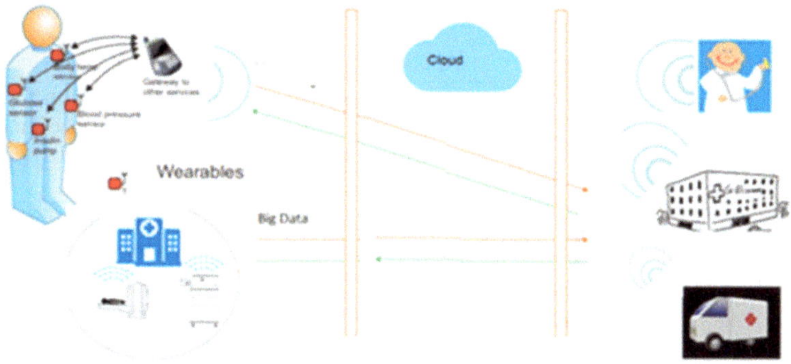

Fig. 1 BSM architecture with fog network

humidity, light, and movement in the home. External data sources include data from magnetic resonance imaging, ultrasound diagnostics, electroencephalographs, etc.

The collected data is transmitted to the control center of the medical institution, which is equipped with significant cloud storage capabilities and computing resources for analysis and presentation of the result.

Today there are dozens of BSM in different countries worldwide, the authors of which use a similar architecture (Table 1) [7–11].

But among them, none are built according to the service-oriented principle. Therefore, it is proposed to implement the service core of the BSM platform with a wide range of applications: from intelligent data analysis to disease diagnosis, from business analytics tasks to participation in smart home management.

In addition to numerous Body Sensor Networks (BSN), there are unique large ecosystems of the world's leading companies that help customers improve their efficiency and delivery of quality healthcare with services that include connections to clinical-grade smart devices, a built-in framework to ingest, manage, and analyze healthcare data, and more (Table 2).

It was demonstrated that the security and privacy of patients and their data would remain the biggest challenges mobile health will face in the nearest future. The eHealth technologies are becoming increasingly mainstream and are present now in primary and secondary healthcare for personal monitoring purposes. Wearable health monitoring systems have highlighted many fundamental security and privacy challenges [11–19].

All e-infrastructure service providers serve the same community of users. It is clear that no single e-infrastructure provider currently serves the full capabilities of e-infra-services the end users need. However, users also want and need a single, easy-to-use interface for all e-Infrastructure services. They need services that are coherent, managed, and, above all, integrated so that they can get on with the disease's treatment. But they also need constant innovation of these services, way ahead of what commercial providers can offer.

Table 1 Appointment of some BSM

No	Name	Applications
1	HEARTFAID	It aims to improve the early diagnosis of the body's condition and, as a result, makes the clinical medical treatment of heart diseases among the elderly more effective
2	ALARM-NET	Human vital signs analysis assistant and human habitat monitoring network based on implemented adaptive medical care
3	CAALYX	It increases the autonomy and self-confidence of older people by implementing devices and sensors to measure certain vital signs, detect falls, and automatically communicate in real-time with a service provider in the event of an emergency, regardless of the person's location
4	Tele-CARE	It integrates mobile agents with the Internet to build a standard flexible and configurable infrastructure for older people to get the maximum effect from the developed technologies
5	CHRONIC	European model focused on care for chronic patients based on an integrated IT environment
6	MyHeart	Intelligent prevention and monitoring system of cardiovascular diseases, which uses smart electronic and textile systems and relevant services that allow patients to monitor their health
7	OLDEST	An innovative low-cost technology platform that provides a wide range of services for the elderly
8	SAPHIRE	It provides patient monitoring based on the technology of distributed agents, supplemented by intelligent systems decision-making support based on clinical practice recommendations
9	@Home	The primary functional purpose is remote monitoring of the patient's vital signs

4 BSN Prototyping

Body Sensor network is one of the principal components in healthcare applications. The most significant advantage of BSNs is real-time monitoring anywhere, at home and outdoors. Stable internet connection at home and mobile network technologies outdoors. This feature allows the patient to have an absolutely normal life while still being monitored. His vital signs are continuously or intermittently transmitted to a remote monitoring center, with health support, and, if needed, inform the patient of his medical status. BSNs are self-organizing networks consisting of many wireless sensor nodes distributed in space and intended to monitor environmental characteristics and manage objects in it. A self-organized network is one in which the number of nodes

Table 2 World's leading companies' ecosystems

No	Name	Applications
1	Amazon.com, Inc.	Amazon Web Services (AWS) for healthcare offers a holistic solution for healthcare cloud computing. It delivers a robust suite of data management tools, enabling facilities to pursue life sciences, genomic application solutions, and client-facing healthcare work. Industry leaders like Moderns for day-to-day operations have trusted this platform
2	Microsoft	Microsoft Azure's cloud services are built around continuous patient monitoring, which helps improve patient outcomes and streamline clinical operations and can integrate with IoT technology. Finally, Azure features AI and machine learning techniques that can help develop personalized preventative care plans
3	Alphabet Inc. (Google)	Google Cloud Computing's healthcare tools equip healthcare providers with the capability to ensure overall population health. Google Cloud's tools accomplish its mission by enabling telehealth appointments, remote work, and easy access to healthcare applications
4	IBM	IBM's main selling point for its cloud services is Merative, previously known as Watson Health—an AI helping revolutionize patient care. Merative is a leading data and analytics AI made to help providers modernize their operations. It offers specialized healthcare consulting services that improve both business and clinical experience

is a random value at each moment, varying from 0 to some value Nmax. Connections between nodes in such networks are also random in time because they are formed to achieve any goal or transfer data to a public communication network or some other types of networks.

The type of disease determines the existing variety of portable diagnostic and measuring devices. It includes, first of all, glucometers, tonometers that measure blood pressure and pulse, or peak flowmeters. This includes devices patients use during home treatment: inhalers, heart rate monitors, and insulin pumps. Now there are smart pills with built-in sensors, allowing doctors to get more accurate information about a person's condition. There are sensors on the patch, with the help of which the system collects data about the patient, and the smart pill simultaneously supplements them with information received inside the body.

The Internet of things (IoT) generates bio-signals which include body temperature, SpO2, blood pressure, blood glucose, electrocardiogram (ECG), electroencephalogram (EEG), electromyography (EMG), and galvanic skin responses from wearable or implantable sensor nodes. In BSN, context awareness information is also required and can be gathered by GPS. BSN is an excellent support in disease management and prevention. Treatment of chronically ill patients is personalized and customized by establishing a Repository of services (web applications with a unified interface) for patient care (care services), for planning and carrying out treatment (treatment services), and ensuring the functioning of the entire system (management services). In this paper, we propose the latest innovative type of cloud BSM based on service-oriented architecture (SOA), using additional fog and edge computing networks as middleware and blockchain technologies to ensure the confidentiality of data received and processed.

Compatibility of medical services for groups of patients with different diseases can be achieved with the help of a medical platform, which consists of the following:

- The block of patient services and portable diagnostic devices.
- The block of services of medical workers.
- The block of cloud platform management services for medical data processing and storage.
- The block of services for Data Analytics.

The block of patient services and portable diagnostic devices implements those services which correspond to the medical specialists' recommendations aimed to provide as close as possible to the best-known treatment strategies for the patient.

That means:

- The possibility of performing measurements of disease indicators at home using devices (glucometer, tonometer, insulin pump, etc.) connected to the patient's smartphone or tablet, data collection and aggregation and their wireless transmission, sending the data to the server, where they will immediately become available to the attending physician through the patient's Electronic Health Records (EHR).
- Remote monitoring of the patient's state, timely reminders to the patient about the need to take medication or undergo the appropriate procedure by the treating doctor (nurse), as well as about his actions in emergencies that arise during an exacerbation of the disease. If the monitored vital signs approach the dangerous borders, the doctor (actually any authorized person as medical personnel or care person) will be notified immediately.
- Integration into social networks or knowledge base where the patient can receive comprehensive information about the treatment methods of his disease and share his experience in this field. The patient can have his account, join interest groups, post his materials and comment on publications of other participants, upload photos, and exchange e-mails with other network members.
- Provision of medical information on the smartphone about the treatment of the disease, organization of proper nutrition, safe and effective use of drugs, and dosage. Advice on actions in urgent situations, when a specialist consultation is required, or it is necessary to contact medical centers and hospitals ready to help, and there is no communication with the attending physician.

The block of services of medical workers provides effective interaction between a medical worker (treating doctor) and a patient through the doctor's smartphone (tablet) and implements the treatment services listed below:

- Remote monitoring of the patient's status in any place and at any time; the ability to see the results of the patient's tests, medical history, and previously prescribed treatment; quickly enter and edit information in the patient's card. Assess the dynamics of changes in the patient's health (is there progress or regression in treatment?) and make changes to his treatment plan. Receive from the server, based on the results of data processing, emergency messages about the approach of the patient's vital signs to dangerous borders.
- Drawing up a treatment plan (roadmap) with constant access to the server with patients' medical records, forming recommendations for proper treatment, nutrition, and leading a healthy lifestyle in the form of a treatment plan and placing them on the server, monitoring their implementation. Advice on the use of other drugs for treatment and actions of the patient in urgent situations, when a specialist's consultation is required, or it is necessary to contact medical centers and clinics ready to come to the aid of the patient.
- The possibility of conducting consultations with colleagues and specialists in voice or video conferences.
- The possibility of maintaining contact with the patient and arranging a personal meeting (visiting the patient at home or in a medical institution). Timely reminders to the patient about the need to take medicine or undergo the appropriate procedure, as well as about his actions in emergencies.

The block of cloud platform management services ensures the functioning of the system of remote monitoring of patients' state and implements the following management services:

- Access to the data of portable diagnostic devices and coordination of their data formats when transferring to the server with medical charts.
- Support of patients' medical records, filling in their fields and reading data from the cards, which contain processed data and recommendations made by the doctor on proper treatment and nutrition, which leads to a healthy lifestyle.
- Composition of repository services and editing of the formal description of the business process of treatment, description of the architecture of the mobile application in the language of the description of business processes.
- Formation of the flow of tasks (sequences of software modules of individual services) for the construction of applications describing the business process of treatment, taking into account the data necessary for the implementation of individual services and using accepted standards and information transfer protocols.
- Transferring data to an external remote system (clinic/network medical professionals) using any of the six available connection options: Wi-Fi, Bluetooth, WiMax, GPRS, 802.15.4, and ZigBee, depending on the application.
- Extreme warnings to the doctor and the patient based on the adverse results of the current data analysis when the patient's vital signs, which are being monitored, are approaching the dangerous limit.
- Call an ambulance based on the patient's current location (GPS) if the server fails to establish emergency contact with the doctor.

- Running algorithms for selecting applied web services from the Repository of services, which in many respects has a universal character (especially in the field of platform management services, initial data processing, measurement, and support of communication with external information systems), and contributing in this way to possible deploy BSM of various purposes with minimal costs.
- Processing of collected patient data at the fog and edge and inter-organizational levels to reduce the load on the cloud.
- Improving the secure exchange of confidential patient data between medical institutions using proof-of-stake blockchain consensus algorithms to reduce the number of resources needed to maintain the platform's infrastructure.
- Support for an electronic prescription (ePrescription) that prescribes the necessary medication to the patient and which was compiled by doctors and medical professionals based on the results of the analysis of data from electronic medical records and measured medical indicators of patients at home.
- Forming an invoice for the services provided to the patient during the treatment, using records of the personal medical card, and sending the invoice to the patient's bank.

The Block of services for Data Analytics. The goals of Personal Health Systems are prediction, prevention, and treatment customized to each patient. Technological capability to aggregate and analyze data from wearable diagnostic devices and information about treatment dynamics contribute to providing patient-specific diagnoses and treatments [29–31]. If patients make home self-measurements over equal time intervals, the time series is formed as a sequence of data points. Time series are explored for a variety of purposes.

In some cases, it is enough to get a description of the time series feathers; in other cases, it is required not only to predict the future values of a time series but also to control its behavior. Method for analysis of the time series is defined, on the one hand, by analysis purposes, and on the other hand, by the probabilistic nature of time series values. They suggest that a time series consists of several system components and random components as the residue. Systematic components usually include periodic trends and cyclical components. The residue is traditionally considered an accidental error or noise. The result of time series analysis is its expansion into simpler components: slow trends, seasonal and other periodic or oscillatory components, and noise components. This expansion can serve as a basis for predicting the original time series and its relative individual components, particularly trends.

- Spectral analysis, which allows finding the periodic components of the time series; Correlation analysis, which deals with the identification of significant periodic dependence and the correspond-ding delay (lag) within one sequence (autocorrelation) and between several series (cross-correlation);
- Seasonal Box-Jenkins model, which allows you to predict the future values of the series;
- Seasonal and non-seasonal exponential smoothing, which is the simplest model of forecasting time series.

- Time series data mining this deal with the identification of temporal patterns for the characterization and prediction of time series events.
- Model of dynamic Bayesian network.
- Neural convolutional network forecasting model based on BSM sensor signal processing.
- Methodologies for the detection and analysis of data outliers.

It can be seen from above that the Repository contains services that belong primarily to Specific (application support) services (SS), which had to be developed. Although between them, there are some Generic (environment-supporting) services (GS): measuring indicators of disease at home and transmitting them, data collection and data aggregation, and remote monitoring. The last ones can be based on Generic Internet of Things (IoT) services databases provided by other service producers (for example, EGI [20], Flatworld [21], FI-WARE [22], SAP [23], ESRC [24], etc. Most Web service for decision support systems is also Specific (application support) services (SS) which had to be developed. Most Web service for medical data analytics is also Genetic or Specific services that had to be developed [33, 34, 38]. A personalized, predictive health profile is then generated, using EHR data, to assist healthcare providers and their patients in working together to improve an individual's health and help prevent the onset of certain diseases whenever possible.

Ontologies are used not only to gather mobile personal device data and behavioral as environmental data together with integrating data formats and software tools as a service collection but also to analyze data processing across the healthcare ecosystem. The last ones may include insights and informative actions for patients, care providers, health payers and insurers, and life science companies. Patients, doctors, and device outcomes can initiate events in the ecosystem with semantically enhanced event-driven service-oriented architecture (SEMSOA) [3, 38].

The proposed BSN platform for supporting the patient and the physician in the treatment and monitoring differs from many available mobile applications for the similar appointment in the following essential features:

(1) Mobile applications for the patient, physician, and service management have a service-oriented architecture. They can be built in on-demand forms by discovering and using reusable services with transparently described interfaces available on the network to perform a specific task.

(2) The Repository of services (web applications with a unified interface) is proposed for patient care (care services), for planning and carrying out treatment (treatment services), and to ensure the functioning of the entire system (services management). Some of these services (Generic Enablers) can be taken from existing service repositories, but many value-added services (so-called Specific Enablers) had to be developed from scratch. Functionalities of Specific Enablers for patients, physicians, and services management were developed.

The experimental verification of some of the above introductory provisions was carried out on the example of the device for patient's respiratory system monitoring in real-time at home, which related to the desire to reduce the risks for sick people in t the conditions of the COVID-19 pandemic. Such devices can detect and prevent the

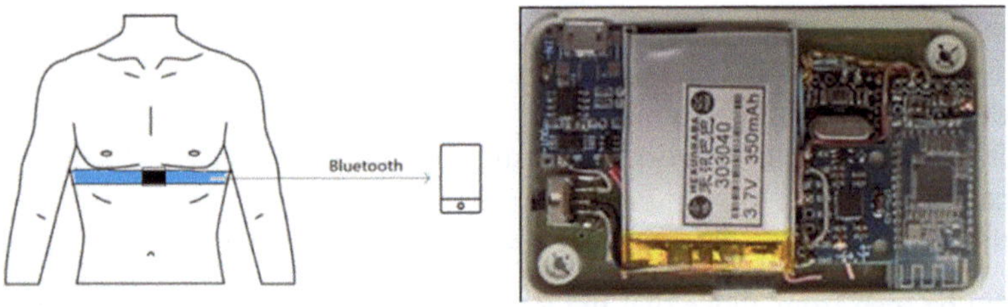

Fig. 2 Respiratory sensor

exacerbation of dangerous respiratory conditions in various daily human activities (such as walking, sleeping, and other physical activities). Their functionality is based on the receipt of breathing signals by fixing the movement of the chest, say, with the help of an accelerometer placed on the patient's body and sending data to the client's smartphone three times per second (Fig. 2).

Instead of traditional statistical and violet-conversion methods for processing respiratory signals, it has implemented alternative louder analytics by applying Deep machine learning methods, which essentially allows the system to be tuned to diagnose and monitor other respiratory diseases [36]. As a result of this experimental study, it was possible to apply CNN to highlight breathing patterns (normal; fast, shallow breathing; deep labored breathing; Cheyn-stokes) using accelerometer signals and demonstrated the usefulness of pre-processing the signal. A method is proposed to convert one-dimensional (1d) accelerometer signals into two-dimensional (2d) graphical images processed by CNN with multiple processing layers. It is better because the accuracy of determining the breathing pattern in different situations for different physical states of patients may be increased compared to one-dimensional conversion accelerometer signals.

For evaluating the proposed method were used the locally gathered and open-source data from PhysioBank and PhysioNet (www.physionet.org.uk) [39].

5 Conclusions

A recent study by Frost & Sullivan on digital health consumer behavior shows that approximately 24% of consumers currently use mobile apps to track health and wellness, 16% use wearable mobile sensors, and 29% use electronic personal health records. This trend is expected to continue as 47% of consumers will consider using Internet-enabled mobile sensors really soon. Analysts from the IDC company estimated that the segment of body sensors alone amounted to $222 billion in 2022 (in 2017, the volume of this market was $113.2 billion). The number of connected medical devices worldwide is expected to grow from 10 billion to 50 billion in the next 10 years.

The main idea proposed is to develop a unified approach to implementing BSM for various purposes based on the use of the SOA paradigm. The use of ontologies of subject areas contributes to the dynamic integration of services from different developers (suppliers) into a single service environment. In contrast, using artificial intelligence methods increases the efficiency and reliability of application software.

The proposed Repository of services has a universal character (especially in the field of platform management services and many services of doctor and patient applications), so on its basis, it is possible to deploy mobile systems of various purposes with minimal costs and adapt them to the tasks of supporting specific patient treatment plans by scaling services: exclusion some of them, adding new ones, replacing some with others of the same purpose.

The social orientation of the proposal should be noted to help those who need help the most: older people with chronic diseases. Remote diagnosis and monitoring of the patient's condition are also crucial for ensuring health care for the rural population, with few doctors and hospitals. Foreign clients may be attracted by domestic medical applications with mathematical forecasting periods of possible crises in the patient's states.

References

1. Healthcare Analytics Adoption Model: A Framework and Roadmap (white paper). https://www.healthcatalyst.com/white-paper/healthcare-analytics-adoption-model/2/
2. GREEN PAPER on mobile health ("mHealth"). http://ec.europa.eu/digital-agenda/en/news/green-paper-mobile-health-mhealth
3. eHealth Action Plan 2012-2020 - Innovative healthcare for the 21st century. http://ec.europa.eu/health/ehealth/docs/com_2012_736_en.pdf
4. Discussion paper on semantic and technical interoperability of eHealth. https://health.ec.europa.eu/system/files/2016-11/ev_20121107_wd02_en_0.pdf
5. eHealth Task Force Report "Redesigning health in Europe for 2020". http://ec.europa.eu/digital-agenda/en/news/eu-task-force-ehealth-redesigning-health-europe-2020
6. FHIR - Fast Healthcare Interoperability Resources (FHIR). https://clinicalarchitecture.com/products/fhir/
7. Yuan, B., Herbert, J.: Web-based real-time remote monitoring for pervasive healthcare. In: Ninth Annual IEEE International Conference on Pervasive Computing and Communications, PerCom 2011, Seattle, WA, USA, Workshop Proceedings. https://doi.org/10.1109/percomw.2011.5766964
8. Naranjo-Hernández, D., Reina-Tosina, J., Roa, L.M.: Special issue body sensors networks for E-health applications. Sensors (Basel) **20**(14), 39–44 (2020). https://doi.org/10.3390/s20143944
9. Ha, I.: Technologies and research trends in wireless body area networks for healthcare: a systematic literature review. Int. J. Distrib. Sens. Netw. **2015**, Article ID 573538, 14. https://doi.org/10.1155/2015/573538
10. Movassaghi, S., Abolhasan, M., Lipman, J., Smith, D., Jamalipour, A.: Wireless body area networks: a survey. Commun. Surv. Tutor. IEEE **16**(3), 1658–1686 (2014)
11. Landi, H.: What Amazon's potential move into at-home medical tests could mean for the market, May 19, 2021. https://www.fiercehealthcare.com/tech/what-amazon-s-potential-move-into-at-home-medical-tests-could-mean-for-market
12. Amazon, Inc. https://www.aws.amazon.com/health
13. Microsoft: https://www.azure.microsoft.com/en-us/solutions/industries/healthcare/

14. Alphabet Inc: https://www.cloud.google.com/solutions/healthcare-life-sciences
15. International Business Machines Corporation: https://www.ibm.com/cloud/healthcare
16. Lindzon, J.: At-home tests put health in your own hands, April 13, 2021
17. https://garage.hp.com/us/en/innovation/telemedicine-consumer-healthcare-devices-at-home.html
18. Competence Centres. https://wiki.egi.eu/wiki/EGI-Engage
19. Horizon Europe Work Programme 2023-2024: Cluster 1 Health. https://research-and-innovation.ec.europa.eu/funding/fundingopportunities/funding-programmes-and-open-calls/horizon-europe/cluster-1-health_en
20. Iqbal, O., Iftakhar, T., Ahmad, S.Z.: Internet of things for in home health based monitoring system: modern advances, challenges and future directions. Quest J Softw Eng Simul **7**(8), 23–37 (2021). ISSN(Online):2321-3795 ISSN (Print):2321-3809. www.questjournals.org
21. Omoogun, M., etc.: When eHealth meets the internet of things: pervasive security and privacy challenges. In: 2017 International Conference on Cyber Security and Protection of Digital Services, 19–20 June 2017. London (2017). https://doi.org/10.1109/CyberSecPODS.2017.8074857
22. Khalid, A., Shahbaz, M.: Using body sensor networks to show that fog computing is more efficient than traditional cloud computing. Int. J. Comput. Sci. Inf. Secur. (IJCSIS), **14**(12) (2016). ISSN 1947-5500. https://sites.google.com/site/ijcsis/
23. Milovanovic, D., Bojkovic, Z.: Cloud-based IoT healthcare applications: requirements and recommendations. Int. J. Internet Things Web Serv. **2**, 60–68 (2017). ISSN:2367-9115
24. Wua, F.-J., Kao, Y.-F., Tseng, Y.-C.: From wireless sensor networks towards cyber physical systems. Pervas. Mobile Comput. **7**, 397–413 (2011). https://doi.org/10.1016/j.pmcj.2011.03.003
25. EGI: Advanced computing for research. https://www.egi.eu/
26. Flatworld. http://www.flatworldsolutions.com
27. FI-WARE. http://catalogue.fi-ware.org/enablers
28. SAP. http://www.sap.com/pc/tech/enterpriseinformation-management
29. ESRC. http://ukdataservice.ac.uk
30. Raghupathi, W., Raghupathi, V.: Big data analytics in healthcare: promise and potential. Health Inf. Sci. Syst. **2**, 3 (2014). https://doi.org/10.1186/2047-2501-2-3
31. IHTT: Transforming Health Care through Big Data Strategies for leveraging big data in the health care industry. 2013, http://ihealthtran.com/wordpress/2013/03/iht%C2%B2-releases-big-data-research-report-download-today/
32. LaValle, S., Lesser, E., Shockley, R., Hopkins, M.S., Kruschwitz, N.: Big data, analytics, and the path from insights to value. MIT Sloan Manag. Rev. **52**, 20–23 (2011)
33. Mouttham, A., Peyton, L., El Saddik, A.: Business process integration and management of next-generation health monitoring systems. J. Emerg. Technol. Web Intell. **1**(2) (2009)
34. Petrenko, O.O.: Comparing the types of service architectures. In: System Research and Information Technologies, Kyiv, №3 (2016). ISSN 1681-6048 (Ukrainian)
35. Petrenko, A., Bulakh, B.: Automatic service orchestration for e-health application, 2019 in advances in science. Technol. Eng. Syst. J. https://doi.org/10.25046/aj040430
36. Pysmennyi, I., Kyslyi, R., Petrenko, A.: Edge computing in multi-scope service-oriented mobile healthcare systems. Syst. Res. Inf. Technol. **1**, 118–127 (2019). https://dx.doi.org/10.20535/SRIT.2308-8893.2019.1.09
37. Petrenko, A., Kyslyi, R., Pysmennyi, I.: Detection of human respiration patterns using deep convolution neural networks. Eastern-Eur. J. Enterprise Technol. **4**(9), 6–13 (2018)
38. Petrenko, A., Kyslyi, R., Pysmennyi, I.: Designing security of personal data in distributed health care platform. In: Apr 2018 in technology audit and production reserves. https://doi.org/10.15587/2312-8372.2018.141299
39. Petrenko, O.O., Petrenko, A.I.: Service-based medical platform for personal health monitoring. Bioinform Proteom Opn Acc J **1**(1), 000106 (2017)
40. Breathmonitor, A.I.: Sleep Apnea Mobile Detector, Studies in Computational Intelligence series: System Analysis and Intelligent Computing: Theory and Applications. Springer Nature, Switzerland AG (2022)

Improving Predictive Models in the Financial Sector Using Fractal Analysis

Alexey Malishevsky

Abstract Fractal theory and analysis have been widely used in many areas, including health care, data mining, economics, urban studies, geology, geography, chemistry, astronomy, and social studies. This chapter proposes advanced attributes based on fractal, spectral, and wavelet analyses for improving predictive models in the financial sector. It provides a brief overview of fractal theory applications and popular methods for computing fractal dimensions. While building predictive models in a financial sector, simple attributes are usually used, including demographic data and simple statistics on transaction activity data. Advanced attributes are presented to enrich the standard set of attributes capturing the complexity and periodic behavior of various client activities. The advanced attributes include fractal-based attributes (head-tail index, Box-counting dimension, compass dimension, and correlation dimension), spectral analysis-based attributes, and wavelet analysis-based attributes. Two case studies are presented using real data from a financial company where classification and regression problems are solved. In these studies, three sources of client activity data were used to capture the dynamics of client behavior: credit card transactions, events from the mobile application, and events from the web portal usage. Proposed advanced attributes improved models for both considered problems. The explanatory power of advanced attributes significantly varied with an actual problem.

Keywords Fractal · Multifractal · Fractal analysis · Power law · Fractal dimension · Hausdorff dimension · Compass dimension · Correlation dimension · Head-tail index · Spectrum analysis · Wavelet · Box-counting method · Data mining · Time series · Attribute · Model · Classification · Regression · Economics · Finance

A. Malishevsky (✉)
Institute for Applied System Analysis of Igor Sikorsky Kyiv Polytechnic Institute, Kyiv, Ukraine
e-mail: alexeym_s@yahoo.com

M. Zgurovsky and N. Pankratova (eds.), *System Analysis and Artificial Intelligence*,
Studies in Computational Intelligence 1107,
https://doi.org/10.1007/978-3-031-37450-0_7

1 Introduction

Mandelbrot invented the term "fractal" to describe "rough or fragmented" objects consisting of parts similar to the whole objects [1]. They can be formally defined as sets whose Hausdorff dimension is greater than the topological dimension. Figure 1 illustrates three fractals as examples. Fractals are not only mathematical curiosity; many natural objects and phenomena can be approximately described as fractals or multifractals, including mountains, trees, blood capillaries, lungs, shorelines, geological fractures, particle paths, boundaries, prices, populations, signal spectra, and many more. Many complex and nonlinear processes have fractal characteristics such as heartbeats, respiration, and market behavior. Fractal theory and fractal analysis have been successfully employed in many areas to study and model various physical objects, processes, and datasets in general.

There are various ways to define the dimension of a fractal set formally. The generalized fractal dimension can be expressed as:

$$D_q = \lim_{r \to 0} \frac{\log \sum_i p_i^q}{(q-1)\log r} \quad (1)$$

In formula (1), we place the data into an n-dimensional lattice having grid size r and calculate the frequency p_i of placing data points in the i-th cell. For different q, frequently used dimensions include the Hausdorff (D_0), information ($D_{q \to 1}$) and correlation (D_2) dimensions. Data mining commonly employs information and correlation dimensions [2, 3]. The dynamics of many complex systems, such as heartbeats, brain activity, turbulence, and meteorology, cannot be sufficiently expressed by a single number representing the fractal dimension. Thus, a multifractal system, described by a continuous spectrum of exponents, generalizes fractal systems [4].

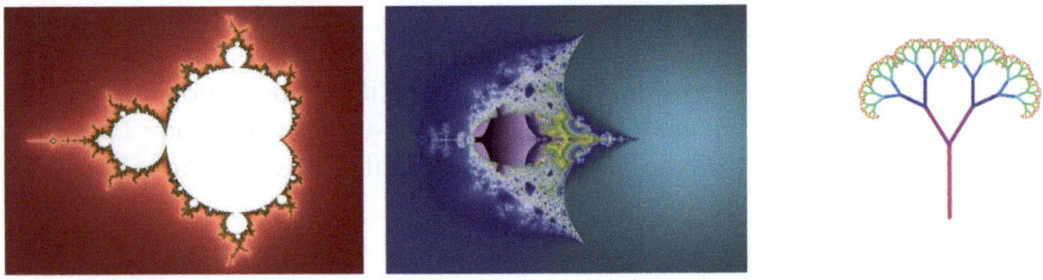

(a) Mandelbrot set (b) Nova fractal (c) Fractal tree

Fig. 1 Examples of fractals

2 Applications of Fractal Analysis

In the last decades, methods of fractal analysis have gained enormous popularity. They are increasingly employed to solve various tasks in medical care, urban studies, ecological research, chemistry, astronomy, data mining, economics, arts, music, etc. Frequently, fractal analysis can represent a dataset by a small set of attributes, measure an object's complexity, determine scaling behavior, and so on. Fractal attributes may uncover hidden characteristics of a data set (ex., its intrinsic dimension) or a process (ex., signal spectrum's complexity). Hidden characteristics may include the state of health, age of geological phenomena, urban land use, market behavior, animal population health, chemical reactivity, and so on. By monitoring the fractal models, changes in hidden characteristics can be detected, such as an onset of an illness, a market crash, or worsening ecology.

Health care has attracted a lot of fractal analysis research because, in most biological systems, shapes (for instance, capillary networks or lungs) or processes (for instance, heartbeats, EEG, or MRI signals) exhibit fractal characteristics. Researchers studied the process of opening peripheral pathways in the lungs to detect lung diseases [5], studied brain degradation due to Alzheimer's disease by computing the fractal dimension of fMRI signals [6], studied ECG signals to identify heart problems [7], studied transthoracic echocardiographic images to identify health problems [8], and measured the fractal dimension of the capillary network in the eye's retina to diagnose diabetes [9].

Cities may be considered fractal systems with self-affinity or self-similarity whose spatial structures conform to scale invariance and whose population density conforms to the inverse power law and has fractal distribution [10, 11]. Monofractal and multifractal models can be applied to street networks, elevations, densities, population densities, street intersections, perimeters, and buildings [11–13]. Fractal analysis was applied to study urban areas, water bodies, and parks of Ukraine and integrated in decision-making processes related to the city development [14, 15].

Fractal analysis was applied to geological structures and processes [16], ecology [17], meteorology [18], astronomy [19], chemistry [20], cultural works [21], music [22].

In recent decades, fractal analysis has been widely used in economics and finance. As early as 1963, Mandelbrot examined cotton prices and showed that financial markets exhibited fractal structure [23]. In 1994, the Fractal Market Hypothesis was proposed by Peters as an alternative theory widely used at that time, the efficient market hypothesis based on standard probability calculus for market modeling [24]. A Fractal Market Hypothesis combines elements of fractals with parts of traditional capital market theory. Commodity prices exhibit fractal properties [23, 25, 26] along with exchange rates and stock markets [27–30]. Studies on transaction intervals [30], distributions of capital, sales, and number of employees [31], distribution of personal incomes, company incomes, and debts owed by bankrupt companies [32], the distribution of bankruptcies [33], and financial market fluctuations [34] confirmed their fractal properties. The multifractal model was used in predicting stock market crashes by detecting pre-cursory signals with a multifractal spectrum analysis [35].

In data mining, fractal analysis is used for clustering, attribute selection, forecasting, compact representation of data sets, including time series, etc. In many works, the concept of the intrinsic "fractal" dimension of a data set is used to facilitate several data mining tasks. The fractal theory was applied to time series mining by using the fractal representation of time series data to help solve clustering problems [36, 37], by monitoring changes in the correlation dimension for anomaly pattern discovery on time series stream [38], by combining fractal-based analysis with clustering to identify intrinsic temporal patterns and trend changes [39], by using the concept of intrinsic dimensionality to do non-linear forecasting, for both periodic and chaotic time series [40] and to improve data stream processing and analysis [41]. Fractal clustering methods were proposed in [3, 36, 37, 42]. By computing the fractal dimension of a given data set, a fast attribute selection can be implemented [43]. Fractal analysis can also be used for recognizing and segmenting textures in images [44].

3 Advanced Attributes

In data mining, an attribute, a feature, a data field, a dimension, or a variable, is the characteristic measured for each observation or record. Attribute types include nominal, binary, ordinal, numeric (interval-scaled or ratio-scaled), and text. In data mining, attributes are used to generate models in various tasks, including classification, regression, clustering, association learning, etc.

When we build predictive models in a financial sector, for example, for predicting client's behavior, demographic data are used, such as age, gender, income, number of dependents, debt, etc. Also, transaction activity data are used to generate several statistical attributes, including transaction frequency, the average amount spent, minimum, maximum, standard deviation, skew, etc. While such attributes exhibit, in many cases, functional dependencies between them and targets of interest, they might miss crucial information. As a result, predictive models may not fully capture the underlying dependencies, thus substantially limiting their predictive power. We suggest enriching the standard set of attributes to capture the complexity and periodic behavior of various activities. To capture the client's behavior dynamics, we suggest utilizing three sources of client activity data: credit card transactions, events from the mobile application, and events from web portal usage. For every activity type, we propose to use fractal, spectral, and wavelet analysis-based attributes in addition to simple statistics-based attributes. Fractal analysis-based attributes capture the scaling behavior and complexity of activity data. Prior research has shown that fractal analysis is a powerful tool for time series analysis [36, 38, 39, 45]. A time series can be compactly represented with several fractal attributes, which allows for comparing and classifying time series data. The spectral analysis can detect periodic patterns in activity data that help capture periodic patterns in a client's activity. The wavelet analysis helps to capture bursts of activity. Together they compactly capture a wealth of information from various activities.

3.1 *Transaction Activity Data*

We propose to generate attributes from a time series of purchase amounts (per client) using fractal, spectral, and wavelet analyses. These attributes help to capture the complexity and shape of the transaction activity. Fractal analysis attributes include the head-tail index, box-counting, correlation, and compass dimensions. Spectral analysis attributes include three dominating frequencies f_1, f_2, and f_3 (sorted by the decreasing amplitude); ratios $\frac{a_2}{a_1}$ and $\frac{a_3}{a_2}$, where a_1, a_2, and a_3 are the amplitudes of the first, second, and third dominating frequencies. Finally, wavelet analysis attributes include three dominating scales.

3.2 *Digital Tracking Activity Data*

Digital tracking activity data consist of events generated by the application and web portal usage. The data are available as a sequence of events with high-resolution timestamps. For each source (the application and web portal), a sequence of events is analyzed using simple statistics and fractal analysis resulting in 15 attributes (30 attributes in total). Statistics-based attributes include average transition time, maximum transition time, average transition time after login, average number of sessions per day, session length (minimum, maximum, average), session time (minimum, maximum, average), number of events per day, and the fraction of errors. Fractal analysis-based attributes include the head-tail index, box-counting, and correlation dimensions.

4 Computing Advanced Attributes

We employed various methods to compute described earlier attributes, including fractal, spectral, and wavelet analyses. A variety of methods exist to compute the fractal dimension, including the grid or box-counting method, the dilation method, the correlation method, the sandbox method, the radius scaling method, the wave spectrum, the walking-divider method, and so on [26, 46–49]. We computed four fractal-based attributes, including the Hausdorf dimension (approximated by the box-counting dimension), compass dimension, correlation dimension, and head-tail index. For spectral analysis, we used the Lomb-Scargle periodogram. For wavelet analysis, we used continuous wavelet transform computed with a convolution.

4.1 Box-Counting Dimension

The box-counting dimension or the Minkowski-Bouligand dimension is used to determine the fractal dimension of a set S in a Euclidean space R^n. The box-counting dimensions of F is defined as $dim_B F = \lim_{\delta \to 0} \frac{\log N_\delta(F)}{-\log \delta}$, where F is any non-empty bounded subset of R^n and $N_\delta(F)$ is the smallest number of sets of diameter at most δ which can cover F [50]. There are many definitions for the box-counting dimension. We use the δ-coordinate mesh of R^n and let $N_\delta(F)$ be the number of mesh cubes of side δ that intersect F. To compute the box-counting dimension of our datasets, we use the box-counting method. This method selects a sequence of decreasing sizes δ. For each δ, we apply a covering of a given set by objects with mesh cubes (intervals in 1d, boxes in 2d, or cubes in 3d cases) while computing the size of each cover $N_\delta(F)$. The resulting $\log N_\delta(F)$ versus $\log \delta$ data is plotted and used to fit the regression line. The regression line's slope (an absolute value) is used as the box-counting dimension. Figure 2 demonstrates an application of the box-counting method to compute the fractal dimension of a Koch snowflake. We generated the object with ten iterations resulting in 3 145 728 lines. Box counting was accomplished using an exponential scale with grids ranging from 256×256 to 4096×4096. By fitting the regression line on $\log(N(s))$ versus $\log(s)$ plot (for $N \times N$ grid, $s = 1/N$), the fractal dimension was calculated to be $D = 1.2610$, which was close to the actual fractal dimension of the Koch snowflake of $\ln(4)/\ln(3) \approx 1.2618595$.

4.2 Compass Dimension

The fractal compass dimension is frequently used for border lines or boundaries. In order to apply it, a border line must be connected. In order to compute this dimension, the Walker's Ruler method is usually employed. The Walker's Ruler method works similarly to the box-counting method; however, instead of "covering" objects with cubes of varying sizes, it traces the boundary using a ruler (or a compass) of varying lengths. Thus, for a given ruler length δ, one ruler's end is fixed at the beginning of the boundary and rotated n times until the other end is reached, resulting in the total length $L(\delta) = n\delta$. Thus, repeating the process M times with varying δ (usually starting with some number and halving each time), we obtain M pairs $< \delta, L(\delta) >$. If these values have the power law relationship $L(\delta) = \delta^{1-D}$, after transformation there is a linear relationship $\log(L(\delta)) = a + (1 - D)\log(\delta)$ for some constant a. D is the fractal dimension being estimated, computed as one minus the slope of the regression line fit on $\log L(\delta)$ versus $\log \delta$ data [48]. We computed the compass dimension for time series data. Figure 3 demonstrates an application of Walker's Ruler method to compute the fractal dimension of a time series for normalized purchase amounts.

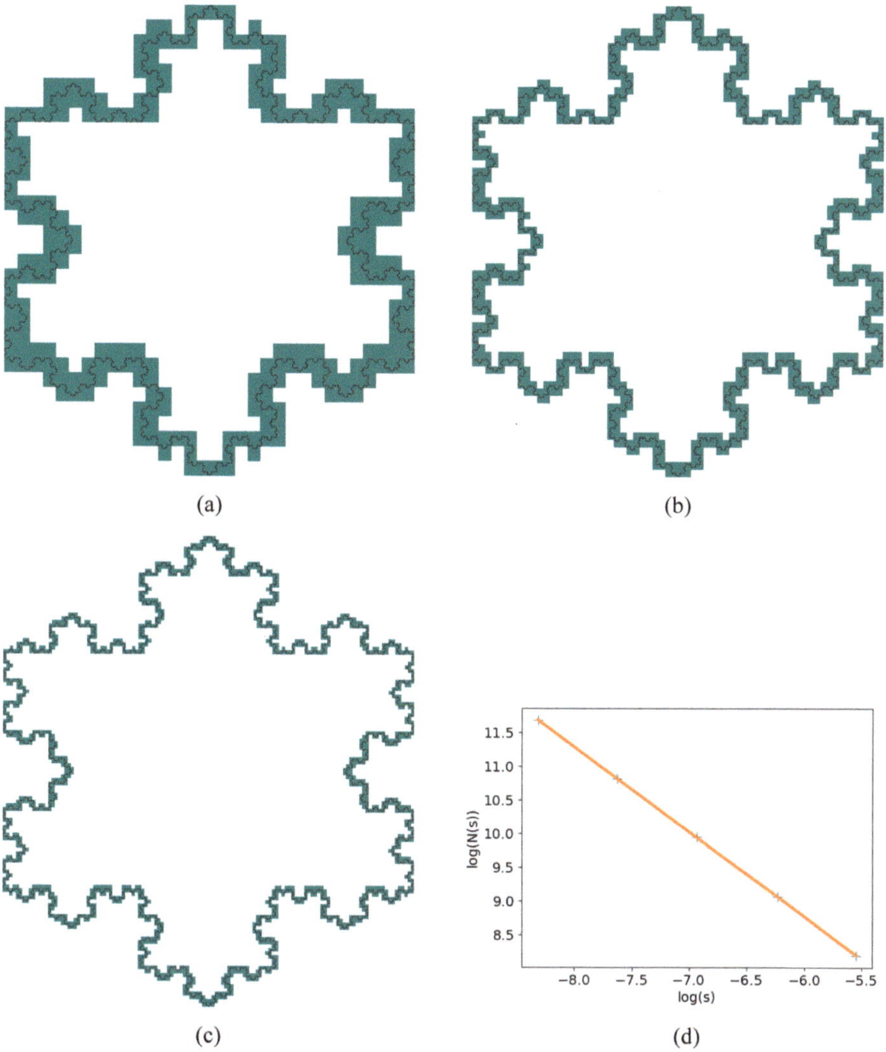

Fig. 2 Koch snowflake: **a**–**c**—box covering using grids of 32x32, 64x64, and 128x128 correspondingly; **d**—$\log(N(\delta))$ versus $\log(\delta)$ plot with regression line from the box-counting method

4.3 Correlation Dimension

The correlation dimension is frequently used to represent the degree of association of data series [51]. To calculate the correlation dimension of a time series, we used G-P algorithm proposed by Grassberger and Procaccia in [52]. The algorithm is based on the embedding theory and phase space reconstruction. During phase space reconstruction, the data series is expanded to a phase space which is a vec-

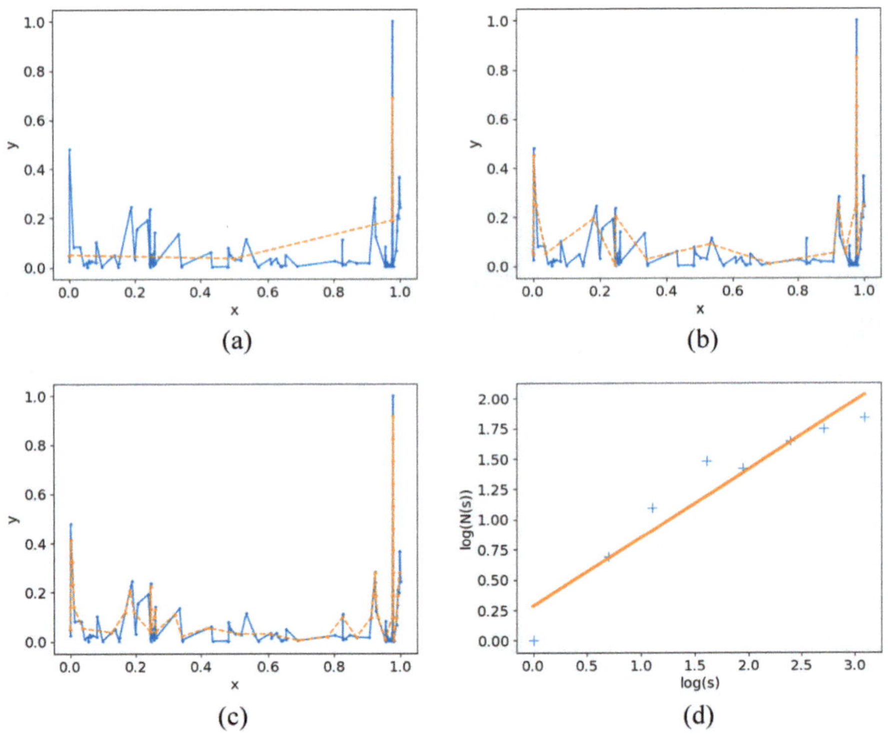

Fig. 3 Time series data: **a–c**—selected steps of Walker's Ruler method; **d**—$\log(N(\delta))$ versus $\log(\delta)$ plot with regression line

tor space, by observing the data series at different times within a specified time. After expanding the data, its correlation dimension is computed. For an increasing sequence of different dimensions N (starting from 3) of phase space, we expand the data and compute the corresponding correlation dimension D stopping when D does not increase with N. The maximum value of D is used as the correlation dimension. To expand the data series $< x_1, x_2, ..., x_M >$ to a phase space, we construct K vectors $y_i = (x_i, x_{i+1}, ..., x_{N+i-1})$, where N is the phase space dimension, M - is the length of the data series, and $i = 1, 2, ..., K$. For a given N, the correlation dimension $D = \lim_{\epsilon \to 0} \frac{\log C(\epsilon)}{\log \epsilon}$, where $C(\epsilon) = \frac{N(\epsilon)}{K^2}$. $N(\epsilon)$ - is the number of vector pairs whose distance is smaller than ϵ. The distance between vectors y_i and y_j is $R_{ij} = \sqrt{\sum_{k=1}^{N}(x_{k+i-1} - x_{k+j-1})^2}$. To compute the correlation dimension for a given N, in practice, we choose several values of scale δ and compute corresponding values $C(\delta)$. Then, we plot a regression line among points $< \log \delta, \log C(\delta) >$. Its slope is the correlation dimension.

4.4 Head-Tail Index

Jiang and Yin suggested the head-tail index (ht-index) in [53] as an indicator that there were "far more small things than large ones." The ht-index measures the extent of a scaling hierarchy. They defined the ht-index as "one plus the recurring times of far more small things than large ones" [53]. A series of data splits are made to compute the ht-index for a dataset. During each iteration, we calculate the mean of the dataset and split it into two parts: smaller than the mean and bigger than the mean. If the number of small elements is much bigger than the number of large elements, we repeat the process by processing only elements bigger than the mean. The value of the ht-index is the number of iterations plus one.

4.5 Spectral Analysis

Spectrum or spectral analysis (frequency domain analysis or spectra density estimation) decomposes a complex signal into simpler parts. Many signals (in our case, time series) can be represented by a sum of many individual frequency components. Spectrum analysis quantifies various amounts versus frequency. Our purpose is to detect any periodicities in the data. It can be done by studying peaks at the corresponding frequencies. To detect periodic behavior, we use the Lomb-Scargle periodogram, a least-squares spectral analysis for estimating a frequency spectrum based on a least-squares fit of sinusoids to data samples [54]. We used this method because it does not boost long-periodic noise in the long and gapped records and allows the data not to be equally spaced.

4.6 Wavelet Analysis

Wavelet analysis is used to analyze localized variations of power within a time series. It transforms a time series into time-frequency space and allows us to determine the dominant modes of variability and their variations in time. Wavelet transform can extract local spectral information along with temporal information. Wavelet transform decomposes a function into a set of wavelets. Usually, a basis wavelet function (the mother wavelet) is chosen, which is scaled and multiplied with the original signal at different locations. Thus, to perform the continuous wavelet transform, a convolution of the original data is computed with a set of functions generated by the mother wavelet. The convolution can be calculated using the fast Fourier transform.

5 Case Studies

We conducted two case studies to evaluate whether advanced attributes can improve predictive models.

In our case studies, we considered the data from a financial company that included credit card transaction data (transaction activity) and event data (digital tracking activity) from the mobile banking application and web portal. In both studies, for each selected client and date (an observation), we computed a set of attributes, including simple statistical and demographic data, along with advanced attributes presented in Sect. 3. Advanced attributes included 12 attributes from transaction activity data and 30 attributes from digital tracking data (the application and web portal). In the first study, we solved a simple classification problem; in the second study, we solved a simple regression problem. Models were trained and evaluated using Azure Machine Learning auto ML. The goal was to find whether advanced attributes could improve models.

We used Python language to implement the required tools.

5.1 Study 1: Predicting the Balance Transfer

In this study, we solved a simple classification problem: predicting whether a client would perform a balance transfer in one week after a promotional message.

We randomly selected 330 000 clients and, for each client, selected 1 to 11 different dates when promotion messages were sent, resulting in 750 000 observations. For each observation (client and date), we computed attributes using six months' activity preceding a given date. We used Azure Machine Learning auto ML twice to build the best model automatically: the first time using only simple statistics-based attributes and the second time using both simple and advanced attributes. The best model was "Extremely randomized trees" when using only simple statistics-based attributes. On the other hand, when using both simple and advanced attributes, the best model was a "Light Gradient Boosting Machine".

Table 1 compares two models. Adding fractal-based attributes improves prediction accuracy.

To better see the impact of advanced attributes, we look at Fig. 4. While simple statistics-based attributes provide the biggest impact, advanced attributes

Table 1 Comparison between models used for the classification problem in the first study

	Accuracy (micro f1)	AUC micro	AUC weighted
Model using only simple attributes	0.69	0.79	0.76
Model using advanced attributes	0.83	0.92	0.78

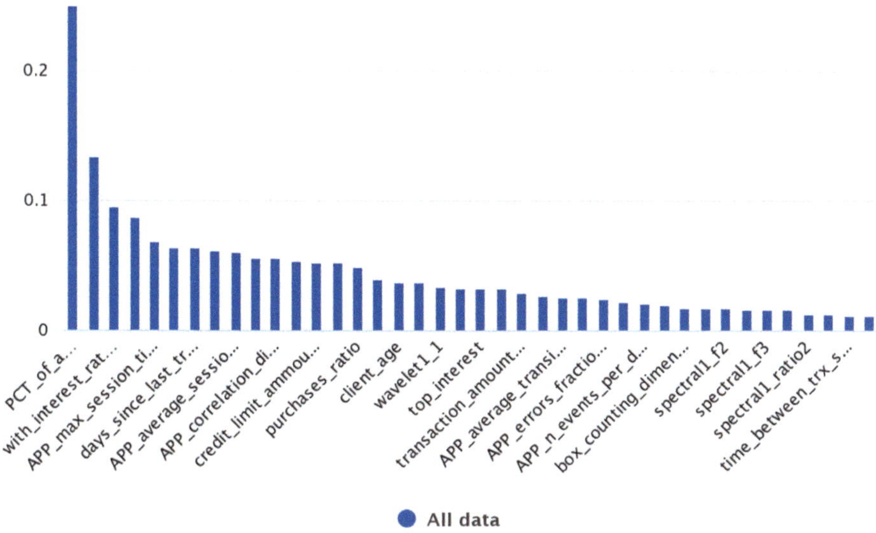

Fig. 4 Attribute importance for the classification problem in the first study

substantially improve a model. For example, the correlation dimension, wavelet attributes, and box-counting dimension have sizable impacts. Table 2 shows combined attribute impact by source. Digital tracking-based attributes have more impact than transaction-based ones. Table 3 shows combined attribute impact by its type. Fractal analysis-based attributes have more impact than spectral and wavelet analysis-based attributes.

Table 2 Attribute importance by source for the classification problem in the first study

Attribute source	Combined attribute importance
Digital tracking	0.58
Transactions	0.16
Other	0.97

Table 3 Attribute importance by type for the classification problem in the first study

Attribute type	Combined attribute importance
Fractal analysis	0.13
Spectral analysis	0.05
Wavelet analysis	0.08
Other digital tracking data statistics	0.49
Other	0.97

5.2 *Study 2: Predicting the Time Between Transactions*

In this study, we solved a simple regression problem: predicting the time between clients' transactions in the next three months after a given date.

We randomly selected 150 000 clients and randomly chose a single date for each client within the 12 months interval, which resulted in 150 000 observations. For each observation (client and date), we computed attributes using six months' activity preceding a given date. We used Azure Machine Learning auto ML twice to build the best model automatically: the first time using only simple statistics-based attributes and the second time using both simple and advanced attributes. The best model was a "Light Gradient Boosting Machine" when using only simple statistics-based attributes. On the other hand, when using both simple and advanced attributes, the best model was "Stacked Ensemble".

Table 4 compares two models. Adding advanced attributes improves prediction accuracy.

To better see the impact of advanced attributes, we look at Fig. 5. Advanced attributes provide the biggest impact on a model. For example, wavelet analysis-

Table 4 Comparison between models used for the regression problem in the second study

	Explained variance	Mean absolute error	Median absolute error
Model using only simple attributes	0.62	14.6	8.0
Model using advanced attributes	0.64	13.8	7.1

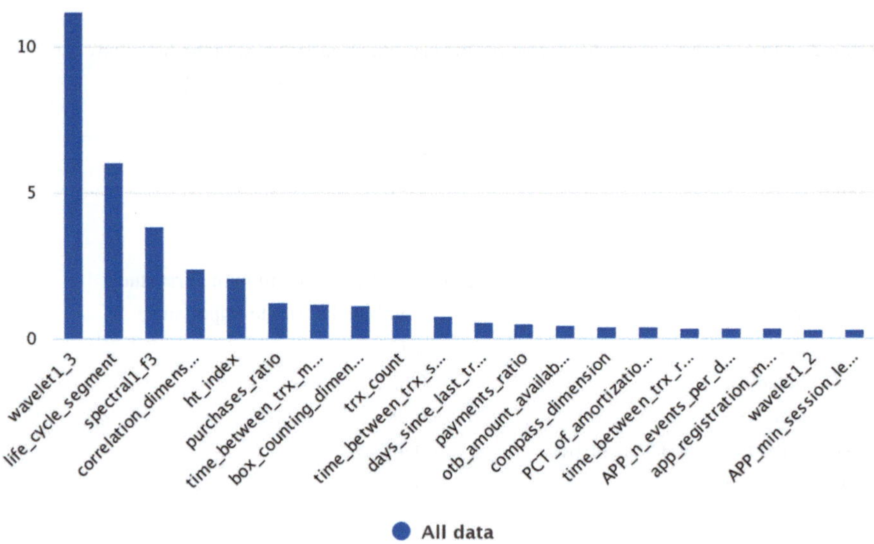

Fig. 5 Attribute importance for the regression problem in the second study

Table 5 Attribute importance by source for the regression problem in the second study

Attribute source	Combined attribute importance
digital tracking	1.74
transactions	21.71
other	14.47

Table 6 Attribute importance by type for the regression problem in the second study

Attribute type	Combined attribute importance
Fractal analysis	6.52
Spectral analysis	4.06
Wavelet analysis	11.60
Other digital tracking data statistics	1.28
Other	14.43

based attributes, correlation, head-tail index, and the box-counting dimension have very high impacts. Table 5 shows combined attribute impact by source. Transaction-based attributes have more impact than digital tracking-based ones. Table 6 shows combined attribute impact by its type. Wavelet analysis-based attributes have more impact than spectral and fractal analysis-based attributes.

6 Conclusions

Fractal theory, along with fractal and multifractal analyses, has been successfully used in many different areas. Fractal analysis has been widely used in the financial sector.

This work aimed to improve predictive models in the financial sector. Transactional and event data were analyzed as time series for calculating advanced attributes, including fractal analysis, spectral analysis, and wavelet analysis-based attributes. These advanced attributes increase prediction models' accuracy. The explanatory power of advanced attributes was demonstrated in several classification and regression problems.

The relative contribution of attributes depends on the problem: while certain attributes are very useful for one problem, they may not be useful for another. In one study, where a simple classification problem was solved, the goal was to predict whether a client would perform a balance transfer in one week after a promotional message. Fractal analysis-based attributes had the biggest contribution among advanced attributes. On the other hand, wavelet analysis-based attributes had the biggest contribution in another study, where a simple regression problem was solved by predicting the time between clients' transactions during the next three months after a given date.

As a result, fractal, spectral, and wavelet analyses can be powerful tools in data mining, especially in improving classification and regression models. These approaches can be valuable tools in financial decision-making.

References

1. Mandelbrot, B.B.: The fractal geometry of nature. W.H. Freeman and Company, San Francisco (1982)
2. Barbara, D.: Chaotic mining: knowledge discovery using the fractal dimension. In: ACM SIGMOD Workshop on Research Issues in Data Mining and Knowledge Discovery, Philadelphia, USA (1999)
3. Barbara, D., Chen, P.: Using the fractal dimension to cluster datasets. In: Proceedings of the Sixth ACM SIGKDD International Conference on Knowledge Discovery and Data Mining, KDD '00, pp. 260–264. Association for Computing Machinery, New York (2000). https://doi.org/10.1145/347090.347145
4. Harte, D.: Multifractals: theory and applications. Chapman and Hall/CRC, New York (2001)
5. Suki, B., Barabasi, A.-L., Hantos, Z., Petak, F., Stanley, H.E.: Avalanches and power-law behaviour in lung inflation. Nature **368**, 615–618 (1994). https://doi.org/10.1038/368615a0
6. M.A. Warsi, The fractal nature and functional connectivity of brain function as measured by BOLD MRI in Alzheimer's disease. Ph.D. thesis, McMaster University (2012)
7. Stanley, H.E., Amaral, L.A.N., Goldberger, A.L., Havlin, S., Ivanov, P.Ch., Peng, C.-K.: Statistical physics and physiology: Monofractal and multifractal approaches. Phys. A: Stat. Mech. Its Appl. **270**(1), 309–324 (1999). https://www.sciencedirect.com/science/article/pii/S0378437199002307
8. Captur, G., Karperien, A.L., Hughes, A.D., Francis, D.P., Moon, J.C.: The fractal heart – embracing mathematics in the cardiology clinic. Nat. Rev. Cardiol. **14**, 56–64 (2017). https://doi.org/10.1038/nrcardio.2016.161
9. Uahabi, K.L., Atounti, M.: Applications of fractals in medicine. Ann. Univ. Craiova Math. Comput. Sci. Ser. **42**, 167–174 (2015)
10. Batty, M.: The size, scale, and shape of cities. Science (New York, N.Y.) **319**, 769–771 (2008). https://doi.org/10.1126/science.1151419
11. Chen, Y.: Fractal analytical approach of urban form based on spatial correlation function. Chaos Solit. Fractals **49**, 47–60 (2013). https://doi.org/10.1016/j.chaos.2013.02.006. https://www.sciencedirect.com/science/article/pii/S0960077913000349
12. Chen, Y., Wang, J., Feng, J.: Understanding the fractal dimensions of urban forms through spatial entropy. Entropy **19**(11) (2017). https://www.mdpi.com/1099-4300/19/11/600
13. Murcio, R., Masucci, A.P., Arcaute, E., Batty, M.: Multifractal to monofractal evolution of the London street network. Phys. Rev. E **92**, 062130 (2015). https://link.aps.org/doi/10.1103/PhysRevE.92.062130
14. Malishevsky, A.: Applications of fractal analysis in science, technology, and art: a case study on geography of Ukraine. In: 2020 IEEE 2nd International Conference on System Analysis and Intelligent Computing (SAIC) (2020), pp. 1–6. https://doi.org/10.1109/SAIC51296.2020.9239196
15. Malishevsky, A.: System analysis & intelligent computing: theory and applications. In: Fractal Analysis and Its Applications in Urban Environment, Springer International Publishing, Cham (2022), pp. 355–376. https://doi.org/10.1007/978-3-030-94910-5_18
16. Chang, Y.-F., Chen, C.-C., Liang, C.-Y.: The fractal geometry of the surface ruptures of the 1999 Chi-Chi earthquake, Taiwan. Geophys. J. Int. **170**(1), 170–174 (2007). https://doi.org/10.1111/j.1365-246X.2007.03420.x

17. Sugihara, G., May, R.M.: Applications of fractals in ecology. Trends Ecol. Evolut. **5**(3), 79–86 (1990). https://www.sciencedirect.com/science/article/pii/0169534790902356
18. Lovejoy, S., Schertzer, D., Ladoy, P.: Fractal characterization of inhomogeneous geophysical measuring networks. Nature **319**, 43–44 (1986). https://doi.org/10.1038/319043a0
19. Ribeiro, M.B., Miguelote, A.Y.: Fractals and the distribution of galaxies. Brazilian J. Phys. **28**(2), 132–160 (1998). https://doi.org/10.1590/S0103-97331998000200007
20. Kaneko, K., Sato, M., Suzuki, T., Fujiwara, Y., Nishikawa, K., Jaroniec, M.: Surface fractal dimension of microporous carbon fibres by nitrogen adsorption. J. Chem. Soc. Faraday Trans. **87**, 179–184 (1991). https://dx.doi.org/10.1039/FT9918700179
21. Vislenko, A.: Possibilities of fractal analysis application to cultural objects. Observat. Cult. 13–19 (2015). https://doi.org/10.25281/2072-3156-2015-0-2-13-19
22. Reljin, N., Pokrajac, D.: Music Performers classification by using multifractal features: a case study. Archiv. Acoust. **42**(2), 223–233 (2017). https://journals.pan.pl/Content/101286/PDF/aoa-2017-0025.pdf
23. Mandelbrot, B.B.: The variation of certain speculative prices. J. Business **36**, 394–419 (1963). https://EconPapers.repec.org/RePEc:ucp:jnlbus:v:36:y:1963:p:394
24. Peters, E.E.: Fractal market analysis: applying chaos theory to investment and economics. Wiley (1994)
25. Mandelbrot, B.B.: The (Mis)behavior of Markets. Basic Books, New York (2004)
26. Richardson, L.F.: The problem of contiguity: an appendix of statistics of deadly quarrels. General Syst. Year. **6**, 139–187 (1961)
27. Cont, R., Bouchaud, J.P.: Herd behaviour and aggregate fluctuations in financial markets. Macroeconomic Dynam. **4**(2), 170–196 (2000). https://doi.org/10.1017/s1365100500015029
28. Gopikrishnan, P., Meyer, M., Amaral, L.A.N., Stanley, H.E.: Inverse cubic law for the distribution of stock price variations. Eur. Phys. J. B **3**(2), 139–140 (1998). https://doi.org/10.1007/s100510050292
29. Mizuno, T., Kurihara, S., Takayasu, M., Takayasu, H.: Analysis of high-resolution foreign exchange data of USD-JPY for 13 years. Phys. A: Stat. Mech. Appl. **324**(1), 296–302 (2003). https://www.sciencedirect.com/science/article/pii/S0378437102018812. Proceedings of the international econophysics conference
30. Takayasu, M., Takayasu, H.: Fractals and economics. In: Meyers, R.A. (eds.) Complex Systems in Finance and Econometrics, pp. 444–463. Springer, New York (2011). https://doi.org/10.1007/978-1-4419-7701-4_25
31. Mizuno, T., Katori, M., Takayasu, H., Takayasu, M.: Statistical laws in the income of Japanese companies. In: Takayasu, H. (ed.) Empirical Science of Financial Fluctuations, pp. 321–330. Springer, Japan (2002)
32. Aoyama, H., Nagahara, Y., Okazaki, M., Souma, W., Takayasu, H., Takayasu, M.: Pareto's law for income of individuals and debt of bankrupt companies. Fractals **08**, 293–300 (2000). https://doi.org/10.1142/S0218348X0000038X
33. Fujiwara, Y.: Zipf law in firms bankruptcy. Phys. A: Stat. Mech. Its Appl. **337**(1), 219–230 (2004). https://doi.org/10.1016/j.physa.2004.01.037. https://www.sciencedirect.com/science/article/pii/S0378437104001165
34. Preis, T., Schneider, J.J., Stanley, H.E.: Switching processes in financial markets. Proc. Natl. Acad. Sci. **108**(19), 7674–7678 (2011). https://www.pnas.org/doi/abs/10.1073/pnas.1019484108
35. Los, C.A., Yalamova, R.M.: Multi-fractal spectral analysis of the 1987 stock market crash. Int. Res. J. Financ. Econ. **1**(4), 106–133 (2006). https://ssrn.com/abstract=1139988
36. Sajjipanon, P., Ratanamahatana, C.A.: Efficient time series mining using fractal representation. In: 2008 Third International Conference on Convergence and Hybrid Information Technology, vol. 2, pp. 704–709 (2008). https://doi.org/10.1109/ICCIT.2008.311
37. Yarlagadda, A., Jonnalagedda, M., Munaga, K.: Clustering based on correlation fractal dimension over an evolving data stream. Int. Arab J. Inf. Technol. **15**, 1–9 (2018)
38. Li, G., Wang, Y., Gu, S., Zhu, X.: Fractal-based algorithm for anomaly pattern discovery on time series stream. J. Converg. Inf. Technol. **6**, 181–187 (2011). https://doi.org/10.4156/jcit.vol6.issue3.20

39. Nunes, S.A., Romani, L.A.S., Avila, A.M.H., Traina, C., Jr., de Sousa, E.P.M., Traina, A.J.M.: Fractal-based analysis to identify trend changes in multiple climate time series. J. Inf. Data Manage. **2**, 51–57 (2011)
40. Chakrabarti, D., Faloutsos, C.: F4: large-scale automated forecasting using fractals. In: Proceedings of the Eleventh International Conference on Information and Knowledge Management, CIKM '02, pp. 2–9. Association for Computing Machinery, New York (2002). https://doi.org/10.1145/584792.584797
41. de Sousa, E.P.M., Traina, A.J.M., Traina, C., Faloutsos, C.: Measuring evolving data streams' behavior through their intrinsic dimension. New Gener. Comput. **25**, 33–60 (2006). https://doi.org/10.1007/s00354-006-0003-3
42. Tasoulis, D.K., Vrahatis, M.N.: Unsupervised clustering using fractal dimension. Int. J. Bifurcat. Chaos **16**(07), 2073–2079 (2006). https://doi.org/10.1142/S021812740601591X
43. Traina Jr., C., Traina, A., Wu, L., Faloutsos, C.: Fast feature selection using fractal dimension. J. Inf. Data Manage. **1**(1), 3–16 (2010). https://sol.sbc.org.br/journals/index.php/jidm/article/view/936
44. Marron, B.: Texture filters and fractal dimension on image segmentation. J. Signal Inf. Process. **9**, 229–238 (2018). https://doi.org/10.4236/jsip.2018.93014
45. Kapecka, A.: Fractal analysis of financial time series using fractal dimension and pointwise holder exponents. Dyn. Econom. Models **13**, 107–126 (2013). https://ideas.repec.org/a/cpn/umkdem/v13y2013p107-126.html
46. Chen, Y.: The solutions to the uncertainty problem of urban fractal dimension calculation. Entropy **21**(5), (2019). https://www.mdpi.com/1099-4300/21/5/453
47. Frankhauser, P.: The fractal approach. A new tool for the spatial analysis of urban agglomerations. Popul.: English Select. **10**(1), 205–240 (1998). http://www.jstor.org/stable/2998685
48. Gonzato, G., Mulargia, F., Marzocchi, W.: Practical application of fractal analysis: problems and solutions. Geophys. J. Int. **132**(2), 275–282 (1998). https://doi.org/10.1046/j.1365-246x.1998.00461.x
49. Mandelbrot, B.B.: How long is the coast of Britain? Statistical self-similarity and fractional dimension. Science **156**(3775), 636–638 (1967)
50. Falconer, K.: Fractal geometry: mathematical foundations and applications. Wiley, Chichester (1990)
51. Mou, D., Wang, Z.-W.: Comparison of box counting and correlation dimension methods in well logging data analysis associate with the texture of volcanic rocks. Nonlinear process. Geophys. Discuss. **2016**, 1–18 (2016). https://npg.copernicus.org/preprints/npg-2014-85/
52. Grassberger, P., Procaccia, I.: Characterization of strange attractors. Phys. Rev. Lett. **50**(5), 346–349 (1983). https://link.aps.org/doi/10.1103/PhysRevLett.50.346
53. Jiang, B., Yin, J.: Ht-index for quantifying the fractal or scaling structure of geographic features. Ann. Assoc. Amer. Geographers **104**(3), 530–540 (2014). https://doi.org/10.1080/00045608.2013.834239
54. Scargle, J.D.: Studies in astronomical time series analysis. II. Statistical aspects of spectral analysis of unevenly spaced data. Astrophys. J. **263**, 835–853 (1982). https://doi.org/10.1086/160554

K-LR Modeling with Neural Economy and Its Utilization in Unclear Data

Glib Mazhara and Kateryna Boiarynova

Abstract In this chapter, we explain the K-LR model, a cognitive hierarchical model that combines linear regression and k-means clustering to improve the interpretability and generalization performance of linear models. We explain the principles of cognitive hierarchy models and provide examples of how K-LR can be applied to logistic regression and supply chain optimization. We also created a dynamic K-LR model with stochastic gradient descent, which allows the model to adapt to changes in the data over time and improve its accuracy and generalization performance. Finally, we propose some possible extensions and alternatives to K-LR, e.g. B. The integration of nonlinear functions and deep learning architectures is discussed, as well as the limitations and trade-offs of these approaches.

1 Cognitive Hierarchy as a Concept in Game Theory

The principle of cognitive hierarchy is a game-theoretic concept that describes how individuals make decisions in strategic interactions. This indicates that people have different cognitive abilities and use different strategies when facing decision-making situations, and these differences may have an important impact on the outcome of the game [1].

According to the principle of cognitive hierarchy, individual cognitive thinking can be divided into different levels, from simple heuristics to complex optimization strategies. The basic idea is that individuals tend to use the simplest strategy they think works in the current situation, but can move to more complex strategies if their initial approach proves ineffective.

Suppose a company is deciding how to price a new product that it plans to introduce to the market. The decision requires a careful evaluation of the preferences and behaviors of consumers, as well as the competitive landscape in the industry. The

G. Mazhara · K. Boiarynova (✉)
Igor Sikorsky Kyiv Polytechnic Institute, Kyiv, Ukraine
e-mail: boyarinovaea@ukr.net

M. Zgurovsky and N. Pankratova (eds.), *System Analysis and Artificial Intelligence*, Studies in Computational Intelligence 1107,
https://doi.org/10.1007/978-3-031-37450-0_8

company can use the cognitive hierarchy model to help identify the most effective pricing strategy for this product.

The cognitive hierarchy model suggests that people make decisions using one of three levels of thinking: level 1 (the most simple and intuitive), level 2 (more complex and analytical), and level 3 (the most complex and strategic). In the context of pricing a new product, level 1 thinking might involve a quick assessment of the product's features and how they compare to competitors' products. Level 2 thinking might involve a more detailed analysis of the costs and benefits of different pricing strategies, such as an evaluation of consumer demand and the potential impact on profits. Level 3 thinking might involve a broader strategic analysis of the company's overall pricing strategy, such as considering the impact of the new product on the company's brand image and long-term goals.

Using the cognitive hierarchy model, the company can identify which level of thinking is most effective for the pricing decision at hand. If the decision is relatively straightforward and the available information is limited, a level 1 thinking strategy may be more appropriate. However, if the decision is complex and the available information is uncertain, a level 2 or level 3 thinking strategy may be more effective.

By applying the cognitive hierarchy model to this decision, the company can improve the quality of its decision-making process and increase the likelihood of setting a pricing strategy that is optimal for both the company and its customers. The principle of cognitive hierarchy has important implications for game theory and economics, as it suggests that individual behavior in strategic interactions may be more complex and varied than traditional models assume [2, 3]. It also has practical applications in fields such as marketing and negotiation, where understanding the cognitive processes of decision-making can help individuals make more effective decisions and achieve better outcomes.

2 K-LR Model as an Algorithm for Classification Tasks

While the CH model provides a framework for understanding the cognitive processes behind a decision, the K-LR model takes a different approach, focusing on the decision-making strategies an individual uses to make a decision. Specifically, the K-LR model suggests that individuals use a range of levels of thought, from more intuitive and automatic processes to more conscious and analytical processes, depending on the complexity of the decision and the information available. By understanding and modeling these layers of thought, K-LR models provide a practical tool for predicting and influencing human behavior in a variety of contexts, from consumer decisions to strategic organizational decisions. Unlike the CH model, which emphasizes the underlying cognitive processes, the K-LR model provides a more concrete and actionable decision-making framework that can be applied to various situations.

The K-LR (k-Logistic Regression) model is a type of machine learning algorithm used for classification tasks. It is similar to standard logistic regression, but it includes a regularization term that prevents overfitting and improves generalization performance [4–6].

In the K-LR model, the regularization term is based on the L1 (Lasso) or L2 (Ridge) norm of the weights of the model. The choice between L1 and L2 regularization is typically determined by the specific problem being solved and the desired properties of the resulting model. L1 regularization tends to produce sparse models by setting some weights to zero, while L2 regularization encourages smaller, more distributed weights.

The K-LR model uses a logistic function to map the input features to a probability of belonging to a certain class. During training, the model adjusts its weights to maximize the likelihood of the observed data, subject to the regularization constraint. The model can then be used to classify new data by computing the probability of belonging to each class and selecting the class with the highest probability.

The K-LR model is widely used in applications such as image classification, natural language processing, and fraud detection, where it is important to classify data accurately and efficiently. It is a simple yet powerful algorithm that can handle high-dimensional data and is relatively easy to interpret.

An example that demonstrates the relationship between K-LR and cognitive hierarchy could be a classification problem in which individuals are asked to classify images of animals into different categories, such as "mammals" or "birds."

In this example, individuals with different levels of cognitive reasoning might use different strategies to make their classifications. For instance, an individual at the lowest level of cognitive reasoning might simply look for obvious features of the image, such as whether it has feathers or fur, and classify it accordingly. This would correspond to a simple heuristic-based strategy.

An individual at a higher level of cognitive reasoning might use a more sophisticated strategy, such as looking for more subtle features of the image, or using prior knowledge about animal taxonomy to guide their classification decision. This might correspond to a more complex optimization-based strategy.

The K-LR model could be used to train a machine learning algorithm to classify these images based on a set of features extracted from the images, such as color, texture, or shape. The K-LR model would use a logistic function to map these features to a probability of belonging to a certain category, and the regularization term would help prevent overfitting.

In this way, the K-LR model provides a way to mathematically model the different levels of cognitive reasoning that individuals might use when making classification decisions, and to learn the most effective classification strategy based on a given set of features. This can help improve the accuracy of classification tasks, such as image recognition, and can also shed light on the underlying cognitive processes that drive human decision-making.

2.1 Example of K-LR Model in Logistics Regression

Example of how the K-LR model can be used in logistic regression for binary classification. Suppose a company decides whether to invest in a new manufacturing plant abroad. This decision requires careful consideration of the costs and benefits of the investment and the potential risks and uncertainties involved. A company can use a K-LR model to determine the most effective decision-making strategy for this complex decision.

In this example, intuitive thinking might include an initial assessment of the opportunity, such as perceptions of the attractiveness of foreign markets, the competitiveness of local industries, and the political stability of the region. A level of thoughtful thinking may involve a more detailed analysis of investment costs and benefits, such as: B. Assessing potential returns, availability of skilled labor, regulatory environment and potential risks of operating in a foreign market.

Using a K-LR model, a company can determine which decision-making strategy is most effective for an upcoming decision. Intuitive reasoning strategies may be more appropriate when decisions are relatively easy and available information is limited. However, when decisions are complex and available information is uncertain, deliberate strategies may be more effective.

By applying the K-LR model to this decision, companies can improve the quality of their decision-making process and increase the likelihood of successful investments in foreign markets.

Or suppose we have a dataset of student grades and want to predict whether a student will pass or fail a course based on their grades in previous courses. We have information about students' grade point average (GPA) and grades (MRG) they earned in their last course.

We can use logistic regression to model the probability of a student passing or failing the course given their GPA and MRG. The K-LR model can be used to select the optimal number of features to include in the model and to regularize the coefficients of the logistic regression model.

We can start by splitting the dataset into a training set and a test set. We can use the training set to fit the K-LR model to select the optimal number of features and to regularize the coefficients of the logistic regression model. We can then use the test set to evaluate the performance of the model.

Here's some example Python code that demonstrates how to use the K-LR model with logistic regression:

Program Code

```
import numpy as np
from sklearn.linear_model import LogisticRegression
from sklearn.model_selection import train_test_split
from sklearn.metrics import accuracy_score
from klr import KLRClassifier
```

```
# Load the dataset
data = np.loadtxt('grades.csv', delimiter=',')
X = data[:, :-1]
y = data[:, -1]
# Split the dataset into a training set and a test set
X_train, X_test, y_train, y_test = train_test_split(X, y,
test_size=0.2)
# Fit the K-LR model to select the optimal number of features
klr = KLRClassifier(n_features=X.shape[1])
klr.fit(X_train, y_train)
# Select the optimal number of features
k = klr.get_optimal_k()
# Fit the logistic regression model with k features
lr = LogisticRegression(penalty='l2', C=1.0)
lr.fit(X_train[:, :k], y_train)
# Make predictions on the test set
y_pred = lr.predict(X_test[:, :k])
# Evaluate the performance of the model
accuracy = accuracy_score(y_test, y_pred)
print('Accuracy:', accuracy)
```

In this example, we first load the dataset and split it into a training set and a test set. We then fit the K-LR model on the training set to select the optimal number of features. Once we have the optimal value of k, we fit a logistic regression model with k features using the training set. Finally, we make predictions on the test set and evaluate the performance of the model using the accuracy metric.

2.2 *Example of K-LR Model in Supply Chains*

Let's say you are the supply chain manager for a manufacturing company and you need to predict the demand for a particular product based on historical sales data, pricing information, and other factors. You have a dataset that includes the following features:

- Date: the date of the sales data;
- Sales: the number of units sold;
- Price: the price of the product;
- Advertising: the amount spent on advertising;
- Competition: the number of competing products in the market.

You want to use this data to build a demand forecasting model that can predict the number of units that will be sold in the future based on these features.

To do this, you can use K-LR to select the most important features and regularize the coefficients of the linear regression model. Here's an example Python code that demonstrates how to use K-LR for this problem:

Program Code

```
import numpy as np
import pandas as pd
import numpy as np
from sklearn.linear_model import LinearRegression
from sklearn.model_selection import train_test_split
from klr import KLRRegressor
# Load the dataset
data = pd.read_csv('sales_data.csv')
# Split the dataset into training and test sets
X_train, X_test, y_train, y_test = train_test_split(data[[
'Price', 'Advertising', 'Competition']],
data['Sales'], test_size=0.2, random_state=42)
# Fit the K-LR model to select the optimal number of features
klr = KLRRegressor(n_features=3)
klr.fit(X_train, y_train)
# Select the optimal number of features
k = klr.get_optimal_k()
# Fit the linear regression model with k features
lr = LinearRegression()
lr.fit(X_train.iloc[:, :k], y_train)
# Make predictions on the test set
y_pred = lr.predict(X_test.iloc[:, :k])
# Evaluate the performance of the model
mse = np.mean((y_test - y_pred) ** 2)
print('Mean Squared Error:', mse)
```

In this example, we first load the sales data and split it into train and test sets. We then fit a K-LR model to the training set to select the optimal number of features. Once we have the best value for k, we can use the training set to fit a linear regression model with k features. Finally, we make predictions on the test set and evaluate the model's performance using the mean squared error (MSE) metric.

The K-LR model can help improve the accuracy and generalization performance of the demand forecasting model by selecting the most important features and regularizing the coefficients of the linear regression model.

2.3 K-LR Dynamic Model as an Instrument for Adaptation to Changes

A dynamic model is a model that can adapt to changes in data over time by updating its parameters and structure in response to new observations. Building a dynamic model using K-LR requires the use of an online learning algorithm that can incrementally update model parameters as new data become available.

One such algorithm is the stochastic gradient descent (SGD) algorithm. It updates the model parameters based on the slope of the loss function with respect to the parameters, using one or several observations at a time. To use SGD with K-LR, you can update the K-LR model for each observation by adding new observations to the training set and refitting the model with the updated set.

Here's an example Python code that demonstrates how to build a dynamic K-LR model using SGD:

Program Code

```
import numpy as np
import pandas as pd
import numpy as np
from sklearn.linear_model import SGDRegressor
from klr import KLRRegressor
# Load the initial training set
data = pd.read_csv('sales_data.csv')
X = data[['Price', 'Advertising', 'Competition']]
y = data['Sales']
# Initialize the K-LR model
klr = KLRRegressor(n_features=3)
klr.fit(X, y)
# Initialize the SGD model
sgd = SGDRegressor(loss='squared_loss', penalty='l2',
                             max_iter=1000, tol=1e-3)
# Train the model on the initial training set
sgd.partial_fit(klr.transform(X), y)
# Start the online learning loop
for i in range(100):
    # Simulate a new observation
    new_observation = pd.DataFrame({'Price': [10],
                'Advertising': [100], 'Competition': [5]})
        # Update the K-LR model with the new observation
    klr.fit(X.append(new_observation), y.append(
                                           pd.Series([50])))
        # Update the SGD model with the new observation
    sgd.partial_fit(klr.transform(new_observation),
                                             pd.Series([50]))
        # Print the predicted value for the new observation
    print('Predicted Value:', sgd.predict(klr.transform(
                                           new_observation)))
```

In this example, we first load the initial training set and fit a K-LR model to select the optimal number of features. Then, we initialize the SGD model with the squared loss function and L2 regularization, and train it on the transformed training set using the partial_fit method.

Next, we simulate new observations and update the K-LR model by adding them to the training set and refitting the model. We then transform the new observations using the updated K-LR model and update the SGD model with the new observations using the partial_fit method. Finally, we print the predicted values for new observations using the updated SGD model.

By updating the K-LR model and the SGD model with new observations, we can create a dynamic model that can adapt to changes in the data over time and improve its accuracy and generalization performance.

2.4 Generalization for Evaluating the Performance of K-LR Modeling

Dataset selection: Choose a dataset that contains both numerical and categorical variables.

Preprocessing: Preprocess data by handling missing values, encoding categorical variables, and scaling numerical variables if necessary.

Training and Test Split: Divide the data into a training set and a test set at 80:20 or other appropriate ratios.

Model Training: A K-LR model is trained on the training set using the K-LR algorithm, which involves clustering the data into k groups using k-means clustering and fitting a linear regression model to each cluster.

Model evaluation: Evaluate the performance of the K-LR model on the test set using metrics such as accuracy, precision, recall, and F1 score. Compare the performance of the K-LR model with that of traditional linear regression models and decision tree models commonly used for similar classification problems.

Hyperparameter Tuning: Experiment with different values of k and other hyperparameters of the K-LR model to optimize its performance on the test set.

Interpretability: Finally, analyze the interpretability of the K-LR model by examining the coefficients and cluster assignments of the linear regression models for each cluster, and compare it with that of the traditional linear regression model.

By following these steps, you can test the performance and interpretability of the K-LR model on a real-world dataset and compare it with other models to see if it provides better results.

2.5 *Determination K-LR Level for Human Thinking (Cognitively Level)*

Determining a person's cognitive level on the K-LR scale would require administering cognitive tasks that measure their ability to reason and solve problems. The K-LR model assumes that individuals at different levels of cognitive ability have different thinking patterns, and these patterns can be captured by clustering their responses to cognitive tasks.

One way to determine a person's level of K-LR is to perform a series of cognitive tasks that measure different types of cognitive abilities, such as: B. Verbal reasoning, spatial reasoning, and mathematical skills. These tasks can be selected from standardized cognitive tests such as Raven's Progressive Matrix or the Wechsler Adult Intelligence Scale.

After the task is completed, the person's responses can be scored and used to determine their cognitive profile. The cognitive profiles can then be used as input to the k-means clustering algorithm to assign individuals to specific clusters or K-LR levels.

It is important to note that using the K-LR model to determine an individual's cognitive level is not a conclusive measure of their overall intelligence or cognitive ability. The K-LR model is just one way to analyze cognitive patterns and can be used in conjunction with other measurements and assessments to gain a more complete picture of an individual's cognitive abilities.

2.6 *Future of K-LR Modeling and Other Possibilities*

The K-LR model has shown promise in improving the interpretability and generalization performance of linear models, and it has been successfully applied in various domains such as finance, marketing, and supply chain optimization.

For the future of K-LR, there are several possible directions for further research and development. One direction is to investigate the effectiveness of K-LR in more complex and high-dimensional data settings, such as text and image data, where linear models may not be sufficient to capture underlying patterns and structures.

Another direction is to explore the potential of K-LR in online and streaming data environments, where data is constantly arriving and models need to be adjusted in real time. In such environments, dynamic K-LR models can be used to detect and respond to changes in data patterns and distributions.

In addition, further research can be conducted to explore the potential of K-LR combined with other machine learning techniques such as deep learning and reinforcement learning to improve the model's performance and applicability to more diverse problems.

Overall, the K-LR model has shown great potential and versatility in improving the interpretability and generalization performance of linear models, and further research and development in this area can lead to more effective and practical solutions for various real-world problems.

Determining a better model than K-LR for a particular problem would depend on the specific characteristics and requirements of that problem. However, there are several models that can potentially outperform K-LR in different scenarios:

1. Random Forest: A type of decision tree-based ensemble model that can handle both classification and regression problems. It is known for its ability to handle high-dimensional data and handle noisy or incomplete data well.
2. Gradient Boosting: A type of boosting algorithm that builds a series of decision trees to make predictions. It is particularly useful for handling imbalanced data sets and can achieve high accuracy in classification problems.
3. Convolutional Neural Networks (CNNs): A type of deep learning model that is particularly useful for image classification tasks. It can automatically learn features from images and achieve state-of-the-art performance on many image classification tasks.
4. Support Vector Machines (SVMs): A type of supervised learning model that can handle both linear and nonlinear classification problems. It works by finding the hyperplane that best separates the data points into different classes.
5. Long Short-Term Memory Networks (LSTMs): A type of recurrent neural network (RNN) that can handle sequential data such as natural language text or time series data. It can capture long-term dependencies between the input data and produce accurate predictions.

In summary, selecting the best model for a particular problem requires understanding the specific characteristics and requirements of that problem. However, the above models can potentially provide better performance than K-LR in different scenarios.

2.7 Examples of K-LR Model Graphs

Here are a few possible graphs related to the K-LR model:

- Scatterplot with Cluster Assignments: A scatterplot can be used to visualize the distribution of the data points, and the cluster assignments from the k-means algorithm can be indicated by different colors or markers.
- Coefficient Plot: A coefficient plot can be used to visualize the coefficients of the linear regression models for each cluster in the K-LR model. This can provide insight into the importance of different features for each cluster and help interpret the model.
- Performance Comparison Plot: A performance comparison plot can be used to compare the performance of the K-LR model with other models such as linear regression, decision trees, and neural networks. This can be done by plotting the accuracy or other performance metrics on the y-axis and the model type on the x-axis.

- Hyperparameter Tuning Plot: A hyperparameter tuning plot can be used to visualize the performance of the K-LR model for different values of k and other hyperparameters. This can help in selecting the optimal values for these hyperparameters and improve the performance of the model.
- Cluster Similarity Plot: A cluster similarity plot can be used to visualize the similarity between the different clusters in the K-LR model. This can be done by using dimensionality reduction techniques such as principal component analysis (PCA) or t-SNE to reduce the high-dimensional feature space to a 2D or 3D plot, and then plotting the cluster assignments on this reduced space. The proximity of the clusters on the plot can provide insight into their similarity and help interpret the model.

Some examples of the graphs are presented on Figs. 1 and 2.

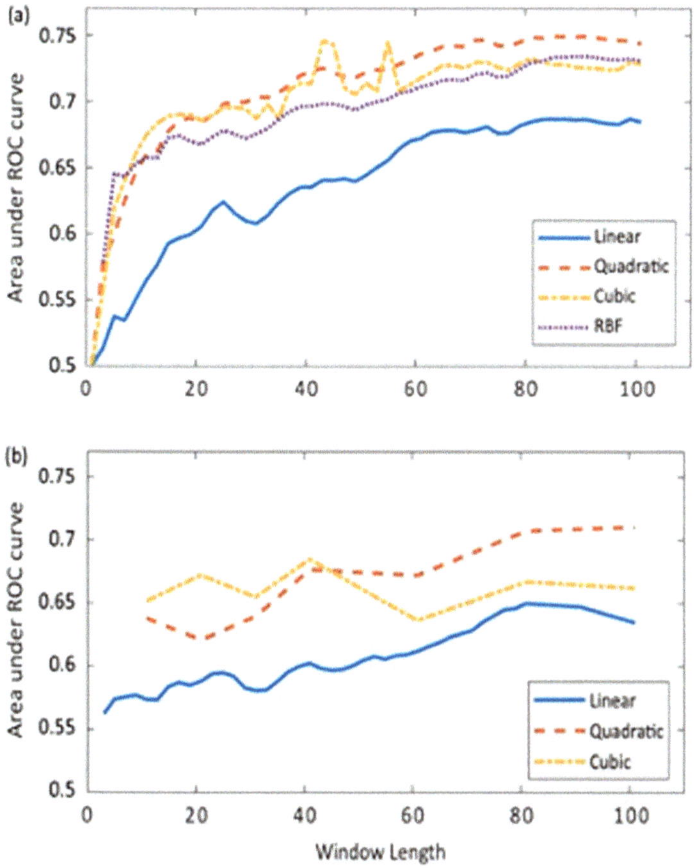

Fig. 1 The performance of KLR models. Results show differences between the linear, quadratic, cubic and RBF models trained across a range of window lengths. **a** Group1 90%; **b** Group2 90% [7]

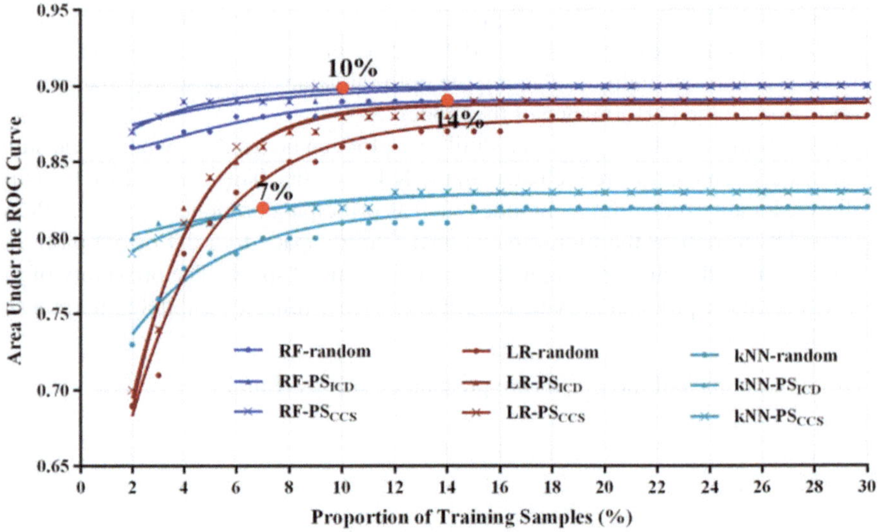

Fig. 2 Predictive performance of random forest (RF), logistic regression (LR), and k-nearest neighbor (kNN) models. For simplicity, only performances of the models built on 2% (200 samples) to 30% (3000 samples) of the 10,000 training sample candidates are displayed in the figure. Blue, cyan, and dark red lines represent RF, kNN, and LR models, respectively. Lines with dot, triangle, and cross markers represent models built on the randomly selected samples and the most similar samples based on patient similarity when the similarity of disease diagnoses feature was calculated using ICD-10 and CCS codes [8]

3 Neural Economy as a Next Step for the Development and Its Utilization

Neuroeconomics and K-LR are two different machine learning models with different approaches to data modeling. K-LR is a linear model that uses clustering to capture nonlinear relationships between features, while Neuroeconomics is a nonlinear model that uses sparse coding to learn compact representations of data.

K-LR can be seen as a simpler and more interpretable version of neuroeconomics, which has been shown to be effective in compressing and reconstructing high-dimensional data such as images and videos. On the other hand, K-LR is mainly used in supervised learning contexts, such as classification and regression, where the goal is to predict an output variable from a set of input features.

One way to combine K-LR and neuroeconomics is to use K-LR as a feature selection mechanism before feeding the selected features into the neuroeconomic model. By using K-LR to identify the most important features, the neural economy model can focus on learning a more compact and efficient representation of the data, which can lead to better generalization performance and faster training times.

Another way to combine K-LR and neural economy is to use the cluster assignments from K-LR as input to a neural economy model, instead of the original features.

By doing so, the neural economy model can learn a more compact and interpretable representation of the data that is based on the underlying clusters, which can be useful in tasks such as unsupervised clustering and anomaly detection.

Overall, combining K-LR and neural economy can lead to more efficient and interpretable machine learning models, and further research can explore the potential of this hybrid approach in various domains and applications.

Here are examples of neural economy:

- Predicting Housing Prices: In this example, we can use a neural network to predict housing prices based on a set of input features, such as location, size, and age of the property. The neural network can be trained on a dataset of historical housing prices and associated features. Once trained, the neural network can be used to make predictions for new properties based on their input features. This approach can help home buyers and sellers make informed decisions about pricing and purchasing properties.
- Fraud Detection: In this example, we can use a neural network to detect fraudulent transactions based on a set of input features, such as transaction amount, location, and time of day. The neural network can be trained on a dataset of historical transactions and associated labels indicating whether each transaction was fraudulent or not. Once trained, the neural network can be used to predict whether new transactions are fraudulent based on their input features. This approach can help financial institutions prevent fraudulent activity and protect their customers.
- Meanwhile K-LR and neural economy are two different concepts, they can be combined to create powerful predictive models for economic forecasting and decision-making. Here are two examples that illustrate the use of K-LR and neural economy together:
  - Forecasting Stock Prices: In this example, we can use K-LR to identify the clusters of similar stocks based on their past performance and use a neural network to forecast their future prices. First, we can use K-LR to cluster similar stocks based on their past performance. Then, for each cluster, we can train a neural network to predict the future prices of the stocks in that cluster. Finally, we can combine the predictions from each neural network to create an overall prediction for the stock market. This approach can help investors make more informed decisions about which stocks to buy and sell.
  - Demand Forecasting: In this example, we can use K-LR to identify the clusters of similar customers based on their purchase history and use a neural network to forecast their future demand. First, we can use K-LR to cluster similar customers based on their past purchases. Then, for each cluster, we can train a neural network to predict the future demand of the customers in that cluster. Finally, we can combine the predictions from each neural network to create an overall prediction for the future demand. This approach can help businesses optimize their inventory management and production planning.

Overall, the combination of K-LR and neural economy can provide a powerful tool for economic forecasting and decision-making, by identifying clusters of similar data points and using neural networks to make accurate predictions.

3.1 Formalization of Principles of Using Neural Networks in Neural Economy

At its core, neural economy involves using neural networks to model and predict economic phenomena. Here is a general formula for neural economy:

$$Y = f(X), \tag{1}$$

where Y is the output variable we are trying to predict (such as house prices, stock prices, or demand) and X is a set of input variables that affect the output (such as the location, size, and age of the house), f is the function that maps the input to the output. In neuroeconomics, f is often implemented as a neural network trained on a dataset of historical examples. Neural networks learn to recognize patterns in input variables that are associated with output variables, which can then be used to make predictions on new examples.

The exact architecture and parameters of a neural network depend on the particular application of neuroeconomics. For example, a neural network designed to predict real estate prices might have a different structure than a neural network designed to detect fraud in financial transactions. However, the basic formula (1) remains the same and the goal is always to use neural networks to model and predict economics phenomena.

3.1.1 Formula That Incorporates Additional Elements for Predicting Different Economic Phenomena

More complex formula for neural economy that incorporates additional elements:

$$Y = f(X, \theta) + \epsilon, \tag{2}$$

where Y is the output variable we are trying to predict, X is a set of input variables that influence the output, θ is a set of parameters that determine the behavior of the neural network, f is a function that maps the inputs to the output, and ϵ is an error term that represents the difference between the predicted output and the true output.

The function f in this formula is typically implemented as a neural network that takes the inputs X and parameters θ as input, and produces a prediction for the output Y. The neural network is trained on a dataset of historical examples, where we know the inputs and true outputs. During training, the neural network adjusts its parameters θ to minimize the error term ϵ and improve the accuracy of its predictions.

The specific architecture and parameters of the neural network depend on the particular application of neural economy, as well as the size and complexity of the dataset. Some common choices for neural network architectures include feedforward networks, recurrent networks, and convolutional networks.

Overall, this formula captures the basic idea of using neural networks to model and predict economic phenomena, while also accounting for the inherent uncertainty and error in the prediction process. By optimizing the parameters of the neural network to minimize the error term, we can create more accurate and reliable predictive models for a wide range of economic applications.

3.1.2 Unclear Data Case as a Part of Modeling

If the data is unclear, it may be difficult to use the formula (2) directly, as the neural network may struggle to find meaningful patterns in the data. However, there are a few strategies that can help to deal with unclear data:

- Feature engineering: Sometimes, the input variables X can be transformed or combined in a way that makes them more informative for the neural network. For example, if we are trying to predict housing prices but only have data on the square footage of the property, we could create a new feature by dividing the square footage by the number of bedrooms to get an average size per bedroom.
- Data cleaning: It is important to check the quality of the data and remove any outliers or errors that could distort the results. For example, if we are trying to predict stock prices and one of the data points has a price that is much higher than all the others, it may be a mistake that needs to be corrected.
- Model selection: If the data is particularly unclear or noisy, it may be necessary to try different types of models or architectures to see which one performs best. For example, we could compare the performance of a feedforward neural network, a recurrent neural network, and a random forest model to see which one produces the most accurate predictions.

Overall, the success of using the formula (2) on unclear data will depend on the quality and nature of the data, as well as the ability to perform feature engineering and model selection to improve the performance of the neural network.

3.1.3 Utilization of the Model with Unclear Data

When dealing with unclear data, the formula (2) can be useful for modeling and predicting the relationship between input variables and output variables. However, this formula relies on the assumption that there is a clear and meaningful relationship between the input variables and output variables, and that any errors or noise in the data can be captured by the error term ϵ.

To use this formula when the data is ambiguous, it is important to carefully preprocess the data and perform feature engineering to identify and extract relevant information. This may include transforming or combining input variables, removing outliers or errors, and choosing appropriate models and architectures to capture potential relationships in the data.

One strategy for dealing with ambiguous data is to use deep learning models, such as convolutional neural networks or recurrent neural networks, which can automatically extract and learn relevant features from data. Another strategy is to use unsupervised learning techniques such as clustering or dimensionality reduction to identify patterns or structures in the data that can be used for modeling and prediction.

Ultimately, the key to utilizing the formula (2) with unclear data is to approach the problem with a flexible and iterative mindset, continually refining the model and data preprocessing steps until a satisfactory level of accuracy and reliability is achieved.

4 Conclusion

We explain the K-LR model and its possible applications, as well as the concept of neuroeconomics and its role in modeling economic phenomena. We examine examples of how K-LR thinking can be applied in different contexts such as logistics and supply chains, and also discuss limitations and potential future directions of K-LR models.

We consider the potential of using neural networks to model economic phenomena and how the formula (2) can be used to capture the relationship between input and output variables. We explore some strategies for dealing with ambiguous data when using neural networks, such as B. feature engineering, data cleaning, and model selection.

Overall, we have addressed many problems related to modeling and forecasting economic phenomena and highlighted some key tools and techniques that can be used to create more accurate and reliable models in this field.

References

1. Camerer, C.F., Teck-Hua, H., Juin-Kuan, Ch.: A cognitive hierarchy model of games. Quart. J. Econ. **119**(3), 861–898 (2004)
2. Mazhara, G.: The principle of cognitive hierarchy in making individual and collective decisions. Acad. Rev. **2**(57), 187–192 (2022). https://doi.org/10.32342/2074-5354-2022-2-57-14
3. Mazhara, G.: "Cognitively hierarchical" modeling approach relationship between consumer and service provider. In: 3rd Eurasian Energy and Sustainability Conference, Baku (2022)
4. Sha, F., Park, Y.A., Saul, L.K.: Multiplicative updates for L 1-regularized linear and logistic regression. In: Berthold, R., Shawe-Taylor, M., Lavrač, J.N., (eds.), Advances in Intelligent Data Analysis VII. IDA. Lecture Notes in Computer Science, vol. 4723. Springer, Berlin (2007). https://doi.org/10.1007/978-3-540-74825-0_2

5. Arad, A., Rubinstein, A.: The 11–20 money request game: a level - k reasoning study. Amer. Econ. Rev. **102**(7), 3561–3573 (2012)
6. Lindner, F., Sutter, M.: Level - k reasoning and time pressure in the 11-20 money request game. In: Working Papers in Economics and Statistics, N 2013-13 (2013)
7. Veevers, R., Cawley, G., Hayward, S.: Investigation of sequence features of hinge-bending regions in proteins with domain movements using kernel logistic regression. BMC Bioinformat. **21**, 137 (2020). https://doi.org/10.1186/s12859-020-3464-3
8. Wang, N., Huang, Y., Liu, H., Fei, X., Wei, L., Zhao, X., Chen, H.: Measurement and application of patient similarity in personalized predictive modeling based on electronic medical records. BioMed. Eng. Line 98 (2019)

Artificial Intelligence Systems: Mathematical Foundations and Applications

Formalization and Development of Autonomous Artificial Intelligence Systems

Pavlo Kasyanov and Liudmyla Borysivna Levenchuk

Abstract This paper explores the problem of formalizing the development of autonomous artificial intelligence systems (AAIS), whose mathematical models may be complex or non-identifiable. Using the Q-value-iterations method, a methodology for constructing of ε-optimal strategies with a given accuracy has been developed. The results allow us to outline classes (including dual-use), for which it is possible to rigorously justify the construction of optimal and ε-optimal strategies even in cases where the models are identifiable but the computational complexity of standard dynamic programming algorithms may not be strictly polynomial.

1 Introduction

Autonomous artificial intelligence systems are systems which can perform tasks assigned to them without external human intervention. Not a full list of autonomous artificial intelligence systems includes [1–3]:

- cars, airplanes, other moving objects that can move under the control of the autopilot: they independently make decisions about the trajectory and program of movement without human intervention;
- flying or floating drones: these are autonomous devices that can perform various tasks, such as delivering certain things to certain points in space, or remote scanning the state of certain territories;

P. Kasyanov (✉)
Institute for Applied System Analysis, National Technical University of Ukraine, "Igor Sikorsky Kyiv Polytechnic Institute", Beresteis'ky Ave., 37, Build, 1, Kyiv 03056, Ukraine
e-mail: kasyanov@i.ua

L. B. Levenchuk
Institute for Applied System Analysis, National Technical University of Ukraine, "Igor Sikorsky Kyiv Polytechnic Institute", Beresteis'ky Ave., 37, Build, 35, Kyiv 03056, Ukraine
e-mail: lusi.levenchuk@gmail.com

M. Zgurovsky and N. Pankratova (eds.), *System Analysis and Artificial Intelligence*, Studies in Computational Intelligence 1107,
https://doi.org/10.1007/978-3-031-37450-0_9

- the Internet of Things (IoT): these are systems which include various sensors and smart devices that can collect data about the environment and perform various actions according to the tasks assigned;
- self-service systems: these are systems which allow customers to buy goods or receive other services without interaction with employees of centers where these services are provided;
- automated medical systems: such systems make it possible to diagnose and provide advice on the treatment of diseases without the participation of doctors;
- robotic financial systems: such systems can automatically allocate assets and manage risks in financial markets;
- automated production systems: these systems include various machines and robots that can automatically perform the tasks of the production process;
- many other systems which can perform the assigned functions without human intervention.

The operating environments for which current and future AAISs are designed and must operate are rapidly evolving. The problems of the functioning of autonomous dual-purpose systems, which must counteract similar AAIS in competitive conditions, are gaining particular relevance [4]. The efficiency of these systems will depend on their ability to provide timely intellectual and reliable support for decision-making, under rapidly changing operational conditions and incomplete information regarding the strategy and program of actions of the adversaries. The operation of such systems should also be based on modern risk management methods and tools.

2 Building AASI in the Paradigm of Markov Decision Process (MDP)

It should be considered that MDP is a model of sequential decision-making under uncertainty, which is based on the theory of sequential statistical decision-making problems developed during the Second World War and studied for more than seventy years in the context of operations research and related fields. Recently, MDPs have been widely used in machine learning problems, as they serve as a standard framework for modeling, developing, and analyzing reinforcement learning algorithms [5]. MDPs model problems where an agent interacts with the environment over time. The current state of the environment can be influenced by both agent actions and factors beyond the agent's control.

An agent has a specific goal (for example, find a hidden goal, navigate to a specific location) and must determine how to interact with the environment (for example, where to search, in which direction to move) to achieve its goal. The goal of the agent is usually expressed in the search for a policy (for example, a mapping that operates from the space of states to the space of actions) that minimizes the expected total costs over the planning horizon [6]. If the dynamics of MDP (it means, transition

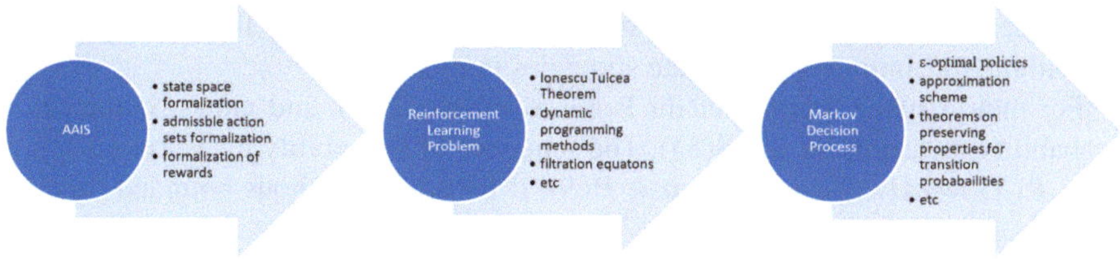

Fig. 1 Formalization of autonomous artificial intelligence systems building

probabilities) are fully known, then optimal policies based on these criteria can be computed using standard methods [6].

Otherwise, approximate simulation versions of these methods may be used to find policies, and this may involve the use of neural networks. Reinforcement learning [5] is used to develop and analyze these and other types of algorithms for MDPs.

Specialists of the artificial intelligence company DeepMind formulated a hypothesis [7]: "Intelligence, and its associated abilities, can be understood as subserving the maximisation of reward by an agent acting in its environment". That is, knowing the rewards for each pair ("state", "action"), theoretically, it is always possible to formalize the problem of building an AAIS as a reinforcement learning problem, which, as a rule, is formalized with the help of partially observable Markov decision-making processes (POMDP), where the spaces of states, admissible controls, and rewards are known.

The model itself, as a rule, is unknown and is identified simultaneously with the search for approximations of optimal actions. In order to develop decision-making support algorithms based on the modern level of mathematical rigor, the problem of step-by-step decision-making must be formulated mathematically rigorously. A key requirement for a mathematically formalized problem is that it must be solved.

For complex sequential decision-making problems, it is often non-trivial to determine whether a model can be solved [8]. There is a need to develop a mathematical apparatus for solving problems of sequential decision-making in order to ensure reliable and robust decision-making in environments with which autonomous systems interact, as well as to develop proof effective algorithms for calculating recommendations for agent actions (see Fig. 1).

3 Search of ε-Optimal Policies

Having formalized the spaces of states and actions, as well as the rewards from certain actions in each of the states (let it be, depending on time), we will set the task of constructing AASI as a task of learning with reinforcement [7]. According to [9, 10], we formalize this problem in the general case of partially observed MDP, where the approximation of the optimal strategies will be found using the so-called Q-functions. In order to delineate the limits of applicability of classes of such problems, we first introduce auxiliary definitions and statements. Then we justify the convergence of algorithms for finding approximate strategies in dynamics.

For metric space S consider thr Borel σ-algebra $\mathcal{B}(S)$, and via $P(S)$ the set of probability measures on $(S, \mathcal{B}(S))$. The sequence of probability measures $\{\mu^{(n)}\}_{n\geq 1}$ from $P(S)$ *weakly converges* to $\mu \in P(S)$, if for any continuous bounded function f on S

$$\int_S f(s)\mu^{(n)}(ds) \to \int_S f(s)\mu(ds), \quad n \to \infty.$$

The sequence of probability measures $\{\mu^{(n)}\}_{n\geq 1}$ from $P(S)$ *converges in total variation* to $\mu \in P(S)$, if

$$\sup\{\int_S f(s)\mu^{(n)}(ds) - \int_S f(s)\mu(ds) : \; f : S \to [-1, 1] \text{ is Borel } \} \to 0, \quad n \to \infty.$$

It must be taken into account that $P(S)$ is separable metric space with respect to the topology of weak convergence of probability measures when S is a separable metric space [11].

For Borel spaces S_1 and S_2, *stochastic kernel* $R(ds_1|s_2)$ on S_1 given S_2 there is such a mapping $R(\cdot|\cdot) : \mathcal{B}(S_1) \times S_2 \to [0, 1]$, if for any $s_2 \in S_2 R(\cdot|s_2)$ is a probability measure on S_1, and $R(B|\cdot)$ is a Borel-measurable function on S_2 for any Borel set $B \in \mathcal{B}(S_1)$.

Stochastic kernel $R(ds_1|s_2)$ on S_1 given S_2 defines the Borel-measurable mapping $s_2 \to R(\cdot|s_2)$ acting from S_2 into metric space $P(S_1)$ with the topology of weak convergence of probability measures. Stochastic kernel $R(ds_1|s_2)$ is called weakly continuous (continuous in total variation) if $R(\cdot|x^{(n)})$ converges weakly (in total variation) to $R(\cdot|x)$, when $x^{(n)}$ converges to x at S_2.

Let $\mathbb{X}$, $\mathbb{Y}$, and $\mathbb{A}$be Borel spaces, $P(dx'|x, a)$ be the stochastic kernel on $\mathbb{X}$ given $\mathbb{X} \times \mathbb{A}$, $Q(dy|a, x)$ be the stochastic kernel on $\mathbb{Y}$ given $\mathbb{A} \times \mathbb{X}$, $Q_0(dy|x)$ be the stochastic kernel on $\mathbb{Y}$ given $\mathbb{X}$, p be probability distribution at $\mathbb{X}$, $c : \mathbb{X} \times \mathbb{A} \to \bar{\mathbb{R}} = \mathbb{R} \cup \{+\infty\}$ be a bounded below Borel function on $\mathbb{X} \times \mathbb{A}$. Partially observed MDPs are determined by $(\mathbb{X}, \mathbb{Y}, \mathbb{A}, P, Q, r)$, where $\mathbb{X}$ is state space, $\mathbb{Y}$ is the space of observations, $\mathbb{A}$ is the set of admissible actions, $P(dx'|x, a)$ is the transition rule for the states, $Q(dy|a, x)$ is observable kernel, $r : \mathbb{X} \times \mathbb{A} \to \bar{\mathbb{R}}$ is the one-step reward. We note that r = -c.

Partially observed MDPs evolve according to the rule: at $t = 0$ initial unobservable state x_0 has a given prior distribution p; initial observation y_0 is generated according to the initial kernel of observation $Q_0(\cdot|x_0)$; at every moment of time $t = 0, 1, \ldots,$ if the system state $x_t \in \mathbb{X}$, the agent chooses an action $a_t \in \mathbb{A}$, and a reward is paid $r(x_t, a_t)$; the system goes into the state x_{t+1} according to the transition rule $P(\cdot|x_t, a_t), t = 0, 1, \ldots;$ observation $y_{t+1} \in \mathbb{Y}$ generated by the kernel $Q(\cdot|a_t, x_{t+1})$, $t = 0, 1, \ldots,$ and observation y_0 at the initial moment of time is generated by the kernel $Q_0(\cdot|x_0)$. Let us define observation histories:$h_0 := (p, y_0) \in H_0$ and

$$h_t := (p, y_0, a_0, \dots, y_{t-1}, a_{t-1}, y_t) \in H_t \text{ for all } t = 1, 2, \dots,$$

where $\mathbb{H}_0 := P(\mathbb{X}) \times \mathbb{Y}$ and $\mathbb{H}_t := \mathbb{H}_{t-1} \times \mathbb{A} \times \mathbb{Y}$ if $t = 1, 2, \dots$. Then the strategy (policy) π for partially observed MDP is defined by a sequence $\pi = \{\pi_t\}_{t=0,1,\dots}$ of stochastic kernels π_t on $\mathbb{A}$ given $\mathbb{H}_t$. The strategy π is called non-randomized if each probability measure $\pi_t(\cdot|h_t)$ is concentrated in one point. A non-randomized strategy is called Markov if all decisions depend only on the current state and time. A Markov strategy is called stationary if all decisions depend only on the current state.

Denote through Π *the set of all strategies.* From the theorem of Ionescu Tulcea ([12], pp. 140–141, [13], p. 178) it follows that strategy $\pi \in \Pi$, initial distribution $p \in P(\mathbb{X})$, initial action a, together with stochastic kernels P, Q, and Q_0, determine an unique probability measure $P^{\pi}_{p,a}$ on the set of all trajectories $\mathbb{H}_{\infty} = P(\mathbb{X}) \times (\mathbb{Y} \times \mathbb{A})^{\infty}$ with the σ-algebra defined by the product of Borel σ-algebras on $P(\mathbb{X})$, $\mathbb{Y}$, and $\mathbb{A}$. The expectation according to this probability measure is denoted by $\mathbb{E}^{\pi}_{p}$. Let us set a performance criterion. For a finite horizon $T = 0, 1, \dots$, determine the expected total discounted rewards (the so-called Q- function):

$$V^{\pi}_{T,\alpha}(p, a) := \mathbb{E}^{\pi}_{p,a} \sum_{t=0}^{T-1} \alpha^n r(x_t, a_t), \ \ \pi \in \Pi, \ \ p \in P(\mathbb{X}), a \in \mathbb{A}, \tag{1}$$

where $\alpha \geq 0$ is the discount factor, $V^{\pi}_{0,\alpha}(p, a) = 0$. Since Q defines the kernel of observations, we will use the notation for the Q-function $V^{\pi}_{T,\alpha}$ – total discounted rewards.

Further, we assume that one of the conditions is always holds:

- **Hypothesis (D).** r is bounded from above on $\mathbb{X} \times \mathbb{A}$, and $\alpha \in [0, 1]$.
- **Hypothesis (P).** r is nonpositive on $\mathbb{X} \times \mathbb{A}$, and $\alpha \in [0, 1]$.

With $T = \infty$ formula (1) determines the expected total discounted rewards with an infinite horizon, which we denote by $V^{\pi}_{\alpha}(p, a)$. For an arbitrary function $g^{\pi}(p, a)$, in particular for $g^{\pi}(p, a) = V^{\pi}_{T,\alpha}(p, a)$ and $g^{\pi}(p, a) = V^{\pi}_{\alpha}(p, a)$, determine the optimal reward

$$g(p, a) = \sup_{\pi \in \Pi} g^{\pi}(p, a)$$

where Π—the set of all strategies.

Strategy π is called optimal for the corresponding criterion if $g^{\pi}(p, a) = g(p, a)$ to all $p \in P(\mathbb{X})$, $a \in \mathbb{A}$. For $g^{\pi} = V^{\pi}_{T,\alpha}$, the optimal strategy is called discounted-optimal with a finite horizon T; for $g^{\pi} = V^{\pi}_{\alpha}$— *discounted-optimal.*

During the analysis similar notations are also used for expected total rewards and target values of Markov decision-making processes and completely observed MDPs. For the sake of certainty, apart from the notations $V^{\pi}_{T,\alpha}$, V^{π}_{α}, $V_{T,\alpha}$ and V_{α}, introduced for partially observed MDP, we will use the notation $v^{\pi}_{T,\alpha}$, v^{π}_{α}, $v_{T,\alpha}$, v_{α} and $\bar{v}^{\pi}_{T,\alpha}$, $\bar{v}^{\pi}_{\alpha}$, $\bar{v}_{T,\alpha}$, $\bar{v}_{\alpha}$ for MDP and completely observed MDP, respectively.

Note that the function r, defined on $\mathbb{X} \times \mathbb{A}$ with values in $\bar{\mathbb{R}}$ *is sup-compact* if the set $\{(x, a) \in \mathbb{X} \times \mathbb{A} : r(x, a) \geq \lambda\}$ is compact for any finite λ. Function r, defined on $\mathbb{X} \times \mathbb{A}$, with values in $\bar{\mathbb{R}}$ is called $\mathbb{K}$-sup-compact on $\mathbb{X} \times \mathbb{A}$, if for any compact set $K \subseteq \mathbb{X}$, function r is sup-compact at $K \times \mathbb{A}$ [14].

According to [14] (lemma 2.5) bounded above function r is $\mathbb{K}$-sup-compact on the product of metric spaces X and A if and only if it satisfy the following two conditions: (a) r upper semi-continuous; (b) if the sequence $\{x^{(n)}\}_{n=1,2,\ldots}$ from $\mathbb{X}$ converges and its limit x belongs to $\mathbb{X}$, then any sequence $\{a^{(n)}\}_{n=1,2,\ldots}$ with $a^{(n)} \in \mathbb{A}$, $n = 1, 2, \ldots,$ such that sequence $\{r(x^{(n)}, a^{(n)})\}_{n=1,2,\ldots}$ is bounded from below, has a limit point $a \in \mathbb{A}$.

For partially observed MDP $(\mathbb{X}, \mathbb{Y}, \mathbb{A}, P, Q, r)$ let us consider MDP $(\mathbb{X}, \mathbb{A}, P, \mathrm{r})$, for which all states are observed. MDP can be considered as partially observed MDP with $\mathbb{Y} = \mathbb{X}$ and $Q(B|a, x) = Q(B|x) = \boldsymbol{I}\{x \in B\}$ for all $x \in \mathbb{X}, a \in \mathbb{A}$, and $B \in \mathcal{B}(\mathbb{X})$. In fact, for this MDP, the action sets of all states are equal.

4 General Conditions for the Existence of Optimal Strategies for MDPs

We define the following general conditions for the existence of optimal strategies for MDPs [15], validity of the optimality equations and convergence of Q-value iterations for total expected rewards.

Formulate these conditions.

Assumption (W^*) [15].

(i) function r is $\mathbb{K}$-sup-compact on $\mathbb{X} \times \mathbb{A}$;
(ii) the stochactic kernel $(x, a) \mapsto P(\cdot|x, a)$ is weakly continuous on $\mathbb{X} \times \mathbb{A}$.

In this case, partially observed MDPs are reduced to completely observed MDPs [12, 16–19]. To simplify the notation, we will sometimes omit the time parameter. For a given posterior distribution z of state x at time $t = 0, 1, \ldots$ and given action a, selected in the period t, denote by $R(B \times C|z, a)$ the cumulative probability that at time $(t + 1)$ the state belongs to $B \in \mathcal{B}(\mathbb{X})$ and observation belongs to $C \in \mathcal{B}(\mathbb{Y})$:

$$R(B \times C|z, a) := \int_{\mathbb{X}} \int_{B} Q(C|a, x')P(dx'|x, a)z(dx), \tag{2}$$

$B \in \mathcal{B}(\mathbb{X})$, $C \in \mathcal{B}(\mathbb{Y})$, $z \in \mathbb{P}(\mathbb{X})$, $a \in \mathbb{A}$, where R is a stochastic kernel on $\mathbb{X} \times \mathbb{Y}$ given $\mathbb{P}(\mathbb{X}) \times \mathbb{A}$. So the probability $R'(C|z, a)$ that observation at time $t + 1$ belongs to the set C is defined as:

$$R'(C|z, a) = \int_{\mathbb{X}} \int_{\mathbb{X}} Q(C|a, x')P(dx'|x, a)z(dx), \tag{3}$$

$C \in \mathcal{B}(\mathbb{Y})$, $z \in \mathbb{P}(\mathbb{X})$, $a \in \mathbb{A}$, where R' is a stochastic kernel on $\mathbb{Y}$ given $\mathbb{P}(\mathbb{X}) \times \mathbb{A}$. According to [20], there is a stochastic kernel H on $\mathbb{X}$ given $\mathbb{P}(\mathbb{X}) \times \mathbb{A} \times \mathbb{Y}$ such that:

$$R(B \times C|z, a) = \int_C H(B|z, a, y) R'(dy|z, a), \tag{4}$$

$B \in \mathcal{B}(\mathbb{X})$, $C \in \mathcal{B}(\mathbb{Y})$, $z \in \mathbb{P}(\mathbb{X})$, $a \in \mathbb{A}$. Stochastic kernel $H(\ \cdot\ |z, a, y)$ defines a measurable mapping $H : \mathbb{P}(\mathbb{X}) \times \mathbb{A} \times \mathbb{Y} \to \mathbb{P}(\mathbb{X})$, where $H(z, a, y)[\ \cdot\ ] = H(\ \cdot\ |z, a, y)$. For each pair $(z, a) \in \mathbb{P}(\mathbb{X}) \times \mathbb{A}$, the mapping $H(z, a, \cdot) : \mathbb{Y} \to \mathbb{P}(\mathbb{X})$ is defined $R'(\ \cdot\ |z, a)$-uniquely in $y \in \mathbb{Y}$ (see [22, Corollary 7.27.1], [21, Appendix 4.4]). For the posterior distribution $z_t \in \mathbb{P}(\mathbb{X})$, action $a_t \in \mathbb{A}$, and observation $y_{t+1} \in \mathbb{Y}$, posterior distribution $z_{t+1} \in \mathbb{P}(\mathbb{X})$ satisfies the filtering equation:

$$z_{t+1} = H(z_t, a_t, y_{t+1}). \tag{5}$$

Note that y_{t+1} is a random variable with distribution $R'(\ \cdot\ |z_t, a_t)$, and the right part of (5) acts from $(z_t, a_t) \in \mathbb{P}(\mathbb{X}) \times \mathbb{A}$ into $\mathbb{P}(\mathbb{P}(\mathbb{X}))$. Hence, z_{t+1} is the random variable with values in $\mathbb{P}(\mathbb{X})$, the distribution of which is uniquely defined by [13]:

$$q(D|z, a) := \int_{\mathbb{Y}} \mathbf{I}\{H(z, a, y) \in D\} R'(dy|z, a), \quad D \in \mathcal{B}(\mathbb{P}(\mathbb{X})),\ z \in \mathbb{P}(\mathbb{X}),\ a \in \mathbb{A}. \tag{6}$$

Selection of stochastic kernel H, satisfying (5) does not affect the definition q from (6), since for each pair $(z, a) \in \mathbb{P}(\mathbb{X}) \times \mathbb{A}$ reflection $H(z, a, \cdot) : \mathbb{Y} \to \mathbb{P}(\mathbb{X})$ is defined $R'(\ \cdot\ |z, a)$-a.s in $y \in \mathbb{Y}$. Similarly, stochastic kernels $R_0, R'_0, H_0,\ q_0$ are defined; see [20]. Therefore $q_0(p)$ is initial distribution, $z_0 = H_0(p, y_0)$ corresponds to the initial state distribution p.

Completely observed MDPs are defined as MDPs with parameters ($\mathbb{P}(\mathbb{X})$,$\mathbb{A}$,q,$\bar{\mathrm{r}}$), where $\mathbb{P}(\mathbb{X})$ is the state space; $\mathbb{A}$ is the set of admissible actions in all states $z \in \mathbb{P}(\mathbb{X})$; one-step rewards function $\bar{\mathrm{r}} : \mathbb{P}(\mathbb{X}) \times \mathbb{A} \to \bar{\mathbb{R}}$, defined as

$$\bar{\mathrm{r}}(z, a) := \int_{\mathbb{X}} r(x, a) z(dx), \qquad z \in \mathbb{P}(\mathbb{X}),\ a \in \mathbb{A}; \tag{7}$$

transition probabilities q on $\mathbb{P}(\mathbb{X})$ given $\mathbb{P}(\mathbb{X}) \times \mathbb{A}$ defined in (6).

For completely observable MDPs, assumptions (W^*) can be presented in the following form: (i) $\bar{\mathrm{r}}$ is $\mathbb{K}$-sup-compact on $\mathbb{P}(\mathbb{X}) \times \mathbb{A}$; (ii) transition probability $q(\ \cdot\ |z, a)$ is weakly continuous at each $(z, a) \in \mathbb{P}(\mathbb{X}) \times \mathbb{A}$.

Denote through $i_t, t = 0, 1, \ldots, t$-horizon history for completely observed MDP, which is called the *information vector* $i_t := (z_0, a_0, \ldots, z_{t-1}, a_{t-1}, z_t) \in I_t$, $t = 0, 1, \ldots,$ where z_0 is an initial posterior distribution and $z_t \in \mathbb{P}(\mathbb{X})$ is recursively defined by Equation (5), $I_t := \mathbb{P}(\mathbb{X}) \times (\mathbb{A} \times \mathbb{P}(\mathbb{X}))^t$ for all $t = 0, 1, \ldots, I_0 := \mathbb{P}(\mathbb{X})$.

Information strategy (I -strategy) is a strategy of completely observed MDP, i.e., I -strategy is such a sequence $\delta = \{\delta_t \, : \, t = 0, 1, \ldots\}$, that $\delta_t(\, \cdot \, |i_t)$ is a stochastic kernel at $\mathbb{A}$ for given I_t and $t = 0, 1$ [21, 22]. Denote through Δ the set of all I-strategies. We will also consider the Markov I-strategy and the stationary I-strategy. For I-strategy $\delta = \{\delta_t \, : \, t = 0, 1, \ldots\}$ let us define a strategy $\pi^\delta = \{\pi_t^\delta \, : \, t = 0, 1, \ldots\}$ on Π:

$$\pi_t^\sigma(\cdot|h_t) = \delta_t(\cdot|i_t(h_t)) \text{ for each } h_t \in H_t, \quad t = 0, 1, \ldots \tag{8}$$

where $i_t(h_t) \in I_t$ is the information vector defined by the history of observations h_t.

Consequently, δ and π^δ are equivalent in the sense that π_t^δ has the same conditional probability on $\mathbb{A}$ given the history of observations h_t, as δ_t for history $i_t(h_t)$.

If δ is the optimal strategy for completely observable MDP, then π^δ is the optimal strategy for it. It follows that $V_{t,\alpha}(p, a) = \bar{v}_{t,\alpha}(q_0(p), \mathrm{a})$, $t = 0, 1, \ldots,$ and $V_\alpha(p, a) = \bar{v}_\alpha(q_0(p), \mathrm{a})$. Let $z_t(h_t)$ be the last element of the information vector $i_t(h_t)$. Using the same notation for a measure concentrated at a point as for a function at this point, we have that if δ is Markov, then it follows from (8) that $\pi_t^\delta(h_t) = \delta_t(z_t(h_t))$.

Moreover, if δ stationary, then $\pi_t^\delta(h_t) = \delta(z_t(h_t))$, $t = 0, 1, \ldots$. Thus, the optimal strategy for completely observed MDP determines the optimal strategy for partially observed MDP. However, very little is known about the conditions for partially observable MDPs under which optimal strategies exist for the corresponding completely observable MDPs. Note that the notation was used to denote the expected total rewards of completely observed MDPs.

The following theorem follows directly from [15], applied to completely observable MDPs $(\mathbb{P}(\mathbb{X}), \mathbb{A}, q, -\bar{\mathrm{r}})$, since the state-action value function is expressed in terms of the state value function and vice versa. From this theorem follows, in particular, the algorithm for searching optimal policy in the case when we can identify the model.

Theorem 1 *Let for completely observable MDPs* $(\mathbb{P}(\mathbb{X}), \mathbb{A}, \mathrm{q}, \bar{\mathrm{r}})$ *Assumption* (W*) *holds, and one of the assumptions (D) or (P) holds. Then:*

1. *functions* $\bar{v}_{t,\alpha}$, $t = 0, 1, \ldots,$ $\bar{v}_\alpha$ *are upper semicontinuous on* $\mathbb{P}(\mathbb{X}) \times \mathbb{A}$, *and* $\bar{v}_{t,\alpha}(z) \to \bar{v}_\alpha(z)$ *as* $t \to \infty$ *to all* $z \in \mathbb{P}(\mathbb{X})$, $a \in \mathbb{A}$;
2. *for any* $z \in \mathbb{P}(\mathbb{X})$, $\mathrm{a} \in \mathbb{A}$ *and* $t = 0, 1, \ldots,$

$$\bar{v}_{t+1,\alpha}(z, a) = \bar{\mathrm{r}}(z, a) + \alpha \int\limits_{\mathbb{P}(\mathbb{X})} \max_{a' \in \mathbb{A}} \bar{v}_{t,\alpha}(z', a') q(dz'|z, a) =$$

$$\int\limits_{\mathbb{X}} r(x, a) z(dx) + \alpha \int_{\mathbb{X}} \int_{\mathbb{X}} \int_{\mathbb{Y}} \max_{a' \in \mathbb{A}} \bar{v}_{t,\alpha}(H(z, a, y), a') Q(dy|a, x') P(dx'|x, a) z(dx)$$

where $\bar{v}_{0,\alpha}(z, a) = 0$ *to all* $z \in \mathbb{P}(\mathbb{X})$, $\mathrm{a} \in \mathbb{A}$, *and nonempty sets*

$$A_{t,\alpha}(z) \; := \{a \in \mathbb{A} : \; \bar{v}_{t,\alpha}(z, a) = \max_{a' \in \mathbb{A}} \bar{v}_{t,\alpha}(z, a')\},$$

$z \in \mathbb{P}(\mathbb{X})$, $t = 0, 1, \ldots$ satisfy properties:

a. *graph $\mathrm{Gr}(A_{t,\alpha}) = \{(z, a) : \; z \in \mathbb{P}(\mathbb{X}), a \in A_{t,\alpha}(z)\}$, $t = 0, 1, \ldots,$ is a Borel subset of $\mathbb{P}(\mathbb{X}) \times \mathbb{A}$,*
b. *if $\bar{v}_{t+1,\alpha}(z, a) = +\infty$, then $A_{t,\alpha}(z, a) = \mathbb{A}$ and if $\bar{v}_{t+1,\alpha}(z, a) < +\infty$, then $A_{t,\alpha}(z, a)$ are compact;*

3. *for any $T = 1, 2, \ldots$, there is an optimal T-horizon Markov I-strategy $(\phi_0, \ldots, \phi_{T-1})$, and if for the T-horizon Markov I-strategy $(\phi_0, \ldots, \phi_{T-1})$ the inclusions hold: $\phi_{T-1-t}(z) \in A_{t,\alpha}(z)$, $z \in \mathbb{P}(\mathbb{X})$, $t = 0, \ldots, T-1$, then this T-horizon I-strategy is optimal;*
4. *for $\alpha \in [0, 1)$ if Assumption (D) holds, and for $\alpha \in [0, 1]$ if Assumption (P) holds,*

$$\bar{v}_\alpha(z, a) = \bar{\mathrm{r}}(z, a) + \alpha \int\limits_{\mathbb{P}(\mathbb{X})} \max_{a' \in \mathbb{A}} \bar{v}_\alpha(z', a') q(dz'|z, a) =$$

$$\int\limits_{\mathbb{X}} r(x, a) z(dx) + \alpha \int_{\mathbb{X}} \int_{\mathbb{X}} \int_{\mathbb{Y}} \max_{a' \in \mathbb{A}} \bar{v}_\alpha(H(z, a, y), a') Q(dy|a, x') P(dx'|x, a) z(dx),$$

$z \in \mathbb{P}(\mathbb{X})$, $a \in \mathbb{A}$, and non-empty sets

$$A_\alpha(z) \; := \{a \in \mathbb{A} : \; \bar{v}_\alpha(z, a) = \max_{a' \in \mathbb{A}} \bar{v}_\alpha(z, a')\},$$

$z \in \mathbb{P}(\mathbb{X})$, satisfy properties (a) graph $\mathrm{Gr}(A_\alpha) = \{(z, a) : \; z \in \mathbb{P}(\mathbb{X}), a \in A_\alpha(z)\}$, $t = 0, 1, \ldots,$ is a Borel subset of $\mathbb{P}(\mathbb{X}) \times \mathbb{A}$, and (b) if $\bar{v}_\alpha(z, a) = +\infty$, then $A_\alpha(z, a) = \mathbb{A}$ and if $\bar{v}_\alpha(z, a) < +\infty$, then $A_\alpha(z, a)$ is compact;
5. *for the problem with an infinite horizon there is a stationary optimal I-strategy ϕ_α and stationary I-strategy $\phi_\alpha^{(*)}$ is optimal if and only if $\phi_\alpha^{(*)}(z) \in A_\alpha(z)$ to all $z \in \mathbb{P}(\mathbb{X})$;*
6. *if $\bar{\mathrm{r}}$ is sup-compact on $\mathbb{P}(\mathbb{X}) \times \mathbb{A}$, then functions $\bar{v}_{t,\alpha}$, $t = 1, 2, \ldots$, and $\bar{v}_\alpha$ are sup-compact on $\mathbb{P}(\mathbb{X})$.*

Remark 1 Let one of Assumptions (D) or (P) hold. According to [20], the Assumption (W^*) on MDP $(\mathbb{P}(\mathbb{X}), \mathbb{A}, \mathrm{q}, \bar{\mathrm{r}})$ in terms of conditions on parameters of partially observable MDP $(\mathbb{X}, \mathbb{Y}, \mathbb{A}, P, Q, r)$ provides the conditions of semi-uniform Fellerness of the convolution of P and Q and $\mathbb{K}$-sup-compactness on $\mathbb{X} \times \mathbb{A}$ of one-step reward function r. Moreover, these conditions are criterion, that is, they are invariant with respect to the transition from partially observed MDP $(\mathbb{X}, \mathbb{Y}, \mathbb{A}, P, Q, r)$ to completely observable MDP $(\mathbb{P}(\mathbb{X}), \mathbb{A}, \mathrm{q}, \bar{\mathrm{r}})$ and vice versa.

Sufficient conditions for semi-uniform Fellerness of the convolution of P and Q are (a) weak continuity of P and continuity in total variation of Q or, for example,

(b) continuity in total variation of P provided that Q is independent of the control parameter a.

In the case when it is impossible to identify the model or the model is complex, paper [21] establishes that the complexity of dynamic programming methods for discounted problems may not be strictly polynomial. Then the algorithm for searching of approximate optimal controls can be modified using Q-learning methods, actor-critic, functional approximations and other deep learning methods of reinforcement learning; see [1, 5, 22, 23].

5 Scope of Potential Applications

Fundamental results on the solvability of sequential decision-making problems [24] are the starting points for the study of solvability problems for general models of decision-making problems. Connections with the theory of stochastic games [25] and complex problems of discrete optimization [26] are especially relevant for dual-purpose applications such as radar/sonar control, logistics and material routing, etc. The expected application of the obtained research results is based on providing a better understanding of risk management issues in the operation of autonomous systems.

Potential applications include control of airborne early warning radar systems [26], multiview sonar target classification with underwater mine countermeasures [27], active sonar control for tracking multiple targets [28], and autonomous underwater vehicle command routing [29]. The application horizons of this class of systems are described in details in [7, 30].

6 Conclusions

1. To formalize the task of development the autonomous artificial intelligence systems, the state and action spaces as well as the rewards of the agent in each pair (state, action) are defined, provided that the model itself may be undefined or complex.
2. By using the Q-value iterations method a mathematically justified methodology for constructing approximate ε-optimal strategies with a given accuracy is developed.
3. It is shown that the conditions of semi-uniform Fellerness on the convolution of transition and observation kernels and K-sup-compactness of the one-step reward function for partially observable MDPs are invariant with respect to the transition to the corresponding completely observable MDPs.
4. The obtained results make it possible to delineate the classes of ASSI (in particular, dual purpose) for which optimal and ε-optimal strategies can be reasonably built at the current level of mathematical rigor even in cases where the model is

identified, however, the computational complexity of standard dynamic programming algorithms may not be strictly polynomial [31].

References

1. Arslan, G., Yuksel, S.: Decentralized Q-learning for stochastic teams and games. IEEE Trans. Autom. Control **62**(4), 1545–1558 (2017)
2. Hernandez-Lerma, O.: Adaptive Markov Control Processes. Springer, New York (1989)
3. Wallis, W.A.: The statistical research group, 1942–1945. J. Am. Stat. Assoc. **75**(370), 320–330 (1980)
4. Department of the Navy, Science and Technology Strategy for Intelligent Autonomous Systems (2021)
5. Sutton, R.S., Barto, A.G.: Reinforcement Learning: An Introduction, 2nd edn. The MIT Press (2020)
6. Puterman, M. L. : Markov Decision Processes: Discrete Stochastic Dynamic Programming. Wiley (2005)
7. Silver, D., Singh, S., Precup, D., Sutton, R.S.: Reward is enough. Artif. Intell. **299**, 103535 (2021)
8. Piunovskiy, A.B.: Examples in Markov Decision Processes. Imperial College Press (2013)
9. Kara, A.D., Saldi, N., Yuksel, S.: Q-learning for MDPs with general spaces: convergence and near optimality via quantization under weak continuity. arXiv preprint arXiv:2111.06781 (2021)
10. Kara, A.D., Yuksel, S.: Convergence of finite memory Q learning for POMDPs and near optimality of learned policies under filter stability. Math. Oper. Res. (2022)
11. Parthasarathy, K.R.: Probability Measures on Metric Spaces. Academic, New York (1967)
12. Bertsekas, D.P., Shreve, S.E.: Stochastic Optimal Control: The Discrete-Time Case. MA. Athena Scientific, Belmont (1996)
13. Hernandez-Lerma, O., Lassere, J.B.: Discrete-Time Markov Control Processes: Basic Optimality Criteria. Springer, New York (1996)
14. Feinberg, E.A., Kasyanov, P.O., Zadoianchuk, N.V.: Berge's theorem for noncompact image sets. J. Math. Anal. Appl. **397**(1), 255–259 (2013)
15. Feinberg, E.A., Kasyanov, P.O., Zadoianchuk, N.V.: Average-cost Markov decision processes with weakly continuous transition probabilities. Math. Oper. Res. **37**(4), 591–607 (2012)
16. Rhenius, D.: Incomplete information in Markovian decision models. Ann. Statist. **2**(6), 1327–1334 (1974)
17. Yushkevich, A.A.: Reduction of a controlled Markov model with incomplete data to a problem with complete information in the case of Borel state and control spaces. Theory Probab. **21**(1), 153–158 (1976)
18. Dynkin, E.B., Yushkevich, A.A.: Controlled Markov Processes. Springer, New York (1979)
19. Feinberg, E.A., Kasyanov, P.O., Zgurovsky, M.Z.: Markov decision processes with incomplete information and semiuniform feller transition probabilities. SIAM J. Control. Optim. **60**(4), 2488–2513 (2022)
20. Sondik E.J: The optimal control of partially observable Markov processes over the infinite horizon: Discounted costs. Oper. Res. **26(2)**, 282–304 (1978)
21. Feinberg, E.A., Huang, J.: The value iteration algorithm is not strongly polynomial for discounted dynamic programming. Oper. Res. Lett. **42**(2), 130–131 (2014)
22. Feinberg, E. A., Kasyanov, P. O., Zgurovsky, M. Z.: Convergence of value iterations for total-cost mdps and pomdps with general state and action sets. In 2014 IEEE Symposium on Adaptive Dynamic Programming and Reinforcement Learning (ADPRL) IEEE, 1–8 (2014)

23. Szepesvari, C.: Algorithms for reinforcement learning. Synthesis Lect. Artif. Intell. Mach. Learn. **4**(1), 1–103 (2010)
24. Feinberg, E.A., Kasyanov, P.O., Zgurovsky, M.Z.: A class of solvable Markov decision models with incomplete information. In: Proceedings of the 60th IEEE Conference on Decision and Control, December 13–15, 2021, Austin, Texas (2021)
25. Feinberg, E.A., Kasyanov, P.O., Zgurovsky, M.Z.: Continuity of equilibria for twoperson zero-sum games with noncompact action sets and unbounded payoffs. Ann. Oper. Res. (2017). https://doi.org/10.1007/s10479-017-2677-y
26. Feinberg, E.A., Bender, M.A., Curry, M.T., Huang, D., Koutsoudis, T., Bernstein, J.L.: Sensor resource management for an airborne early warning radar. In: Proceedings of SPIE 4728, Signal and Data Processing of Small Targets, August 7, Orlando, Florida (2002)
27. Myers, V., Williams, D.P.: Adaptive multiview target classification in synthetic aperture sonar images using a partially observable Markov decision process. IEEE J. Oceanic Eng. **37**(1), 45–55 (2012)
28. Wakayama, C.Y., Zabinsky, Z.B.: Simulation-driven task prioritization using a restless bandit model for active sonar missions. In: Proceedings of the 2015 Winter Simulation Conference, December 6–9, 2015, Huntington Beach, California (2015)
29. Yordanova, V., Griffiths, H., Hailes, S.: Rendezvous planning for multiple autonomous underwater vehicles using a Markov decision process. IET Radar Sonar Navig. **11**(12), 1762–1769 (2017)
30. Rempel, M., Cai, J.: A review of approximate dynamic programming applications within military operations research. Oper. Res. Perspect. **8**, 100204 (2021)
31. Zgurovsky M.Z. et al.: On formalization of methods for the development of autonomous artificial intelligence systems. Cybern. Syst. Anal. **59**(5) (2023)

Attractors for Differential Inclusions and Their Applications for Evolution Algorithms

Nataliia Gorban, Oleksiy Kapustyan, Pavlo Kasyanov, and Bruno Rubino

Abstract In this paper we study the issues of global solvability and existence of global attractors for differential inclusion of parabolic type in the phase space of continuous functions with sup-norm. Assuming an upper semicontinuity of the multi-valued right-hand side and sign condition of dissipativity we prove that all mild-solutions generate a multi-valued semiflow, and establish the existence of global attractor for the semiflow. We present also an evolutional algorithm for finding attractor elements.

1 Introduction

Qualitative studying various problems that are incorrect in the classical sense was significantly developed by V. Melnik. His works [1, 2] allowed to study these problems using methods of the global attractors theory. The concept of the global attractor is one of the key instruments of the qualitative theory of infinite-dimensional dynamical systems [3]. Partial differential equations without uniqueness of solution for the cor-

N. Gorban
Fakultät Bauingenieurwesen, Bauhaus-Universität Weimar, 99423 Weimar, Germany
e-mail: nataliia.gorban@uni-weimar.de

O. Kapustyan (✉)
Chair of Integral and Differential Equations, Taras Shevchenko National University of Kyiv, Volodymyrska Street, 64, Kyiv 01601, Ukraine
e-mail: kapustyan@knu.ua

P. Kasyanov
Institute for Applied System Analysis, National Technical University of Ukraine “Igor Sikorsky Kyiv Polytechnic Institute”, Beresteis’ky Ave., 37, Build, 1, Kyiv 03056, Ukraine
e-mail: kasyanov@i.ua

B. Rubino
Department of Information Engineering, Computer Science and Mathematics, University of L’Aquila, Via Vetoio, L’Aquila 67100, Italy
e-mail: bruno.rubino@univaq.it

M. Zgurovsky and N. Pankratova (eds.), *System Analysis and Artificial Intelligence*, Studies in Computational Intelligence 1107,
https://doi.org/10.1007/978-3-031-37450-0_10

responding initial value problem, impulsive and stochastically perturbed evolutional systems, variational inequalities, optimal control problems are only particular types of objects for which the existence of a compact uniformly attracting set (uniform attractor) in the phase space was provided [4–9] and references therein. For evolutional inclusions with upper semicontinuous right-hand side, the classical results on the solvability in the reflexive Banach spaces were provided in [10, 11]. These results were significantly extended in [12–14]. The existence of the global attractors was proven in [15–17]. Based on these results we prove the global solvability and convergence to the attractor in non-reflexive phase space with sup-norm for the differential inclusion of parabolic type. We propose also the evolutional algorithm a for finding attractor elements using its topological properties.

2 Setting of the Problem

Let $d = 1, 2, \ldots$, $\Omega \subset \mathbb{R}^d$ be a bounded domain, $f : \mathbb{R} \to 2^{\mathbb{R}}$ be a set-valued mapping satisfying the following assumptions:

$$f(u) \in C_v(\mathbb{R}), \text{ that is,} \quad f(u) \text{ is a nonempty, convex, and compact set for each } u \in \mathbb{R}; \tag{1}$$

$$0 \in f(0); \tag{2}$$

$$\mathrm{Gr}(f) := \{(p, u) : p \in f(u)\} \text{ is closed;} \tag{3}$$

$$\text{there exists } M > 0 \text{ such that } \xi \cdot u \le 0 \text{ for each } u \in \mathbb{R}, \text{ and } \xi \in f(u). \tag{4}$$

Remark 10.1 Note that Assumptions (1) and (3) imply that

$$f : \mathbb{R} \to 2^{\mathbb{R}} \text{ is upper semi-continuous,} \tag{5}$$

see, for example, [18] and references therein. Moreover, [6] and (1), (2), and (5) imply that $f(u) = [\underline{f}(u), \overline{f}(u)]$, $u \in \mathbb{R}$, where $\underline{f}(0) \le 0 \le \overline{f}(0)$, $\underline{f}$ is lower semi-continuous, $\overline{f}$ is upper semi-continuous, and $\underline{f}, \overline{f}$ satisfy (4).

Consider the problem: find $u = u(x, t)$ satisfying

$$\begin{cases} \frac{\partial u}{\partial t} - \Delta u \in f(u), & t > 0,\ x \in \Omega, \\ u|_{\partial\Omega} = 0, \\ u|_{t=0} = u_0, \end{cases} \tag{6}$$

where $u_0 = u_0(x)$ is the initial condition that satisfy the certain assumption, $\partial\Omega$ is the boundary of Ω.

We consider Problem (6) in the following phase space:

$$X := C_0(\Omega) = \{v \in C(\overline{\Omega}) \ : \ v|_{\partial\Omega} = 0\}$$

endowed with the norm:

$$\|v\| := \max_{x \in \overline{\Omega}} |v(x)|.$$

So, the statement of Problem (6) implies that considering some diffusion (heat) process we know only lower and upper boundary for the interaction function (heat source function). In this paper we perform the following.

1. Define the solution of Problem (6) and prove the global solvability of (6) in the phase space X.
2. Construct the semiflow generated by the solutions of Problem (6), prove the existence of a uniform attractor in the phase space X with sup-norm, and establish the sufficient conditions which guarantee that the uniform attractor is a global attractor.
3. Using topological properties of a global attractor construct the procedure based on a genetic algorithm for finding attractor elements.

3 Existence of Solution

Let $T(t)$ be a C_0-semigroup generated by Δ in X.

Definition 10.1 We say that solution of Problem (6) on $[0, T]$ is a function $u \in C\,(0, T)$ such that $u(0) = u_0$, and there exists $l \in L^1(0, T; X)$ such that $\forall t \in [0, T]$

$$u(t) = T(t)u_0 + \int_0^T T(t-s)l(s)ds \text{ in } X, \tag{7}$$

$$l(s, x) \in f(u(s, x)) \text{ a.e. on } (0, T) \times \Omega \tag{8}$$

Remark 10.2 Note that $l(s, \cdot)$, $u(s, \cdot) \in X$, and f has a closed graph [19]. So, embedding (8) is satisfied by all $x \in \Omega$.

Remark 10.3 In [11] it was proved that the additional assumption

$$\exists D_1, D_2 \geq 0 : \ \forall u \in \mathbb{R} \ |f(u)|_+ := \sup_{\xi \in f(u)} |\xi| \leq D_1 + D_2|u| \tag{9}$$

guarantee the solvability of Problem (6) in sense of Definition 10.1 in the phase space $X = L^2(\Omega)$.

The main problem which does not allow us to apply to Problem (6) classical theorems on solvability of evolutional inclusions is the non-existence of a continuous selector of $f : \mathbb{R} \to C_v(\mathbb{R})$. For this reason we establish the solvability using special approximations of f. For $R > M$ consider

$$f_R(u) = \begin{cases} f(u), & |u| \le R, \\ f\left(\frac{Ru}{|u|}, \quad |u| > R.\right) \end{cases} \tag{10}$$

It's easy to see that $\forall R > M$ map f_R satisfies the following conditions:

$$f_R : \mathbb{R} \to C_v(\mathbb{R}), \quad 0 \in f_R(0);$$

$$f_R \text{ is upper semicontinuous and satisfies (10.4);}$$

$$|f_R(u)|_+ \le D(R) \ \forall u \in \mathbb{R}.$$

Let's choose $R > M$ such that $\|u_0\| < R$. Consider the problem

$$\begin{cases} \frac{\partial u}{\partial t} - \Delta u \in f_R(u), \quad t > 0, \ x \in \Omega, \\ u|_{\partial\Omega} = 0, \\ u|_{t=0} = u_0(u). \end{cases} \tag{11}$$

Assume now that we have established the existence of solution for Problem (11) in sense of Definition 10.1, i.e. $\exists u_R \in C([0, T]; X)$, $\exists l_R \in L^1(0, T; X)$ such that

$$\forall t \in [0, T] \ u_R = T(t)u_0 + \int_0^t T(t-s)l_R(s)ds,$$

$$l_R(s, x) = f_R\left(u_R(s, x)\right) \text{ a.e. on } (0, T) \times \Omega.$$

Following the arguments in [20] remark that a solution in sense of Definition 10.1 is a weak solution [21] of problem

$$\begin{cases} \frac{du}{dt} = Au + l, \\ u|_{t=0} = u_0, \end{cases}$$

where the equation is considered in sense of scalar distributions on $(0, T)$, A is a generator of C_0-semigroup generated by Δ in X. Multiply the equation for u_R by $(u_R - R)_+$ in $L^2(\Omega)$, where

$$v_+ = \begin{cases} v, & v \ge 0, \\ 0, & v < 0. \end{cases}$$

Then we obtain that due to (4)

$$\frac{1}{2}\frac{d}{dt}\|(u_R - R)_+\|^2_{L^2} + \|(u_R - R)_+\|^2_{H^1_0} = \int\limits_{\Omega} l_R(s,x)(u_R(s,x) - R)^2_+ dx \leq 0.$$

Hence, using Poincare inequality we obtain that for some $\lambda > 0$

$$\int\limits_{\Omega} (u_R(t,x) - R)^2_+ dx \leq e^{-2\lambda t} \int\limits_{\Omega} (u(0,x) - R)^2_+ dx. \tag{12}$$

Recall that $\|u_0\| < R$, so, from (12) it follows that

$$u_R(t,x) \leq R \text{ for a.a. } (t,x).$$

Repeating these considerations for $(u_R - R)_-$ we get that for $\|u_0\| < R$ the solution of Problem (11) satisfies the estimate

$$\forall t \in [0,T] \ \ \|u_R(t)\| \leq R.$$

It implies that for a.a. $(\xi, x) \in (0,T) \times \Omega$

$$l_R(\xi, x) \in f(u_R(\xi, x)).$$

So, u_R is a solution of Problem (6). Moreover, for any solution of Problem (6) $u \in C([0,T]; X)$ there exists $R > 0$ such that $\max\limits_{t\in[0,T]} \|u(t)\| < R$. Therefore, each solution of Problem (6) is a solution of Problem (11) with corresponding $R > 0$. So, we need to prove the existence of problem (11). Let's consider $\|u_0\| < R$, and f_R defined by (10). According to [22], there exists a sequence of locally Lipschitz maps $\left\{f_R^N : \mathbb{R} \to C_v(\mathbb{R})\right\}_{N=1}^{\infty}$ such that

$$|f_R^N(u)|_+ \leq C(R) \ \ \forall u \in \mathbb{R}, \tag{13}$$

$$f_R(u) \subset \ldots \subset f_R^{N+1}(u) \subset f_R^N(u) \subset \ldots \quad \forall u \in \mathbb{R}, \tag{14}$$

$$\forall \varepsilon > 0 \ \forall u \in \mathbb{R} \ \exists N_0 \text{ such that } \forall N > N_0 \ f_R^N(u) \subset O_\varepsilon(f_R(u)). \tag{15}$$

Let g_R^N be a locally Lipschitz selector of map f_R^N. Generally, the existence of such a selector follows from Michael selection theorem [19]. In our case $f_R^N(u) = [\underline{f}_R^N(u), \overline{f}_R^N(u)]$, where $\underline{f}_R^N(u), \overline{f}_R^N(u)$ are single-valued locally Lipschitz maps. So, we can define g_R^N by following

$$g_R^N(u) = \alpha \underline{f}_R^N(u) + (1-\alpha)\overline{f}_R^N(u). \tag{16}$$

Then, due to (13) we obtain that

$$|g_R^N(u)| \le C(R) \ \ \forall N > 1 \ \forall u \in \mathbb{R}.$$

Therefore,

$$g_R^N(u) \cdot u \le \varepsilon u^2 + C_1,$$

where $\varepsilon \in (0, \lambda)$, λ is from Poincare inequality.

Using the inequality $\|T(t)\| \le M_1 e^{-\lambda t}$ we can proceed to the differential operator $\widehat{A} = Au + \widehat{\varepsilon} u$, $\widehat{\varepsilon} u \in (\varepsilon, \lambda)$. Repeating the previous considerations (estimation (12)) we obtain that solution u_R^N of Problem

$$\begin{cases} \frac{\partial u}{\partial t} = \Delta u + g_R^N(u), \\ u|_{\partial\Omega} = 0, \\ u|_{t=0} = u_0 \end{cases} \tag{17}$$

which satisfies the following estimate

$$\exists M_\varepsilon \ge M \ \ \forall R > M_\varepsilon \ \ \forall u_0 \in X \ \ \|u_0\| < R.$$

exists on $[0, T]$, and

$$\{ \, |u_R^N(t, x)| \le R \ \forall (t, x) \in [0, T] \times \overline{\Omega} \tag{18}$$

Denote $l_N = l_N(t, x) = g_R^N(u_R^N(t, x))$. From (13) we have that $\{l_N\} \subset L^1(0, T; X)$ is uniformly integrable. So, sequence $\{u_R^N\}$ is precompact in $C([0, T]; X)$ [10]. Therefore, we have the following convergence (up to subsequence)

$$u_R^N \to u_R \text{ in } C([0, T]; X) \tag{19}$$

Estimate (18), inequality

$$\|T(t)u_0\|_{C^{1+\alpha}} \le \frac{C}{t^\delta}\|u_0\|, \ \ \alpha \in (0, 1), \ \ \delta \in \left(\frac{1}{\varepsilon}, 1\right),$$

and equality

$$u_R^N = T(t) + \int_0^T T(t-s) l_N(s) ds$$

guarantee us the following inequality

$$\forall t \in (0, T) \ \|u_R^N\|_{C^{1+\alpha}} \le C(R), \tag{20}$$

where $C(R) > 0$ does not depend on N. Due to local Lipschitz g_R^N from (20) we obtain that for $t \in (0, T)$

$$\|l_N(t)\|_{C^\alpha} \le C_1(R), \tag{21}$$

Inequality (21) and compact embedding $C^\alpha \Subset X$ implies pre-compactness of convergence $\{l_N(t)\}$ in X for any $t \in (0, T)$. Hence, according to [10]

$$l_N \to l \text{ weakly in } L^1(0, T; X). \tag{22}$$

Due to Masur theorem

$$l(t, x) \in f_R(u(t, x)) \text{ a.e. on } (0, T) \times \Omega.$$

Therefore, u is a solution of (11). So, the following theorem is proved.

Theorem 10.1 *Let conditions (1), (2), (4), (5) holds. Then* $\forall T > 0 \forall u_0 \in X$ *Problem (6) has a solution* $u \in C([0, T]; X)$ *in sense of Definition 10.1.*

4 Existence of Attractor

Taking into account the result of Theorem 10.1 we can conclude that $\forall u_0 \in X$ Problem (6) has a solution defined on $[0, +\infty)$, i.e. there exists $u \in C([0, +\infty); X)$ such that $u(0) = u_0$, and $\exists l \in L^1_{loc}(0, +\infty; X)$ such that

$$\forall t \ge 0 \quad u(t) = T(t)u_0 + \int_0^t T(t-s)l(s)ds; \tag{23}$$

$$l(s, x) \in f(u(s, x)) \text{ a.e. on } (0, +\infty) \times \Omega \tag{24}$$

In this section we study the behavior of solutions of Problem (6) as $t \to \infty$. For this reason, we need to consider a multi-valued map $G : \mathbb{R} \times X \to 2^X$ defined as follows:

$$G(t, u_0) = \{u(t) : \; u(\cdot) \text{ is a solution of (10.6)}, u(0) = u_0\}. \tag{25}$$

From [23] it follows that map (25) is a multi-valued semiflow, i.e. $\forall u_0 \in X, \forall t, s \ge 0$

$$G(0, u_0) = u_0, \quad G(t + s, u_0) = G(t, G(s, u_0)).$$

Definition 10.2 A compact set $\Theta \subset X$ is called a uniform attractor for a multi-valued semiflow G if the following conditions hold.

1. for any $r > 0$ $\sup_{\|u_0\| \le r} dist\,(G(t, u_0), \Theta) \to 0, t \to \infty$;
2. Θ is a minimal closed set which satisfies (6).

Here, $dist(A, B)$ for $A, B \subset X$ means

$$dist(A, B) = \sup_{a \in A} \inf_{b \in B} \|b - a\|.$$

If, additionally, $\Theta = G(t, \Theta)$ $\forall t \ge 0$ then the uniform attractor is called a global attractor.

It was proved in [15, 24] that the following conditions are sufficient for the existence of a uniform attractor.

1. multi-valued semiflow G is dissipative, i.e. $\exists R_0 > 0\, \forall r > 0\, \exists T = T(r)\, \forall u_0 \in X$ $(\|u_0\| \le r)\, \forall t \ge T$ $\|G(t, u_0)\|_+ \le R_0$;
2. for all $r > 0$ the set $\{G(1, u_0) : \ \|u_0\| \le r\}$ is pre-compact in X.

If, additionally, for all $t > 0$ the map $u \mapsto G(t, u)$ has a closed graph, i.e.

$$\begin{cases} u_0^n \to u_0, \\ \xi_n \in G(t, u_0^n), \quad \Rightarrow \xi \in G(t, u_0), \\ \xi_n \to \xi \end{cases}$$

then the uniform attractor is a global attractor.

Let's check these conditions (6), (1) for multi-valued semiflow (25). Let's prove the second condition. Suppose $\|u_0\| < r$. Then, due to (4) for any solution of Problem (10.1) $u = u(t, x)$ we have that

$$\frac{1}{2}\frac{d}{dt}\|(u - M)_+\|_{L^2}^2 + \|(u - M)_+\|_{H_0^1}^2 = \int_\Omega l(s, x)\,(u(s, x) - M)_+\,dx \le 0.$$

Therefore,

$$\int_\Omega (u(s, x) - M)_+^2\,dx \le e^{-2\lambda t} \int_\Omega (u_0(x) - M)_+\,dx. \tag{26}$$

Hence, we have that

$$\exists C(r) > 0\ \forall t \ge 0\ \|u(t)\| \le C(r). \tag{27}$$

Then from (23), we obtain that for $t = 1$

$$\|u(1)\|_{C^{1+\alpha}} \le K_1 r + \frac{K_2}{1 - \delta} \sup_{|u| \le C(r)} |f(u)|_+, \tag{28}$$

where K_1, K_2 do not depend on $u(\cdot)$, and sup in the right-hand side is a finite value which does not depend on r because of the upper semicontinuity of f. Then, from (28), it follows that $\exists K(r) > 0$ such that

$$G(1, u_0) \subset \{u : \ \|u\|_{C^{1+\alpha}} \le K(r)\}. \tag{29}$$

From compactness of embedding $C^{1+\alpha} \Subset X$, we have precompactness of $G(1, u_0)$. So, the second condition is proved.

Let's prove the first condition. Consider $\|u_0\| \le r$. Let $u(\cdot)$ be a solution of Problem (6) with initial data u_0. Remark that $u(t) \in G(1, u(t-1))$ $\forall t \ge 1$. So, due to (27), we obtain that the set

$$A(r) = \left\{\bigcup_{t\ge1} G(t, u_0) : \ \|u_0\| \le r\right\}$$

is bounded in $C^{1+\alpha}$, and precompact in X. Note that

$$\sup_{\|u_0\|\le r} dist\,(G(t, u_0), A(r)) \to 0, \ t \to \infty.$$

Hence, from [7], it follows that for $B_r = \{u : \ \|u\| \le r\}$, the set

$$\omega(B_r) = \bigcap_{T>0} \overline{\bigcup_{t\ge T} G(t, B_r)}$$

is non-empty, compact, and

$$dist\,(G(t, B_r), \omega(B_r)) \to 0, \quad t \to \infty. \tag{30}$$

Moreover, $\forall \xi \in \omega(B_r)$ $\exists t_n \nearrow \infty$, and $\exists \xi_n \in G(t_n, B_r)$ such that $\xi = \lim_{n\to\infty} \xi_n$.

It follows from (26) that for $\xi_n = u_n(t_n)$

$$\int_\Omega (\xi_n(x) - M)_+^2 \, dx \le e^{-2\lambda t_n} \int_\Omega 2(r^2 + M^2) dx.$$

As $n \to \infty$, we obtain that

$$\xi(x) \le M \ \ \forall x \in \Omega.$$

Carrying out the same considerations for $(\xi_n + M)_-$, we obtain that

$$\xi(x) \ge -M \ \ \forall x \in \Omega.$$

Therefore, we have that

$$\forall r > 0 \ \ \|\omega(B_r)_+\| \le M. \tag{31}$$

Then from (30) and (31), for $R_0 = M + 1$, we obtain that

$$\forall r > 0 \ \exists T = T(r) \ \forall u_0 \in X, \ \ \|u_0\| \geq r \ \forall t \geq T : \ \ \|G(t, u_0)\|_+ \leq R_0.$$

So, the following theorem is proved.

Theorem 10.2 *Let conditions (1), (2), (4), (5) holds. Then multi-valued semiflow (25) generated by solutions of Problem (6), has a uniform attractor* Θ *in the sense of Definition 10.2. Moreover,* $\|\Theta\|_+ \leq M$.

Remark 10.4 Assume additionally that $f : \mathbb{R} \to C_v(\mathbb{R})$ is a continuous map. Then, using the considerations from [20], we can deduce that multi-valued semiflow (25) has a closed graph. So, it has a global attractor in this case.

5 Algorithm for Finding Attractor Elements

A global attractor of semigroup generated by (6) can have a significantly complicated structure even in the case of smooth single-valued map $f : \mathbb{R} \to \mathbb{R}$ [3], [25]. In the simplest case, when $f'(u) \leq \beta$, an attractor is a connected union of equilibrium states and complete trajectories, the order of its dimension is $\beta^{\frac{1}{2}}$. But, even for the simplest inclusion of type (6) with Lipschitz map $f : \mathbb{R} \to C_v(\mathbb{R})$ it is known [26] that under the assumption

$$\exists a > 0 : \ [-a, a] \subset f(0),$$

for all sufficiently small $\delta > 0$, functions of the type

$$u(x) = \sum_{k=1}^{\infty} \alpha_k \sin(kx), \ \text{ where } \sum_{k=1}^{\infty} k^2 |\alpha_k| < \delta,$$

belongs to attractor, and its dimension is ∞.

The following result allows to identify the simplest elements of the attractor.

Lemma 10.1 *Let's the multi-valued semiflow* G *has a uniform attractor* Θ*, and for a bounded set* $\Gamma \subset X$ *there exists* $\tau > 0$ *such that*

$$\Gamma \subset G(\tau, \Gamma). \tag{32}$$

Then $\Gamma \subset \Theta$.

Proof It follows from (32) that $\forall n \geq 1$

$$\Gamma \subset G(t, \Gamma) \subset \ldots \subset G(n\tau, \Gamma).$$

So, $\forall \varepsilon > 0$ there exists $N \in \mathbb{N}$ such that $\forall n \geq N$

$$dist(\Gamma, \Theta) \leq dist\,(G(n\tau, \Gamma), \Theta) < \varepsilon.$$

Hence, the lemma is proved.

The simplest example of a bounded set for which condition (32) holds, is stationary solutions of (6) $u(t,x) \equiv \overline{u}(x)$ for which $\forall t \geq 0\, \overline{u} \in G(t, \overline{u})$. In particular, condition (2) implies that $\overline{u}(x) \equiv 0$ is a stationary solution of (6). Therefore, $0 \in \Theta$. Periodic trajectories also satisfy condition (32). So, the use of (32) looks promising for the search the elements of attractor.

To present the algorithm consider $\Omega = (0, \pi)$, $0 < x_1 < x_2 < \ldots < x_n < \pi$, $f(u) = \left[\underline{f}(u), \overline{f}(u)\right]$, where $\underline{f}, \overline{f}$ are locally lipschitz functions which satisfy (4).

1. *Step 1.* Let's take a set of numbers $\{\widehat{u}_1, \ldots, \widehat{u}_n\}$, $|\widehat{u}_i| < M$, and consider the corresponding piecewise linear function $\widehat{u}(x), \widehat{u} \in X$ such that $\widehat{u}(0) = \widehat{u}(\pi) = 0$, $\widehat{u}(x_i) = \widehat{u}_i$.
2. *Step 2.* Let's take any $\alpha \in (0, 1)$, and consider the problem

$$\begin{cases} \frac{\partial u}{\partial t} - \frac{\partial^2 u}{\partial x^2} = \alpha \underline{f}(u) + (1-\alpha)\overline{f}(u), \quad x \in (0, \pi), \ t > 0, \\ |_{x=0} = u|_{x=\pi} = 0, \\ |_{t=0} = \widehat{u}(x). \end{cases} \tag{33}$$

   Note that the right-hand side of the equation in Problem (33) is a Lipschitz function. Using convergent numerical algorithms for Problem (33) we can obtain the values of a solution of Problem (33) at $t = 1$, $x = x_i$: $\{\widehat{u}(1, x_1), \ldots, \widehat{u}(1, x_n)\}$.
3. *Step 3.* Find the value of the criteria $J(\widehat{u}) = \sum\limits_{i=1}^{n} |\widehat{u}(1, x_i) - \widehat{u}_i|^2$.

   Note that a solution of the inclusion $\overline{u} \in G(1, \overline{u})$ solves the minimization problem:

$$\begin{cases} \max\limits_{x \in [0,\pi]} |u(x) - \xi(x)| \to \inf, \\ \in G(1, u). \end{cases}$$

   So, the the minimal value of J implies that $\widehat{u}(x)$ is close to attractor. Then, use the evolutional crossover and mutation operators for $\{\alpha, \widehat{u}_1, \ldots, \widehat{u}_n\}$. As a result, we obtain new values of $\{\widehat{u}(1, x_1), \ldots, \widehat{u}(1, x_n)\}$, and a new value of J. It remains only to determine the stopping moment.

References

1. Melnik, V.S., Valero, J.: Set-Valued Anal. **6**(1), 83 (1998)
2. Melnik, V.S., Valero, J.: Set-Valued Anal. **8**(4), 375 (2000)
3. Temam, R.: Infinite-Dimensional Dynamical Systems in Mechanics and Physics, vol. 68. Springer Science & Business Media (2012)
4. Cui, H., Langa, J.A.: J. Differ. Equ. **263**(2), 1225 (2017)
5. Gorban, N.V., Kapustyan, O.V., Kasyanov, P.O.: Nonlinear analysis: theory. Methods Appl. **98**, 13 (2014)
6. Kapustyan, O.V., Kasyanov, P.O., Valero, J., Zgurovsky, M.Z.: Continuous and Distributed Systems: Theory and Applications, pp. 163–180 (2014)
7. Zgurovsky, M.Z., Kasyanov, P.O.: In Advances in Global Optimization, pp. 283–294. Springer (2015)
8. Kapustyan, O., Kasyanov, P., Valero, J.: J. Math. Anal. Appl. **373**(2), 535 (2011)
9. Kapustyan, O.V., Kasyanov, P.O., Valero, J., Zgurovsky, M.Z.: Discret. Contin. Dyn. Syst.-Ser. B **24**(3) (2019)
10. Tolstonogov, A.: Differential Inclusions in a Banach Space, vol. 524. Springer Science & Business Media (2012)
11. Denkowski, Z., Migórski, S., Papageorgiou, N.S.: An Introduction to Nonlinear Analysis: Theory. Springer Science & Business Media (2013)
12. Zgurovsky, M.Z., Melnik, V.S., Kasyanov, P.O.: Evolution Inclusions and Variation Inequalities for Earth Data Processing II: Differential-Operator Inclusions and Evolution Variation Inequalities for Earth Data Processing, vol. 25. Springer Science & Business Media (2010)
13. Gluzman, M.O., Gorban, N.V., Kasyanov, P.O.: Appl. Math. Lett. **39**, 19 (2015)
14. Kasyanov, P.O., Toscano, L., Zadoianchuk, N.V.: In: Abstract and Applied Analysis, vol. 2012. Hindawi (2012)
15. Dashkovskiy, S., Kapustyan, O., Romaniuk, I.: Discret. Contin. Dyn. Syst.-Ser. B **22**(5) (2017)
16. Kapustyan, O., Shkundin, D.: Ukrainian Math. J. **55**(4), 535 (2003)
17. Dashkovskiy, S., Feketa, P., Kapustyan, O., Romaniuk, I.: J. Math. Anal. Appl. **458**(1), 193 (2018)
18. Feinberg, E.A., Kasyanov, P.O., Voorneveld, M.: J. Math. Anal. Appl. **413**(2), 1040 (2014)
19. Aubin, J.P., Frankowska, H.: Set-Valued Analysis. Springer Science & Business Media (2009)
20. Feketa, P., Kapustyan, O., Kapustian, O., Korol, I.: Appl. Math. Lett. **135**, 108435 (2023)
21. Ball, J.: In: Mechanics: From Theory to Computation: Essays in Honor of Juan-Carlos Simo, pp. 447–474. Springer (2000)
22. Aubin, J.P., Cellina, A.: Differential Inclusions: Set-Valued Maps and Viability Theory, vol. 264. Springer Science & Business Media (2012)
23. Kapustyan, A., Valero, J.: J. Math. Anal. Appl. **323**(1), 614 (2006)
24. Kapustyan, A.V., Melnik, V.S., Valero, J.: Int. J. Bifurc. Chaos **13**(07), 1969 (2003)
25. Robinson, J.C.: Dimensions, Embeddings, and Attractors, vol. 186. Cambridge University Press (2010)
26. Kapustyan, O., Melnik, V., Valero, J., Kyiv, V.Y.: Naukova Dumka (2008)

System Analysis and Method of Ensuring Functional Sustainability of the Information System of a Critical Infrastructure Object

Oleg Barabash, Valentyn Sobchuk, Andrii Musienko, Oleksandr Laptiev, Volodymyr Bohomia, and Serhii Kopytko

Abstract Modern society is characterized by the intensive development of information technologies with a high degree of autonomy. This issue is particularly acute for the critical infrastructure facilities that operate under the influence of extreme factors. Therefore, an effective approach is to apply a new method of ensuring the functional stability of the information system of critical infrastructure objects, through the presentation of the functioning of the system in the form of a formalized process, in which the main types of the procedures are: accumulation of checks, analysis of check links, diagnosis of the failed module and restoration of the system's functioning. The paper examines the development of a new methodology of ensuring the functional stability of the critical infrastructure objects, the functioning of which is presented as a continuous process. The main types of procedures of the formalized process are accumulation of checks, analysis of check links, diagnosis of the module that failed and restoration of the subsystems of the generalized information system of the enterprise. The procedure for accumulating checks and analyzing the structure of check connections can be presented in the form of a graph of transition states, based on which the main parameters of this procedure are determined, such as: the minimum time to issue the information about the state of the information system

O. Barabash (✉) · A. Musienko
National Technical University of Ukraine "Igor Sikorsky Kyiv Polytechnic Institute", Kyiv, Ukraine
e-mail: bar64@ukr.net

V. Sobchuk · O. Laptiev
Taras Shevchenko National University of Kyiv, Kyiv, Ukraine
e-mail: sobchuk@knu.ua

O. Laptiev
e-mail: olaptiev@knu.ua

V. Bohomia
Kyiv University of Intellectual Property and Rights of the National University "Odesa Law Academy", Odesa, Ukraine
e-mail: bog2603@ukr.net

S. Kopytko
State University of Intellectual Technologies and Communication, Odesa, Ukraine

M. Zgurovsky and N. Pankratova (eds.), *System Analysis and Artificial Intelligence*, Studies in Computational Intelligence 1107,
https://doi.org/10.1007/978-3-031-37450-0_11

of a critical infrastructure object with a given reliability, the optimal value for the execution time of the main and additional check accumulation cycles. Since the study of the state transition graph, which describes the method of ensuring the functional stability of the critical infrastructure objects in a form of a continuous process, is complicated, a new approach is proposed. According to the proposed approach, the state graph is used sequentially. Each time the graph of states is used, its states are strengthened. Thus, each graph of states corresponds to a certain level of strengthening. The method of determining the main probabilities of transition to a state of complete failure without pre-calculating all transition probabilities in the state graph is a characteristic feature of the first level of strengthening of the state transition graph is. The purpose of the article is to develop a new methodology for ensuring the functional stability of critical infrastructure objects.

Keywords Functional stability · Graph model · State transition graph · Restoration of the information system

1 Introduction

The main achievement of the transformation of the global information infrastructure and large-scale production automation is the actual merging the automated production, the data exchange, and the production technologies into a single, self-regulating system with minimal or no human intervention in the production process.

Currently, there is a massive introduction of information systems in production and breakthroughs in the fields of artificial intelligence, robotics, and others. We are experiencing a real boom in the development of information technologies and their widespread introduction into various spheres of human activity, including their integration into the production processes of modern enterprises. This issue is especially acute for the critical infrastructure facilities that operate under the influence of extreme factors.

The main goal of the functional stability of the enterprise managing is to ensure the stability of the operations and the enterprise's development in the long term within the limits of the permissible level of risk. Ensuring a high level of functional stability of the enterprise in the process of its development and functioning is realized through the formation of an optimal and flexible structure of the functioning of all the links of the enterprise. Approaches to ensuring financial stability are inextricably linked with the general strategy of the enterprise. Management of the functional stability of critical infrastructure objects is a system of principles and methods for the development and implementation of management solutions. With the help of the structural-functional model, it is possible to formalize the sequence of the management process of critical infrastructure objects, to illustrate the course of action in absentia, and to observe the intermediate results.

2 Analysis of Literary Data and Statement of the Problem

The development of the modern society requires an intensive development of the information technologies with a high degree of autonomy. This issue is especially acute for critical infrastructure facilities that operate under the influence of extreme factors. Among such enterprises, enterprises of metallurgy, energy, chemical industry, etc. stand out.

In works [1, 2], the functioning of production enterprises operating under the influence of extreme factors was studied. An algorithm for managing the production processes of such enterprises is proposed, which is the basis of the operation of the information systems of corresponding types. Systems are being constantly modernized due to the intensification of the capital investments in the production process. With the help of such systems, planning and control of production processes is carried out. They work autonomously under the influence of external and internal destabilizing factors. The term functional stability is considered as a category that characterizes specific forms of system manifestation (state, development, finances, use of resources, etc.). In the modern understanding, the sustainability of the enterprise as an economic system is a functionally limited parameter. Based on the analogy of thermodynamic and economic processes, the second law of thermodynamics, a criterion for assessing the stability of an enterprise (economic system) is proposed as the inverse of the change in its entropy, which characterizes the degree of stability of the system to resist crisis processes. A structural-functional model for optimizing the process of managing the financial stability of a metallurgical enterprise is proposed. But the model is very limited and does not apply to other critical infrastructure facilities.

The article [3] studies the problem of ensuring the properties of functional stability of production processes based on the use of the neural networks. Examples of information systems helping to ensure the functional stability of all production centers through the use of the neural networks are given. The works [4, 5] built a system of indicators and criteria for assessing the level of functional stability of the information heterogeneous networks and gave an assessment of the quality of functional stability of an automated control system with a hierarchical structure. In articles [6, 7], the problems of designing stable periodic modes for a certain class of hybrid systems are studied and the conditions for the existence of optimal control for a parabolic system with perturbations of the coefficients on the semi-axis are obtained. The results obtained in the works can serve as mathematical models of relevant processes for critical infrastructure systems.

Articles [8, 9] emphasize that information systems of the enterprises are functioning under the influence of external and internal destabilizing factors. Under negative influence, system modules may fail. However, the systems must function offline for a specified time. Such a condition of functioning can be fulfilled by ensuring the property of functional stability. Functional stability is the key to the functioning of the information system, possibly with a decrease in quality, during the specified time under the influence of external and internal destabilizing factors. External and internal destabilizing factors mean failures, failures of system modules, mechanical damage, thermal effects, errors of service personnel. The main stages of ensuring the functional stability are the detection of the module that failed during the control, the diagnosis of the module that failed, and the restoration of the functioning of the enterprise's information system. However, the methodology is given in a general form, without paying attention to the development of the mathematical apparatus.

The article [10] outlines the priorities of functional strategies by stages of the enterprise's life cycle. The methodical foundations for researching the possibility of functional subsystems of management to ensure risk protection and dynamic stability of machine-building enterprises in the process of their innovative development have been formed. But the methods and techniques of their provision are not given.

In the article [11], the prerequisites for ensuring functional stability are given. Therefore, one of the most important prerequisites is ensuring the functional stability of the enterprise's generalized information system. At the same time, a rather effective approach is the application of a certain method of ensuring the property of the functional stability of the information system of the enterprise, due to the presentation of the functioning of the system in the form of a formalized process, where the main types of procedures are the accumulation of checks, the analysis of test links, the diagnosis of the failed module and the restoration of functioning systems.

Therefore, the development of a new methodology for ensuring the functional stability of critical infrastructure objects is an urgent scientific task to ensure the reliable functioning of the information system of critical infrastructure objects.

3 Research Goals and Objectives

The purpose of this study is to develop a methodology for ensuring the functional stability of critical infrastructure objects.

To achieve the goal, the following tasks were set:

- determining the probability of transition to a state of complete failure of the facility control procedure;
- determining the probability of transition to a state of complete failure of the diagnostic procedure;
- determining the probability of transition to a state of complete failure of the recovery procedure.

4 Research Materials and Methods

4.1 *Parameters Describing the Process of Functioning of the Information System*

It is known that the mathematical model that describes the formalized process of the functioning of a complex technical system, in particular, the information system of an enterprise, includes only the main characteristics [12]. Additional characteristics are not considered in such models. When considering the functional stability of systems, the main task is to determine the failure state of the system. Therefore, the process of functioning is considered as a process of the successive changes of states. Usually, the main components of mathematical models are formed when such operations are performed:

- determining the types of faults (permanent, flashing or failure);
- determining the probability distribution for the time of malfunctions in the system;
- formalization of the concept of failure in the information system.

The parameters describing the process of the information system's functioning are:

- time of finding a malfunction in the system $t_{\mathcal{D}}$;
- reliability of the fault detection $\mathfrak{D}$;
- the probability of the system recovery $P_{\mathcal{R}}$;
- system recovery time $T_{\mathcal{R}}$.

For multi-module systems with passive and active redundant resources, there are several options for disclosing the parameters that describe the functioning of the system. For enterprise information systems, the time to find a malfunction $t_{\mathcal{D}}$ plays a significant role. Therefore, the model that describes the functioning of the information system of the enterprise from the point of view of its functional stability must take into account the parameter $t_{\mathcal{D}}$. In addition, the fault detection and the system recovery procedures are implemented by the system itself. It follows that the parameters $t_{\mathcal{D}}$, $\mathfrak{D}$, $P_{\mathcal{R}}$, and $T_{\mathcal{R}}$ are functions of the state of the system, i.e. each time malfunction is detected and the system is restored, the specified parameters will change.

The model of a functionally stable information system of an enterprise must take into account all the parameters describing the functioning and the dependence of these parameters on the state of the system. The basis of such a model is the presentation of the functioning of the system in the form of a process $\mathcal{Q}_t$:

$$\mathcal{Q}_t = \langle N,\ N_s,\ TP,\ n_0,\ \rho,\ \varphi\rangle \tag{1}$$

where N—the number of active modules in the system; N_s—the number of modules that failed but continue to work (that is, those that are not yet detected or detected but participate in elementary checks); TP—type of procedure; n_0—operation number; $\rho,\ \varphi$—relevant process parameters.

The process $\mathcal{Q}_t$ has the following procedure types:

- accumulation of inspections for system control (AC);
- analysis of the structure of verification relations (AR);
- accumulation of checks for system diagnostics (AD);
- analysis of the set of inspection results to determine the failed module (FM);
- restoration of system operation (RO).

The operation number n_0 corresponds to the cycle number within the procedure; the parameter ρ characterizes the moment of failure in relation to the operation number; the parameter φ characterizes the moment of failure detection relative to the operation number.

4.2 *Restoration of the Information System of the Enterprise After Diagnosing the Failed Module*

Restoration of the subsystems of the enterprise's generalized information system occurs by disconnecting the failed module or by disconnecting the suspicious pair. In the case when, before performing the diagnostic procedure, the tasks performed by the modules of the generalized information system of the enterprise [13, 14] (GISE) were not redistributed, then the recovery procedure also includes the selection of the module and transferring of the task performed by the failed module to it.

It is assumed that the means of recovery, with the probability of q_R, themselves may turn out to be faulty. Such a situation leads to a complete failure of GISE. The recovery time is chosen on the condition that the probability t_{RD} of exceeding the set value T_0 by a random variable must be within the set limits. The value t_{RD} is determined in the following way

$$t_{\mathcal{RD}} = t_{\mathcal{D}} + t_{\mathcal{R}} \tag{2}$$

where $t_{\mathcal{D}}$ is a random value of system diagnostics time; $t_{\mathcal{R}}$ is a random value of recovery time GISE.

Just as when performing diagnostics, the parameters of the recovery procedure depend on which approach is used after detecting a system failure, that is, after detecting a suspicious pair of modules. Two approaches are possible:

- an approach related to redistribution of tasks;
- tasks are not redistributed.

For the first approach, the value T_0 equals $t_{z\ \max}$, for the second approach $T_0 = t_{w\ \max}$. Time of diagnosis $t_{\mathcal{RD}}$ for these approaches is equal to $t_{\mathcal{D}} = t_z$ and $t_{\mathcal{D}} = t_x + t_z$ in accordance.

In the case when the value $t_{\mathcal{RD}}$ exceeds the value T_0, the system goes into a state of complete failure. Thus, *the system recovery time* $t_{\mathcal{R}}$ can be defined as follows:

for the first approach

$$t_{\mathcal{R}} \leq T_0 - t_z^{(1)}, \tag{3}$$

where $t_z^{(1)}$ is the diagnosis time calculated for the diagnosis procedure with task redistribution; *for the second approach*

$$P\{t_{\mathcal{RD}} < T_0\} \geq C \tag{4}$$

or

$$P\left\{t_x + t_z^{(2)} < T_0 - t_{\mathcal{R}}\right\} \geq C \tag{5}$$

where $t_z^{(2)}$ is the diagnosis time calculated for the diagnosis procedure without task redistribution; C is the constant obtained basing on the study of the indicators of the functional stability of the information system of the enterprise [15].

The maximum value of the diagnosis time is selected

$$t_{z\ \max}^{(2)} = t_{\mathcal{D}w}. \tag{6}$$

In the present case $t_{\mathcal{D}w}$ is diagnosis time, at which the probability of the system transitioning to a state of complete failure due to exceeding the limit $t_{w\ \max}$ less than V [16]. Then

$$P\{t_x < T_0 - t_{\mathcal{R}} - t_{\mathcal{D}w}\} \geq C$$

or

$$\Phi_{t_x}\{T_0 - t_{\mathcal{R}} - t_{\mathcal{D}w}\} \geq C$$

or

$$1 - \left(T_0 - t_R - t_{\mathcal{D}w} + \frac{1}{\lambda}\right) \lambda \exp\left[-\lambda\left(T_0 - t_{\mathcal{R}} - t_{\mathcal{D}w}\right)\right] \geq C. \tag{7}$$

The recovery time in this case is defined as the largest value $t_{\mathcal{R}}$ in which the inequality holds (7).

When solving practical problems, it is necessary to consider the inverse problem.

Problem 1 According to the known (specified) recovery time $t_{\mathcal{R}}$, it is necessary to determine the probability of the information system of the enterprise going into a state of complete failure due to exceeding the specified limit T_0.

This probability can be determined in the following way

$$P_{\mathcal{RD}} = P\{t_{\mathcal{RD}} > T_0\} = 1 - P\{t_{\mathcal{RD}} < T_0\} = 1 - \Phi_{t_x}\{T_0 - t_{\mathcal{R}} - t_{\mathcal{D}w}\} =$$

$$= \left(T_0 - t_{\mathcal{R}} - t_{\mathcal{D}w} + \frac{1}{\lambda}\right) \lambda \exp\left[-\lambda\left(T_0 - t_{\mathcal{R}} - t_{\mathcal{D}w}\right)\right]. \tag{8}$$

Improvement of the method of ensuring the property of functional stability of the information system of the enterprise.

The methodology for ensuring a functionally stable information system of the enterprise is based on the researched state transition graph, which describes the process Q_t. However, this research is difficult due to a number of reasons:

Firstly, the random value of the time the process stays in the selected states does not obey the exponential law of distribution. Therefore, characteristic models cannot be used.

Secondly, the probabilities of transition from state to state are functions of the time the system stays in the state from which the transition occurs. In addition, these probabilities depend on the states in which the system was previously. Therefore, it is not possible to use the apparatus of semi-Markov processes.

In order to use a state transition graph for the research of the process Q_t it is necessary to increase the number of states in the graph. Such an increase in the number of states of the graph will allow to bypass the indicated difficulties. However, such a graph model is difficult to depict due to the large number of states and the impossibility of determining all transition parameters [17].

Therefore, a new approach is proposed. According to the new approach, sequential use of the state graph is performed. It should be noted that with each use of the graph, its states are strengthened. Thus, each graph of states corresponds to a certain level of strengthening.

Consider the graphical model of the process Q_t, which consists of several levels of strengthening, in reverse order, starting from the upper third level.

The distribution of integral combined states into the components of their states leads to the second level of strengthening.

In turn, each state of the second level of strengthening of a functionally stable GISE can be broken down into more detailed states that can be considered separately. At the first level, GISE states can no longer be represented as a collection of more detailed states.

A distinctive feature of the first level of strengthening is the following circumstance. For the state transition graph corresponding to the first level, determining all the probabilities of transitions to the state of complete failure (PF) is a difficult task. Knowledge of all the transition probabilities is necessary to determine the selected three probabilities R_1, R_2, R_3. Therefore, for the first level, an approach is considered, thanks to which probabilities R_1, R_2, R_3 are allocated are determined without prior determination of all transition probabilities in the state graph.

Let's consider the methods of determining probabilities R_1, R_2, R_3.

Determination of the probability of transition to a state of complete failure of the control procedure R_1.

After performing the control procedure, the transition to the state of complete failure is possible for two reasons:

- the failed module will be controlled but not sufficiently (with a second type error);
- the failed module will remain unmonitored for the set time t_M.

The probability of transition to a state of complete failure due to the first reason is discussed above. Here we assume that this probability is known and equal $P_{M_1}(t_M)$.

The probability of transition to a state of complete failure due to the second reason $P_{M_2}(t_M)$ can be determined using the following technique.

Assume that repeated control blocks are not allowed and the optimal value of the number of additional cycles M is known and probabilities $\mathcal{S}_{OC}, \mathcal{S}_{AC_i}, i = 1, \ldots, k$, where $\mathcal{S}_{AC_i}$, is the probability that the i-th additional cycle will be required. Then the probability $P_{M_2}(t_M)$ is from the condition

$$P_{M_2} = \mathcal{S}_{OC}\Omega_{OC} + \mathcal{S}_{AC_1}\Omega_{AC_1} + \cdots + \mathcal{S}_{AC_k}\Omega_{AC_k}, \tag{9}$$

where Ω_{AC_i} the probability that the failed module will not be controlled in time t_s. In the future instead of $\mathcal{S}_{OC}$ let's use the notation $\mathcal{S}_{AC}$. Probabilities Ω_{AC_i} can be represented as

$$\Omega_{AC_1} = f(t_s) = f(t_M - t_0),$$

where t_0 is a random value of the time interval between the last moment when the control result was issued and the moment when the failure appeared.

Function $\Omega_{AC_i} = f(t_M - t_0)$ has an exponential character and is presented in the form

$$\Omega_{AC_i} = e^{-\mu t_s} = e^{-\mu(t_{M_i} - t_0)}.$$

Because t_0 is a random variable, then and Ω_{AC_i} is random variable. Suppose, first, that on the time interval $[0, t_{M_i}]$ the density of the distribution of a random variable t_0 has a constant value, equal to $\frac{1}{t_{M_i}}$. In this case, the distribution density is also a constant value

$$f_{\Omega_{AC_i}} = \left(1 - e^{-\mu t_{M_i}}\right)^{-1}.$$

According to (9), the value P_{M_2} is the sum of uniformly distributed independent random variables. In this case, the density of the distribution is a random variable P_{M_2} is equal to

$$f_{P_{M_2}}(p) = f_{x_{OC}} * f_{x_{AC_1}} * \cdots * f_{x_{AC_k}}, \qquad x_{AC_i} = \mathcal{S}_{AC_i}\Omega_{AC_i},$$

where $f_{x_{AC_i}}$ is the density of the distribution x_{AC_i}; $*$ is symbol that means a convolution of functions.

The density of the distribution of a random variable x_{AC_i}

$$f_{x_{AC_i}} = \frac{1}{\mathcal{S}_{AC_i}}\left(1 - e^{-\mu t_{M_i}}\right)^{-1}.$$

Using the Laplace transform of the function $f_{x_{AC_i}}$, we will get

$$F_{z_{AC_i}}(z) = \frac{1}{\mathcal{S}_{AC_i}\left(1 - e^{-\mu t_{M_i}}\right)z}. \tag{10}$$

From (10) for the Laplace transform of a random variable P_{M_2}, we will get

$$F_{P_{M_2}}(z) = \frac{1}{z^M} \prod_{i=0}^{k} \frac{1}{\mathcal{S}_{AC_i}\left(1 - e^{-\mu t_{M_i}}\right)}.$$

After the inverse transformation for the density distribution of a random variable P_{M_2} the following equality holds

$$f_{P_{M_2}}(p) = \frac{p^{M-1}}{(M-1)!} \prod_{i=0}^{k} \frac{1}{\mathcal{S}_{AC_i}\left(1 - e^{-\mu t_{M_i}}\right)},$$

where P is a value of a random variable P_{M_2}, $P = [0, 1]$. Thus, with probability equal to the value $\mathfrak{D}$, it can be argued that the value P_{M_2} will not be greater than the value $\mathcal{A}$, where

$$\mathfrak{D} = \frac{\mathcal{A}^M}{M!} \prod_{i=0}^{k} \frac{1}{\mathcal{S}_{AC_i}\left(1 - e^{-\mu t_{M_i}}\right)}.$$

For the mean value P_{M_2} we have

$$P_{M_2 m}(p) = \frac{M}{(M+1)!} \prod_{i=0}^{k} \frac{1}{\mathcal{S}_{AC_i}\left(1 - e^{-\mu t_{M_i}}\right)}.$$

Now consider the case when the density of the distribution of a random variable t_0 on the interval $\left[0, t_{M_i}\right]$ is a monotonically decreasing function of the form

$$f_{t_0} = \frac{2 - t_0/t_{M_i} - \sigma t_0}{t_{M_i}}, \tag{11}$$

where σ is the value which shows by how much value $t_0 = 0$ is more likely than $t_0 = t_M$.

For this case, the density of the distribution of a random variable Ω_{AC_i} is defined as follows:

$$f_{\Omega_{AC_i}}(\Omega) = f_{t_0}\left[t_0\left(\Omega_{AC}\right)\right] \times \left|\frac{dt_0}{d\Omega_{AC}}\right|_{\Omega_{AC}=\Omega}. \tag{12}$$

Formula (11) contains

$t_0\left(\Omega_{AC}\right) = \frac{\mu t_{M_i} + \ln \Omega_{AC}}{\mu}$ and $\left|\frac{dt_0}{d\Omega_{AC}}\right|_{\Omega_{AC}=\Omega} = \frac{1}{\mu\Omega}$.

From (11) and (4.2), relation (12) will take the form

$$f_{\Omega_{AC_i}}(\Omega) = \frac{\mu t_{M_i} + \ln \Omega - \sigma \mu t_{M_i}^2 - \sigma t_{M_i} \ln \Omega}{\mu^2 t_{M_i}^2 \Omega}, \tag{13}$$

where $\Omega = \left[e^{-\mu t_{M_i}}, 1\right]$.

Let's denote random variable $S_{AC_i}\Omega_{AC_i}$ as Z_{AC_i}. Based on the relation (13), it is possible to determine the distribution density of a random variable Z_{AC_i}.

$$f_{Z_{AC_i}}(z) = \frac{\mu t_{M_i} + \ln\frac{z}{S_{AC_i}} - \sigma\mu t_{M_i}^2 - \sigma t_{M_i}\ln\frac{z}{S_{AC_i}}}{\mu^2 t_{M_i}^2 z}.$$

Searched value P_{M_2} will be the sum of k mutually independent values Z_{AC_0}, Z_{AC_1}, ... , Z_{AC_k}, the distribution densities of which, respectively, equal f_{AC_0}, f_{AC_1}, ... , f_{AC_k}.

The density of the distribution of a random variable P_{M_2} is found as a convolution of the corresponding densities of the distribution of random variables f_{AC_0}, f_{AC_1}, ... , f_{AC_k} that is

$$f_{P_{M_2}}(p) = f_{Z_{AC_0}} * f_{Z_{AC_1}} * \cdots * f_{Z_{AC_k}}.$$

So, the final expression for the probability of transition of the information system of the enterprise to a state of complete failure of the control procedure R_1 looks like

$$R_1 = P_{M_1} + P_{M_2m} - P_{M_1}P_{M_2m}.$$

Determination of the probability of transition to a state of complete failure of the diagnostic procedure R_2.

Earlier, two cases of performing the diagnostic procedure were considered. In the first case, after disabling a suspicious pair of modules, the tasks they performed are transferred to the other modules. In this case, the complete failure of the company's information system is possible as a result of incorrect diagnosis, that is, the failure of the failed module. The probability of such a refusal is considered in the third chapter.

More complicated is the case when, after disconnecting a suspected pair of modules, the redistribution of tasks is not carried out. A complete failure of the system becomes possible due to exceeding the set limit $t_{w\ \max}$. Thus, determining the probability of transition to a state of complete failure is mainly reduced to finding the probability that the total time of detection and diagnosis of the failed module will exceed the value $t_{w\ \max}$ that is, $P_0 = P\{t_x + t_z > t_{w\ \max}\}$.

In this case, one of the summed random variables t_x is a continuous random variable, and the second t_z is a discrete random variable. Since the diagnostic procedure is divided into the main and additional diagnostic cycles, then the random variable t_z can take values:

$$\gamma_0 = t_{OC}^{\partial},$$

$$\gamma_1 = t_{OC}^{\partial} + t_{AC_1}^{\partial},$$

$$\gamma_2 = t_{OC}^{\partial} + t_{AC_1}^{\partial} + t_{AC_2}^{\partial},$$

$$\cdots\cdots\cdots\ \cdots\cdots$$

$$\gamma_k = t^{\partial}_{OC} + t^{\partial}_{AC_1} + t^{\partial}_{AC_2} + \cdots + t^{\partial}_{AC_k}.$$

Assume that the probabilities of the specified values are random t_z are known and are equal $\mathcal{M}_{OC}, \mathcal{M}_{AC_1}, \mathcal{M}_{AC_2}, \ldots, \mathcal{M}_{AC_k}$. Then, for the distribution function of a random variable $Y = t_x + t_z$, we will get

$$F_y(Y) = \mathcal{M}_{OC}(\gamma_0) \int_0^{y-\gamma_0} f_{t_x}(t_x)\, dt_x + \mathcal{M}_{AC_1}(\gamma_1) \int_0^{y-\gamma_1} f_{t_x}(t_x)\, dt_x + \cdots +$$

$$+\mathcal{M}_{AC_k}(\gamma_k) \int_0^{y-\gamma_k} f_{t_x}(t_x)\, dt_x.$$

In the case when the random variable t_x has a density distribution of the form

$$f_{t_x}(t_x) = \lambda^2 t_x e^{-\lambda t_x}.$$

The distribution function can be represented as

$$\begin{aligned} F_y(Y) = {} & \mathcal{M}_{OC}\left[1 - e^{-\lambda(Y-\gamma_0)}\left(1 + \lambda(Y - \gamma_0)\right)\right] + \\ & + \mathcal{M}_{AC_1}\left[1 - e^{-\lambda(Y-\gamma_1)}\left(1 + \lambda(Y - \gamma_1)\right)\right] + \ldots \\ & + \mathcal{M}_{AC_k}\left[1 - e^{-\lambda(Y-\gamma_k)}\left(1 + \lambda(Y - \gamma_k)\right)\right]. \end{aligned} \tag{14}$$

Knowing $F_y(Y)$ we find the probability P_0:

$$P_0 = 1 - F_y(Y = t_{w\ \max}) == 1 - \mathcal{M}_{OC}\left[1 - e^{-\lambda(t_{w\ \max}-\gamma_0)}\left(1 + \lambda(t_{w\ \max} - \gamma_0)\right)\right] -$$

$$-\mathcal{M}_{AC_1}\left[1 - e^{-\lambda(t_{w\ \max}-\gamma_1)}\left(1 + \lambda(t_{w\ \max} - \gamma_1)\right)\right] - \cdots -$$

$$-\mathcal{M}_{AC_k}\left[1 - e^{-\lambda(t_{w\ \max}-\gamma_k)}\left(1 + \lambda(t_{w\ \max} - \gamma_k)\right)\right].$$

If, after completing the diagnostic procedure, it is possible to disconnect a pair of modules (with insufficient diagnostic reliability), then the sought probability R_2 is equal to the probability value P_0. In the case when only one module is disabled, the probability R_2 is located as

$$R_2 = P_0 + P_{\mathcal{D}} - P_0 P_{\mathcal{D}}. \tag{15}$$

where $P_{\mathcal{D}}$ is a reliability of diagnosing the failed module. A calculation algorithm has been developed for the case when a pair of modules is turned off R_2.

Determination of the probability of transition to a state of complete failure of the recovery procedure R_3.

There are two reasons to go from a system recovery state to a full failure state $\mathcal{A}_1$ and $\mathcal{A}_2$. In this case $\mathcal{A}_1$ is recovery failure event, $\mathcal{A}_1$ is a event of exceeding a random variable $t_{\mathcal{RD}}$ the established limit T_0, where $t_{\mathcal{RD}} = t_{\mathcal{D}} + t_{\mathcal{R}}, t_{\mathcal{D}}, t_{\mathcal{R}}$ is a independent random variables. It is believed that the probability of an event $\mathcal{A}_1$ is known and is equal to $q_{\mathcal{R}}$ and the distribution functions of random variables $t_{\mathcal{D}}$, $t_{\mathcal{R}}$ equal F_{t_z} and $F_{t_{\mathcal{R}}}$ in accordance. Then, the probability of transition of the information system of the enterprise to a state of complete failure from the recovery procedure R_3 is defined as

$$R_3 = q_{\mathcal{R}} + P_{\mathcal{RD}} - q_{\mathcal{R}} P_{\mathcal{RD}}, \tag{16}$$

where

$$P_{\mathcal{RD}} = 1 - F_{t_{\mathcal{RD}}} = 1 - \int_0^{\infty} F_{t_{\mathcal{R}}}(t - t_z)\, dF_{t_z}(t_z). \tag{17}$$

Thus, finally, the block diagram of the new method of determining the probability of transition to a state of complete failure from the recovery procedure. With the help of the given methodology, it is possible to calculate the probabilities of the transition to a state of complete failure from the recovery procedure.

5 Research Results

We will simulate the process of determining the probability of transition to a state of complete failure of a critical infrastructure object from the recovery procedure R_3.

Results of simulation of the probability of transition to the state of complete failure of the critical infrastructure object from the recovery procedure R_3 shows that with the growth of failures, the probability of transition to the state of complete failure also increases. But the growth rate during the recovery procedure is lower than the growth rate without recovery, which fully corresponds to functionally stable critical infrastructure objects.

After the probabilities of transitions into states of complete failure have been determined R_1, R_2, R_3 (parameters of the second level), the parameters of the third level of a functionally stable information system of the enterprise can be found. So, for example, probability P_1 is determined in this way

$$P_1 = 1 - (1 - R_1)(1 - R_2)(1 - R_3).$$

When the probabilities are determined R_1, R_2 and R_3 parameters of the first level are used, which are determined as a result of simulation modeling.

6 Discussion of Research Results

The obtained simulation results show that the proposed methods and algorithms show that the procedure for accumulating checks and analyzing the structure of check connections can be presented in the form of a state transition graph, on the basis of which the main parameters of the procedure are determined, such as: the minimum time for issuing information about the state of the information systems of the critical infrastructure object with a given reliability, the optimal value for the execution time of the main and additional cycles of accumulation of checks.

The study of the state transition graph, which describes the methodology for ensuring the functional stability of critical infrastructure objects in the form of a continuous process, is complicated. Therefore, an approach is proposed in which the sequential use of the graph of states is performed. Each time the graph of states is used, its states are strengthened. Therefore, each graph of states corresponds to a certain level of strengthening.

7 Conclusions

The section presents a new method of ensuring the functional stability of critical infrastructure objects. The basis of the methodology for ensuring the functional stability of the enterprise's generalized information system, the functioning of which is presented in the form of a process Q_t The main types of procedures of the formalized process are the accumulation of checks, the analysis of check connections, the diagnosis of the failed module and the restoration of the enterprise information system.

The procedure for accumulating checks and analyzing the structure of check connections can be presented in the form of a transition state graph, on the basis of which the main parameters of this procedure are determined, such as: the minimum time for issuing information about the state of the information system of the enterprise with a given reliability, the optimal value for the execution time of the main and additional check accumulation cycles.

Since the study of the state transition graph, which describes the method of ensuring the functional stability of the information system of the enterprise in the form of a process Q_t complicated, a new approach is proposed. According to the proposed approach, the state graph is used sequentially. Each time the graph of states is used, its states are strengthened. Thus, each graph of states corresponds to a certain level of strengthening. A feature of the first level of strengthening the state transition graph is the method of determining the main transition probabilities of the process Q_t into a state of complete failure without pre-calculating all transition probabilities in the state graph.

References

1. Pichkur, V., Sobchuk, V.: Mathematical models and control design of a functionally stable technological process. J. Optim. Diff. Equ. Appl. (JODEA) **29**, 1, 1–11 (2021)
2. Sobchuk, V., Pichkur, V., Barabash, O., Laptiev, O., Kovalchuk, I., Zidan, A.: Algorithm of control of functionally stable manufacturing processes of enterprises. In: 2020 IEEE 2nd International Conference on Advanced Trends in Information Theory (ATIT), Kyiv, Ukraine, pp. 206–210 (2020)
3. Sobchuk, V., Olimpiyeva, Y., Musienko, A., Sobchuk, A.: Ensuring the properties of functional stability of manufacturing processes based on the application of neural networks. CEUR Workshop Proc. **2845**, 106–116 (2021)
4. Maksymuk, O., Sobchuk, V., Salanda, I., Sachuk, Yu.: A system of indicators and criteria for evaluation of the level of functional stability of information heterogenic networks. Math. Model. Comput. **7**(2), 285–292 (2020)
5. Barabash, O., Tverdenko, H., Sobchuk, V., Musienko, A., Lukova-Chuiko, N.: The assessment of the quality of functional stability of the automated control system with hierarchic structure. In: 2020 IEEE 2nd International Conference on System Analysis and Intelligent Computing (SAIC). Conference Proceedings. 05–09 Oct. 2020, Kyiv, Ukraine, pp. 158–161. Igor Sikorsky Kyiv Polytechnic Institute (2020)
6. Sobchuk, V., Kapustyan, O., Pichkur, V., Kapustyan, O.: Design of stable periodic regimes for one class of hybrid planar systems. In: II International Scientific Symposium "Intelligent Solutions", 28–30 Sept. 2021, Kyiv, pp. 89–100 (2021)
7. Kapustyan, O.A., Kapustyan, O.V., Ryzhov, A., Sobchuk, V.: Approximate optimal control for a parabolic system with perturbations in the coefficients on the half-axis. Axioms **11**, 175 (2022)
8. Sobchuk, V.V.: The method of creating a single information space at a production enterprise with a functionally stable production process. Control Navig. Commun. Syst. **6**(58), 84–91 (2019)
9. Laptiev, O., Sobchuk, V., Shcheblanin, Y., Barabash, O., Musienko, A., Kozlovskyi, V.: Evaluation of efficiency of application of functionally sustainable generalized information system of the enterprise. In: 4th International Congress on Human-Computer Interaction, Optimization and Robotic Applications, 9–11 June 2022, Ankara, Turkey (2022)
10. Sobchuk, V., Zamrii, I., Vlasyk, H., Tsvietkova, Y.: Strategies for management of operation of production centers to provide functionally sustainable technological processes of production. In: 2021 IEEE 3 nd International Conference on Advanced Trends in Information Theory (ATIT), 15–18 Dec. 2021, Kyiv, Ukraine, pp. 61–66 (2021)
11. Sobchuk, V.V., Zamrii, I.V., Barabash, O.V., Musienko, A.P.: The method of ensuring the property of functional stability of the intellectual information system of a production enterprise. Bulletin of Taras Shevchenko Kyiv National University. Seri. Phys. Math. Sci. **4**, 116–127 (2021)
12. Yevseiev, S., Ponomarenko, V., Laptiev, O., Milov, O., others: Synergy of building cybersecurity systems: monograph. In: Kharkiv (ed.) PC Technology Center, 2021. 188 p. (2021). http://monograph.com.ua/pctc/catalog/book/64
13. Pichkur, V., Laptiev, O., Polovinkin, I., Barabash, A., Sobchuk, A., Salanda, I.: The method of managing man-generated risks of critical infrastructure systems based on ellipsoidal evaluation. In: 2022 IEEE 4th International Conference on Advanced Trends in Information Theory (ATIT), pp. 133–137 (2022)
14. Zamrii, I., Haidur, H., Sobchuk, A., Hryshanovych, T., Zinchenko K., Polovinkin, I.: The method of increasing the efficiency of signal processing due to the use of harmonic operators. In: 2022 IEEE 4th International Conference on Advanced Trends in Information Theory (ATIT), pp. 138–141 (2022)
15. Salanda, I.P., Barabash, O.V., Musienko, A.P.: System of indicators and criteria for formalization of processes of ensuring local functional stability of extensive information networks. Syst. Control Navig. Commun. **1**(41), 122–126 (2017)

16. Petrivskyi, V., Shevchenko, V., Yevseiev, S., Milov, O., Laptiev, O., Bychkov, O., Fedoriienko, V., Tkachenko, M., Kurchenko, O., Opirsky, I.: Development of a modification of the method for constructing energy-efficient sensor networks using static and dynamic sensors. Eastern-European J. Enterp. Technol. **1**(9), 15–23 (2022)
17. Lenkov, S., Zhyrov, G., Zaytsev, D., Tolok, I., Lenkov, E., Bondarenko, T., Gunchenko, Y., Zagrebnyuk, V., Antonenko, O.: Features of modeling failures of recoverable complex technical objects with a hierarchical constructive structure. Eastern-European J. Enterp. Technol. **4**(4), 34–42 (2017)
18. Yevseiev, S., Rzayev, K., Laptiev, O., Hasanov, R., Milov, O., Asgarova, B., Camalova, J., Pohasii, S.: Development of a hardware cryptosystem based on a random number generator with two types of entropy sources. Eastern-European J. Enterp. Technol. **5**(9), 6–16 (2022)
19. Laptiev, O., Lukova-Chuiko, N., Laptiev, S., Laptieva, T., Savchenko, V., Yevseiev, S.: Development of a method for detecting deviations in the nature of traffic from the elements of the communication network. In: International Scientific and Practical Conference "Information Security and Information Technologies": Conference Proceedings. 13–19 Sept. 2021. Kharkiv—Odesa, Ukraine, pp. 1–9 (2021)
20. Mashkov, V.A., Barabash, O.V.: Engineering Simulation, pp. 43–51 (1998). Amsterdam: OPA
21. Mashkov, V.A., Barabash, O.V.: Self-Testing of multimodule systems based on optimal check-connection structures. Eng. Simul. **13**, 479–492 (1996). Amsterdam: OPA
22. Mashkov, V.A., Barabash, O.V.: Self-checking of modular systems under random performance of elementary checks. Eng. Simul. **12**, 433–445 (1995). Amsterdam: OPA

Intellectual Data Analysis and Machine Learning Approaches for Car Insurance Rates Reconstruction

Pavlo Pustovoit and Olga Kupenko

Abstract The compulsory obligation for car insurance has led to a significant increase in the number of offers, that may not differ qualitatively from each other but still be considerably varying in price. Companies operating in the highly saturated insurance market have to constantly monitor their competitiveness. As the rates of insurance offers are of great importance to consumers, the reconstruction of leading competitors' rates can help an insurance company to determine its rates policy in an efficient way. In this paper, the authors provide the rate reconstruction on competitors' data using feature engineering techniques and machine learning algorithms, such as XGBoost, CatBoost, and LightGBM. The results of these algorithms are easy to interpret, all models perform similarly, with a reliability of about 80% and a relative error of 4–5%.

1 Introduction

According to de.Statista.com, as of January 1, 2023, there were 48.76 million vehicles registered in Germany—the highest number of cars ever registered. According to the same resource, but based on 2021 statistics, 93 companies in Germany provide car insurance, with a net income of 29.1 billion euros. Undoubtedly, this is a pretty small number of insurance companies in one large and quite rich market, and therefore the question of their competitiveness is crucial for any of them.

How popular is the company, who are its main competitors, in which parts of the market is it losing its opportunities, and does it need to make more efforts there? Is its offer competitive in terms of service and price? All these and many other questions are deeply investigated by departments of big insurance companies. For many years of work, they have established their own algorithms for solving mentioned problems, and the most important thing is that every year these algorithms are replenished with

P. Pustovoit · O. Kupenko (✉)
National Technical University of Ukraine "Igor Sikorsky Kyiv Polytechnic Institute", 37, Beresteisky Avenue (former Peremohy), Kyiv 03056, Ukraine
e-mail: kupenko.olga@gmail.com

M. Zgurovsky and N. Pankratova (eds.), *System Analysis and Artificial Intelligence*, Studies in Computational Intelligence 1107,
https://doi.org/10.1007/978-3-031-37450-0_12

new methods [1–4]. Amongst them are the Intelligent data analysis and Machine learning that we use to predict competitors' prices. The problems we encountered on the way and their solutions we will describe in this article. We will work with data taken from the websites of aggregators and directly from the websites of insurance companies, which include 31 necessary features: the age of the driver, the driving class, the type of driver's profession, the presence and age of another driver, the presence of an escort of a young driver, the type of fuel of the car, its mileage, vehicle risk class, vehicle age, regional class (risk class relative to zip code), garage availability, and others.

2 Statement of the Problem

Our goal is to create a generalized model able to reconstruct the car insurance rate of any company competing with us. It is worth specifying that this is only possible only due to the two conditions:

1. We are assuming that all tariffs are similar and have the same structure
2. We work in a specific economic market, as the questions asked when applying for insurance, which are our direct initial attributes to work with, may vary depending on the country and its legislation (in our case, Germany).

As the required data collecting takes place on aggregator websites as well as competitors' websites directly, it takes a lot of effort. Mostly, all such websites have protection systems against data collectors, which, in addition to simple blocking, can also deliberately give incorrect information. With this regard, we set another condition: the desired model must produce confidence-level results based on the minimum amount of data available.

3 Determination of Evaluation Metrics

First, to determine the appropriate evaluation function for the model, we will take into account the competitive environment of the insurance market.

The majority of existing insurance companies are represented at any popular aggregator site, and their offers for a specific combination of input parameters are sorted by price in ascending order. Therefore, for further convenience, we will introduce the concept of rank—the position of an offer in the list of all offers, where the first rank corresponds to the lowest price.

Based on this concept, for most clients, the first several ranks will be the most appealing. For every insurance company, it is vital to know what position it is in and who its main competitors are. This is why it is crucial for us to be confident that the competitor's price prediction for a specific combination of parameters stays within the same rank where the competitor's offer can be located.

Moreover, for the stated problem metrics based on absolute values, such as RMSE, are not appropriate us.

Let's imagine there is a combination of features valued at \$2. Suppose our model predicts a price of \$4 for this combination. Let's also imagine there is a second combination valued at \$500 and predicted at \$550. If we use a metric based on absolute values for estimation, the error in the first case is only \$2, while in the second case, it is much larger, \$50. Due to this fact, the model penalization will occur for the second case and not for the first. In fact, the relative error in the first case is as much as 100%, while in the second case, it is only 10%.

Of course, when working in a competitive environment, the relative error for the first case is much more dangerous because the forecast can go considerably beyond the assumed rank and, therefore, cannot be trusted.

From what was mentioned above, it makes sense if we take the ratio of our forecast to the actual price instead of their difference, and our primary metric for the evaluation of a single case will look like this:

$$E^* = \sqrt{(\frac{prediction}{true} - 1)^2}$$

and for a general assessment of the model:

$$E = \frac{1}{N}\sum_{n=1}^{N}\sqrt{(\frac{prediction_n}{true_n} - 1)^2},$$

where *prediction* is the model's prediction for a specific combination, *true* is true price value, and N is a dataset size. For a better understanding of the desirable model accuracy let's calculate the value of the maximum allowable error. For this purpose, let's take the value of the first rank as accurate and the value of the second rank as predicted. As a result, the maximum permissible error for our problem is 4.7%. It is worth keeping in mind the disadvantages of chosen error function:

1. It is sensitive to negative values.
2. The overall model estimate, similar to RMSE, will be significantly distorted by outliers, which, however, can be quite useful for analyzing such combinations.

Therefore we cannot fully trust only one averaging metric to adequately train our model because only the 50th quantile of all errors can get a value equal to the maximum acceptable error. Hence, the second important way to estimate our model will be the calculation of its "reliability", which is the ratio between the number of cases whose error is within the desired limits to the total number of combinations. For the model's results to be trustworthy, its "reliability" must be equal to or greater than 80%.

4 Feature Selection

As was mentioned earlier, all the parameters entered on the aggregator website or insurance company website must be used to create a working model, and they can be divided into the following conditional groups:

1. Main parameters that affect the formation of the main price, such as insurance start date (date feature), insurance holder (categorical feature), purchase/production date of the car (date feature), driver's and holder's date of birth (date feature), mileage (numeric feature), etc. Between all these features there are many interactions and dependencies. Also in this group, there is the car's model, the insurance holder's, and the driver's postal code, which require replacement. The fact is that for each car model as well as postal code, there is a risk group where the pair model/area is located. Those risk coefficients are important parameters that highly affect the process of price formation. The problem, however, is that exact reconstruction of their values with a limited amount of data and the presence of many other parameters, affecting the result, is simply impossible when there are over thirty thousand car models or twenty-eight thousand postal codes. Those data will mostly create noise and therefore will not be used for the model input. The risk group values we need are publicly available online, so clients can independently check them for their cases. We will replace the parameters mentioned with these values. Many companies may have their own risk group coefficients, but we will not attempt to reconstruct them due to the difficulties listed. We should also note that these rates for full insurance coverage and mandatory coverage will be different.
2. Some parameters provide a discount when taking out insurance, such as the policyholder's garage ownership, property ownership, and other insurance policies held by the same policyholder.
3. Additional services such as protection against loss of driver's license, protective letter for travel abroad, and delivery of the car to a repair shop in case of an accident may be related to some basic parameters. It is also important to note that there are parameters, such as seasonal license plates when the base price is multiplied by a certain coefficient, which we use before sending our data to the model.

Each parameter that is entered when applying for insurance is essential and affects the price, regardless of which group we put it.

5 Feature Processing

An essential step in model preparation is feature pre-processing and extracting the important information from them. For our model, we calculated the driver's age at the insurance coverage time rather than using the less relevant birth year. Such processing is necessary for all dates among input features. Another vital pre-processing step was dividing the car mileage values into intervals. Boundary values of these intervals are publicly available but not obvious and are not included in our data. Dividing the policyholder's age into intervals is also important, but unlike mileage, it depends more on the policyholder and cannot be unambiguously established for all companies. In this case, age can be left unchanged as a categorical feature, as working with sufficient data from one company showed that the lack of this processing does not critically affect the results. More important is the feature processing that protects against a reduction in the driver's class. This feature has different coefficients depending on the driver's age, the policyholder's age (as this can be another person in general), and correspondence between them. Mentioned interactions must be shown directly to work correctly with this feature. Other processing steps are less critical and theoretically one can go on with the model training without these.

6 Data Processing Issues

Features' format was one of the first and most essential issues in working with data. The specifics of insurance rates formation force us to convert all features, even numerical ones such as driver's age, policyholder's age, regional class, and car age, to categorical type. The reason for this is that coefficients for any such features in an insurance tariff do not have a direct linear relationship with their numerical values but are related to the risk groups where these values are located, interacting with other features. For example, the presence of a young driver affects coefficient values, and driving under the supervision of a more experienced driver can lower them. The model is not able to detect the relationship between these two features unless they are in a categorical format. Similar reasoning can be also applied to many other input variables. Another issue is the imbalance of essential parameters, such as age values or protection against driver's class loss. Protection against loss of a driver class is a binary feature coming from additional service client can go for. It may not be selected by many people, as the driver's class will be recalculated according to the presence of car accidents when changing the policyholder. As a result, during training, the model often receives negative values and ignores positive ones, which prevents it from predicting prices correctly. The same applies to age feature, where young and elderly drivers are much less represented in the insurance data than drivers aged 25–55. This is natural, but as we have seen, analysing the results, the model makes the largest errors predicting cases, that it doesn't see very often during it's training process. The problem of unbalanced data is much more common in classification tasks, so we used popular existing techniques to solve it [5]. The first one is to increase the number of data points for smaller classes to the level of larger ones. To do this, we took the value ranges for our parameters corresponding to drivers ages 18–23 and 58+ years old, randomly generated new cases from these ranges, and then requested prices for them. Unfortunately, random oversampling of our data did not improve the model's performance but on the contrary, it even got worse. This had most likely happened due to the fact that the randomly created combinations weren't natural. Tree algorithms, which are known to be one of the most interpretable and are able to explain the logic of the tariff, were significantly distorted by unnatural cases [6]. Therefore oversampling solution can only be effective when generating data with many accompanying constraints. The second method involved reducing the number of elements in the larger class for its size to become similar to the smaller class's size—the so-called undersampling technique. Of course, this approach is not possible to apply to a small dataset. Still, when working with a large amount of data from a single company (around 50,000 cases), we obtained much better results after balancing dataset than before, even though there were less data to train the model on.

7 Correlation and Multicollinearity

There is a strong correlation and multicollinearity between many features. For example, the correlation between a young driver's age and the car driver's binary feature, which has the values "insurance holder" and "other driver", is—. This happens because the young driver's age only exists when there is an "other" driver. It would make sense to combine these two features into one, but "driver" is also related to "insurance holder", "loss of driver's class protection", and many other features. This increases the multicollinearity coefficient for the "driver"—more than two hundred units. The following table, based on VIF analysis (Variance Inflation Factor analysis), shows the 10 parameters with the highest multicollinearity values (Table 1).

A large number of interactions between features make their combination extremely complex and impractical. Multicollinearity does not threaten the accuracy of tree-based algorithms, as they are immune to it, but it can significantly distort Feature importance.

Table 1 Results of multicollinearity analysis

N	Feature	VIF
1	Driver	590.29
2	Driver-Owner-Age interaction	177.61
3	Youngest driver's age	122.26
4	Accompanying a young driver	102.79
5	Beginning of insurance	84.15
6	Second car	68.62
7	Police holder	43.76
8	Own house	30.64
9	Profession group	22.21
10	Protection against driver's class loss	15.63

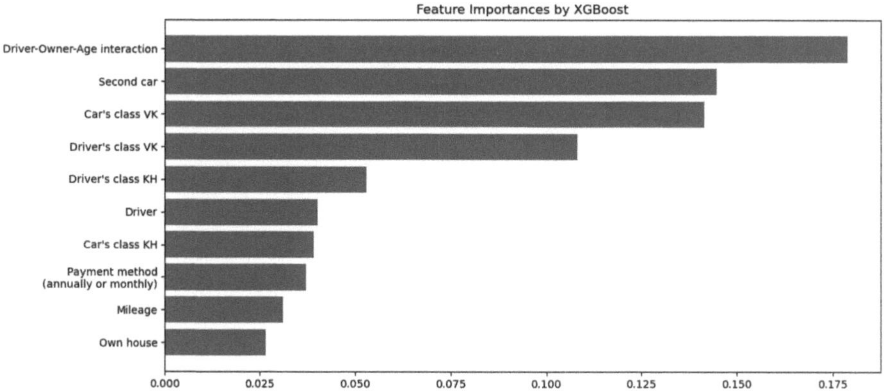

Fig. 1 Feature importance based on XGBoost

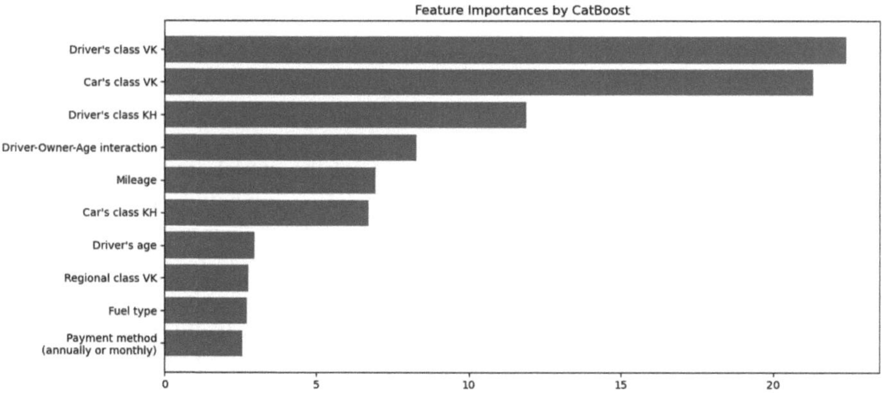

Fig. 2 Feature importance based on CatBoost

Figures 1 and 2 show the feature importance based on XGBoost and CatBoost. As we can see, the analysis given by XGBoost does not correspond to the actual tariff logic. The Second car parameter belongs to our conditional second category of parameters that give additional discounts, but are not essential when determining the main price. CatBoost gives a much more realistic result, which we can see later.

We used the Shapley values methodology to check the features significance for the model. It is important to note that for our model, due to a large number of features, variations in their values, and the size of the training dataset, the standard algorithm of the SHAP library could take more than 8 hours to compute the values, which was not suitable for us. Instead, we used the algorithm of a specialized library for working with tree-based algorithms, FastTreeShap, which took around a minute to complete. As a result, we obtained the desired answer, which entirely coincides with the logic of the insurance tariff.

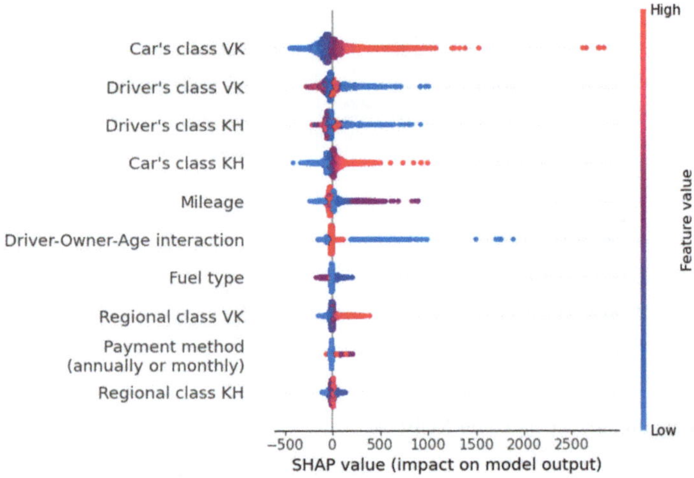

Fig. 3 Most significant features based on Shapley values, where VK and KH are data for full insurance coverage and mandatory, respectively

According to the Fig. 3, the essential features of the model are the regional and car classes, driver's class, age, car mileage, and relationship between a policyholder and a driver. The graph also shows that all features perform their task: their common values are around zero Shapley value. Any changes lead to a corresponding negative or positive increase of influence to the model's output. Thus, Shapley values analysis solves the problem of highlighting the most significant features caused by the distortion effect coming from multicollinearity for the Feature importance method in tree-based algorithms.

8 Analysis of Results

By the interpretability condition set, as the company must see the logic behind insurance rates of its competitors, we used only tree-based algorithms, namely XGBoost, CatBoost, and LightGBM. The average relative error they give is about the same—around 4–5%, but in terms of Reliability and average precision, CatBoost performs best. This result makes sense, as CatBoost algorithm was designed directly for training models on data where a lot of features are categorical, which is exactly our case. The following table shows the results of chosen models, trained on a dataset with 78.000 cases (Table 2).

As we can see, the most effective estimator is CatBoost, which gives us the highest value of Reliability with the lowest average relative error. The next plot shows the model's error distribution on the test dataset (Fig. 4).

Table 2 Results of used models

Estimator	AVG error (%)	Reliability (%)	Biggest error (%)
XGBoost	4.69	81.49	49.69
LightGBM	4.49	81.33	52.55
CatBoost	4.26	82.03	36.25

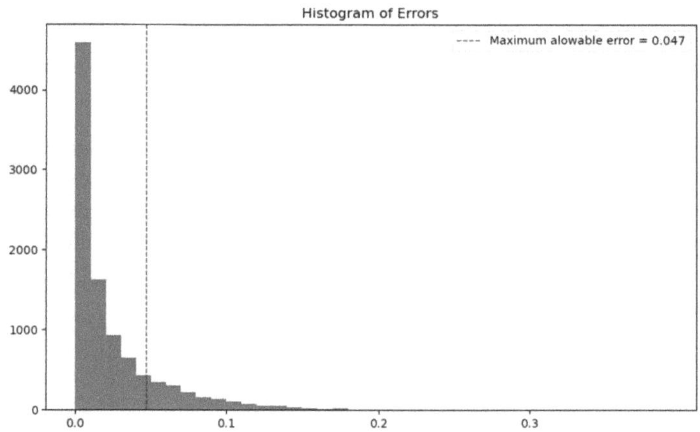

Fig. 4 Histogram of the errors the CatBoost model makes on the test set

Table 3 Results of CatBoost model on the datasets with different train size

Train size	AVG error (%)	Reliability (%)	Biggest error (%)	AVG cross-valid. error (%)	AVG cross-valid. reliability (%)
5,588	8.24	59.89	49.27	8.00	61.38
14,352	6.38	68.81	62.78	6.53	69.00
20,151	5.51	77.50	63.51	5.54	77.02
21,045	5.51	78.08	98.03	5.39	76.28
31,000	5.79	65.77	55.47	6.08	64.54
59,410	5.55	64.20	34.08	5.79	63.92
65,406	4.58	78.17	86.93	4.57	78.04
78,756	4.62	75.28	37.83	4.79	74.69
98,144	4.49	81.78	48.39	4.49	81.98

We are still interested in the following question: what is the minimum required amount of data to feed tree models with for the best results. To analyse this, we cite the following table based on CatBoost's results for different tariffs and their size (Table 3).

As we can see, the result does not have a linear dependence on the size of the dataset, but it definitely depends more on the quality of the training data. The difference between reliability when working with data of minimum and maximum sizes is only 20%, while the ratio of the sizes themselves is 17.5 times! This is primarily due to the previously discussed problem—the presence of unbalanced features—which cannot be simply solved for regression tasks.

9 Conclusion

In this project, we were able to achieve our goal—to reconstruct the tariffs of our competitors. Its logical continuation is the following two studies: optimization of work with unbalanced datasets for regression tasks and reconstruction of the minimum price in the entire insurance market. Obviously, without solving the first problem, we will not be able to effectively solve the second, because the biggest snag lies precisely in the amount, quality, and availability of data.

References

1. Hanafy, M., Ming, R.: Machine Learning Approaches for Auto Insurance Big Data/Mohamed Hanafy, Ruixing Ming Risks, vol. 9, p. 42 (2021). https://doi.org/10.3390/risks9020042
2. Fang, Kuangnan, Jiang, Yefei, Song, Malin: Customer profitability forecasting using Big Data analytics: a case study of the insurance industry. Comput. Ind. Eng. **101**, 554–64 (2016)
3. Henckaerts, R., Antonio, K., Clijsters, M., Verbelen, R.: A data driven binning strategy for the construction of insurance tariff classes. Scand. Actuar. J. 1–25 (2018). https://doi.org/10.1080/03461238.2018.1429300
4. Salcedo-Sanz, S., Mario DePrado-Cumplido, M.J.S.-V., Pèrez-Cruz, F., Bousoño-Calzòn, C.: Feature selection methods involving support vector machines for prediction of insolvency in non-life insurance companies intelligent systems in accounting, finance and management **12**(4), 261–281 (2004). https://doi.org/10.1002/isaf.255
5. Huyen, C.: Designing Machine Learning Systems: O'Reilly, 463 p (2022)
6. Chen, T., Guestrin, C.: XGBoost: a scalable tree boosting system. In: Paper presented at the 22nd ACM SIGKDD International Conference on Knowledge Discovery and Data Mining, San Francisco, CA, USA, 13–17 Aug. 2016
7. Anzahl zugelassener Pkw in Deutschland von 1960 bis 2023. https://de.statista.com/statistik/daten/studie/12131/umfrage/pkw-bestand-in-deutschland/
8. Anzahl der Versicherungsunternehmen in der Kfz-Versicherung in Deutschland von 1991 bis 2021. https://de.statista.com/statistik/daten/studie/38848/umfrage/versicherungsunternehmen-in-der-kraftfahrtversicherung-seit-2000/
9. Beiträge und Leistungen in der Kfz-Versicherung in Deutschland von 1991 bis (2021). https://de.statista.com/statistik/daten/studie/38847/umfrage/beitraege-und-leistungen-in-der-kraftfahrtversicherung-seit-2000/

On Generation of Daily Cloud-Free Satellite Images at High Resolution Level

Natalya Ivanchuk, Peter Kogut, and Petro Martyniuk

Abstract In this paper we discuss a new variational approach to the Date Fusion problem of multi-spectral satellite images from Sentinel-2 and MODIS that have been captured at different resolution level and, arguably, on different days. The crucial point of this problem is that the MODIS images are cloud-free whereas the images from Sentinel-2 can be corrupted by clouds or noise. We consider this problem in three different statements—restoration, interpolation and prediction. We show that each of these subtask can be stated as a constrained minimization problem that lives in variable Sobolev-Orlicz spaces. We discuss the consistency of the proposed models, give the schemes for their regularization, and derive the corresponding optimality systems. The numerical experiments confirm the efficacy of the proposed approach and its accuracy for agro-scenes with rather complicated texture of background surface.

1 Introduction

Following [26], the Image Fusion is a process of combining the relevant information from a set of images of the same scene into a single image and the resultant fused image must be more informative and complete than any of the input images. At the

N. Ivanchuk · P. Martyniuk
National University of Water and Environmental Engineering, Soborna str., 11, Rivne 33028, Ukraine
e-mail: n.v.medvid@nuwm.edu.ua; natalya.ivanchuk@eosda.com

P. Martyniuk
e-mail: p.m.martyniuk@nuwm.edu.ua; petro.martyniyk@eosda.com

EOS Data Analytics Ukraine, Desyatynny lane, 5, Kyiv 01001, Ukraine

P. Kogut (✉)
Oles Honchar Dnipro National University, Gagarin av., 72, Dnipro 49010, Ukraine
e-mail: p.kogut@i.ua; peter.kogut@eosda.com

EOS Data Analytics Ukraine, Gagarin av., 103a, Dnipro, Ukraine

M. Zgurovsky and N. Pankratova (eds.), *System Analysis and Artificial Intelligence*, Studies in Computational Intelligence 1107,
https://doi.org/10.1007/978-3-031-37450-0_13

same time, when we deal with the data fusion problem for satellite images, such images, as a rule, are multi sensor, multi-modal, multi-focus and multi temporal. Moreover, the data fusion problem is often exacerbated by cloud contamination. In some cloudy areas, researchers are fortunate to get 2–3 cloud-free satellite scenes per year, what is insufficient for many applications that require dense temporal information, such as crop condition monitoring and phenology studies [20, 40]. In view of this, we can indicate the following general requirements for the satellite image fusion process:

(i) The fused image should preserve all relevant information from the input images;
(ii) The image fusion should not introduce artifacts which can lead to wrong inferences.

In spite of the fact that the first requirement (item (i)) sounds rather vague, we give a precise treatment for it in Sect. 5, making use of a collection of special constrained minimization problems (see (35)–(36)). As for the second item, it is important to emphasize that we are mainly interesting by satellite images that can be useful from agricultural point of view (land cover change mapping, crop condition monitoring, yield estimation, and many others). Because of this an important option in the image data fusion is to preserve the precise geo-location of the existing crop fields and avoid an appearance of the so-called false contours and pseudo-boundaries on a given territory.

In this paper we mainly focus on the image fusion problem coming from two satellites—Sentinel-2 and Moderate Resolution Imaging Spectroradiometer (MODIS). Since each band (spectral channel) in Sentinel images has 10, 20, or 60 meters in pixel size, it gives an ideal spatial resolution for vegetation mapping at the field scale. Moreover, taking into account that Sentinel-2 has 3–5 revisit cycle over the same territory, it makes its usage for studying global biophysical processes, which allows to evolve rapidly during the growing season, essentially important. The unique problem that drastically restricts its practical implementation, is the fact that the satellite images, as a rule, are often contaminated by clouds, shadows, dust, and other atmospheric artifacts.

One of possible solutions for practical applications is to make use of frequent coarse-resolution data of the MODIS. Taking into account that the MODIS data can be delivered with the daily repeat cycle and 500-m surface reflectance, the core idea is to use the Sentinel and cloud-free MODIS data to generate synthetic 'daily' surface reflectance products at Sentinel spatial resolution [32].

The problem we consider in this paper can be briefly described as follows. We have a collection of multi-band images $\{S_1, S_2, \dots, S_N : G_H \to \mathbb{R}^m\}$ from Sentinel-2 that were captured at some time instances $\{t_1, t_2, \dots, t_N\}$, respectively, and we have a MODIS image $M : G_L \to \mathbb{R}^n$ from some day t_M. It is assumed that all of these images are well co-registered with respect to the unique geographic location. We also suppose that the MODIS image is cloud-free and the day t_M may does not coincide with any of time instances $\{t_1, t_2, \dots, t_N\}$. Meanwhile, the Sentinel images $\{S_2, \dots, S_N : G_H \to \mathbb{R}^m\}$ can be corrupted by clouds. The main question is how to generate a new synthetic 'daily' multi-band image of the same territory from the day

t_M at the Sentinel-2 spatial resolution G_H, utilizing for that the above mentioned data.

In principle, this problem is not new in the literature (see, for instance, [20, 27, 40]). For nowadays the spatial and temporal adaptive reflectance fusion model (STARFM) is one of the most popular model where the idea to generate a new synthetic 'daily' satellite images at high resolution level has been realized. Basing on a deterministic weighting function computed by spectral similarity, temporal difference, and spatial distance, this model (as many other generalizations) allows to predict daily surface reflectance at high spatial resolution and MODIS temporal frequency. However, its performance essentially depends on the characteristic patch size of the landscape and degrades somewhat when used on extremely heterogeneous fine-grained landscapes [20].

As it was mentioned in [5], the majority image interpolation techniques are essentially based on certain assumptions regarding the data and the corresponding imaging devices. First, it is often assumed that the image of high spatial resolution is a linear combination of the spectral channels with known weights [35]. Second, the loss of resolution is usually modeled as a linear operator which consists of a subsampled convolution with known kernel (point spread function). While both assumptions may be justified in some applications, it may be difficult to measure or estimate the weights and the convolution kernel in a practical situation.

Instead of this, we mainly focus on the variational approach to the satellite image data fusion. In order to eliminate the above mentioned restrictions, we formulate the data fusion problem as the two-level optimization problem. At the first level, following a simple iterative procedure, we generate the so-called structural prototype for a synthetic Sentinel image from the given day t_M. The main characteristic feature of this prototype is the fact that, it must have a similar geometrical structure (namely, precise location of contours and field boundaries) to the nearest in time 'visible' Sentinel images, albeit they may have rather different intensities in all bands. Since the revisit cycle of Sentinel-2 is 3–5 days, such prototype can be easily generated (for the details, we refer to Sect. 4). In fact, we consider the above mentioned structural prototype as a reasonable input data for 'daily' prediction problem that we formulate in the form of a special constrained minimization problem, where the cost functional has a nonstandard growth and the edge information for restoration of MODIS cloud-free images at the Sentinel resolution is accumulated both in the variable exponent of nonlinearity and in the directional image gradients which we derive from the predicted structural prototype.

Our approach is based on the variational model in Sobolev-Orlicz space with a non-standard growth condition of the objective functional and on the assumption that, to a large extent, the image topology in each spectral channel is similar to the topographic map of its structural prototype. It is worth to emphasize that this model is considerably different from the variational model for P+XS image fusion that was proposed in [2]. We discuss the well thoroughness of the above approach and consistency of the corresponding variational problem, and we show that this problem admits a unique solution. We also derive some optimality conditions and supply our approach by results of numerical simulations with the real satellite images.

The paper is organized as follows. Section 2 contains some preliminaries, auxiliary results, and a non-formal statement of the data fusion problem. In Sect. 3 we begin with some key assumptions and after we give a precise statement of the satellite image data fusion in the form of two-level constrained optimization problem with a nonstandard growth energy functional. We show that, in principle, we can distinguish three different statements of the data fusion problem. Namely, it is the so-called restoration problem (when the main point is to restore the information in the cloud-corrupted zone for Sentinel images), the interpolation problem (when the day t_M is intermediate for some time instances of $\{t_1, t_2, \dots, t_N\}$), and the so-called extrapolation problem (when $t_M > t_N$). We also discuss the specific of each of these problems and their rigorous mathematical description. Section 4 is devoted to the study of a particular model for the prediction of structural prototypes. We also illustrate this approach by some numerical simulations.

Consistency issues of the proposed minimization problems, optimality conditions, and their substantiation are studied in Sects. 5 and 6. For illustration of this approach, we refer to [25] where some results of numerical experiments with real satellite images are given. The experiments confirmed the efficacy of the proposed method and revealed that it can acquire plausible visual performance and satisfactory quantitative accuracy for agro-scenes with rather complicated texture of background surface. In Appendix we give the main auxiliary results concerning the Orlicz spaces and the Sobolev-Orlicz spaces with variable exponent.

2 Non-Formal Statement of the Problem

Let $\Omega \subset \mathbb{R}^2$ be a bounded connected open set with a sufficiently smooth boundary $\partial\Omega$ and nonzero Lebesgue measure. In majority cases Ω can be interpreted as a rectangle domain. Let G_H and G_L be two sample grids on Ω such that $G_H = \widehat{G}_H \cap \Omega$ and $G_H = \widehat{G}_H \cap \Omega$, where

$$\widehat{G}_H = \left\{ (x_i, y_j) \left| \begin{array}{l} x_1 = x_H,\ x_i = x_1 + \Delta_{H,x}(i-1),\ i = 1, \dots, N_x, \\ y_1 = y_H,\ y_j = y_1 + \Delta_{H,y}(j-1),\ j = 1, \dots, N_y, \end{array} \right. \right\},$$

$$\widehat{G}_L = \left\{ (x_i, y_j) \left| \begin{array}{l} x_1 = x_L,\ x_i = x_1 + \Delta_{L,x}(i-1),\ i = 1, \dots, M_x, \\ y_1 = y_L,\ y_j = y_1 + \Delta_{L,y}(j-1),\ j = 1, \dots, M_y, \end{array} \right. \right\},$$

with some fixed points (x_H, y_H) and (x_L, y_L). Here, it is assumed that $N_x >> M_x$ and $N_y >> M_y$.

Let $[0, T]$ be a given time interval. Normally, by T we mean a number of days. Let t_M and $\{t_k\}_{k=1}^N$ be moments in time (particular days) such that $0 \le t_1 < t_2 < \dots < t_N \le T$ and $t_1 < t_M < T$. Let $\{S_1, S_2, \dots, S_N : G_H \to \mathbb{R}^m\}$ be a collection of multispectral images of some territory, delivered from Sentinel-2, that were taken at time instances $t_1, t_2, \dots, t_N$, respectively. Hereinaftre, $m = 13$ and it stands for the number of spectral channels in images from Sentinel-2. Let $M : G_L \to \mathbb{R}^n$, with

Table 1 Spectral channels of J_1

Band	Resolution m	Central wavelenth m	Description
B_2	10	490	Blue
B_3	10	560	Green
B_4	10	665	Red
B_{8a}	20	865	Visible and near infrared (VNIR)
B_{11}	20	1610	Short wave infrared (SWIR)
B_{12}	20	2190	Short wave infrared (SWIR)

Table 2 Spectral channels of J_2

Band	Resolution m	Central wavelenth m	Description
B_1	60	443	Ultra blue (Coastal and Aerosol)
B_5	20	705	Visible and near infrared (VNIR)
B_6	20	740	Visible and near infrared (VNIR)
B_7	20	783	Visible and near infrared (VNIR)
B_8	10	842	Visible and near infrared (VNIR)
B_9	60	940	Short wave infrared (SWIR)
B_{10}	60	1375	Short wave infrared (SWIR)

$n = 6$, be a MODIS image of the same territory and this image has been captured at time $t = t_M$. It is assumed that (Tables 1 and 2):

(i) The Sentinel-2 images $S_k : G_H \to \mathbb{R}^m, k = 2, \dots, N$ can be corrupted by some noise, clouds and blur. However, the first one $S_1 : G_H \to \mathbb{R}^m$ is a cloud-free image;
(ii) For further convenience we divide the set of all bands for Sentinel images onto two parts J_1 and J_2 with $\dim(J_1) = 6$ and $\dim(J_2) = 7$
(iii) Each spectral channel of the MODIS image $M = [M_1, M_2, \dots, M_6] : G_L \to \mathbb{R}^6$ has the similar spectral characteristics to the corresponding channel of J_1-group $\{B_2, B_3, B_4, B_{8a}, B_{11}, B_{12}\}$, respectively;
(iv) The principle point is that the MODIS image $M : G_L \to \mathbb{R}^6$ is visually sufficiently clear and does not corrupted by clouds or its damage zone can be neglected;
(v) The MODIS image $M : G_L \to \mathbb{R}^6$ and the images $\{S_1, S_2, \dots, S_N : G_H \to \mathbb{R}^m\}$ from Sentinel-2 are rigidly co-registered. This means that the MODIS image after arguably some affine transformation and each Sentinel images after the resampling to the grid with low resolution G_L, could be successfully matched according to the unique geographic location.
In practice, the co-registration procedure can be realized using, for instance, the open-source LSReg v2.0.2 software [38, 41] that has been used in a number

of recent studies [19, 39], or the rigid co-registration approach that has been recently developed in [21, 22]. However, in both cases, in order to find an appropriate affine transformation, we propose to apply this procedure not to the original images, but rather to the contour's map of their spectral energies $Y_M : G_L \to \mathbb{R}$ and $Y_{S_j} : G_H \to \mathbb{R}$, where the last ones should be previously resampled to the grid of the low resolution G_L. Here,

$$\begin{aligned} Y_M(z) &:= \alpha_1 M_1(z) + \alpha_2 M_2(z) + \alpha_3 M_3(z), \quad \forall z = (x, y) \in G_L, \\ Y_{S_i}(z) &:= \alpha_1 S_{i,1}(z) + \alpha_2 S_{i,2}(z) + \alpha_3 S_{i,3}(z), \quad \forall z = (x, y) \in G_H \end{aligned}$$

with $\alpha_1 = 0.114$, $\alpha_2 = 0.587$, and $\alpha_3 = 0.299$.

2.1 Functional Spaces

Let us recall some useful notations. For vectors $\xi \in \mathbb{R}^2$ and $\eta \in \mathbb{R}^2$, $(\xi, \eta) = \xi^t \eta$ denotes the standard vector inner product in $\mathbb{R}^2$, where ^{t} stands for the transpose operator. The norm $|\xi|$ is the Euclidean norm given by $|\xi| = \sqrt{(\xi, \xi)}$. Let $\Omega \subset \mathbb{R}^2$ be a bounded open set with a Lipschitz boundary $\partial\Omega$. For any subset $E \subset \Omega$ we denote by $|E|$ its 2-dimensional Lebesgue measure $\mathcal{L}^2(E)$. Let $\overline{E}$ denote the closure of E, and ∂E stands for its boundary. We define the characteristic function χ_E of E by

$$\chi_E(x) := \begin{cases} 1, & \text{for } x \in E, \\ 0, & \text{otherwise.} \end{cases}$$

Let X denote a real Banach space with norm $\|\cdot\|_X$, and let X' be its dual. Let $\langle \cdot, \cdot \rangle_{X' X}$ be the duality form on $X' \times X$. By $\rightharpoonup$ and $\overset{*}{\rightharpoonup}$ we denote the weak and weak* convergence in normed spaces, respectively.

For given $1 \le p \le +\infty$, the space $L^p(\Omega; \mathbb{R}^2)$ is defined by

$$L^p(\Omega; \mathbb{R}^2) = \left\{ f : \Omega \to \mathbb{R}^2 \ : \ \|f\|_{L^p(\Omega;\mathbb{R}^2)} < +\infty \right\},$$

where $\|f\|_{L^p(\Omega;\mathbb{R}^2)} = \left(\int_\Omega |f(x)|^p \, dx \right)^{1/p}$ for $1 \le p < +\infty$. The inner product of two functions f and g in $L^p(\Omega; \mathbb{R}^2)$ with $p \in [1, \infty)$ is given by

$$(f, g)_{L^p(\Omega;\mathbb{R}^2)} = \int_\Omega (f(x), g(x)) \, dx = \int_\Omega \sum_{k=1}^{2} f_k(x) g_k(x) \, dx.$$

We denote by $C_c^\infty(\mathbb{R}^2)$ a locally convex space of all infinitely differentiable functions with compact support in $\mathbb{R}^2$. We recall here some functional spaces that will be used throughout this paper. We define the Banach space $H^1(\Omega)$ as the closure of $C_c^\infty(\mathbb{R}^2)$ with respect to the norm

$$\|y\|_{H^1(\Omega)} = \left(\int_\Omega \left(y^2 + |\nabla y|^2\right) dx\right)^{1/2}.$$

We denote by $\left(H^1(\Omega)\right)'$ the dual space of $H^1(\Omega)$.

Given a real Banach space X, we will denote by $C([0,T];X)$ the space of all continuous functions from $[0,T]$ into X. We recall that a function $u : [0,T] \to X$ is said to be Lebesgue measurable if there exists a sequence $\{u_k\}_{k\in\mathbb{N}}$ of step functions (i.e., $u_k = \sum_{j=1}^{n_k} a_j^k \chi_{A_j^k}$ for a finite number n_k of Borel subsets $A_j^k \subset [0,T]$ and with $a_j^k \in X$) converging to u almost everywhere with respect to the Lebesgue measure in $[0,T]$.

Then for $1 \le p < \infty$, $L^p(0,T;X)$ is the space of all measurable functions $u : [0,T] \to X$ such that

$$\|u\|_{L^p(0,T;X)} = \left(\int_0^T \|u(t)\|_X^p\, dt\right)^{\frac{1}{p}} < \infty,$$

while $L^\infty(0,T;X)$ is the space of measurable functions such that

$$\|u\|_{L^\infty(0,T;X)} = \sup_{t\in[0,T]} \|u(t)\|_X < \infty.$$

The full presentation of this topic can be found in [17].

For our purpose X will mainly be either the Lebesgue space $L^p(\Omega)$ or $L^p(\Omega;\mathbb{R}^2)$ or the Sobolev space $W^{1,p}(\Omega)$ with $1 \le p < \infty$. Since, in this case, X is separable, we have that $L^p(0,T;L^p(\Omega)) = L^p(Q_T)$ is the ordinary Lebesgue space defined in $Q_T = (0,T) \times \Omega$. As for the space $L^p(0,T;W^{1,\alpha}(\Omega))$ with $1 \le \alpha, p < +\infty$, it consists of all functions $u : [0,T] \times \Omega \to \mathbb{R}$ such that u and $|\nabla u|$ belongs to $L^p(0,T;L^\alpha(\Omega))$.

2.2 Topographic Maps and Geometry of Satellite Multispectral Images

Following the main principle of the Mathematical Morphology, a scalar image $u : \Omega \to \mathbb{R}$ is a representative of an equivalence class of images v obtained from u via a contrast change, i.e., $v = F(u)$, where F is a continuous strictly increasing function. Under this assumption, a scalar image can be characterized by its upper (or lower) level sets $Z_\lambda(u) = \{x \in \Omega : u(x) \ge \lambda\}$ (resp. $Z'_\lambda(u) = \{x \in \Omega : u(x) \le \lambda\}$). Moreover, each image can be recovered from its level sets by the reconstruction formula $u(x) = \sup\{\lambda : x \in Z_\lambda(u)\}$. Thus, according to the Mathematical Morphology Doctrine, the reliable information in the image contains in the level sets, independently of their actual levels (see [7] for the details). So, we can suppose that the entire geometric information about a scalar image is contained in those level sets.

In order to describe the level sets by their boundaries, $\partial Z_\lambda(u)$, we assume that $u \in W^{1,1}(\Omega)$, where $W^{1,1}(\Omega)$ stands for the standard Sobolev space of all functions $u \in L^1(\Omega)$ with respect to the norm

$$\|u\|_{W^{1,1}(\Omega)} = \|u\|_{L^1(\Omega)} + \|\nabla u\|_{L^1(\Omega)^2},$$

where the distributional gradient $\nabla u = \left(\frac{\partial u}{\partial x_1}, \frac{\partial u}{\partial x_2}\right)$ is represented as follows

$$\int_\Omega u \frac{\partial \phi}{\partial x_i}\, dx = -\int_\Omega \phi \frac{\partial u}{\partial x_i}\, dx, \quad \forall \phi \in C_0^\infty(\Omega),\ i = 1, 2.$$

It was proven in [1] that if $u \in W^{1,1}(\Omega)$ then its upper level sets $Z_\lambda(u)$ are sets of finite perimeter. So, the boundaries $\partial Z_\lambda(u)$ of level sets can be described by a countable family of Jordan curves with finite length, i.e., by continuous maps from the circle into the plane $\mathbb{R}^2$ without crossing points. As a result, at almost all points of almost all level sets of $u \in W^{1,1}(\Omega)$ we may define a unit normal vector $\theta(x)$. This vector field formally satisfies the following relations

$$(\theta, \nabla u) = |\nabla u| \quad \text{and} \quad |\theta| \leq 1 \text{ a.e. in} \Omega.$$

In the sequel, we will refer to the vector field θ as the vector field of unit normals to the topographic map of a function u. So, we can associate θ with the geometry of the scalar image u.

In the case of multi-band satellite images $I : \Omega \to \mathbb{R}^m$, we will impose further the following assumption: $I \in W^{1,1}(\Omega; \mathbb{R}^m)$ and each spectral channel of a given image I has the same geometry. We refer to [8] for the experimental discussion.

Remark 1 In practice, at the discrete level, the vector field $\theta(x, y)$ can be defined by the rule $\theta(x_i, y_j) = \frac{\nabla u(x_i, y_j)}{|\nabla u(x_i, y_j)|}$ when $\nabla u(x_i, y_j) \neq 0$, and $\theta = 0$ when $\nabla u(x_i, y_j) = 0$. However, as was mentioned in [2], a better choice for $\theta(x, y)$ would be to compute it as the ration $\frac{\nabla U(t,\cdot)}{|\nabla U(t,\cdot)|}$ for some small value of $t > 0$, where $U(t, x, y)$ is a solution of the following initial-boundary value problem with $1D$-Laplace operator in the principle part

$$\frac{\partial U}{\partial t} = \operatorname{div}\left(\frac{\nabla U}{|\nabla U|}\right), \quad t \in (0, +\infty),\ (x, y) \in \Omega, \tag{1}$$

$$U(0, x, y) = u(x, y), \quad (x, y) \in \Omega, \tag{2}$$

$$\frac{\partial U(0, x, y)}{\partial \nu} = 0, \quad t \in (0, +\infty),\ (x, y) \in \partial\Omega. \tag{3}$$

As a result, for any $t > 0$, there can be found a vector field

$$\xi \in L^\infty(\Omega; \mathbb{R}^2) \quad \text{with} \quad \|\xi(t)\|_{L^\infty(\Omega;\mathbb{R}^2)} \leq 1$$

such that

$$(\xi(t), \nabla U(t, \cdot)) = |\nabla U(t, \cdot)| \text{ in } \Omega, \quad \xi(t) \cdot \nu = 0 \text{ on } \partial\Omega, \tag{4}$$

and $U_t(t, x, y) = \operatorname{div} \xi(t, x, y)$ in the sense of distributions on Ω for a.a. $t > 0$.

We notice that in the framework of this procedure, for small value of $t > 0$, we do not distort the geometry of the function $u(x, y)$ in an essential way. Moreover, it can be shown that this regularization of the vector field $\theta(x, y) = \frac{\nabla U(x,y)}{|\nabla U(x,y)|}$ satisfies condition $\operatorname{div} \theta \in L^2(\Omega)$.

2.3 *Texture Index of a Gray-Scale Image*

Let $u \in C([0, T]; L^2(\Omega))$ be a given function. For each $t \in [0, T]$, we associate the real-valued mapping $u(t, \cdot) : \Omega \mapsto \mathbb{R}$ with a gray-scale image, and the mapping $u : (0, T) \times \Omega \to \mathbb{R}$ with an optical flow. A widely-used way to smooth $u(t, \cdot)$ is by calculating the convolution

$$u_\sigma(t, x) := (G_\sigma * \widetilde{u}(t, \cdot))\,(x) = \int_{\mathbb{R}^2} G_\sigma(x - y)\widetilde{u}(t, y)\, dy,$$

where $\widetilde{u}$ denotes zero extension of u from $Q_T = (0, T) \times \Omega$ to $\mathbb{R}^3$, and G_σ stands for the two-dimentional Gaussian of width (standard deviation) $\sigma > 0$:

$$G_\sigma(x) = \frac{1}{2\pi\sigma^2} \exp\left(-\frac{|x|^2}{2\sigma^2}\right).$$

Definition 1 We say that a function $p_u : (0, T) \times \Omega \to \mathbb{R}$ is the texture index of a given optical flow $u : (0, T) \times \Omega \to \mathbb{R}$ if it is defined by the rule

$$p_u(t, x) := 1 + g\left(\frac{1}{h}\int_{t-h}^{t} |(\nabla G_\sigma * \widetilde{u}(\tau, \cdot))\,(x)|^2\, d\tau\right), \ \forall\,(t, x) \in Q_T, \tag{5}$$

where $g{:}[0, \infty) \to (0, \infty)$ is the edge-stopping function which we take in the form of the Cauchy law $g(s) = \frac{a}{a+s}$ with $a > 0$ small enough, and $h > 0$ is a small positive value.

Since $G_\sigma \in C^\infty(\mathbb{R}^2)$, it follows from (5) and absolute continuity of the Lebesgue integral that $1 < p_u(t, x) \leq 2$ in Q_T and $p_u \in C^1([0, T]; C^\infty(\mathbb{R}^2))$ even if u is just an absolutely integrable function in Q_T. Moreover, for each $t \in [0, T]$, $p_u(t, x) \approx 1$

in those places of Ω where some edges or discontinuities are present in the image $u(t,\cdot)$, and $p_u(t,x) \approx 2$ in places where $u(t,\cdot)$ is smooth or contains homogeneous features. In view of this, $p_u(t,x)$ can be interpreted as a characteristic of the sparse texture of the function u that can change with time. The following result plays a crucial role in the sequel (for the proof we refer to [31, Lemma 1]).

Lemma 1 *Let $u \in C([0,T]; L^2(\Omega))$ be a measurable function extended by zero outside of Q_T. Let*

$$p_u = 1 + g\left(\frac{1}{h}\int_{t-h}^{t} |(\nabla G_\sigma * \widetilde{u}(\tau,\cdot))|^2\, d\tau\right)$$

be the corresponding texture index. Then there exists a constant $C > 0$ depending on Ω, G, and $\|u\|_{C([0,T];L^2(\Omega))}$ such that

$$\alpha := 1 + \delta \le p_u(t,x) \le \beta := 2, \quad \forall\,(t,x) \in Q_T, \tag{6}$$

$$\begin{gathered} p_u \in C^{0,1}(Q_T),\ |p_u(t,x) - p_u(s,y)| \le C\left(|x-y| + |t-s|\right), \\ \forall\,(t,x),(s,y) \in \overline{Q_T}, \end{gathered} \tag{7}$$

where

$$\delta = ah\left[ah + \|G_\sigma\|^2_{C^1(\overline{\Omega-\Omega})}|\Omega|\|u\|^2_{L^2(0,T;L^2(\Omega))}\right]^{-1}, \tag{8}$$

$$\begin{aligned} \|G_\sigma\|_{C^1(\overline{\Omega-\Omega})} &:= \max_{\substack{z=x-y\\ x\in\overline{\Omega},\, y\in\overline{\Omega}}} \left[|G_\sigma(z)| + |\nabla G_\sigma(z)|\right] \\ &= \frac{e^{-1}}{\left(\sqrt{2\pi}\sigma\right)^2}\left[1 + \frac{1}{\sigma^2}\operatorname{diam}\Omega\right]. \end{aligned} \tag{9}$$

3 Data Fusion Problem. Main Requirements to the Formal Statement

Let $\{S_1, S_2, \ldots, S_N : G_H \to \mathbb{R}^m\}$, with $m = 13$, be a collection of multispectral images of some territory from Sentinel-2 that were taken at time instances $\{t_1, t_2, \ldots, t_N\} \subset [0,T]$, respectively. We admit that these images can be corrupted because of poor weather conditions, such as rain, clouds, fog, and dust conditions. Typically, the measure of degradation of optical satellite images can be such that we cannot even rely on some reliability of pixel's values inside the damaged regions for each of spectral channels. As a result, some subdomains of such images become absolutely invisible. Let $\{D_1, D_2, \ldots, D_N\} \subset 2^\Omega$ be a collection of damage regions for the corresponding Sentinel-images. So, in fact, we deal with the set of images

$$\left\{S_i : G_H \backslash D_i \to \mathbb{R}^m,\ i = 1,\ldots,N\right\}.$$

Let $M : G_L \to \mathbb{R}^n$, with $n = 6$, be a MODIS image of the same territory and this image has been captured at time $t = t_M \in (t_1, T)$.

Before proceeding further, we begin with the following assumption:

(a) $D_1 = \emptyset$ and the damage zones for the rest images from Sentinel-2 are such that each D_i, $i = 2, \dots, N$, is a measurable closed subset of Ω with property $\mathcal{L}^2(D_i) \le 0.6\, \mathcal{L}^2(\Omega)$, where $\mathcal{L}^2(D_i)$ stands for the 2-D Lebesgue measure of D_i;
(b) The MODIS image $M : G_L \to \mathbb{R}^n$ is assumed to be cloud-free;
(c) The images $M : G_L \to \mathbb{R}^6$ and $\{S_1, S_2, \dots, S_N : G_H \to \mathbb{R}^m\}$ are rigidly co-registered. This means that the MODIS image after arguably some affine transformation and each Sentinel images after the resampling to the grid with low resolution G_L, could be successfully matched according to the unique geographic location;
(d) There exists an impulse response $\mathcal{K}$ such that, for any multi-band image with high resolution $I : G_H \to \mathbb{R}^m$, its resampling to the grid with low resolution G_L can be expressed as follows

$$I(x_i, y_j) = [\mathcal{K} * I]\,(x_i, y_j), \quad \forall\, i = 1, \dots, M_x, \ \forall\, j = 1, \dots, M_y,$$

where $\mathcal{K} * I$ stands for the convolution operator. For instance, setting $\mathcal{K} = [k_{p,q}]_{p,q=1,\dots,K}$, we have

$$[\mathcal{K} * I]\,(x_i, y_j) = \sum_{p=1}^{K}\sum_{q=1}^{K} k_{p,q} I(x_{i-p+1}, y_{j-q+1})$$

provided $I(x, y) = 0$ if $(x, y) \notin \Omega$. In majority of cases, it is enough to set up $k_{p,q} = \frac{1}{K^2}, \forall\, p, q = 1, \dots, K$, with an appropriate choice of $K \in \mathbb{N}$. However, if we deal with satellite images containing some agricultural areas with medium sides fields of various shapes, then the more efficient way for the choice of kernel $\mathcal{K} = [k_{p,q}]_{p,q=1,\dots,K}$ is to define it using the weight coefficients of the Lanczos interpolation filters.

For our further analysis, we make use of the following notion. We say that the multi-band images $\left\{\widehat{S}_i : G_H \to \mathbb{R}^m\right\}_{i=1}^N$ are structural prototypes of the corresponding cloud-corrupted ones $\{S_i : G_H \setminus D_i \to \mathbb{R}^m\}_{i=1}^N$ if they are defined as follows:

$$\begin{aligned} &\widehat{S}_{1,k}(z) = S_{1,k}(z), \\ &\widehat{S}_{i,k}(z) = \left\{ \begin{array}{ll} S_{i,k}(z), & z \in G_H \setminus D_i, \\ \gamma_{i,k} \widehat{S}_{i-1,k}(z), & z \in G_H \cap D_i, \end{array} \right\}, \ i = 2, \dots, N, \ k = 1, \dots, m, \end{aligned} \tag{10}$$

where

$$\gamma_{i,k} = \frac{\left[\chi_{\Omega \setminus D_i} S_{i,k}\right] \cdot \left[\chi_{\Omega \setminus D_i} \widehat{S}_{i-1,k}\right]}{\left\| \chi_{\Omega \setminus D_i} \widehat{S}_{i-1,k} \right\|^2_{\mathcal{L}(\mathbb{R}^2, \mathbb{R}^2)}},$$

$A \cdot B$ stands for the scalar product of two matrices A and B, and $\|\cdot\|_{\mathcal{L}(\mathbb{R}^2,\mathbb{R}^2)}$ denotes the Euclidean norm of a matrix. Moreover, each structural prototype $\widehat{S}_i : G_H \to \mathbb{R}^m$ is rigidly related to the corresponding day t_i when the image $S_i : G_H \setminus D_i \to \mathbb{R}^m$ had been captured.

Remark 2 As follows from the rule (10), this iterative procedure should be applied to each spectral channel of all multi-band images from Sentinel-2. Since the revisit time for Sentinel-2 is 3–5 days and the collection of images $\{S_i : G_H \setminus D_i \to \mathbb{R}^m\}_{i=1}^N$ is rigidly co-registered, it follows from (10) that the structural prototypes $\left\{\widehat{S}_i\right\}_{i=1}^N$ are also well co-registered and they have the similar topographic maps with respect to their precise space location, albeit some false contours can appear along the boundaries of the damage zones D_i. In fact, in order to avoid the appearance of the false contours, the weight coefficients $\gamma_{i,k}$ have been introduced.

Since the MODIS image has been captured at a time instance $t_M \in (t_1, T)$, we can have three possible cases:

(A1) there exists an index $i^* \in \{1, 2, \dots, N\}$ such that $t_M = t_{i^*}$;
(A2) there exists an index $i^* \in \{1, 2, \dots, N-1\}$ such that $t_{i^*} < t_M < t_{i^*+1}$;
(A3) $t_N < t_M < T$.

In view of this, we will distinguish three different statements of the data fusion problem:

Case (A1) (Restoration Problem) The problem (A1) consists in restoration of the damaged multi-band optical image $S_{i^*} : G_H \setminus D_{i^*} \to \mathbb{R}^m$ using result of its fusion with the cloud-free MODIS image $M : G_L \to \mathbb{R}^6$ of the same territory. It means that, we have to create a new image $S_{i^*}^{rest} : G_H \to \mathbb{R}^m$, which would be well defined on the entire grid G_H, such that

$$S_{i^*}^{rest}(z) = S_{i^*}(z), \quad \forall z = (x, y) \in G_H \setminus D_{i^*}, \tag{11}$$

$$\sum_{z \in G_L \cap D_{i^*}} \left(\left(\mathcal{K} * S_{i,k}^{rest}\right)(z) - M_k(z) \right)^2 = \inf_{I \in \mathcal{I}} \sum_{z \in G_L \cap D_{i^*}} \left(\left(\mathcal{K} * I\right)(z) - M_k(z) \right)^2, \quad \forall k \in J_1, \tag{12}$$

$$S_{i^*,k}^{rest}(z) = \widehat{S}_{i^*,k}(z), \quad \forall z \in G_H, \ \forall k \in J_2. \tag{13}$$

The precise description of the class of admissible (or feasible) images $\mathcal{I}$ will be given in the next section.

Case (A2) (Interpolation Problem) The problem (A2) consists in generation of a new multi-band optical image $S_{t_M}^{int} : G_H \to \mathbb{R}^m$ at the Sentinel-level of resolution using result of the fusion of cloud-free MODIS image $M : G_L \to \mathbb{R}^6$ with the predicted structural prototype $\widehat{S}_{t_M} : G_H \to \mathbb{R}^m$ from the given day t_M. In fact, in this case we deal with the two-level problem. At the first level, having the collection of structural prototypes $\left\{\widehat{S}_i : G_H \to \mathbb{R}^m\right\}_{i=1}^N$ which is associated with the time instances $\{t_1, t_2, \ldots, t_N\} \subset [0, T]$, we create a new 'intermediate' image $\widehat{S}_{t_M} : G_H \to \mathbb{R}^m$ that can be considered as daily prediction of the topographical map of a given territory from the day t_M. Then, at the second level, we realize the fusion procedure of this predicted image with the cloud-free MODIS image $M : G_L \to \mathbb{R}^6$ of the same territory. As a result, we have to create a new image $S_{t_M}^{interp} : G_H \to \mathbb{R}^m$ with properties:

$$\sum_{z \in G_L} \left(\left(\mathcal{K} * S_{t_M,k}^{int} \right)(z) - M_k(z) \right)^2 = \inf_{I \in \mathcal{I}} \sum_{z \in G_L} \left(\left(\mathcal{K} * I \right)(z) - M_k(z) \right)^2, \quad \forall k \in J_1, \tag{14}$$

$$S_{t_M,k}^{int}(z) = \widehat{S}_{t_M,k}(z), \quad \forall z \in G_H, \ \forall k \in J_2. \tag{15}$$

Case (A3) (Extrapolation Problem) The problem (A3) consists in generation of a new multi-band optical image $S_{t_M}^{ext} : G_H \to \mathbb{R}^m$ using result of the data assimilation from the cloud-free MODIS image $M : G_L \to \mathbb{R}^6$ into the structural prototype $\widehat{S}_N : G_H \to \mathbb{R}^m$ of the Sentinel-image $S_N : G_H \setminus D_N \to \mathbb{R}^m$. Here, it is assumed that the level sets of the given territory (or topographical map) for each Sentinel spectral channel from the day t_M have the same geo-location as they have in $\widehat{S}_N : G_H \to \mathbb{R}^m$. So, we can set $\widehat{S}_{t_M} = \widehat{S}_N$. Thus, in the framework of this problem, we have to retrieve a new image $S_{t_M}^{ext} : G_H \to \mathbb{R}^m$, which would be well defined on the entire grid G_H, such that

$$\sum_{z \in G_L} \left(\left(\mathcal{K} * S_{t_M,k}^{ext} \right)(z) - M_k(z) \right)^2 = \inf_{I \in \mathcal{I}} \sum_{z \in G_L} \left(\left(\mathcal{K} * I \right)(z) - M_k(z) \right)^2, \quad \forall k \in J_1, \tag{16}$$

$$S_{t_M,k}^{ext}(z) = \widehat{S}_{t_M,k}(z), \quad \forall z \in G_H, \ \forall k \in J_2. \tag{17}$$

To provide the detailed analysis of the above mentioned problems, we begin with some auxiliaries.

4 The Model for Prediction of Structural Prototypes

To begin with, we notice that, due to the iterative procedure (10), we can define the so-called structural prototypes $\left\{\widehat{S}_i : G_H \to \mathbb{R}^m\right\}_{i=1}^N$ for each cloud-corrupted Sentinel image $\{S_i : G_H \setminus D_i \to \mathbb{R}^m\}_{i=1}^N$. Let i^* be an integer such that $t_{i^*} < t_M < t_{i^*+1}$. Let $\widehat{S}_{i^*,j}$ and $\widehat{S}_{i^*+1,j}$ be structural prototypes of the corresponding images from given days t_{i^*} and t_{i^*+1}. Since $\widehat{S}_{i^*,j}$ and $\widehat{S}_{i^*+1,j}$ are well co-registered images, it is reasonable to assume that they have the similar geometric structure albeit they may have rather different intensities.

The main question we are going to discuss in this section is: how to correctly define the 'intermediate' image $\widehat{S}_{t_M} : G_H \to \mathbb{R}^m$ that can be considered as daily prediction of the topographical map of a given territory from the day t_M. With that in mind, for each spectral channel $j \in \{1, 2, \ldots, m\}$, we make use of the following model

$$\frac{\partial u}{\partial t} - \operatorname{div}\left(|\nabla u|^{p_u(t,x)-2}\nabla u\right) = v \quad \text{in } (t_{i^*}, t_{i^*+1}) \times \Omega, \tag{18}$$

$$\partial_\nu u = 0 \quad \text{on } (t_{i^*}, t_{i^*+1}) \times \partial\Omega, \tag{19}$$

$$u(t_{i^*}, \cdot) = \widehat{S}_{i^*,j}(\cdot) \quad \text{in } \Omega, \tag{20}$$

where $p_u(t, x)$ stands for the texture index of the scalar image u (see Definition 1), and $v \in L^2(t_{i^*}, t_{i^*+1}; L^2(\Omega))$ is an unknown source term that has to be defined in the way to guarantee the fulfillment (with some accuracy) of the relation

$$u(t_{i^*+1}, \cdot) \approx \widehat{S}_{i^*+1,j}(\cdot) \quad \text{in } \Omega. \tag{21}$$

Here, we assume that the images $\widehat{S}_{i^*,j}$ and $\widehat{S}_{i^*+1,j}$ in (20) and (21) are well defined into the entire domain Ω.

Remark 3 The main characteristic feature of the proposed initial-boundary value problem (IBVP) is the fact that the exponent p_u depend not only on (t, x) but also on a solution $u(t, x)$ of this problem. It is well-known that the variable character of the exponent p_u causes a gap between the monotonicity and coercivity conditions. Because of this gap, equations of the type (18) can be termed as equations with nonstandard growth conditions. So, in fact, we deal with the Cauchy-Neumann IBVP for a parabolic equation of $p_u = p(t, x, u)$-Laplacian type with variable exponent of nonlinearity. It was recently shown that the model (18)–(20) naturally appears as the Euler-Lagrange equation in the problem of restoration of cloud contaminated satellite optical images [15, 28]. Moreover, the above mentioned problem can be considered as a model for the deblurring and denoising of multi-spectral images. In particular, this model has been proposed in [16, 32] in order to avoid the blurring of edges and other localization problems presented by linear diffusion models in images processing. We also refer to [30], where the authors study some optimal control problems associated with a special case of the model (18)–(20) and show that the given class of optimal control problems is well posed.

Before proceeding further, we note that the distributed control v in the right hand side of (18) describes the fictitious sources or sinks of the intensity u that may have a tendency to change at most pixels even for co-registered structural prototypes $\widehat{S}_{i^*,j}(\cdot)$ and $S_{i^*+1,j}(\cdot)$. As for the Neumann boundary condition $\partial_\nu u = 0$ on $\partial\Omega$, this condition corresponds to the reflection of the image across the boundary and has the advantage of not imposing any value on the boundary and not creating 'edges' on it. So, it is very natural conditions if we assume that the boundary of the image is an arbitrary cutoff of a larger scene in view.

In order to characterize the solvability issues of the IBVP (18)–(20), we adopt the following concept.

Definition 2 We say that, for given $v \in L^2(\Omega)$ and $\widehat{S}_{i^*,j} \in L^2(\Omega)$, a function u is a weak solution to the problem (18)–(20) if

$$\begin{gathered} u \in L^2(t_{i^*}, t_{i^*+1}; L^2(\Omega)), \;\; u(t,\cdot) \in W^{1,1}(\Omega) \text{ for a.a.} t \in [t_{i^*}, t_{i^*+1}], \\ \int_{t_{i^*}}^{t_{i^*+1}} \int_\Omega |\nabla u|^{p_u(t,x)}\, dxdt < +\infty, \end{gathered} \tag{22}$$

and the integral identity

$$\begin{aligned} \int_{t_{i^*}}^{t_{i^*+1}} \int_\Omega \left(-u\frac{\partial \varphi}{\partial t} + \left(|\nabla u|^{p_u} \nabla u, \nabla\varphi\right)\right) dxdt \\ = \int_{t_{i^*}}^{t_{i^*+1}} \int_\Omega v\varphi\, dxdt + \int_\Omega \widehat{S}_{i^*,j}\varphi|_{t=t_{i^*}}\, dx \end{aligned} \tag{23}$$

holds true for any function $\varphi \in \Phi$, where

$$\Phi = \left\{\varphi \in C^\infty([t_{i^*}, t_{i^*+1}] \times \overline{\Omega}) \;:\; \varphi|_{t=t_{i^*+1}} = 0\right\}.$$

The following result highlights the way in what sense the weak solution takes the initial value $u(t_{i^*}, \cdot) = \widehat{S}_{i^*,j}(\cdot)$.

Proposition 1 ([31]) *Let $v \in H^1(\Omega)$ and $\widehat{S}_{i^*,j} \in L^2(\Omega)$ be given distributions. Let u be a weak solution to the problem* (18)–(20) *in the sense of Definition 2. Then, for any $\eta \in C^\infty(\overline{\Omega})$, the scalar function $h(t) = \displaystyle\int_\Omega u(t,x)\eta(x)\, dx$ belongs to $W^{1,1}(t_{i^*}, t_{i^*+1})$ and $h(0) = \displaystyle\int_\Omega \widehat{S}_{i^*,j}(x)\eta(x)\, dx$.*

Utilizing the perturbation technique and the classical fixed point theorem of Schauder [36], it has been recently proven the following existence result.

Theorem 1 ([31]) *Let $v \in L^2(\Omega)$ and $\widehat{S}_{i^*,j} \in L^2(\Omega)$ be given distributions. Then the initial-boundary value problem* (18)–(20) *admits at least one weak solution $u = u(t,x)$ with the following higher inegrability properties*

$$\begin{gathered} u \in L^\infty(t_{i^*}, t_{i^*+1}; L^2(\Omega)), \;\; u \in W^{1,\alpha}((t_{i^*}, t_{i^*+1}) \times \Omega), \\ u \in L^{2\alpha}\left(t_{i^*}, t_{i^*+1}; L^{2\alpha}(\Omega)\right), \end{gathered} \tag{24}$$

where the exponent α is given by the rule

$$\alpha = ah\left[ah + \|G_\sigma\|^2_{C^1(\overline{\Omega - \Omega})}|\Omega|\left(\|v\|^2_{L^2(Q_T)} + 2\|\widehat{S}_{i^*,j}\|^2_{L^2(\Omega)}\right)\right]^{-1}$$

In order to satisfy the condition (21) and define an appropriate source term $v = v(t,x)$, we utilize some issues coming from the well-known method of Horn and Schunck [23] that has been developed in order to compute optical flow velocity from spatiotemporal derivatives of image intensity. Following this approach, we define the function v^* as a solution of the problem

$$\int_\Omega \left(\frac{\partial Y}{\partial t}\Big|_{t=(t_{i^*}+t_{i^*+1})/2} - \operatorname{div}\left(|\nabla Y|^{p_Y} \nabla Y\right)\Big|_{t=(t_{i^*}+t_{i^*+1})/2} - v \right)^2 dx + \lambda_1^2 \int_\Omega |\nabla v|^2 \, dx \to \inf_{v \in H^1(\Omega)}, \quad (25)$$

where $\lambda_1 > 0$ is a tuning parameter (for numerical simulations we take $\lambda_1 = 0.5$), and the spatiotemporal derivatives are computed by the rules

$$\frac{\partial Y}{\partial t}\Big|_{t=(t_{i^*}+t_{i^*+1})/2} = \frac{1}{t_{i^*+1} - t_{i^*}} \left(\widehat{S}_{i^*+1,j} - \widehat{S}_{i^*,j}\right),$$

$$\operatorname{div}\left(|\nabla Y|^{p_Y} \nabla Y\right)\Big|_{t=(t_{i^*}+t_{i^*+1})/2} = \frac{1}{2}\Big[\operatorname{div}\left(|\nabla \widehat{S}_{i^*,j}|^{p_{\widehat{S}_{i^*,j}}} \nabla \widehat{S}_{i^*,j}\right) + \operatorname{div}\left(|\nabla \widehat{S}_{i^*+1,j}|^{p_{\widehat{S}_{i^*+1,j}}} \nabla \widehat{S}_{i^*+1,j}\right) \Big].$$

It is clear that a minimum point $v^* \in H^1(\Omega)$ to unconstrained minimization problem (25) is unique and satisfies necessarily the Euler-Lagrange equation

$$\lambda_1^2 \Delta v^* + \left(\frac{\partial Y}{\partial t}\Big|_{t=(t_{i^*}+t_{i^*+1})/2} - \operatorname{div}\left(|\nabla Y|^{p_Y} \nabla Y\right)\Big|_{t=(t_{i^*}+t_{i^*+1})/2} - v^* \right) = 0 \quad (26)$$

with the Nuemann boundary condition $\partial_\nu v^* = 0$ on $\partial\Omega$.

Setting $v = v^*$ in (18), we can define a function $u^* = u^*(t, x)$ as the weak solution of the IBVP (18)–(20). Numerical experiments show that, following this way, we obtain a function u^* with properties (22) and (24) such that

$$u^*(t_{i^*}, x) = \widehat{S}_{i^*,j}(x) \quad \text{and} \quad u^*(t_{i^*+1}, x) \approx \widehat{S}_{i^*+1,j}(x) \quad \text{in } \Omega,$$

where the peak signal-to-noise ratio (PSNR) between images $u^*(t_{i^*+1}, x)$ and $\widehat{S}_{i^*+1,j}(x)$ is sufficiently large, $PSNR > 46$.

This observation leads us to the following conclusion: the 'intermediate' image $\widehat{S}_{t_M} : G_H \to \mathbb{R}^m$ can be defined as follow:

$$\widehat{S}_{t_M}(x) = u^*(t_M, x), \quad \forall x \in G_H. \quad (27)$$

To illustrate how the proposed model (18)–(20) works, we refer to the full version of our paper (see [25]) where it was considered as an input data two images of some region that represent a typical agricultural area in Australia the resolution $20m/pixel$. These images have been delivered from Sentinel-2 and captured at the time instances $t_1 =' July, 08'$ and $t_2 =' August, 25'$, respectively.

5 Variational Statements of the Data Fusion Problems (A1)–(A3)

Coming back to the principle cases (A1)–(A3), that have been described in Sect. 3, we can suppose that a structural prototype $\widehat{S}_{t_M} : G_H \to \mathbb{R}^m$ from the given day t_M is well defined. As it was emphasized in Sect. 3, this prototype coincides either with one of the images $\left\{\widehat{S}_i : G_H \to \mathbb{R}^m\right\}_{i=1}^N$ in cases (A1) and (A3), or it is defined using the solutions of the problem (18)–(20), (26) for each $j = 1, \dots, m$ in the case (A2)-problem (see the rule (27)). For further convenience, we assume that $\widehat{S}_{t_M} : G_H \to \mathbb{R}^m$ is zero-extended outside of Ω.

Let $j \in \{1, 2, \dots, m\}$ be a fixed index value (the number of spectral channel). Let $q_j : \Omega \to \mathbb{R}$ be the texture index of the j-th band for the structural prototype $\widehat{S}_{t_M} : G_H \to \mathbb{R}^m$, i.e.

$$q_j(x) := 1 + g\left(\left|\left(\nabla G_\sigma * \widehat{S}_{t_M,j}\right)(x)\right|^2\right), \ \forall x \in \Omega, \tag{28}$$

where g:$[0, \infty) \to (0, \infty)$ is the edge-stopping function which we take in the form of the Cauchy law $g(s) = \frac{a}{a+s}$ with $a > 0$ small enough. Let $\eta \in (0, 1)$ be a given threshold. Let $\theta_j = \begin{bmatrix} \theta_{j,1} \\ \theta_{j,2} \end{bmatrix} \in L^\infty(\Omega, \mathbb{R}^2)$ be a vector field such that

$$|\theta_j(x)|_{\mathbb{R}^2} \leq 1 \ \text{ and } \ \left(\theta_j(x), \nabla \widehat{S}_{t_M,j}(x)\right)_{\mathbb{R}^2} = |\nabla \widehat{S}_{t_M,j}(x)|_{\mathbb{R}^2} \quad \text{a.e. in}\Omega.$$

As it was mentioned in Sect. 2.2, for each spectral channel $j \in \{1, \dots, m\}$ this vector-field can be defined by the rule $\theta_j(x) = \frac{\nabla U_j(t,x)}{|\nabla U_j(t,x)|}$ with $t > 0$ small enough, where $U_j(t, x)$ is a solution the following initial-boundary value problem

$$\frac{\partial U}{\partial t} = \operatorname{div}\left(\frac{\nabla U}{|\nabla U| + \varepsilon}\right), \quad t \in (0, +\infty), \ x \in \Omega, \tag{29}$$

$$U(0, x) = \widehat{S}_{j,t_M}(x), \quad x \in \Omega, \tag{30}$$

$$\frac{\partial U(0, x)}{\partial \nu} = 0, \quad t \in (0, +\infty), \ x \in \partial\Omega \tag{31}$$

with a relaxed version of the $1D$-Laplace operator in the principle part of (29). Here, $\varepsilon > 0$ is a sufficiently small positive value.

Taking into account the definition of the Directional Total Variation (see [5]), we define a linear operator $R_{j,\eta} : \mathbb{R}^2 \to \mathbb{R}^2$ as follows

$$R_{j,\eta} \nabla v := \nabla v - \eta^2 \left(\theta_j, \nabla v\right)_{\mathbb{R}^2} \theta_j, \quad \forall v \in W^{1,1}(\Omega). \tag{32}$$

It is clear that $R_{j,\eta} \nabla v$ reduces to $(1 - \eta^2)\nabla v$ in those regions where the gradient ∇v is co-linear to θ_j, and to ∇v, where ∇v is orthogonal to θ_j. It is important to emphasize that this operator does not enforce gradients in the direction θ_j.

Let $\delta_{(x_i,y_j)}$ be the Dirac's delta at the point (x_i, y_j). Then $\Pi_{S_L} = \sum_{(x_i,y_j)\in S_L} \delta_{(x_i,y_j)}$ stands for the Dirac's comb defined onto the sample grid G_L.

We are now in a position to give a precise meaning of the solutions to Problems (A1)–(A3). We say that:

(A1) A multi-band image $S^{rest}_{i^*} : G_H \to \mathbb{R}^m$, where $t_{i^*} = t_M$, is a solution of the Restoration Problem, if it is given by the rule

$$S^{rest}_{i^*,j}(x) = \begin{cases} S_{i^*,j}(x) \ , \ \forall x \in G_H \setminus D_{i^*}, \\ \beta_j u^0_j(x) \ , \ \forall x \in G_H \cap D_{i^*}, \end{cases} \forall j \in J_1, \tag{33}$$

$$S^{rest}_{i^*,j}(x) = \widehat{S}_{t_M,j}(x), \quad \forall x \in G_H, \ \forall j \in J_2. \tag{34}$$

Here, β_j is the weight coefficient and we define it as follows

$$\beta_j = \left[\int_{\Omega\setminus D_{i^*}} S_{i^*,j}(x) u^0_j(x)\, dx\right] / \left[\int_{\Omega\setminus D_{i^*}} |u^0_j(x)|^2\, dx\right]$$

and u^0_j is a solutions of the following constrained minimization problem

$$(\mathcal{P}) \qquad \mathcal{F}_j\left(u^0_j\right) = \inf_{u\in\Xi_j} \mathcal{F}_j(u), \tag{35}$$

where

$$\begin{aligned} \mathcal{F}_j(u) := &\int_\Omega \frac{1}{q_j(x)} |R_{j,\eta} \nabla u(x)|^{q_j(x)}\, dx + \frac{\mu}{2} \int_\Omega \left|\nabla u(x) - \nabla \widehat{S}_{t_M,j}(x)\right|^2 dx \\ &+ \frac{\gamma}{2} \int_{\Omega\setminus D_{i^*}} \left|u(x) - S_{t_M,j}(x)\right|^2 dx + \frac{\vartheta}{2} \int_\Omega \Pi_{S_L}\left(\left[\mathcal{K} * u - M_{t_M,j}\right]\right)^2 dx, \end{aligned} \tag{36}$$

$\Xi_j = \left\{u \in W^{1,q_j(\cdot)}(\Omega) \ : \ 0 \le u(x) \le C_j \text{ a.e. in } \Omega\right\}$ stands for the set of feasible solutions, $\mu > 0, \gamma > 0, \vartheta > 0$ are some weight coefficients, and $W^{1,q_j(\cdot)}(\Omega)$ denotes the Sobolev space with variable exponent (for the details we refer to Appendix C). As for the constants C_j, their choice depends on the format of signed integer numbers in which the corresponding intensities $S_{i,j}(x)$ are represented. In particular, it can be $C_j = 2^8 - 1$, $C_j = 2^{16} - 1$, and so on.

(A2) A multi-band image $S^{int}_{t_M} : G_H \to \mathbb{R}^m$, with $t_{i^*} < t_M < t_{i^*+1}$, is a solution of the Interpolation Problem, if it is given by the rule

$$S^{int}_{t_M,j}(x) = \begin{cases} \beta_j u^0_j(x) \ , \ \forall j \in J_1, \\ \widehat{S}_{t_M,j}(x) \ , \ \forall j \in J_2, \end{cases} \forall x \in G_H, \tag{37}$$

where

$$\beta_j = \left[\int_\Omega \widehat{S}_{t_M,j}(x) u^0_j(x)\, dx\right] / \left[\int_\Omega |u^0_j(x)|^2\, dx\right] \tag{38}$$

and u_j^0 is a solutions of the constrained minimization problem (35)–(36) with $D_{i^*} = \Omega$.

(A3) A multi-band image $S_{t_M}^{ext} : G_H \to \mathbb{R}^m$, with $t_N < t_M < T$, is a solution of the Extrapolation Problem, if it is given by the rule

$$S_{t_M,j}^{ext}(x) = \begin{cases} \beta_j u_j^0(x) \ , \ \forall\, j \in J_1, \\ \widehat{S}_{t_M,j}(x) \ , \ \forall\, j \in J_2, \end{cases} \quad \forall\, x \in G_H, \tag{39}$$

where u_j^0 is a solutions of the constrained minimization problem (35)–(36) with $D_{i^*} = \Omega$, and β_j is defined as in (38).

Let us briefly discuss the relevance of the proposed minimization problem $(\mathcal{P})$. We begin with the motivation to the choice of the energy functional in the form (35)–(36).

The first term in (36) can be considered as the regularization in the Sobolev-Orlicz space $W^{1,q_j(\cdot)}(\Omega)$ because, for each spectral channel, we have

$$(1-\eta^2)|\nabla u| \le |R_{j,\eta}\nabla u| \le |\nabla u| \quad \text{in}\Omega \tag{40}$$

with a given threshold $\eta \in (0,1)$. Hence,

$$\int_\Omega \frac{1}{q_j(x)}|R_{j,\eta}\nabla u(x)|^{q_j(x)}\,dx \ge (1-\eta^2)^2\int_\Omega |\nabla u(x)|^{q_j(x)}\,dx, \quad \forall\, u \in W^{1,q_j(\cdot)}(\Omega) \tag{41}$$

and, therefore, if

$$u \in \Xi_j \subset W^{1,q_j(\cdot)}(\Omega) \cap L^\infty(\Omega) \quad \text{and} \quad \mathcal{F}_j(u) < +\infty, \tag{42}$$

then

$$\begin{aligned}
\|u\|^\alpha_{W^{1,q_j(\cdot)}} &= \left(\|u\|_{L^{q_j(\cdot)}(\Omega)} + \|\nabla u\|_{L^{q_j(\cdot)}(\Omega;\mathbb{R}^2)}\right)^\alpha \\
&\le C\left(\|u\|^\alpha_{L^{q_j(\cdot)}(\Omega)} + \|\nabla u\|^\alpha_{L^{q_j(\cdot)}(\Omega;\mathbb{R}^2)}\right) \\
&\overset{\text{by (13.58)}}{\le} C\left(\int_\Omega |u(x)|^{q_j(x)}\,dx + \int_\Omega |\nabla u(x)|^{q_j(x)}\,dx + 2\right) \\
&\overset{\text{by (13.42)}}{\le} C\left(|\Omega| C_j^2 + \int_\Omega |\nabla u(x)|^{q_j(x)}\,dx + 2\right) \\
&\overset{\text{by (13.41)}}{\le} C\left(|\Omega| C_j^2 + 2 + \frac{1}{(1-\eta^2)^2}\int_\Omega \frac{1}{q_j(x)}|R_{j,\eta}\nabla u(x)|^{q_j(x)}\,dx\right) \\
&\overset{\text{by (13.36)}}{\le} C\left(|\Omega| C_j^2 + 2 + \frac{1}{(1-\eta^2)^2}\mathcal{F}_j(u)\right) < +\infty.
\end{aligned} \tag{43}$$

On the other side, this term plays the role of a spatial data fidelity. Indeed, what we are going to achieve in this interpolation problem, it is to preserve the following property for the retrieved images at the Sentinel-2 resolution level: the geometry of each spectral channel of the retrieved image has to be as close as possible to the geometry of the predicted structural prototype $\widehat{S}_{t_M} : G_H \to \mathbb{R}^m$ that we obtain either as a solution of the problem (18)–(20), (26), or as a result of the iterative procedure (10). Formally, it means that relations

$$\left(\theta_j^{\perp}, \nabla u\right) = 0 \quad \text{a.e. in}\Omega, \ \forall j \in J_1, \tag{44}$$

have to be satisfied. Hence, the magnitude $\int_\Omega \left| \left(\theta_j^{\perp}, \nabla u\right) \right| dx$ must be small enough for each spectral channel, where θ_j stands for the vector field of unit normals to the topographic map of the predicted band $\widehat{S}_{t_M,j} : G_H \to \mathbb{R}$. In order to achieve this property, we observe that the expression $R_{j,\eta}\nabla u$ can be reduced to $(1-\eta^2)\nabla u$ in those places of Ω where ∇u is co-linear to the unit normal θ_j, and to ∇u if ∇u is orthogonal to θ_j.

Thus, gradients of the intensities u that are aligned/co-linear to θ_j are favored as long as $|\theta_j| > 0$. Moreover, this property is enforced by the special choice of the exponent $q_j(x)$. Since $q_j(x) \approx 1$ in places in Ω where edges or discontinuities are present in the predicted band $\widehat{S}_{t_M,j}$, and $q_j(x) \approx 2$ in places where $\widehat{S}_{t_M,j}$ is smooth or contains homogeneous features, the main benefit of the energy functional (36) is the manner in which it accommodates the local image information. For the places where the gradient of $\widehat{S}_{t_M,j}$ is sufficiently large (i.e. likely edges), we deal with the so-called directional TV-based diffusion [5, 6], whereas in the places where the gradient of $\widehat{S}_{t_M,j}$ is close to zero (i.e. homogeneous regions), the model becomes isotropic. Specifically, the type of anisotropy at these ambiguous regions varies according to the strength of the gradient. Apparently, the idea to involve the norm of $W^{1,q_j(\cdot)}(\Omega)$ with a variable exponent $q_j(x)$ was firstly proposed in [4] in order to reduce the staircasing effect in the TV image restoration problem.

As for the second term in (36), it reflects the fact that the topographic map of the retrieved image should be as close as possible to the topographic map of predicted structural prototype $\widehat{S}_{t_M} : G_H \to \mathbb{R}^m$. We interpret this closedness in its simplified form, namely, in the sense of L^2-norm of the difference of the corresponding gradients.

It remains to say a few words about the last term in (36). Basically, this term represents an L^2-distortion between a j-th spectral channel in the MODIS image $M = [M_1, M_2, \dots, M_6] : G_L \to \mathbb{R}^6$ and the corresponding channel of the retrieved image u_j^0 which is resampled to the grid of low resolution G_L.

6 Existence Result and Optimality Conditions for Constrained Minimization Problem $(\mathcal{P}_k)$

Our main intention in this section is to show that, for each $j \in J_1$, constrained minimization problem (35)–(36) is consistent and admits at least one solution. Because of the specific form of the energy functional $\mathcal{F}_j(u)$, the minimization problem (35)–(36) is rather challenging and we refer to [3–5, 9] for some specific details.

Following in many aspects the recent studies [12, 29] (see also [11, 13–15, 24, 33]), we can give the following existence result.

Theorem 2 *Let $\widehat{S}_{t_M} : G_H \to \mathbb{R}^m$ be a given structural prototype for unknown image S_{t_M} from Sentinel-2. Then for any given $j \in J_1$, $\mu > 0$, $\gamma > 0$, $\vartheta > 0$, and $\eta \in (0, 1)$, the minimization problem* (35)–(36) *admits a unique solution $u_j^0 \in \Xi_j$.*

In order to derive some optimality conditions to the problem (35)–(36) and characterize its solution $u_j^0 \in W^{1,q_j(\cdot)}(\Omega)$, we show that the cost functional $\mathcal{F}_j : \Xi_j \to \mathbb{R}$ is Gâteaux differentiable. To this end, we note that, for arbitrary $v \in W^{1,q_j(\cdot)}(\Omega)$, the following assertion

$$\frac{|R_{j,\eta}\nabla u_j^0(x) + tR_\eta \nabla v(x)|^{q_j(x)} - |R_{j,\eta}\nabla u_j^0(x)|^{q_j(x)}}{q_j(x)t}$$
$$\to \left(|R_{j,\eta}\nabla u_j^0(x)|^{q_j(x)-2} R_{j,\eta}\nabla u_j^0(x), R_{j,\eta}\nabla v(x)\right) \quad \text{as } t \to 0$$

holds almost everywhere in Ω. Indeed, by convexity,

$$|\xi|^{q_j(x)} - |\eta|^{q_j(x)} \le 2q_j(x)\left(|\xi|^{q_j(x)-1} + |\eta|^{q_j(x)-1}\right)|\xi - \eta|,$$

it follows that

$$\left|\frac{|R_{j,\eta}\nabla u_j^0(x) + tR_{j,\eta}\nabla v(x)|^{q_j(x)} - |R_{j,\eta}\nabla u_j^0(x)|^{q_j(x)}}{q_j(x)t}\right|$$
$$\le 2\left(|R_{j,\eta}\nabla u_j^0(x) + tR_{j,\eta}\nabla v(x)|^{q_j(x)-1} + |R_{j,\eta}\nabla u_j^0(x)|^{q_j(x)-1}\right)|R_{j,\eta}\nabla v(x)|$$
$$\le \text{const}\left(|R_{j,\eta}\nabla u_j^0(x)|^{q_j(x)-1} + |R_{j,\eta}\nabla v(x)|^{q_j(x)-1}\right)|R_{j,\eta}\nabla v(x)|. \quad (45)$$

Taking into account that

$$\int_\Omega |R_{j,\eta}\nabla u_j^0(x)|^{q_j(x)-1}|R_{j,\eta}\nabla v(x)|\,dx$$
$$\overset{\text{by (13.67)}}{\le} 2\||R_{j,\eta}\nabla u_j^0(x)|^{q_j(x)-1}\|_{L^{(q_j)'(\cdot)}(\Omega)}\||R_{j,\eta}\nabla v(x)|\|_{L^{q_j(\cdot)}(\Omega)}$$
$$\le 2\||\nabla u_j^0(x)|^{q_j(x)-1}\|_{L^{(q_j)'(\cdot)}(\Omega,\mathbb{R}^2)}\|\nabla v(x)\|_{L^{q_j(\cdot)}(\Omega,\mathbb{R}^2)},$$

and

$$\int_\Omega |R_{j,\eta}\nabla v(x)|^{q_j(x)}\,dx \overset{\text{by (13.40)}}{\leq} \int_\Omega |\nabla v(x)|^{q_j(x)}\,dx \leq \|\nabla v\|^2_{L^{q_j(\cdot)}(\Omega,\mathbb{R}^2)} + 1,$$

we see that the right hand side of inequality (45) is an $L^1(\Omega)$ function. Therefore,

$$\int_\Omega \frac{|R_{j,\eta}\nabla u_j^0(x) + tR_{j,\eta}\nabla v(x)|^{q_j(x)} - |R_{j,\eta}\nabla u_j^0(x)|^{q_j(x)}}{q_j(x)t}\,dx$$
$$\to \int_\Omega \left(|R_{j,\eta}\nabla u_j^0(x)|^{q_j(x)-2}R_{j,\eta}\nabla u_j^0(x),\, R_{j,\eta}\nabla v(x)\right)dx \text{ as } t \to 0$$

by the Lebesgue dominated convergence theorem.

Since the cost functional $\mathcal{F}_j : \Xi_j \to \mathbb{R}$ can be cast in the form

$$P_k(u) = \mathcal{A}_j(u) + \mu\mathcal{B}_j(u) + \gamma\mathcal{C}_j(u) + \vartheta\mathcal{D}_j(u),$$

where

$$\mathcal{A}_j(u) = \int_\Omega \frac{1}{q_j(x)}|R_{j,\eta}\nabla u(x)|^{q_j(x)}\,dx,\ \mathcal{B}_j(u) = \frac{1}{2}\int_\Omega \left|\nabla u(x) - \nabla\widehat{S}_{t_M,j}(x)\right|^2 dx,$$
$$\mathcal{C}_j(u) = \frac{1}{2}\int_{\Omega\setminus D_{i*}} \left|u(x) - S_{t_M,j}(x)\right|^2 dx,\ \mathcal{D}_j(u) = \frac{1}{2}\int_\Omega \Pi_{S_L}\left(\left[\mathcal{K} * u - M_{t_M,j}\right]\right)^2 dx,$$

we deduce that

$$\mathcal{A}_j'(u_j^0)[v] = \int_\Omega \left(|R_{j,\eta}\nabla u_j^0(x)|^{q_j(x)-2}R_{j,\eta}\nabla u_j^0(x),\, R_{j,\eta}\nabla v(x)\right)_{\mathbb{R}^2} dx, \tag{46}$$

for each $v \in W^{1,q_j(\cdot)}(\Omega)$.

As for the rest terms $\mu\mathcal{B}_j(u)$, $\gamma\mathcal{C}_j(u)$, and $\lambda\mathcal{D}_j(u)$ in the cost functional $\mathcal{F}_j : \Xi_j \to \mathbb{R}$, utilizing the similar arguments, we have the following representation for their Gâteaux derivatives.

Proposition 2 *For a given MODIS image $M : G_L \to \mathbb{R}^6$, the functionals $\mathcal{B}_j, \mathcal{C}_j, \mathcal{D}_j : L^2(\Omega) \to \mathbb{R}$ are convex and Gâteaux differentiable in $L^2(\Omega)$ with*

$$\mathcal{B}_k'(u_j^0)[v] = \int_\Omega \left(\nabla u_j^0(x) - \nabla\widehat{S}_{t_M,j}(x),\, \nabla v(x)\right) dx, \tag{47}$$

$$\mathcal{C}_k'(u_j^0)[v] = \int_{\Omega\setminus D_{i*}} \left(u_j^0(x) - S_{t_M,j}(x)\right) v(x)\,dx, \tag{48}$$

$$\begin{aligned}\mathcal{D}_k'(u_j^0)[v] &= \int_\Omega \Pi_L\left(\left[\mathcal{K} * u_j^0\right] - M_{t_M,j}\right)\left[\mathcal{K} * v\right] dx \\ &= \int_\Omega \Pi_L\left[\mathcal{K}^* * \left(\left[\mathcal{K} * u_j^0\right] - M_{t_M,j}\right)\right] v\,dx,\end{aligned} \tag{49}$$

for all $v \in W^{1,q_j(\cdot)}(\Omega)$.

Thus, in order to derive some optimality conditions for the minimizer $u_j^0 \in W^{1,q_j(\cdot)}(\Omega)$ to the problem $\inf_{u \in \Xi_j} \mathcal{F}_j(u)$, we note that Ξ_j is a nonempty convex subset of $W^{1,q_j(\cdot)}(\Omega) \cap L^\infty(\Omega)$ and the objective functional $\mathcal{F}_j : \Xi_j \to \mathbb{R}$ is strictly convex. Hence, the well known classical result (see [34, Theorem 1.1.3]) leads us to the following conclusion.

Theorem 3 *Let $\widehat{S}_{t_M} : G_H \to \mathbb{R}^m$ be a given structural prototype for unknown image S_{t_M} from Sentinel-2. Let $M : G_L \to \mathbb{R}^6$ be a given MODIS image. Let q_j stands for the texture index of the j-th band for the predicted structural prototype $\widehat{S}_{t_M}$ (see (28)). Then the unique minimizer $u_j^0 \in \Xi_j$ to the minimization problem $\inf_{u \in \Xi_j} \mathcal{F}_j(u)$ is characterized by the following variational inequality*

$$\begin{aligned}\int_\Omega \left(\left| R_{j,\eta} \nabla u_j^0(x) \right|^{q_j(x)-2} R_{j,\eta} \nabla u_j^0(x),\, R_{j,\eta} \nabla v(x) - R_{j,\eta} \nabla u_j^0(x) \right) dx \\ + \mu \int_\Omega \left(\nabla u_j^0(x) - \nabla \widehat{S}_{t_M,j}(x),\, \nabla v(x) \right) dx \\ + \gamma \int_{\Omega \setminus D_{i*}} \left(u_j^0(x) - S_{t_M,j}(x) \right) v(x)\, dx \\ + \vartheta \int_\Omega \Pi_L \left[\mathcal{K}^* * \left(\left[\mathcal{K} * u_j^0 \right] - M_{t_M,j} \right) \right] v\, dx \ge 0, \quad \forall v \in \Xi_j. \quad (50)\end{aligned}$$

Remark 4 In practical implementation, it is reasonable to define an optimal solution $u_j^0 \in \Xi_j$ using a 'gradient descent' strategy. Indeed, observing that

$$\begin{aligned}\int_\Omega \left(\left| R_{j,\eta} \nabla u_j^0(x) \right|^{q_j(x)-2} R_{j,\eta} \nabla u_j^0(x),\, \nabla v(x) \right) dx \\ = - \int_\Omega \operatorname{div} \left(\left| R_{j,\eta} \nabla u_j^0(x) \right|^{q_j(x)-2} R_{j,\eta} \nabla u_j^0(x) \right) v\, dx \\ + \int_{\partial\Omega} \left(\left| R_{j,\eta} \nabla I_{k,i}(x) \right|^{q_j(x)-2} R_{j,\eta} \nabla u_j^0(x),\, \nu \right) v\, d\mathcal{H}^1,\end{aligned}$$

and

$$\begin{aligned}\int_\Omega \left(\left| R_{j,\eta} \nabla u_j^0(x) \right|^{q_j(x)-2} R_{j,\eta} \nabla u_j^0(x),\, (\theta, \nabla v)\, \theta \right) dx \\ = - \int_\Omega \operatorname{div} \left(\left(\left| R_{j,\eta} \nabla u_j^0(x) \right|^{q_j(x)-2} R_{j,\eta} \nabla u_j^0(x), \theta \right) \theta \right) v\, dx \\ + \int_{\partial\Omega} \left(\left| R_{j,\eta} \nabla u_j^0(x) \right|^{q_j(x)-2} R_{j,\eta} \nabla u_j^0(x), \theta \right) (\theta, \nu)\, v\, d\mathcal{H}^1,\end{aligned}$$

we see that

$$\int_\Omega \left(\left| R_{j,\eta} \nabla u_j^0(x) \right|^{q_j(x)-2} R_{j,\eta} \nabla u_j^0(x), \, R_{j,\eta} \nabla v(x) \right) dx$$

$$= \int_\Omega \left(\left| R_{j,\eta} \nabla u_j^0(x) \right|^{q_j(x)-2} R_{j,\eta} \nabla u_j^0(x), \, \nabla v - \eta^2 \left(\theta, \nabla v\right) \theta \right) dx$$

$$= - \int_\Omega \operatorname{div} \left(\left| R_{j,\eta} \nabla u_j^0(x) \right|^{q_j(x)-2} R_{j,\eta} \nabla u_j^0(x) \right) v \, dx$$

$$+ \eta^2 \int_\Omega \operatorname{div} \left(\left(\left| R_{j,\eta} \nabla u_j^0(x) \right|^{q_j(x)-2} R_{j,\eta} \nabla u_j^0(x), \theta \right) \theta \right) v \, dx$$

provided

$$\left(\left| R_{j,\eta} \nabla u_j^0(x) \right|^{q_j(x)-2} R_{j,\eta} \nabla u_j^0, \, \nu \right) = 0 \quad \text{on } \partial\Omega.$$

Thus, following the standard procedure and starting from the initial image $\widehat{S}_{t_M,j}$, we can pass to the following initial value problem for the quasi-linear parabolic equations with Nuemann boundary conditions

$$\frac{\partial u_j^0}{\partial t} - \operatorname{div} \left(|R_{j,\eta} \nabla u_j^0(x)|^{q_j(x)-2} R_{j,\eta} \nabla u_j^0(x) \right)$$

$$= -\eta^2 \operatorname{div} \left(\left(\left| R_{j,\eta} \nabla u_j^0(x) \right|^{q_j(x)-2} R_{j,\eta} \nabla u_j^0(x), \theta \right) \theta \right)$$

$$+ \mu \operatorname{div} \left(\nabla u_j^0(x) - \nabla \widehat{S}_{t_M,j}(x) \right) - \gamma \left(u_j^0(x) - S_{t_M,j}(x) \right)$$

$$- \vartheta \Pi_L \left[\mathcal{K}^* * \left(\left[\mathcal{K} * u_j^0 \right] - M_{t_M,j} \right) \right], \tag{51}$$

$$\left(\left| R_{j,\eta} \nabla u_j^0(x) \right|^{q_j(x)-2} R_{j,\eta} \nabla u_j^0, \, \nu \right) = 0 \quad \text{on } \partial\Omega, \tag{52}$$

$$0 \le u_j^0(x) \le C_j \text{ a.a. in } \Omega, \tag{53}$$

$$u_j^0(0, x) = \widehat{S}_{t_M,j}(x), \quad \forall x \in \Omega. \tag{54}$$

In principle, instead of the initial condition (54) we may consider other image that can be generated from $\widehat{S}_{t_M,j}$ and the bicubic interpolation of the MODIS band $M_{t_M,j}$ onto the entire domain Ω. For instance, it can be one of well-known simple data fusion methods (for the details, we refer to [37]).

In order to illustrate the proposed approach for the restoration of satellite multi-spectral images, we refer to the full version of this paper (see [25]) where we have used a series of Sentinel-2 images (725×600 in pixels) over the South Dakota area (USA) with resolution $10m/pixel$ that were captured at different time instances in period from July 10 to July 15, 2021.

7 Conclusion

We propose a novel model for the restoration of satellite multi-spectral images. This model is based on the solutions of special variational problems with nonstandard growth objective functional. Because of the risk of information loss in optical images, we do not impose any information about such images inside the damage region, but instead we assume that the texture of these images can be predicted through a number of past cloud-free images of the same region from the time series. So, the characteristic feature of variational problems, which we formulate for each spectral channel separately, is the structure of their objective functionals. On the one hand, we involve into consideration the energy functionals with the nonstandard growth $p(x)$, where the variable exponent $p(x)$ is unknown a priori and it directly depends on the texture of an image that we are going to restore. On the other hand, the texture of an image $\mathbf{u}$, we are going to restore, can have rather rich structure in the damage region D. In order to identify it, we push forward the following hypothesis: the geometry of each spectral channels of a cloud corrupted image in the damage region is topologically close to the geometry of the total spectral energy that can be predicted with some accuracy by a number of past cloud-free images of the same region. As a result, we impose this requirement in each objective functional in the form of a special fidelity term. In order to study the consistency of the proposed collection of non-convex minimization problems, we develop a special technique and supply this approach by the rigorous mathematical substantiation.

Appendix 1. On Orlicz Spaces

Let $p(\cdot)$ be a measurable exponent function on Ω such that $1 < \alpha \le p(x) \le \beta < \infty$ a.e. in Ω, where α and β are given constants. Let $p'(\cdot) = \frac{p(\cdot)}{p(\cdot)-1}$ be the corresponding conjugate exponent. It is clear that

$$1 \le \underbrace{\frac{\beta}{\beta - 1}}_{\beta'} \le p'(x) \le \underbrace{\frac{\alpha}{\alpha - 1}}_{\alpha'} \quad a.e.\ in\ \Omega,$$

where β' and α' stand for the conjugates of constant exponents. Denote by $L^{p(\cdot)}(\Omega)$ the set of all measurable functions $f(x)$ on Ω such that $\int_\Omega |f(x)|^{p(x)}\,dx < \infty$. Then $L^{p(\cdot)}(\Omega)$ is a reflexive separable Banach space with respect to the Luxemburg norm (see [10, 18] for the details)

$$\|f\|_{L^{p(\cdot)}(\Omega)} = \inf\left\{\lambda > 0 \,:\, \rho_p(\lambda^{-1} f) \le 1\right\}, \tag{55}$$

where $\rho_p(f) := \int_\Omega |f(x)|^{p(x)}\,dx$.

It is well-known that $L^{p(\cdot)}(\Omega)$ is reflexive provided $\alpha > 1$, and its dual is $L^{p'(\cdot)}(\Omega)$, that is, any continuous functional $F = F(f)$ on $L^{p(\cdot)}(\Omega)$ has the form (see [43, Lemma 2]) $F(f) = \int_\Omega fg\,dx$, with $g \in L^{p'(\cdot)}(\Omega)$. As for the infimum in (55), we have the following result.

Proposition 3 *The infimum in (55) is attained if $\rho_p(f) > 0$. Moreover*

$$if\ \lambda_* := \|f\|_{L^{p(\cdot)}(\Omega)} > 0, \ \ then\ \ \rho_p(\lambda_*^{-1} f) = 1. \tag{56}$$

Taking this result and condition $1 \le \alpha \le p(x) \le \beta$ into account, we see that

$$\begin{aligned} &\frac{1}{\lambda_*^\beta}\int_\Omega |f(x)|^{p(x)}\,dx \le \int_\Omega \left|\frac{f(x)}{\lambda_*}\right|^{p(x)} dx \le \frac{1}{\lambda_*^\alpha}\int_\Omega |f(x)|^{p(x)}\,dx,\\ &\frac{1}{\lambda_*^\beta}\int_\Omega |f(x)|^{p(x)}\,dx \le 1 \le \frac{1}{\lambda_*^\alpha}\int_\Omega |f(x)|^{p(x)}\,dx. \end{aligned}$$

Hence, (see [10, 18, 42] for the details)

$$\begin{aligned} &\|f\|^\alpha_{L^{p(\cdot)}(\Omega)} \le \textstyle\int_\Omega |f(x)|^{p(x)}\,dx \le \|f\|^\beta_{L^{p(\cdot)}(\Omega)}, \ \ if\ \|f\|_{L^{p(\cdot)}(\Omega)} > 1,\\ &\|f\|^\beta_{L^{p(\cdot)}(\Omega)} \le \textstyle\int_\Omega |f(x)|^{p(x)}\,dx \le \|f\|^\alpha_{L^{p(\cdot)}(\Omega)}, \ \ if\ \|f\|_{L^{p(\cdot)}(\Omega)} < 1, \end{aligned} \tag{57}$$

and, therefore,

$$\|f\|^\alpha_{L^{p(\cdot)}(\Omega)} - 1 \le \int_\Omega |f(x)|^{p(x)}\,dx \le \|f\|^\beta_{L^{p(\cdot)}(\Omega)} + 1, \quad \forall f \in L^{p(\cdot)}(\Omega), \tag{58}$$

$$\|f\|_{L^{p(\cdot)}(\Omega)} = \int_\Omega |f(x)|^{p(x)}\,dx, \ \ if\ \ \|f\|_{L^{p(\cdot)}(\Omega)} = 1. \tag{59}$$

The following estimates are well-known (see, for instance, [10, 18, 42]): if $f \in L^{p(\cdot)}(\Omega)$ then

$$\|f\|_{L^\alpha(\Omega)} \le (1 + |\Omega|)^{1/\alpha}\,\|f\|_{L^{p(\cdot)}(\Omega)}, \tag{60}$$

$$\|f\|_{L^{p(\cdot)}(\Omega)} \le (1 + |\Omega|)^{1/\beta'}\,\|f\|_{L^\beta(\Omega)}, \quad \beta' = \frac{\beta}{\beta - 1}, \quad \forall f \in L^\beta(\Omega). \tag{61}$$

Let $\{p_k\}_{k\in\mathbb{N}} \subset C^{0,\delta}(\overline{\Omega})$, with some $\delta \in (0, 1]$, be a given sequence of exponents. Hereinafter in this subsection we assume that

$$\begin{aligned} &p,\, p_k \in C^{0,\delta}(\overline{\Omega})\ \ for\ k = 1, 2, \dots, \ \text{and}\\ &p_k(\cdot) \to p(\cdot)\ \text{uniformly in}\ \overline{\Omega}\ \text{as}\ k \to \infty. \end{aligned} \tag{62}$$

We associate with this sequence the following collection $\left\{f_k \in L^{p_k(\cdot)}(\Omega)\right\}_{k\in\mathbb{N}}$. The characteristic feature of this set of functions is that each element f_k lives in the corresponding Orlicz space $L^{p_k(\cdot)}(\Omega)$. We say that the sequence $\left\{f_k \in L^{p_k(\cdot)}(\Omega)\right\}_{k\in\mathbb{N}}$

is bounded if

$$\limsup_{k\to\infty} \int_\Omega |f_k(x)|^{p_k(x)}\,dx < +\infty. \tag{63}$$

Definition 3 A bounded sequence $\left\{f_k \in L^{p_k(\cdot)}(\Omega)\right\}_{k\in\mathbb{N}}$ is weakly convergent in the variable Orlicz space $L^{p_k(\cdot)}(\Omega)$ to a function $f \in L^{p(\cdot)}(\Omega)$, where $p \in C^{0,\delta}(\overline{\Omega})$ is the limit of $\{p_k\}_{k\in\mathbb{N}} \subset C^{0,\delta}(\overline{\Omega})$ in the uniform topology of $C(\overline{\Omega})$, if

$$\lim_{k\to\infty} \int_\Omega f_k\varphi\,dx = \int_\Omega f\varphi\,dx, \quad \forall\,\varphi \in C_0^\infty(\mathbb{R}^N). \tag{64}$$

We make use of the following result (we refer to [43, Lemma 3] for comparison) concerning the lower semicontinuity property of the variable $L^{p_k(\cdot)}$-norm with respect to the weak convergence in $L^{p_k(\cdot)}(\Omega)$.

Proposition 4 *If a bounded sequence* $\left\{f_k \in L^{p_k(\cdot)}(\Omega)\right\}_{k\in\mathbb{N}}$ *converges weakly in* $L^\alpha(\Omega)$ *to* f *for some* $\alpha > 1$, *then* $f \in L^{p(\cdot)}(\Omega)$, $f_k \rightharpoonup f$ *in variable* $L^{p_k(\cdot)}(\Omega)$, *and*

$$\liminf_{k\to\infty} \int_\Omega |f_k(x)|^{p_k(x)}\,dx \ge \int_\Omega |f(x)|^{p(x)}\,dx. \tag{65}$$

Remark 5 Arguing in a similar manner and using, the estimate

$$\liminf_{k\to\infty} \int_\Omega \frac{1}{p_k(x)}\|f_k(x)|^{p_k(x)}\,dx \ge \int_\Omega f(x)\varphi(x)\,dx - \int_\Omega \frac{1}{p_k'(x)}|\varphi(x)]|^{p'(x)}\,dx,$$

it can be shown that the lower semicontinuity property (65) can be generalized as follows

$$\liminf_{k\to\infty} \int_\Omega \frac{1}{p_k(x)}|f_k(x)|^{p_k(x)}\,dx \ge \int_\Omega \frac{1}{p(x)}|f(x)|^{p(x)}\,dx. \tag{66}$$

The following result can be viewed as an analogous of the Hölder inequality in Lebesgue spaces with variable exponents (for the details we refer to [10, 18]).

Proposition 5 *If* $f \in L^{p(\cdot)}(\Omega)^N$ *and* $g \in L^{p'(\cdot)}(\Omega)^N$, *then* $(f, g) \in L^1(\Omega)$ *and*

$$\int_\Omega (f, g)\,dx \le 2\|f\|_{L^{p(\cdot)}(\Omega)^N}\|g\|_{L^{p'(\cdot)}(\Omega)^N}. \tag{67}$$

Appendix 2. Sobolev Spaces with Variable Exponent

We recall here well-known facts concerning the Sobolev spaces with variable exponent. Let $p(\cdot)$ be a measurable exponent function on Ω such that $1 < \alpha \le p(x) \le \beta < \infty$ a.e. in Ω, where α and β are given constants. We associate with it the so-called Sobolev-Orlicz space

$$W^{1,p(\cdot)}(\Omega) := \left\{ u \in W^{1,1}(\Omega) : \int_\Omega \left[|u(x)|^{p(x)} + |\nabla u(x)|^{p(x)} \right] dx < \infty \right\} \quad (68)$$

and equip it with the norm $\|u\|_{W^{1,p(\cdot)}(\Omega)} = \|u\|_{L^{p(\cdot)}(\Omega)} + \|\nabla u\|_{L^{p(\cdot)}(\Omega;\mathbb{R}^N)}$.

It is well-known that, in general, unlike classical Sobolev spaces, smooth functions are not necessarily dense in $W = W_0^{1,p(\cdot)}(\Omega)$. Hence, with variable exponent $p = p(x)$ $(1 < \alpha \leq p \leq \beta)$ we can associate another Sobolev space,

$$H = H^{1,p(\cdot)}(\Omega) \text{ as the closure of the set } C^\infty(\overline{\Omega}) \text{ in } W^{1,p(\cdot)}(\Omega)\text{-norm.}$$

Since the identity $W = H$ is not always valid, it makes sense to say that an exponent $p(x)$ is regular if $C^\infty(\overline{\Omega})$ is dense in $W^{1,p(\cdot)}(\Omega)$.

The following result reveals the important property that guarantees the regularity of exponent $p(x)$.

Proposition 6 *Assume that there exists $\delta \in (0, 1]$ such that $p \in C^{0,\delta}(\overline{\Omega})$. Then the set $C^\infty(\overline{\Omega})$ is dense in $W^{1,p(\cdot)}(\Omega)$, and, therefore, $W = H$.*

Proof Let $p \in C^{0,\delta}(\overline{\Omega})$ be a given exponent. Since

$$\lim_{t\to 0} |t|^\delta \log(|t|) = 0 \quad with\ \delta \in (0, 1], \quad (69)$$

it follows from the Hölder continuity of $p(\cdot)$ that

$$|p(x) - p(y)| \leq C|x-y|^\delta \leq \left[\sup_{x,y\in\Omega} |x-y|^\delta \log(|x-y|^{-1}) \right] \omega(|x-y|), \quad \forall x, y \in \Omega,$$

where $\omega(t) = C/\log(|t|^{-1})$, and $C > 0$ is some positive constant.

Then property (69) implies that $p(\cdot)$ is a log-Hölder continuous function. So, to deduce the density of $C^\infty(\overline{\Omega})$ in $W^{1,p(\cdot)}(\Omega)$ it is enough to refer to Theorem 10 in [43].

References

1. Ambrosio, L., Caselles, V., Masnou, S., Morel, J.M.: The connected components of sets of finite perimeter. European J. Math. **3**, 39–92 (2001)
2. Ballester, C., Caselles, V., Igual, L., Verdera, J., Rougé, B.: A variational model for P+XS image fusion. Int. J. Comput. Vis. **69**, 43–58 (2006)
3. Blomgren, P.: Total variation methods for restoration of vector valued images, Ph.D. thesis, pp. 384–387 (1998)
4. Blomgren, P., Chan, T.F., Mulet, P., Wong, C.: Total variation image restoration: numerical methods and extensions. In: Proceedings of the 1997 IEEE International Conference on Image Processing, vol. III, III, pp. 384–387 (1997)
5. Bungert, L., Coomes, D.A., Ehrhardt, M.J., Rasch, J., Reisenhofer, R., Schönlieb, C.-B.: Blind image fusion for hyperspectral imaging with the directional total variation. Inverse Probl. **34**(4), Article 044003 (2018)

6. Bungert, L., Ehrhardt, M.J.: Robust image reconstruction with misaligned structural information. IEEE Access **8**, 222944–222955 (2020)
7. Caselles, V., Coll, B., Morel, J.M.: Topographic maps and local contrast changes in natural images. IEEE Trans. Image Process. **10**(8), 5–27 (1999)
8. Caselles, V., Coll, B., Morel, J.M.: Geometry and color in natural images. J. Math. Imaging Vis. **16**, 89–107 (2002)
9. Chen, Y., Levine, S., Rao, M.: Variable exponent, linear growth functionals in image restoration. SIAM J. Appl. Math. **66**(4), 1383–1406 (2006)
10. Cruz-Uribe, D.V., Fiorenza, A.: Variable Lebesgue Spaces: Foundations and Harmonic Analysis. Birkhäuser, New York (2013)
11. D'Apice, C., De Maio, U., Kogut, P.I.: Suboptimal boundary control for elliptic equations in critically perforated domains. Ann. Inst. H. Poincaré Anal. Non Lineaire **25**, 1073–1101 (2008)
12. D'Apice, C., De Maio, U., Kogut, P.I.: An indirect approach to the existence of quasi-optimal controls in coefficients for multi-dimensional thermistor problem. In: Sadovnichiy, V.A., Zgurovsky, M. (eds.) Contemporary Approaches and Methods in Fundamental Mathematics and Mechanics, pp. 489–522. Springer (2020)
13. D'Apice, C., Kogut, P.I., Kupenko, O., Manzo, R.: On variational problem with nonstandard growth functional and its applications to image processing. J. Math. Imaging Vis. (2022). https://doi.org/10.1007/s10851-022-01131-w
14. D'Apice, C., Kogut, P.I., Manzo, R.: On coupled two-level variational problem in Sobolev-Orlicz space. Differ. Integral Equ. (2023). (in press)
15. D'Apice, C., Kogut, P.I., Manzo, R., Uvarov, M.V.: Variational model with nonstandard growth conditions for restoration of satellite optical images using synthetic aperture radar. Europian J Appl. Math. **34**(1), 77–105 (2023)
16. D'Apice, C., Kogut, P.I., Manzo, R., Uvarov, M.V.: On variational problem with nonstandard growth conditions and its applications to image processing. In: Proceeding of the 19th International Conference of Numerical Analysis and Applied Mathematics, ICNAAM 2021, Rhodes, Greece (2021)
17. Dautray, R., Lions, J.L.: Mathematical Analysis and Numerical Methods for Science and Technology. Springer, Berlin, Heidelberg (1985)
18. Diening, L., Harjulehto, P., Hästö, P., Růžička, M.: Lebesgue and Sobolev Spaces with Variable Exponents. Springer, New York (2011)
19. Frantz, D.: Landsat + Sentinel-2 analysis ready data and beyond. Remote Sens. **11**, 1124 (2019)
20. Gao, F., Masek, J., Schwaller, M., Hall, F.: On the blending of the Landsat and MODIS surface reflectance: predicting daily Landsat surface reflectance. IEEE Trans. Geosci. Remote Sens. **44**(8), 2207–2218 (2006)
21. Hnatushenko, V.V., Kogut, P.I., Uvarov, M.V.: On flexible co-registration of optical and SAR satellite images. In: Lecture Notes in "Computational Intelligence and Decision Making". Series 'Advances in Intelligent Systems and Computing', pp. 515–534. Springer (2021)
22. Hnatushenko, V.V., Kogut, P.I., Uvarov, M.V.: Variational approach for rigid co-registration of optical/SAR satellite images in agricultural areas. J. Comput. Appl. Math. **400**, Id 113742 (2022)
23. Horn, B.K., Schunck, B.G.: Determining optical flow. Artif. Intell. **17**, 185–203 (1981)
24. Horsin, T., Kogut, P.: Optimal L^2-control problem in coefficients for a linear elliptic equation. I. Existence result. Math. Control Relat. Fields **5** (1), 73–96 (2015)
25. Ivanchuk, N., Kogut, P., Martyniuk, P.: Data fusion of satellite imagery for generation of daily cloud-free images at high resolution level (2023). arXiv:2302.12495 [math.OC]
26. Joshi, M.V., Upla, K.P.: Multi-resolution Image Fusion in Remote Sensing. Cambridge University Press, Cambridge (2019)
27. Ju, J., Roy, D.P.: The availability of cloud-free Landsat ETM+ date over the conterminous united states and globally. Remote Sens. Environ. **112**(3), 1196–1211 (2008)
28. Khanenko, P., Kogut, P., Uvarov, M.: On variational problem with nonstandard growth conditions for the restoration of clouds corrupted satellite images. In: CEUR Workshop Proceedings, the 2nd International Workshop on Computational and Information Technologies for Risk-Informed Systems, CITRisk-2021. Kherson, Ukraine, vol. 3101, pp. 6–25 (2021)

29. Kogut, P.I.: On optimal and quasi-optimal controls in coefficients for multi-dimensional thermistor problem with mixed Dirichlet-Neumann boundary conditions. Control Cybern. **48**(1), 31–68 (2019)
30. Kogut, P., Kohut, Y., Manzo, R.: Fictitious controls and approximation of an optimal control problem for Perona-Malik Equation. J. Optim. Diff. Equ. Appl. **30** (1), 42–70 (2022)
31. Kogut, P., Kohut, Y., Parfinovych, N.: Solvability issues for some noncoercive and nonmonotone parabolic equations arising in the image denoising problems. J. Optim. Diff. Equ. Appl. **30** (2), 19–48 (2022)
32. Kogut, P.I., Kupenko, O.P., Uvarov, M.V.: On increasing of resolution of satellite images via their fusion with imagery at higher resolution. J. Optim. Diff. Equ. Appl. **29**(1), 54–78 (2021)
33. Kogut, P.I., Manzo, R.: On vector-valued approximation of state constrained optimal control problems for nonlinear hyperbolic conservation laws. J. Dyn. Control Syst. **19**(2), 381–404 (2013)
34. Lions, J.-L.: Optimal Control of Systems Governed by Partial Differential Equations. Springer, Berlin (1971)
35. Loncan, L., De Almeida, L.B., Bioucas-Dias, J.V., Briottet, X., Chanussot, J., Dobigeon, N., Fabre, S., Liao, W., Licciardi, G.A., Simoes, M., Tourneret, J.Y., Veganzones, M.A., Vivone, G., Wei, Q., Yokoya, N.: Hyperspectral pansharpening: a review. IEEE Geosci. Remote Sens. Mag. **3**(3), 27–46 (2015)
36. Nirenberg, L.: Topics in Nonlinear Analysis. Lecture Notes, New York University, New York (1974)
37. Rani, K., Sharma, R.: Study of different image fusion algorithm. Int. J. Emerging Technol. Adv. Eng. **3**(5), 288–290 (2013)
38. Roy, D.P., Li, J., Zhang, H.K., Yan, L.: Best practices for the reprojection and resampling of Sentinel-2 Multi Spectral Instrument Level 1C data. Remote Sens. Lett. **7**, 1023–1032 (2016)
39. Roy, D.P., Huang, H., Boschetti, L., Giglio, L., Zhang, H.K., Li, J.: Landsat-8 and Sentinel-2 burned area mapping—a combined sensor multi-temporal change detection approach. Remote Sens. Environ. **231**, 111254 (2019)
40. Wang, P., Gao, F., Masek, J.G.: Operational data fusion framework for building frequent Landsat-like imagery. IEEE Trans. Geosci. Remote Sens. **52**(11), 7353–7365 (2014)
41. Yan, L., Roy, D.P., Zhang, H., Li, J., Huang, H.: An automated approach for sub-pixel registration of Landsat-8 Operational Land Imager (OLI) and Sentinel-2 Multi Spectral Instrument (MSI) imagery. Remote Sens. **8**, 520 (2016)
42. Zhikov, V.V.: Solvability of the three-dimensional thermistor problem. Proc. Steklov Inst. Math. **281**, 98–111 (2008)
43. Zhikov, V.V.: On variational problems and nonlinear elliptic equations with nonstandard growth conditions. J. Math. Sci. **173**(5), 463–570 (2011)

Computational Intelligence for Digital Healthcare Informatics

Abdel-Badeeh M. Salem

Abstract The term "digital healthcare informatics" has been coined to represent a move forward in information and communication technology-enhanced healthcare. It is a multidisciplinary field of research, at the intersection of medical sciences, biology sciences, biochemistry neurosciences, cognitive sciences and informatics. In the last years, various computational intelligence (CI) techniques and methodologies have been proposed by the researchers in order to develop digital knowledge-based systems(DKBS) for different medical and healthcare tasks. These systems are based on the knowledge engineering (KE) paradigms and artificial intelligence (AI) concepts and theories. Many types of DKBS are in existence today and are applies to different healthcare domains and tasks. The objective of the paper is to presents a comprehensive and up-to-date research in the area of digital medical decision making covering a wide spectrum of CI methodological and intelligent algorithmic issues, discussing implementations and case studies, identifying the best design practices, assessing implementation models and practices of AI paradigms in digital healthcare systems. This paper presents some of the CI techniques for managing and engineering knowledge in digital healthcare systems(DHS). Some of the research results and applications of the author and his colleagues that have been carried out in last year's are discussed.

Keywords Machine learning · Computational intelligence · AI · Knowledge engineering · Digital healthcare informatics

---

Abdel-Badeeh M. Salem (✉)
Faculty of Computer and Information Sciences, Head of Artificial Intelligence and Knowledge Engineering Research Labs, Ain Shams University, Abbassia, Cairo, Egypt
e-mail: absalem@cis.asu.edu.eg; abmsalem@yahoo.com
URL: http://staff.asu.edu.eg/Badeeh-Salem; http://aiasulab.000webhostapp.com/

M. Zgurovsky and N. Pankratova (eds.), *System Analysis and Artificial Intelligence*,
Studies in Computational Intelligence 1107,
https://doi.org/10.1007/978-3-031-37450-0_14

1 Introduction

Computational intelligence (CI) is the study of intelligent computer algorithms that improve automatically through experience. CI aims to enable computers to learn from data and make improvements without any dependence on commands in a program. This learning could eventually help computers in building models such as those used in the prediction of weather. CI is inherently interdisciplinary, includes: neurobiology, information theory, probability, statistics, AI, control theory, Bayesian methods, physiology and philosophy [1, 2]. In the recent years, various AI paradigms and computational intelligence (CI) techniques have been proposed by the researchers in order to develop efficient and smart systems in the areas of health informatics and health monitoring systems. AI and CI offers robust, intelligent algorithms and smart methods that can help to solve problems in a variety of health and life sciences areas [3–5].

Life sciences, including biology and medicine, are a growing application area of computational intelligence [6, 7]. CI is successfully used on a wide variety of medical problems and data. Medicine is largely an evidence-driven discipline where large quantities of relatively high-quality data are collected and stored in databases. The medical data are highly heterogeneous and are stored in numerical, text, image, sound, and video formats. Medical data includes; (a) Clinical data (symptoms, demographics, biochemical tests, diagnoses and various imaging, video, vital signals, etc.), (b) logistics data (e.g. charges and cost policies, guidelines, clinical trials), (c) bibliographical data, and (d) molecular data. CI techniques are also used to modify medical procedures in order to reduce cost and improve perceived patient's experience and outcomes [4–7].

Bioinformatics concerns with biological data, conceptualizes biology in terms of molecules and applies ML techniques, to understand and organize the information associated with these molecules on a large scale. Bioinformatics encompasses analysis of molecular data expressed in the form of nucleotides, amino acids, DNA, RNA, peptides, and proteins. The huge amount and breadth of biological data requires development of efficient methods for knowledge/information extraction that can cope with the size and complexity of the accumulated data [6, 7].

There are numerous examples of successful applications of CI in areas of diagnosis and prevention, prognosis and therapeutic decision making. CI algorithms are used for the following tasks; (a) discovering new diseases, (b) finding predictive and therapeutic biomarkers, and (c) detecting relationships and structure among the clinical. CI contributes to the enhancement of management and information retrieval processes leading to development of intelligent (involving ontologies and natural language processing) and integrated (across repositories) literature searches. Moreover applications of CI in bioinformatics include the following areas of research; (a) microarray analysis, chromosome and proteome databases, modeling of inhibition of metabolic networks (b) Signal analysis (echocardiograph images and electroencephalograph time series) and (c) Drug delivery, Information retrieval, software for pattern recognition in biomedical data [8–10].

This paper discusses the potential role of the AI and CI approaches, techniques, and theories which are used in developing the intelligent health informatics and health monitoring systems. The paper discusses the following CI paradigms: (a) bio-inspired computing, (b) analogical reasoning computing, (c) vagueness and fuzzy computing, and (d) deep learning. Strengths and weaknesses of these approaches were enunciated. Also, the paper presents the challenges as well as the current research directions in the areas of digital health informatics. Examples of the research performed by the author and his associates for developing knowledge-based systems for cancer, heart, brain tumor, and thrombosis diseases are discussed. The paper is organized as follows: Sect. 2 presents the subareas of the intelligent of healthcare informatics. Section 3 discusses the smart health monitoring paradigms and models Sect. 4 reviews the most common bio-inspired computing approaches. In Sect. 5, we discuss the analogical Reasoning paradigms as cognitive models for humans. While Sect. 6 discusses the vagueness and Fuzzy computing, namely rough sets and fuzzy logic. Section 7 presents deep learning methods classified by subfields and applications of Health Informatics. Section 8 presents our cases and applications, Sect. 9 presents our future research directions and then we conclude in Sect. 10.

2 Intelligent Health Informatics

Figure 1 shows the subareas of the intelligent health informatics. From the figure it can be seen that, it is a multidisciplinary field of research and covered many digital healthcare areas, namely; dental, neuro, biological, medical, nursing, and clinical. From the informatics perspective, each are composed of many fields of research, e.g. the medical informatics contains, knowledge engineering, medical imaging, expert systems, robotic surgery, education, learning and training [4–6, 12, 13].

3 Smart Health Monitoring Systems (SHMS) Paradigms and Models

SHMS involves deploying intelligent computing, information, and networking technologies to aid in preventing disease, improving the quality of care and lowering overall cost [9, 10, 13]. Figure 2 shows the different models of the most common SHMS.

1. Real-Time Monitoring Model
   In this model, sophisticated sensors and mobile devices can feed real-time medical data directly to patients and doctors via secure computing networks (Fig. 2a).
2. Computer-Aided Surgery Model
   In this model, advanced robotic devices make surgery more accurate and potentially less invasive (Fig. 2b).

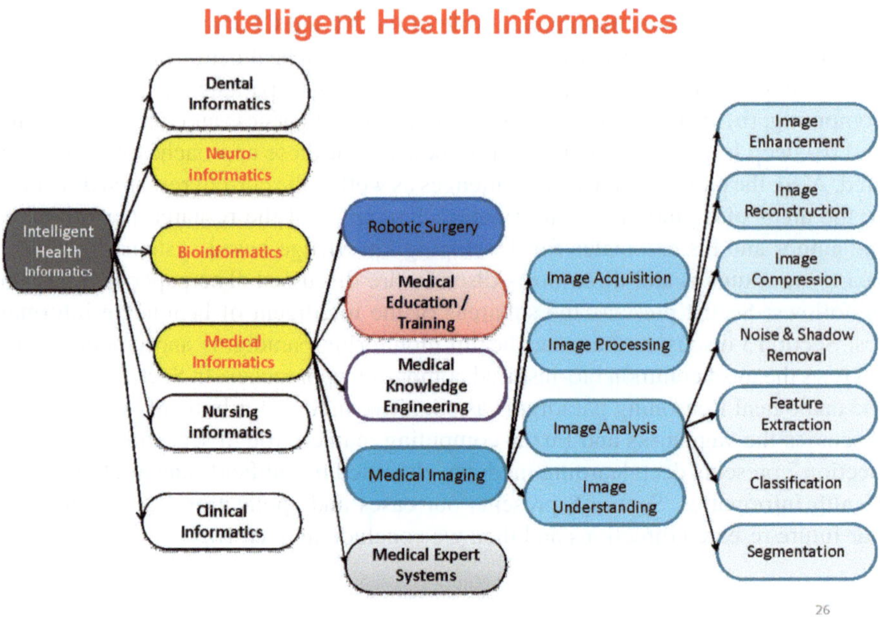

Fig. 1 Intelligent bio-medical and healthcare informatics

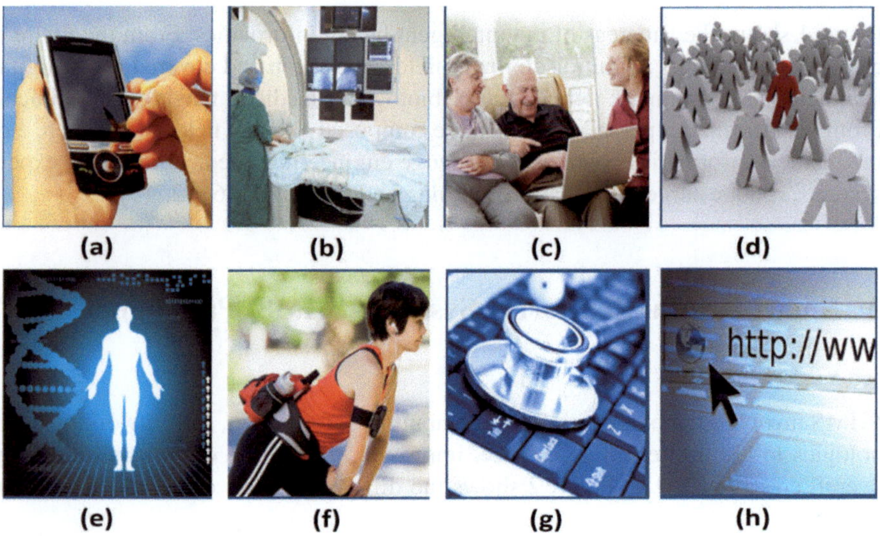

Fig. 2 The most common models of the Smart Health Monitoring Systems

3. Telemedicine Model
   In which, automated tools in the home and on mobile devices are able help patients interact with providers remotely, enabling the patients to adjust their daily lives by better managing their own care (Fig. 2c).
4. Population-Based Care Model
   In which, inexpensive monitoring devices enable collection of data from large populations with lower administrative and research costs than current method (Fig. 2d).
5. Personalized Medicine Model
   In this system, machine learning and predictive modeling will identify trends and causal relationships in medical data—leading to improved understanding of disease, development of new cures, and more accurate treatments tailored to each patient's specific needs (Fig. 2e).
6. Ubiquitous Computing Model
   This model is characterized by improved security and privacy ensure the integrity of data stored in the "cloud," allowing stakeholders—patients, providers, and relatives—to access the right information at the right time from anywhere in the world (Fig. 2f).
7. Decision Support Model
   In which, computer systems offer possible diagnoses and recommend treatment approaches, allowing doctors to quickly assess situations and viable options (Fig. 2g).
8. Health 2.0 Model
   In which, smart Web-based tools such as wikis and social networks connect patients and clinicians to shared experiences, symptoms and treatments (Fig. 2h).

4 Bio-inspired Computing

Figure 3 shows the different biological techniques of the bio-inspired computing this section presents a brief overview about the different bio-inspired techniques, namely; artificial neural networks, support vector machines, genetic algorithms, evolutionary computing, DNA computing.

4.1 Artificial Neural Networks (ANN)

ANN is: a class of learning algorithm consisting of multiple nodes that communicate through their connecting synapses. Neural networks imitate the structure of biological systems. ANN are inspired in biological models of brain functioning [11, 14]. They are capable of learning by examples and generalizing the acquired knowledge. Due to these abilities the neural networks are widely used to find out nonlinear relations which otherwise could not be unveiled due to analytical constraints. The learned

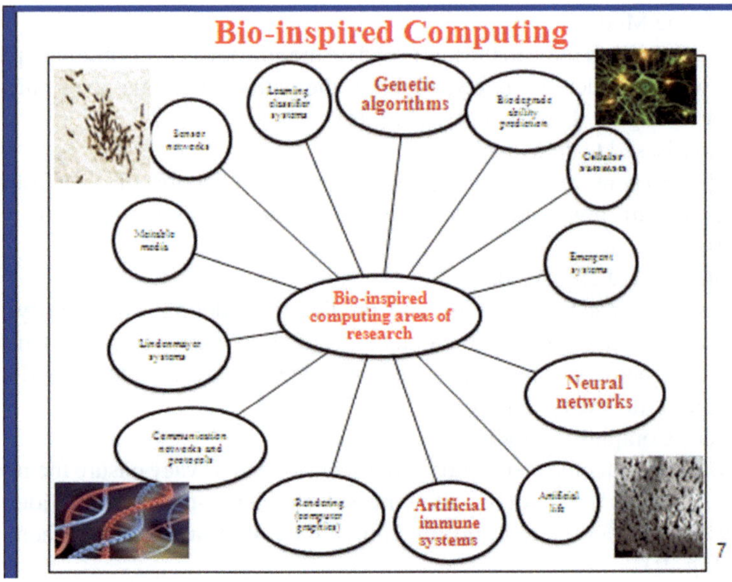

Fig. 3 Bio-inspired computing

knowledge is hidden in their structure thus it is not possibly to be easily extracted and interpreted. The structure of the multilayered perception, i.e. the number of hidden layers and the number of neurons, determines its capacity, while the knowledge about the relations between input and output data is stored in the weights of connections between neurons. The values of weights are updated in the supervised training process with a set of known and representative values of input—output data samples.ANN can be used in the following medical purposes;

(a) Modelling: Simulating and modeling the functions of the brain and neurosenory organs.
(b) Signal processing: Bioelecteric signal filtering and evaluation.
(c) System control and checking: Intelligent artificial machine control and checking based on responses of biological or technical systems given to any signals.
(d) Classification tasks: Interpretation of physical and instrumental findings to achieve more accurate diagnosis.
(e) Prediction: Provide prognostic information based on retrospective parameter analysis.

4.2 Support Vector Machines (SVM)

SVM are new learning-by example paradigm for classification and regression problems [11, 15]. SVM have demonstrated significant efficiency when compared with neural networks. Their main advantage lies in the structure of the learning algorithm which consists of a constrained quadratic optimization problem (QP), thus avoiding the local minima drawback of NN. The approach has its roots in statistical learning theory (SLT) and provides a way to build "optimum classifiers" according to some optimality criterion that is referred to as the maximal margin criterion. An interesting development in SLT is the introduction of the Vapnik-Chervonenkis (VC) dimension, which is a measure of the complexity of the model. Equipped with a sound mathematical background, support vector machines treat both the problem of how to minimize complexity in the course of learning and how high generalization might be attained. This trade-off between complexity and accuracy led to a range of principles to find the optimal compromise. Vapnik and co-authors' work have shown the generalization to be bounded by the sum of the training error and a term depending on the Vapnik-Chervonenkis (VC) dimension of the learning machine leading to the formulation of the structural risk minimization (SRM) principle. By minimizing this upper bound, which typically depends on the margin of the classifier, the resulting algorithms lead to high generalization in the learning process [11].

4.3 Genetic Algorithms (GA)

GA follows the lead of genetics, reproduction, evolution and the survival for the fittest theory by Darwin. GA is a class of machine learning algorithm that is based on the theory of evolution [11, 16]. Genetic Algorithms (GA) provide an approach to learning that based loosely on simulated evolution. The GA methodology hinges on a population of potential solutions, and as such exploits the mechanisms of natural selection well known in evolution. Rather than searching from general to specific hypothesis or from simple to complex GA generates successive hypotheses by repeatedly mutating and recombining parts of the best currently known hypotheses. The GA algorithm operates by iteratively updating a poll of hypotheses (population). One each iteration, old members of the population are evaluated according a fitness function. A new generation is then generated by probabilistically selecting the fittest individuals form the current population. Some of these selected individuals are carried forward into the next generation population others are used as the bases for creating new offspring individuals by applying genetic operations such as crossover and mutation.

4.4 Evolutionary Computing (EC)

EC is an approach to the design of learning algorithms that is structured along the lines of the theory of evolution. A collection of potential solutions for a problem compete with each other. The best solutions are selected and combined with each other according to a kind of 'survival of the fittest' strategy. Genetic algorithms are a well known variant of evolutionary computation [11, 16].

4.5 DNA Computing

DNA computing is essential computation using biological molecules rather than traditional silicon chips. In recent years, DNA computing has been a research tool for solving complex problems. Despite this, it is still not easy to understand. The main idea behind DNA (Deoxyribo Nucleic Acid) computing is to adopt a biological (wet) technique as an efficient computing vehicle, where data are represented using strands of DNA. Even though a DNA reaction is much slower than the cycle time of a silicon-based computer, the inherently parallel processing offered by the DNA process plays an important role. This massive parallelism of DNA processing is of particular interest in solving NP-complete or NP-hard problems [17–19].

It is not uncommon to encounter molecular biological experiments which involve 6×1016/ml of DNA molecules. This means that we can effectively realize 60,000 Tera Bytes of memory, assuming that each string of a DNA molecule expresses one character. The total execution speed of a DNA computer can outshine that of a conventional electronic computer, even though the execution time of a single DNA molecule reaction is relatively slow. A DNA computer is thus suited to problems such as the analysis of genome information, and the functional design of molecules (where molecules constitute the input data) [20].

DNA computing will solve that problem and serve as an alternative technology. DNA computing is also known as molecular computing. It is computing using the processing power of molecular information instead the conventional digital components. It is one of the non-silicon based computing approaches. DNA has been shown to have massive processing capabilities that might allow a DNA-based computer to solve complex problems in a reasonable amount of time. DNA computing was proposed by Leonard Adleman, who demonstrated in 1994 that DNA could be applied in computations [21, 22].

5 Analogical Reasoning Computing (ARC) as Cognitive Models for Humans

CBR is an analogical reasoning method provides both a methodology for problem solving and a cognitive model of people [23–26]. CBR means reasoning from experiences or "old cases" in an effort to solve problems, critique solutions and explain anomalous situations. The case is a list of features that lead to a particular outcome; e.g. The information on a patient history and the associated diagnosis. We feel more comfortable with older doctors because they have seen and treated more patients

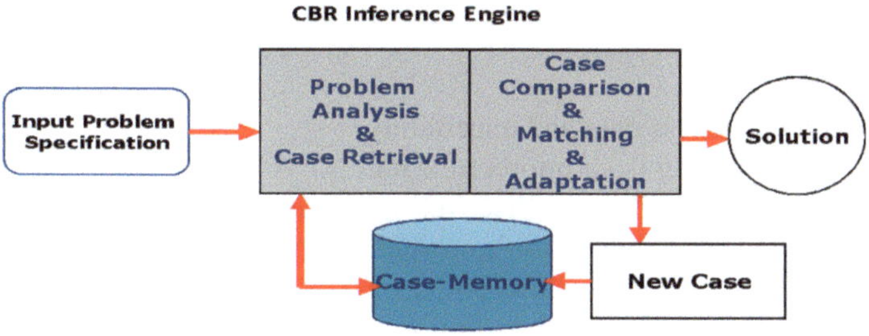

Fig. 4 Case-based reasoning methodology

who have had illnesses similar to our own. CBR is an analogical reasoning method provides both: a methodology for building an efficient knowledge-based reasoning systems (CBRS), and a cognitive model for People. CBR is a preferred method of reasoning in dynamically changing situations and other situations where solutions are not clear cut [27].

Most commonly application of CBR used in developing expert systems technology. In CBR expert systems, the system can reason from analogy from the past cases. This system contains what is called "case-memory" which contains the knowledge in the form of old cases (experiences). CES solves new problems by adapting solutions that were used for previous and similar problems [22]. The methodology of CBR directly addresses the problems found in rule-based technology, namely: knowledge acquisition, performance, adaptive solution, maintenance.

According to Kolonder [24], CBR from the computational perspective refers to a number of concepts and techniques (e.g. data structures and intelligent algorithms) that can be used to perform the following operations; (a) Record and index cases, (b) Search cases to identify the ones that might be useful in solving new cases when they are presented, (c) Modify earlier cases to better match new cases, and (d) Synthesize new cases when they are needed (Fig. 4).

6 Vagueness and Fuzzy Computing

6.1 Rough Sets

Rough set theory was proposed as a new approach to vague concept description from incomplete data. The rough set theory is one of the most useful techniques in many real life applications such as medicine, pharmacology, engineering, banking and market analysis. This theory provides a powerful foundation to reveal and discover important structures in data and to classify complex objects. The theory is very useful in practice, e.g. in medicine, pharmacology, engineering, banking, financial and market analysis [28]. In what follows, we can summarize the benefits and advantages of rough set theory

(a) Deals with vagueness data and uncertainty.
(b) Deals with reasoning from imprecise data.
(c) Used to develop a method for discovering relationships in data.
(d) Provides a powerful foundation to reveal and discover important structures in data and to classify complex objects.
(e) Do not need any preliminary or additional information about data.
(f) Concerned with three basics: granularity of knowledge, approximation of sets and data mining.

6.2 *Fuzzy Rules*

In the rich history of rule-based reasoning in AI, the inference engines almost without exception were based on Boolean or binary logic. However, in the same way that neural networks have enriched the AI landscape by providing an alternative to symbol processing techniques, fuzzy logic has provided an alternative to Boolean logic based systems [29]. Unlike Boolean logic, which has only two states, true or false, fuzzy logic deals with truth values which range continuously from 0 to 1. Thus something could be half true 0.5 or very likely true 0.9 or probably not true 0.1. The use of fuzzy logic in reasoning systems impacts not only the inference engine but the knowledge representation itself [11]. For, instead of making arbitrary distinctions between variables and states, as is required with Boolean logic systems, fuzzy logic allows one to express knowledge in a rule format that is close to a natural language expression.

The difference between this fuzzy rule and the Boolean-logic rules we used in our forward- and backwardchaining examples is that the clauses "temperature is hot" and "humidity is sticky" are not strictly true or false. Clauses in fuzzy rules are real-valued functions called membership functions that map the fuzzy set "hot" onto the domain of the fuzzy variable "temperature" and produce a truth-value that ranges from 0.0 to 1.0 (a continuous output value, much like neural networks).

Reasoning with fuzzy rule systems is a forward-chaining procedure. The initial numeric data values are fuzzified, that is, turned into fuzzy values using the membership functions. Instead of a match and conflict resolution phase where we select a triggered rule to fire, in fuzzy systems, all rules are evaluated, because all fuzzy rules can be true to some degree (ranging from 0.0 to 1.0). The antecedent clause truth values are combined using fuzzy logic operators (a fuzzy conjunction or and operation takes the minimum value of the two fuzzy clauses). Next, the fuzzy sets specified in the consequent clauses of all rules are combined, using the rule truth values as scaling factors. The result is a single fuzzy set, which is then defuzzified to return a crisp output value. More technical details and applications can be found in the recent book of Voskoglou [30].

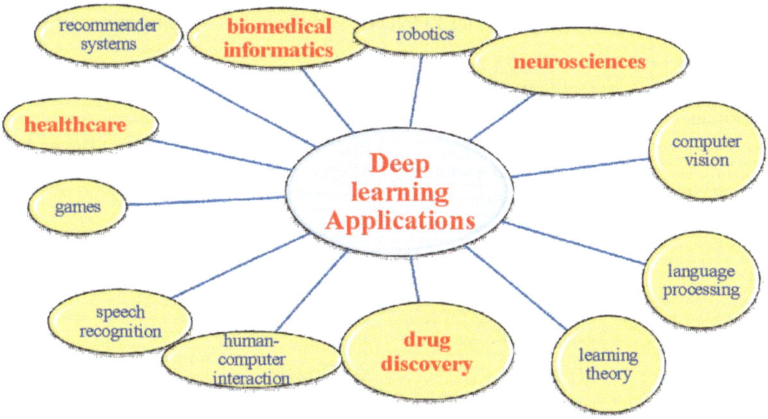

Fig. 5 Deep learning applications

7 Deep Learning

Deep learning is a branch of AI covering a spectrum of current exciting research and industrial innovation that provides more efficient algorithms to deal with large-scale data in healthcare, recommender systems, learning theory, robotics, games, neurosciences, computer vision, speech recognition, language processing, human-computer interaction, drug discovery, biomedical informatics, act [31–33]. DL is a branch of AI covering a spectrum of current exciting research and industrial innovation that provides more efficient algorithms to deal with large-scale data in many areas, see Fig. 5. In the last decade, with the development of ANN, many researchers have tried to develop further studies using DL methods. In this section, DL studies are investigated in popular areas to illuminate the paths of researchers working in DL Tables 1 and 2. Summarize the different DL methods by subfields and applications of Health Informatics.

8 Case Studies and Applications

8.1 CI for Thyroid Cancer Diagnosis

Cancer is a group of more than 200 different diseases. From the medical point of view, Cancer occurs when cells become abnormal and keep dividing and forming either benign or malignant tumors. Cancer has initial signs or symptoms if any is observed, the patient should perform complete blood count and other clinical examinations. To specify cancer type, patient needs to perform special lab-tests. In our research Labs, we have performed an interesting research for developing intelligent systems

Table 1 The different deep learning (DL) methods by subfields and applications of health informatics

Subfield	Application	Input data	DL Method
Bioinformatics	Cancer diagnosis Gene selection/classification Gene variants	Gene expression Micro RNA Microarray data	Deep Autoencoders Deep Belief Network Deep Neural Network
	Drug design	Molecule compounds	Deep Neural Network
	Compound-Protein interaction RNA binding protein DNA methylation	Protein structure Molecule compounds Genes/RNA/DNA sequences	Deep Belief Network Deep Neural Network
Medical imaging	3D brain reconstruction Neural cells classification Brain tissues classification Alzheimer/MCI diagnosis	MRI/fMRI Fundus images PET scans	Deep Autoencoders Convolutional Neural Network Deep Belief Network Deep Neural Network
	Tissue classification Organ segmentation Cell clustering Hemorrhage detection Tumor detection	MRI/CT images Endoscopy images Microscopy Fundus images X-ray images Hyperspectral images	Convolutional Deep Belief Network Convolutional Neural Network Deep Autoencoders Group Method of Data Handling Deep Neural Network

for thyroid cancer diagnoses. This research are based on using case-based and rule-based reasoning, ontological engineering, and artificial neural networks [33–36]. Figure 6a, b show an examples for the encoded rules for cancer diagnosis and the liver cancer case of old Egyptian women respectively.

8.2 *Heart Disease Expert System*

Heart disease is a vital health care problem affecting millions of people. Figure 7 shows Types of Heart Diseases. Expert system (ES) is a consultation intelligent system that contains the knowledge and experience of one or more experts in a specific domain that anyone can tap as an aid in solving problems [37, 38]. At our research unit, we have developed two versions of expert systems for heart diseases diagnosis.

The first one uses the rule-based reasoning while the second one uses case-based reasoning. The first version is composed of three components: knowledge base, user interface and computational model. The knowledge was gathered from expert doctors in EL-Maadi Military Egyptian hospital, Egyptian Health Insurance Institute and

Table 2 Health informatics subfields and applications areas of DL

Subfield	Application	Input data	DL Method
Medical informatics	Prediction of disease Human behavior monitoring Data mining	Electronic health records Big medical dataset Blood/Lab tests	Deep Autoencoders Deep Belief Network Convolutional Neural Network Recurrent Neural Network Convolutional Deep Belief Network Deep Neural Network
Public health	Predicting demographic info Lifestyle diseases Infectious disease epidemics Air pollutant prediction	Social media data Mobile phone metadata Geo-tagged images Text messages	Deep Autoencoders Deep Belief Network Convolutional Neural Network Deep Neural Network
Pervasive sensing	Anomaly detection Biological parameters monitoring	EEG ECG Implantable device	Deep Belief Network
	Human activity recognition	Video Wearable device	Convolutional Neural Network Deep Belief Network Deep Neural Network
	Hand gesture recognition Obstacle detection Sign language recognition	Depth camera RGB-D camera Real-Sense camera	Convolutional Neural Network Deep Belief Network
	Food intake Energy expenditure	Wearable device RGB image Mobile device	Convolutional Neural Network Deep Neural Network

medical books. We have built the system's knowledge base for the 24 heart diseases and it is composed of 24 facts and 65 rules. The system is implemented in Visual Prolog and has been tested for 13 real experiments (patients). The experimental results have shown 76.9% accuracy in estimating the right conclusion [39].

The Second version of the expert system uses CBR methodology [40]. We have represented the knowledge in the form of frames and built the case memory for 4 heart diseases namely; mistral stenosis, left-sided heart failure, left-sided heart failure, stable angina pectoris and essential hypertension. The system has been implemented in visual prolog for Windows and has trained set of 42 cases for Egyptian cardiac patients and has been tested by another 13 different cases. Each case contains 33 significant attributes resettled from the statistical analysis performed to 110 cases. The system has been tested for 13 real cases. The experimental results have shown

Rule1: **IF the symptom is changes in bowel habits,**
or The symptom is unusual bleeding,
or The symptom is thickening lymph
Then cancer is infecting

Rule2: **IF cancer is infecting,**
and The symptom is severe headaches,
and The symptom is seizures
Then brain cancer

(a) Example of encoded rules for cancer diagnosis

Patient: 65-years old, female, not working, with nausea and vomiting.

Medical history: cancer head of pancreas

Physical Exam: tender hepatomgaly liver, large amount of inflammatory about 3 liters, multiple liver pyogenic abcesses and large pancreatic head mass.

Laboratory findings: total bilrubin 1.3 mg/dl, direct bilrubin 0.4 mg/dl, sgot (ast) 28 IU/L. sgpt (alt) 26 IU/L.

(b) Example of liver cancer case of old Egyptian women

Fig. 6 Examples

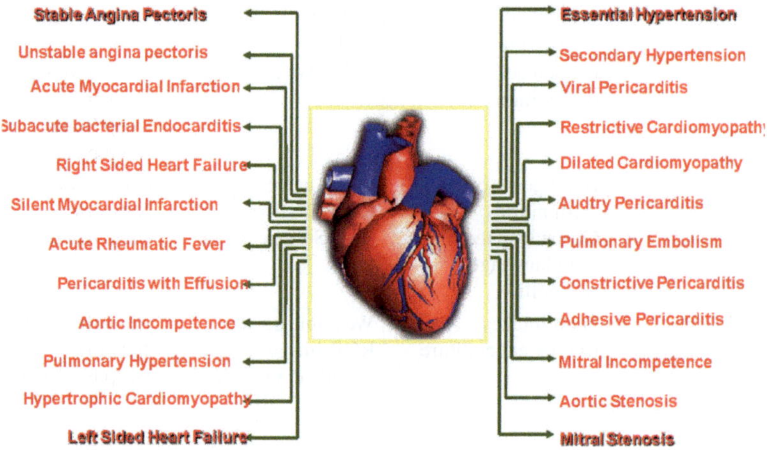

Fig. 7 Types of Heart Diseases

100% accuracy in estimating the correct results for using nearest neighbor algorithm and this percentage is dropped to 53.8% in case of using the induction algorithm. The systems are able to give an appropriate diagnosis for the presented symptoms, signs and investigations done to a cardiac patient with the corresponding certainty factor. It aims to serve as doctor diagnostic assistant and support the education for the undergraduate and postgraduate young physicians.

8.3 Mining Patient Date for Determining Thrombosis Disease Using Rough Sets

At our research unit at Ain Shams, a rough set-based medical system for mining patient data for predictive rules to determine thrombosis disease was developed [41, 42] this system aims to search for patterns specific/sensitive to thrombosis disease. This system reduced the number of attributes that describe the thrombosis disease from 60 to 16 significant attributes in addition to extracting some decision rules, through decision applying decision algorithms, which can help young physicians to predict the thrombosis disease.

8.4 Genetic Algorithms Approach for Mining Medical Data

In our research group we developed a hybrid classifier that integrates the strengths of genetic algorithms and decision trees. The algorithm was applied on a medical database of 20 MB size for predicting thrombosis disease [43]. The results show that our classifier is a very promising tool for thrombosis disease prediction in terms of predictive accuracy.

8.5 An Agent-Based Modeling for Pandemic Influenza in Egypt

Pandemic influenza has great potential to cause large and rapid increases in deaths and serious illness. The first major pandemic influenza H1N1 is recorded in 1918–1919, which killed 20–40 million people and is thought to be one of the most deadly pandemics in human history. In 1957, a H2N2 virus originated in China, quickly spread throughout the world and caused 1–4 million deaths world wide. In 1968, an H3N2 virus emerged in Hong Kong for which the fatalities were 1–4 million [44]. In recent years, novel H1N1 influenza has appeared. Novel H1N1 influenza is a swine-origin flue and is often called swine flue by the public media. The novel H1N1 outbreak began in Mexico, with evidence that there had been an ongoing epidemic for months before it was officially recognized as such.

In the absence of reliable pandemic detection systems, computational intelligence techniques and paradigms have become an important smart software tools for both policy-makers and the general public [45]. Intelligent computer models can help in providing a global insight of the infectious disease outbreaks' behavior by analyzing the spread of infectious diseases in a given population, with varied geographic and demographic features [46]. Agent-based modeling of pandemics recreates the entire populations and their dynamics through incorporating social structures, heterogeneous connectivity patterns, and meta-population grouping at the scale of the single individual [47].

In our application, we propose a stochastic multi-agent model to mimic the daily person-to-person contact of people in a large scale community affected by a pandemic influenza (novel H1N1) in Egypt. The developed multi-agent model is based on the modeling of individuals' interactions in a space time context. The model involves different types of parameters such as: social agent attributes, distribution of Egypt population, and patterns of agents' interactions. Analysis of the results leads to understanding the characteristics of the modeled pandemic, transmission patterns, and the conditions under which an outbreak might occur. In addition, model is used to measure the effectiveness of different control strategies to intervene the pandemic spread. More technical and computing aspects can be found in reference [48, 49].

8.6 Daily Meal Planner Expert System for Diabetics Type-2 in Sudan

Diabetes is a serious, life-threatening and chronic disease. It is estimated that this figure will reach 366 million by 2030 [50], with 81% of these diabetics being in developing Countries, where medical care remains severely limited. Actually, recent estimates place the diabetes population in Sudan at around one million—around 95% of whom have type 2 diabetes and patients with diabetes make up around 10% of all hospital admissions in Sudan [51]. The types of food eaten in Sudan vary according to climate, although the Sudanese diet has plenty of carbohydrate rich items some patients believe sugar is the only source of energy, therefore on hot days people consume large amount of sugary carbonated drinks [50]. Fortunately, diabetes can be managed very effectively through healthy lifestyle choices, primarily diet and exercise. Mostly, Type 2 diabetes is strongly connected with obesity, age, and physical inactivity [52]. Most medical resources reported that 90 to 95% of diabetic is diagnosed as type-2. So, a successful intelligent control of patient food for treatment purpose must combines patient interesting food list and doctors' efficient treatment food list. In addition, many rural communities in Sudan have extremely limited access to diabetic diet centers. People travel long distances to clinics or medical facilities, and there is a shortage of medical experts in most of these facilities. This results in slow service, and patients end up waiting long hours without receiving any attention. Hence the expert systems paradigm can play a very important role in such cases where medical experts are not readily available.

In our research Labs at Ain Shams University, Egypt, we design and implement of an intelligent medical expert system for diabetes diet that intended to be used in Sudan [51, 52]. The expert system provides the patients with medical advice and basic knowledge on diabetes diet. The development of such system went through a number of technical stages, namely; requirements analysis, food knowledge acquisition, formalization, design and implementation. Visual prolog was used for designing the graphical user interface and the implementation of the system. The system is a promising helpful smart tool that reduces the workload for physicians and provides a more comfort for diabetic patients.

8.7 *CI in Medical Volume Visualization (MVV)*

MVV brings profound changes to personal health programs and clinical healthcare delivery. It's seeks to reveal internal structures hidden by the skin and bones, as well as to diagnose and treat disease. Visual representation of the interior body is a key element in medicine. There are many techniques for creating it; such as magnetic resonance imaging, computed tomography, and ultra-sound [53]. The past few decades have witnessed an increasing number of new techniques being developed for practical clinical image display. Regarding physician diagnosis and therapy monitoring, medical imaging is one of the most important tools in the field and it also comes up handy in other fields like remote emergency assistance and surgical planning. The main problem with this large number of medical images lies in processing the enormous amount of obtained data (slice resolution with 16 bits/voxel precision can be provided by modern CT scanner). One approach is to render the data interactively using a specialized multi-processor hardware support. Since these devices are not cheap, they are not widely used in practice. Another alternative is to use volume visualization [54]. In our research labs, we performed a comprehensive study for the recent intelligent techniques and algorithms used for medical data visualization [55]. These techniques cover filtering, segmentation, classification and visualization as well as the intelligent software supporting medical volume visualization. The study reveals hybrid techniques are the best approach for medical image segmentation is the best. In our future, work we are looking to develop mobile based intelligent system using direct volume rendering texture mapping technique with bones data sets.

9 Future Research Directions

In this section we describe some of the research directions in the area of healthcare informatics at Artificial Intelligence and Knowledge Engineering Research Labs (The ASU-AIKE lab). This lab was founded in 2005 and belongs to the Computer Science department, Faculty of Computer and Information Sciences, Ain Shams University, Cairo, Egypt. The Lab is currently composed of 3 full Profs., 3 Assistant Prof., 15 Drs (Ph.D. holders) 5 of them from Arabic countries, 10 Ph.D. students and 10 M.Sc. students.

We are working in the following disciplines:

1. Intelligent Information Processing, Data Mining.
2. Knowledge Engineering and Expert Systems.
3. Image Processing and Pattern Recognition.
4. Data Science and Big Data Analytics.
5. Intelligent Education and Smart Learning.

6. Medical Informatics, Bio-informatics and e-Health.
7. Computing Intelligence and Machine learning.
8. Internet of Things(IoT), Technologies for HealthCare.
9. Space Science and Cosmic Rays Big Data Analytics.

9.1 CI for the Diagnoses of Alzheimer's Disease

Alzheimer's disease (AD) is one of the most serious diseases that destroy the brain and classified as the most widespread type of dementia. According to Alois Alzheimer, Alzheimer's disease is a physical disease that affects the brain [56]. In Alzheimer's disease procedure, proteins are created in the brain in order to form two main components: plaques and tangles. This process may leads to damage the connections between nerve cells. In addition to, Alzheimer's patient have a shortage of some significant chemicals in their brain. When there is some problems of them, the chemical messengers are not send the signals through the region of the brain [57]. Currently, we working in determining the appropriate computational intelligent paradigm and model capable to classify AD and to build computer aided diagnostic system capable of detecting Alzheimer's at any stage [58].

9.2 Big Data Analytics and IoT in Healthcare

Data science confronts the challenges of extracting a fountain of knowledge from "mountains" of big data. Data analytics is the science of drawing conclusions and making sense of raw data/information. On the other side, in the digital healthcare industry, a huge amount of medical data is already being collected to generate insights on emerging conditions and to improve patient care. The healthcare sector continues to grapple with issues such as data silos and security challenges that stand in the way of harnessing insights. Predictive analytics are revolutionizing healthcare, as wearable IoT and sophisticated mobile apps collect data on individuals to inform them and their caregivers about their health. This data is invaluable for spotting incipient conditions of concern and nipping them in the bud. But IoT, wearables and mobile apps aren't the only sources of data that can predict health issues: Old healthcare claims data is a great source, and potentially a richer one. Currently, we working in intensive care unit data analytics using machine learning and computational intelligence techniques [59, 60].

9.3 CI for Detecting Novel Corona's Virus

The importance of using computational intelligence techniques in the field of medical imaging technologies have increased in both research and clinical care over the recent years. So, detecting novel corona's virus in precise way and early phase is essential for patient care. Currently, we are working in developing an intelligent medical imaging evaluation system to help doctors diagnose pneumonia caused by the novel corona. The system capable of detecting corona's at any stage.

10 Conclusions

The development of robust intelligent medical decision support systems is a very difficult and complex process that raises a lot of technological and research challenges that have to be addressed in an interdisciplinary way. The development of robust intelligent healthcare systems is a very difficult and complex process that raises a lot of technological and research challenges that have to be addressed in an interdisciplinary way. This paper analyzes the main paradigms and applications of the computational intelligence (CI) in healthcare from the artificial intelligence perspective.

The main results based on our analysis, CI offer potentially powerful tools for the development a novel digital healthcare systems. The variety of such techniques enabling the design of a robust and efficient IHS. The key to the success of such systems is the selection of the CI technique that best fits the domain knowledge and the problem to be solved. That choice is depends on the experience of the knowledge engineer. And, digital medical decision support systems can benefit from systematic knowledge engineering and structure using techniques from the different disciplines artificial intelligence. AI technologies and techniques play an key role in developing intelligent tools for medical tasks and domains. CI Techniques (e.g. CBR, Data Mining, Rough Set, Ontology) can cope with medical noisy data, sub symbolic data, and complex structure data. In addition, CI offer intelligent computational methods for accumulating, changing and updating medical knowledge in IHS, and in particular learning mechanisms that will help us to induce knowledge from medical information or data.

From our comprehensive analysis, one can recommend the following recommendations:

(a) The cooperation between physicians and AI communities is essential to produce efficient computing systems for medical purposes. The physicians will have more information to deliver a better service and dynamic guidelines to improve quality and reduce risks.
(b) The industry of intelligent medical decision support systems is a promising area of research for developing successful telemedicine projects.

(c) Mobile devices can feed real-time medical data directly to patients and doctors via secure computing networks and IoT. The web-based and IoT medical systems can enhance the online education/learning/training processes.
(d) The use of ICT technologies also improves the quality of patient care and reduces clinical risk. At the same time, the patient will be part of the healthcare process, having more information about diseases and access to his/her electronic health record.
(e) The pharmaceutical industries can get more accurate information about drug's effects and supply chain delivery systems.
(f) Public health authorities can get more accurate information and develop dashboards to make better and fast decisions.
(g) Hospital management benefits from a more updated meaningful data. This data is used by management systems to delivery KPI (Key Performance Indicators).

References

1. Russell, S.J., Norvig, P.: Artificial Intelligence: A Modern Approach, 2nd ed. Prentice Hall (2003)
2. Borzemski, L., Światek, J., Wilimowska, Z. (eds): Information Systems Architecture and Technology: Proceedings of 39th International Conference on Information Systems Architecture and Technology—ISAT 2018. ISAT 2018. Advances in Intelligent Systems and Computing, vol. 85. Publisher Name Springer, Cham. https://doi.org/10.1007/978-3-319-99981-4_1, Print ISBN 978-3-319-99980-7, Online ISBN 978-3-319-99981-4, eBook Packages Intelligent Technologies and Robotics Reprints and Permissions
3. Cook, D.: Health monitoring and assistance to support aging in place. JUCS **12**(1), 15–29 (2006)
4. Salem, A.-B.M., Shendryk, V., Shendryk, S.: Exploiting the knowledge computing and engineering in medical informatics and healthcare. Int. J. Comput. **3**, 130–139 (2018). http://www.iaras.org/iaras/journals/ijc, ISSN: 2367-8895
5. Salem, A.-B.M.: Knowledge engineering paradigms in the intelligent health informatics and medical knowledge-based systems. J. Glob. Res. Comput. Sci. ISSN: 2229-371X
6. Salem, A.-B.M., Aref, M.M., Moawad, I.F., Alfonse, G.A.M.: Ontological engineering in medical informatics. Int. J. Genom. Proteomics Metab. Bioinform. (IJGPMB). SciDoc Publishers, Infer, Interpret & Inspire Science **1**(2), 9–13 (2016)
7. Said, H.M., Salem, A.-B.M.: Exploiting computational intelligence paradigms in e-technologies and activities. Elsevier Procedia Comput. Sci. **65**, iii–vii, 396–405 (2015)
8. Cook, D.J., Das, S. (eds.): Smart Environments: Technologies Protocols and Applications. Wiley (2004)
9. Dix, A., Finlay, J., Abowd, G.D., Beale, R.: Human-Computer Interaction, 3d edn. Prentice Hall (2003)
10. Cios, K.J., Kurgan, L.A., Reformat, M.: Machine learning in the life sciences "How it is used on a wide variety of medical problems and data". In: IEEE Engineering in Medicine and Biology Magazine, pp. 14–16 (2007)
11. Mitchell, T.: Machine Learning. McGraw-Hill, New York (1997)
12. Salem, A.B.M.: Artificial intelligence technology in intelligent health informatics. In: Borzemski, L., Światek, J., Wilimowska, Z. (eds.), Information Systems Architecture and Technology: Proceedings of 39th International Conference on Information Systems Architecture and

Technology—ISAT 2018. ISAT 2018. Advances in Intelligent Systems and Computing, vol. 85. Springer (2019). https://doi.org/10.1007/978-3-319-99981-4_1
13. Noorbakhsh-Sabet, N., et al.: Artificial intelligence transforms the future of healthcare. Am. J. Med. **132**(7), 795–801 (2019)
14. Simon, S.: Neural Networks and Learning Machines, 3rd ed. Originally Published (1993). http://dai.fmph.uniba.sk/courses/NN/haykin.neural-networks.3ed.2009.pdf
15. Cortes, C., Vapnik, V.: Support vector networks. Mach. Learn. **20**, 273–297 (1995)
16. Salem, A.-B.M., Mahmoud, A.M.: A hybrid genetic algorithm-decision tree classifier. In: Proceedings of the 3rd International Conference on New Trends in Intelligent Information Processing and Web Mining, Zakopane, Poland, pp. 221–232, 2–5 June 2003
17. Adleman, L.M.: Computing with DNA. Science **279**, 54–61 (1998)
18. Paun, G.H., Rozenberg, G., Salomaa, A.: DNA Computing: New Computing Paradigms, Yokomori T (Translated Ed). Springer (1999)
19. Regalado, A.: DNA computing. Technol. Rev. 80–84 (2000)
20. Lewin, D.I.: DNA computing. Comput. Sci. Eng. 5–8 (2002)
21. Steele, G., Stojkovic, V.: Agent-oriented approach to DNA computing. In: Proceedings of Computational Systems Bioinformatics Conference, pp. 546–551 (2004)
22. Huang, Y., He, L.: DNA computing research progress and application. In: The 6th International Conference on Computer Science and Education, pp. 232–235 (2011)
23. Kolodner, J.L.: Improving human decision making through Case based decision aiding. AI Magazine, pp. 52–69 (1991)
24. Kolonder, J.: Case-Based Reasoning. Morgan Kaufmann (1993)
25. Lopez, B., Plaza, E.: Case-Based Planning for Medical Diagnosis (1993)
26. Voskoglou, 1 M.G., Salem, A.-B.M.: Analogy-based and case-based reasoning: two sides of the same coin. Int. J. Appl. Fuzzy Sets Artif. Intell. **4**, 5–51 (2014). (ISSN 2241-1240)
27. Aamodt, A., Plaza, E.: Case-based reasoning: foundational issues, methodological variations, and system approaches. A. I. Commun. **7**(1), 39–52 (1994)
28. Pawlak, Z.: Rough Sets: Theoretical Aspects of Reasoning about Data. Kluwer Academic Publishers, Dordrecht (1991)
29. Klir, G.J., Folger, T.A.: Fuzzy Sets, Uncertainty and Information. Prentice Hall Int, London (1988)
30. Voskoglou, M.: Special Issue Fuzzy Sets, Fuzzy Logic and Their Applications. Published in Mathematics) (2020). ISBN 978-3-03928-520-4 (Pbk); ISBN 978-3-03928-521-1 (PDF). https://doi.org/10.3390/books978-3-03928-521-1
31. Miotto, R., et al.: Deep learning for healthcare: review, opportunities and challenges. Brief. Bioinform. **19**(6), 1236–1246 (2017)
32. Rajkomar, A., et al.: Scalable and accurate deep learning with electronic health records. NPJ Digit. Med. **1**(1), 18 (2018)
33. Kwak, G.H.-J., Hui, P.: DeepHealth: deep Learning for Health Informatics reviews, challenges, and opportunities on medical imaging, electronic health records, genomics, sensing, and online communication health (2019). arXiv:1909.00384
34. Salem, A.-B.M., Nagaty, K.A., El-Bagoury, B.M.: A hybrid case-based adaptation model for thyroid cancer diagnosis. In: Proceedings of 5th International Conference on Enterprise Information Systems, ICEIS 2003, Angres, France, pp. 58–65 (2003)
35. Salem, A.-B.M., El-Bagoury, B.M.: A comparison of case-based adaptation method for thyroid cancer diagnosis using neural networks. In: Proceeding of the 9th International Conference on Soft Computing, MENDEL 2003, pp. 263–269. Brno University of Technology, Czech Republic (2003)
36. Salem, A.-B.M., El-Bagoury, B.M.: A case-based adaptation model for thyroid cancer diagnosis using neural networks. In: Proceedings of the 16th International FLAIRS Conference, Florida, US, pp. 155–159 (2003)
37. Abdrabou, E.A.M., Salem, A.B.: A breast cancer classifier based on a combination of case-based reasoning and ontology approach. In: Proceedings of 2nd International Multi-conference on Computer Science and Information Technology (IMCSIT 2010), Wisła, Poland (2010)

38. Salem, A.-B.M., Alfonse, M.: Ontological engineering approach for breast cancer knowledge management. In: Proceeding of Med-e-Tel the International eHealth, Telemedicine and Health ICT for Education, Networking and Business. Luxembourg (2009)
39. Salem, A-B.M., Roushdy, M., HodHod, R.A.: A rule-base expert system for diagnosis of heart diseases. In: Proceedings of 8th International Conference on Soft Computing MENDEL, Brno, Czeeh Republic, 5–7 June 2002, pp. 258–263 (2002)
40. Salem, A.-B.M., Roushdy, M., HodHod, R.A.: A case based expert system for supporting diagnosis of heart diseases. Int. J. Artif. Intell. Mach. Learn. AIML, Tubungen, Germany **1**, 33–39 (2004)
41. Salem, A.-B.M., Roushdy, M., Mahmoud, S.A.: Mining patient data based on rough set theory to Determine Thrombosis Disease. Int. J. Artif. Intell. Mach. Learn. AIML, Tubungen, Germany **1**, 27–31 (2004)
42. Salem, A.M., Mahmoud, S.A.: Mining patient data based on rough set theory to determine thrombosis disease. In: Proceedings of first intelligence conference on intelligent computing and information systems, pp. 291–296. ICICIS 2002, Cairo, Egypt, 24–26 June 2002
43. Salem, A.-B.M., Mahmoud, AM.: Applying the genetic algorithms approach for data mining classification task. In: IFIP WG12.6, First IFIP Conference on Artificial Intelligence Applications and Innovations, Toulouse France, 22–27 Aug. 2004
44. World Health Organization. http://www.who.int/en/
45. Timpka, T., Eriksson, H., Gursky, E.A., Nyce, J.M., Morin, M., Jenvald, J., Strömgren, M., Holm, E., Ekberg, J.: Population based simulations of influenza pandemics: validity and significance for public health policy. Bull World Health Organ. **87**(4), 305–311 (2009). ISSN 0042-9686
46. Perez, L., Dragicevic, S.: An agent-based approach for modeling dynamics of contagious disease spread. Int. J. Health Geograph. **8**, 50 (2009)
47. Chowell, G., Hyman, J.M., Eubank, S., Castillo-Chavez, C.: Scaling laws for the movement of people between locations in a large city. Phys. Rev. E **68**, 066102 (2003)
48. Khalil, K.M., Abdel-Aziz, M., Nazmy, T.T., Salem, A.-B.M.: An agent-based modeling for pandemic influenza in Egypt. In: INFOS2010: 7th International Conference on Informatics and Systems, Cairo, Egypt, AOM, pp. 24–30 (2010)
49. Khalil, K.M., Abdel-Aziz, M., Nazmy, T.T., Salem, A.-B.M.: Artificial immune system metaphor for agent based crisis response operations. In: MATES2010: Eighth German Conference on Multi Agents System Technologies, Karlsruhe, Germany, 21–23 Sept. 2010
50. Ahmed, I.M., Mahmoud, A.M.: Development of an expert system for diabetic Type-2 Diet. Int. J. Comput. Appl. (IJCA) **107**(1) (2014)
51. Ahmed, I.M., Alfonse, M., Mahmoud, A.M., Aref, M.M., Salem, A.-B.M.: Knowledge acquisition for developing knowledge-base of diabetic expert system. In: Proceeding Abstract of the 7th International Conference on Information Technology(ICIT15), Jordan, p. 17, May 2015
52. Ahmed, I.M., Alfonse, M., Aref, M.M., Salem, A.-B.M.: Daily meal planner expert system for Diabetics Type-2. E-Leader Int. J. New York, USA **10**(2) (2015)
53. Balazs, C.: Interactive volume-rendering techniques for medical data visualization. In: Balazs, C. (ed.) Interactive Volume-Rendering Techniques for Medical Data Visualization. Institut fur Computergraphik und Algorithmen (2001)
54. Lundstrom, C.: Efficient medical volume visualization: an approach based on domain knowledge. In: Lundstrom, C. (ed.) Efficient Medical Volume Visualization: An Approach Based on Domain Knowledge. Institutionen for teknikoch naturvetenskap (2007)
55. Abdallah, Y., Abdelhamid, A., Elarif, T., Salem, A.-B.M.: Intelligent techniques in medical volume visualization. Procedia Comput. Sci. **65**, 546–555 (2015)
56. Wilson, R.S., Segawa, E., Boyle, P.A., Anagnos, S.E., Hizel, L.P., Bennett, D.A.: The natural history of cognitive decline in Alzheimer's disease. Psychol. Aging **27**(4), 1008–17 (2012)
57. Kruthika, K.R., et al.: Multistage classifier-based approach for Alzheimer's disease prediction and retrieval. Inf. Med. Unlocked (2018). https://doi.org/10.1016/j.imu.12.003
58. Solimana, S.A., Mohsenb, H., El-Dahshanc, E.-S.A., Salem, A.-B.M.: Exploiting of machine learning paradigms in Alzheimer's disease. Int. J. Psychiatr. Psychotherap. **4**, 1–11 (2019). http://www.iaras.org/iaras/journals/ijpp, ISSN: 2535-0994

59. Rayan, Z., Alfonse, M., Salem, A.-B.M.: Machine learning approaches in smart health. Procedia Comput. Sci. **154**, 361–368 (2019). ScienceDirect. www.sciencedirect.com, www.elsevier.com/locate/prodedia
60. Rayan, Z., Alfonse, M., Salem, A.-B.M.: Intensive care unit (ICU) data analytics using machine learning techniques. ITHEA® Int. J. Inf. Theor. Appl. **26**(1), 69–82 (2019). http://www.foibg.com/ijita/vol26/ijita-fv26.htm

Continuous and Convex Extensions Approaches in Combinatorial Optimization

Sergiy Yakovlev and Oksana Pichugina

Abstract The paper is dedicated to extensions of functions and applications in combinatorial optimization. We present a review of the main contributions in the area of continuous and smooth extensions of functions, on the basis of which we formulate the singularities of such extensions for finite sets. We consider a class of finite sets coinciding with their convex hull. For such sets, the existence of convex extensions of functions is proved, which makes it possible to apply the methods of convex analysis to solve relaxation problems. For combinatorial sets, we formulated an equivalent statement of the discrete optimization problem with a convex objective function and proposed methods for estimating the optimum.

1 Introduction

Many discrete and combinatorial optimization methods are based on decomposing and relaxing original problems. At the same time, the relaxation of the discreteness condition makes it possible to propose original approaches to solving relaxation problems. One such approach is constructing continuous extensions for functions defined on discrete sets. There are an infinite number of possibilities for constructing these extensions. Therefore, it is interesting to utilize properties of discrete sets and

S. Yakovlev (✉) · O. Pichugina
National Aerospace University "Kharkiv Aviation Institute", 17 Chkalova Street, Kharkiv 61070, Ukraine
e-mail: svsyak7@gmail.com

O. Pichugina
e-mail: o.pichugina@khai.edu

S. Yakovlev
Institute of Information Technology, Lodz University of Technology, 116 Zeromskiego Str, 90-924 Lodz, Poland

O. Pichugina
Department of Mechanical and Industrial Engineering, University of Toronto, 27 King's College Cir, Toronto M5S1A1, Canada

M. Zgurovsky and N. Pankratova (eds.), *System Analysis and Artificial Intelligence*, Studies in Computational Intelligence 1107,
https://doi.org/10.1007/978-3-031-37450-0_15

functions defined on them in optimization. The paper proposes approaches to constructing convex extensions of functions on the class of discrete sets that coincide with the set of vertices of their convex hull. Thus, as a result of relaxation, a convex optimization problem arises, for which many efficient solution methods exist. Note that this approach is especially interesting for combinatorial optimization problems since it allows one to additionally use the properties of the corresponding combinatorial polytopes.

2 Related Work

In mathematics, a concept of a restriction of a function from its domain to a subset of this domain is commonly used.

Definition 1 Let $F : E' \to D, E \subset E'$, then a function $f : E \to D$ coinciding with F on E is called *a restriction* of function F to set E.

For a function f, being a restriction of function F to E, a notation $f = F|_E$ is commonly used. Respectively, function F is called *an extension* of function f to set E'.

For indicating that, $\forall x \in E : f(x) = F(x)$, i.e., F coincides with f on E, the following [1] notation is used:

$$F(x) \underset{E}{=} f(x). \tag{1}$$

An extension of a function can now be defined as follows.

Definition 2 An extension of a function $f(x)$ from E to $E' \supset E$ is a function $F(x)$, defined on E' and satisfied the condition (1).

The domain E from which the extension is carried out is called *an extension domain* [2].

With its help, the definition (2) can be reformulated.

Definition 3 An extension of a function $f : E \to \mathbb{R}^1$ from E to $E' \supset E$ is a function $F : E' \to \mathbb{R}^1$ satisfying the constraint (1).

Extensions of various types—convex, concave, smooth, Lipschitz ones, etc.—are known in scientific literature [3–16]. Besides, various types of an extension domain E and a domain E', to which they are carried out, are considered in research sources. For instance, as E, the boundary of some domain or its part is chosen. For example, it cat be a set of extreme points of the domain, convex, finite and countable sets, etc. At the same time, for E' the whole space $\mathbb{R}^n$, the convex hull of E and another proper superset of E can be chosen.

There are three main areas of study on extensions of functions—continuous, differentiable, including analytic ones, convex extensions.

For instance, in the seminal paper [9], H. Whitney establishes that the problem of constructing a continuous extension from a set $E \subset \mathbb{R}^n$ to the entire Euclidean space is always solvable if the extension domain E is a closed set. Namely, under a certain set of constraints, such an extension is a solution to the Dirichlet problem being a harmonic function that on $\overline{E} = \mathbb{R}^n \backslash E$ and such that it takes given values on E.

Based on this fact, H. Whitney raises new questions: (a) it is possible to construct such a continuous extension on $\overline{E}$, which is differentiable and even analytic; (b) if the answer is positive, then it is possible to make this extension smooth to a certain order if f has this order of smoothness on E.

The fundamental result was formulated in the so-called H.Whitney extension theorem [9]. This theorem provides sufficient conditions on the existence of a differentiable (including infinitely differentiable) extension of a function, which is analytic on $\overline{E}$ and ensures fulfillment of (1) and similar conditions on derivatives. Besides, in [9], peculiarities of extensions from a set of isolated points were studied, and the result is formulated in the form of H. Whitney's extension theorem for sets of isolated points (finite or countable).

Let us move to the consideration of convex extensions. They play a special role among all classes of function extensions. On the one hand, it is related to the problem of constructing extensions of a given order of smoothness because, according to [17], convex functions are twice continuously differentiated almost everywhere. On the other hand, convex extensions have many practical applications in geometric analysis, nonlinear dynamics, quantum computing and economics (see references in [16]) and in optimization [1, 8, 10, 14, 16, 18–20].

Usually, speaking about the construction of convex extensions, one means that the domain E' to which they are built and the extension domain E are convex. In this case, we can say that a convex extension always exists because to construct $F(x)$, it is sufficient to redefine $f(x)$ by the value $+\infty$ on $\overline{E}$, and then move to the convex envelope of the epigraph of the resulting function [21]. However, as the author of [16] notes, this method is suitable in some applications such as optimization, in particular, convex analysis, and theoretical areas such as duality theory. Still, in the calculus of variations, more analytic approaches to constructing convex extensions are required. That is why special attention is paid to solving these issues [6, 7, 21] in the research literature.

Meanwhile, issues of constructing convex extensions from non-convex domains [4] are of particular interest. Especially it concerns parts of the boundaries of convex domains as extension domains [10]. Among the domain are sets of vertices of polytopes [1, 8].

3 Convex Extensions of Functions Defined on Some Finite Sets

Thus, in work [15], several fundamental theorems on the existence of convex extensions of functions from finite sets are formulated. As an extension domain, point configurations, particularly, sets coinciding with vertex sets of the convex hull of polytopes were considered.

Let us list some of the results needed for the further presentation and discussion. Introduce E, E' are sets such that

$$E' = conv\ E,\ E = \text{vert}\ E',\ n_E = |E|.$$

Theorem 1 *For any function* $f : E \to \mathbb{R}^1$*, there exists a convex function* $\varphi : E' \to \mathbb{R}^1$ *such that* $\varphi(x) \underset{E}{=} f(x)$.

The function $\varphi(x)$ is built in the form of

$$\varphi(x) = \min_{\alpha \in \Lambda(x)} \sum_{i=1}^{n_E} \alpha_i f(x_i), \tag{2}$$

$$\Lambda(x) = \left\{ (\alpha_1, ..., \alpha_{n_E}) : x = \sum_{i=1}^{n_E} \alpha_i x_i;\ \sum_{i=1}^{n_E} \alpha_i = 1;\ \alpha_i \in [0, 1],\ x_i \in E,\ i \in \overline{1, n_E} \right\}.$$

According to [22], $\Lambda(x)$ is the convex envelope of $f(x)$, therefore, $\varphi(x) = conv\ f(x)$ for $x \in E'$.

The convex hull of a function f is called a uniquely defined convex extension φ satisfying a constraint $\varphi(x) \geq \phi(x)$ for all convex extensions ϕ of function f and $\forall x \in E'$ [8].

The following theorems develop and generalize the assertions of Theorem 1.

Theorem 2 *For any function* $f : E \to \mathbb{R}^1$ *and arbitrary* $\rho > 0$*, there exists a strongly convex function* $\Psi : E \to \mathbb{R}^1$ *with a parameter of at least* ρ*, such that* $\Psi(x) \underset{E}{=} f(x)$.

Theorem 3 *For any function* $f : E \to \mathbb{R}^1$*, there exists a differentiable convex function* $\xi : E \to \mathbb{R}^1$*, so* $\xi(x) \underset{E}{=} f(x)$.

If $|E| > 1$, then $E \subset E'$ holds. Respectively, Theorems 1–3 establish an existence of convex, strongly convex and differentiable extensions of an arbitrary function $f(x)$ from any vertex set E of a polytope E' nondegenerate at a point. These theorems provide sufficient conditions for the existence of not only convex but strongly convex and differentiable convex extensions to the polytope E'. The question arises, what are sufficient conditions for this and how to formulate a criterion on the existence of convex extensions of a function, depending on the type of an extension domain and function f. Another issue is extending the domain to which the extension is carried out, from $conv\ E$ to a convex set $E' \supset conv E$, particularly to $E' = \mathbb{R}^n$. If φ

is convex on K, which is a convex superset of E, its lower bound on E can be found according to the results presented in [15, 22].

Statements of the following theorems are based on the properties of differentiable convex functions on a set E under consideration.

Assume that $K \subseteq \mathbb{R}^n$ is a closed convex set, $E \subset K$, $\varphi : K \to \mathbb{R}^1$.

Theorem 4 *If $\varphi(x)$ is convex and differentiable on K, then $\forall\, x \in K$*

$$\min_{y\in E} \varphi(y) \geq \varphi(x) - (grad\, \varphi(x), x) + \min_{y\in E} \sum_{i=1}^{n} \frac{\partial \varphi(x)}{\partial x_i} y_i. \tag{3}$$

Theorem 5 *Let $\varphi(x)$ be strongly convex with parameter $\rho > 0$ on K, then*

$$\min_{x\in E} \varphi(x) \geq \varphi(y^*) + \rho \min_{x\in E} \|x - y^*\|^2, \tag{4}$$

where $y^* = \arg\min_{y\in K} \varphi(y)$.

Theorem 6 *If $\varphi(x)$ is strongly convex with parameter $\rho > 0$ and differentiable on K, then $\forall x \in K$*

$$\min_{y\in E} \varphi(y) \geq \varphi(x) - \frac{1}{4\rho} \|grad\, \varphi(x)\|^2 + \rho \min_{y\in E} \left\| y - x + \frac{1}{2\rho} grad\, \varphi(x) \right\|^2. \tag{5}$$

Theorem 7 *In order for the point $y^{**} \in E$ to deliver a minimum on E of a convex function φ differentiable on K, it is sufficient that*

$$\min_{y\in E} \sum_{i=1}^{n} \frac{\partial \varphi(x)}{\partial x_i} y_i = (grad\, \varphi(y^*), y^*). \tag{6}$$

Theorem 1 generalizes a contribution introduced in [4]. It utilizes an authors' observation that, in decision-making problems under risk, it is often required to extend functions convexly from non-convex domains. In their study [4], H. Peters and P.'Wakke gave the necessary and sufficient conditions for extending a function on a non-convex domain to a convex function on the convex hull of the domain.

This led to an introduction of a concept of a convex function defined on a domain, which is not necessarily convex. This need was caused by the fact that often observations serving as input data in mathematical models are given on non-convex, particularly discrete, parts of space. At the same time, using all available data in the model processing is highly desirable. In order to distinguish this concept from the classical one of a convex function, the term the generalized convex function (GCF) was introduced.

Definition 4 Let $E \subseteq \mathbb{R}^n$. A function $f : E \to \mathbb{R}^1$ is a GCF on E if, for all convex linear combinations of elements $\{x^i\}_{i\in I} \subseteq E$, it holds:

$$\sum_{i\in I}\lambda_i x^i \in E \Rightarrow f(\sum_{i\in I}\lambda_i x^i) \leq \sum_{i=1}^{|I|}\lambda_i f(x^i). \tag{7}$$

Thus, a GCF is a function that satisfies Jensen's inequality only for those $\{x^i\}_{i\in I} \subseteq E$ whose convex linear combination is also a point of E [23].

Theorem 8 (extension theorem) *[4] Let V be a linear space over $\mathbb{R}$, $E \subset V$, $f : E \to \overline{\mathbb{R}}^1 = \mathbb{R}^1 \cup \{-\infty, \infty\}$ is GCF. Then there exists a convex function $F : V \to \overline{\mathbb{R}}^1$, which extends f to V.*

Theorem 9 *Let $E \subset \mathbb{R}^n$, $f : E \to \mathbb{R}^1$ is a GCF. Then there exists a convex function $F : \mathbb{R}^n \to \mathbb{R}^1$, which is a convex extension of f to V.*

Since the proof of Theorem 8 utilizes a convex extension obtained from (2) by replacement min by inf, this corollary establishes the existence of a convex extension of any GCF f in the form (2). Since the function given on a vertex-located set is a GCF, the following corollary is true.

Theorem 10 *For an arbitrary VLS E, a function $f : E \to \mathbb{R}^1$, and a convex set $E' \supset E$, there exists a convex extension f from E to E'.*

Finally, [10] establishes a necessary and sufficient condition for the existence of a convex extension, which exploits the fact that for points representable as a convex linear combination of other points of E, the condition (7) is a necessary one for the existence of a convex function' extension.

Theorem 11 *A convex extension $F : E' \to \mathbb{R}^1$ of a function $f : E \to \mathbb{R}^1$ to a convex set $E' \supset E$ exists if and only if $\forall x \in E$*

$$f(x) \leq \min\left\{\sum_{i\in I}\lambda_i f(x^i) : \sum_{i\in E}\lambda_i x^i = x; \sum_{i\in I}\lambda_i = 1; \lambda_i \in (0, 1), x^i \in E\right\}, \tag{8}$$

where the summation is over all $I \subset \mathbb{N}$, $|I| < \infty$.

Remark 1 In the proof of Theorem 11 presented in [10], a convex extension $F(x)$ is formed in the form of

$$F(x) = conv\ \phi(x), \text{ where } \phi(x) = \begin{cases} f(x), \text{ if } x \in E, \\ \infty, \text{ if } x \in E' \backslash E. \end{cases} \tag{9}$$

Such a convex extension turns into (2) when choosing $E = \text{vert } conv\ E$, $E' = conv\ E$ and to the one suggested in [21] when $E \subset \mathbb{R}^n$ is convex, $E' = \mathbb{R}^n$. This implies that it is a generalization of the above two ways of designing convex extensions of functions.

4 Convex Extensions for Polyhedral-Spherical Sets

Regarding the existence of convex differentiable extensions, generalizing Theorem 6, we mention a recent study [24], where a convex compact is considered as an extension domain E, and there are given a convex function $f : E \to \mathbb{R}^1$ and a set of polynomials satisfying the Whitney conditions (see [9]). Thus, according to [9], provided that certain conditions are satisfied,a continually differentiable extension exists up to the mth order of smoothness, which satisfies the Whitney conditions. Authors of [24] study an issue under what conditions convexity of resulting extensions can be guaranteed. They also point to an interest that Whitney-type extensions have attracted in recent years in constructing linear extension operators with almost optimal norms, designing the extensions in non-Euclidean spaces such as Sobolev ones. Generalizing Whitney's results to the class of convex functions is profitable from the theoretical point of view and, according to the authors of [24], has wide applications in differential geometry, partial differential equations, etc.

Concerning strictly and strongly convex extensions, the theorem in [25] formulates a fundamental result on their existence and a search for constructive ways of their construction.

Theorem 12 *If E is located on a hypersphere $S_r(x^0)$ centered at a point x^0 of radius r, a function $f \in \mathbf{C}^2(\mathbb{R}^n)$ has bounded second partial derivatives, then $\exists M \in \mathbb{R}^1_+$, such that*

$$F(x, \mu) = f(x) + \mu f_1(x) \tag{10}$$

is convex for any $\mu \geq M$, where

$$f_1(x) = (x - x^0)^2 - r^2. \tag{11}$$

According to [26], the value (11) is called the *regularization term*, and μ is a *regularization parameter*.

Corollary 1 *Let $\rho > 0$, then, under the conditions of Theorem 12, function* (10) *is strictly convex for any $\mu > M$ and strongly convex with parameter ρ for $\mu \geq M + \rho$.*

Theorem 12 and Corollary 1 establish necessary conditions on the existence of convex (strictly/strongly convex) twice continuously differentiable extensions in the form of (10), (11). In addition, in [25], the lower bound M on the threshold value μ is obtained, ensuring convexity of function (10).

Corollary 2 *If E is located on the hypersphere $S_r(x^0)$, $f \in \mathbf{C}^2(K)$, where $K \subset \mathbb{R}^n$ is a convex compact, then $\forall \varepsilon > 0\, \exists M(\varepsilon) \in \mathbb{R}^1_{>0}$, such that for any $\mu \geq M(\varepsilon)/M(\varepsilon) + \varepsilon$ function $F(x, \mu)$ in the form of* (10), *function* (11) *is convex/strongly convex on K with parameter ε.*

The possibility of constructing $f(x)$ with the help of $f_1(x)$ exists due to a strong convexity of the latter. The question arises whether it is possible to generalize Theorem 12 onto convex regularization terms of general form.

After the existence of a convex extension is established, a question arises of finding effective ways to construct the extension, which avoids involving all elements of E in consideration, as was assumed in Theorems 1 and 11. Indeed, for combinatorial sets with exponential cardinality, convex extensions presented in the study are of theoretical interest only.

Together with a choice of a regularization parameter according to [25], the formulas (10), (11) provides a constructive way of constructing a convex extension from a hypersphere and its arbitrary subset, discrete or continuous.

We emphasize that the construction of convex extensions for the class of quadratic functions. Let $E \subset \mathbb{R}^n$ be a polyhedral-spherical set with parameters (τ, r). Then the convex extension of the quadratic function $f(x) = (Cx, x) + (b, x)$ to the space $\mathbb{R}^n$ is the function $\widetilde{f}(x) = (\widetilde{C}x, x) + (\widetilde{b}, x) + \widetilde{d}$, where $C = \left[c_{ij}\right]_{n \times n}$ is arbitrary $n \times n$-matrix, b – n-dimensional vector, $\widetilde{C} = C + \beta I$, $\beta = \sum\limits_{i:c_{ii}<0} |c_{ii}| + \sum\limits_{i<j} \left|c_{ij}\right|$, $\widetilde{b} = b - 2\beta\tau$, $\widetilde{d} = \beta\left(\tau^2 - r^2\right)$, I is a unit matrix.

5 Convex Extensions in Discrete Optimization

Let $E \subset \mathbb{R}^n$ be a finite set of cardinality $|E| < \infty$. We formulate the discrete optimization problem as following:

$$f(x) \to \min, x \in E', \tag{12}$$

where E' is given by constraints

$$x \in E, \tag{13}$$

$$g_i(x) \leq 0, \ i \in J_k, \tag{14}$$

$$g_i(x) = 0, \ i \in J_m \backslash J_k; \tag{15}$$

and functions $f(x), \ g_i(x), \ i \in J_m$ are defined on E'.

Specifics and properties of problem (12)–(15) are determined by the properties of a set X as well as of functions $f(x), \ g_i(x), \ i \in J_m$ involved in optimization and given on X.

One of the main results of the convex extensions theory is the approach to equivalent reformulating an optimization problem on vertex-located sets proposed in [27] and further formalized in [18–20]. Consider optimization problem (12)–(15) under the assumption that the set E is vertex-located. We represent equalities (15) in the form of inequalities and construct convex extensions of the original functions involved in these new constraints to a convex set $X \supseteq conv E$:

$$\widetilde{f}(x) \underset{E}{=} f(x), \tag{16}$$

$$\tilde{g}_i(x) \underset{E}{=} g_i(x), \; i \in J_m, \tag{17}$$

$$\tilde{g}_i(x) \underset{E}{=} \; - g_{i-l}(x), \; i \in J_r \backslash J_m, \tag{18}$$

where $r = 2m - k$. Taking into account (16)–(18), we can form the optimization problem

$$\tilde{f}(x) \to \min, x \in \tilde{X}, \tag{19}$$

where a set $\tilde{X}$ is given by constraints

$$x \in X, \tag{20}$$

$$\tilde{g}_i(x) \le 0, \; i \in J_r, \tag{21}$$

while functions $\widetilde{f}(x), \; \tilde{g}_i(x), \; i \in J_r$ are convex on X.

The coincidence of the original functions and their convex extensions on vertex-located sets X allows formulating the following optimization problem on such sets, which is equivalent to the original problem (12)–(15).

Let a set $E \subset \mathbb{R}^n$ is finite and $E = \text{vert}\, convE$. Introduce sets

$$G = \{x \in E : \; g_i(x) \le 0, \; i \in J_k, \;\; g_i(x) = 0, \; i \in J_m \backslash J_k\};$$

$$\widetilde{G} = \{x \in E : \; \tilde{g}_i(x) \le 0, \; i \in J_r\}.$$

Then

$$\arg\min_{x \in G} \; f(x) = \arg\min_{x \in \tilde{G}} \; \widetilde{f}(x), \tag{22}$$

In equality (22), one can use differentiable and strongly convex extensions of the functions $\widetilde{f}(x), \; \tilde{g}_i(x), \; i \in J_r$ onto the set $\Pi = convE$, as well as its convex supersets (if such a convex extension is constructed).

We use the above approach for polyhedral and convex surface relaxations to construct an equivalent optimization problem on vertex-located sets.

Let us form the following optimization problem

$$\tilde{f}(x) \to \min, x \in \widetilde{G}, \tag{23}$$

where $\widetilde{G} = \{x \in X : \; \tilde{g}_i(x) \le 0, \; i \in J_r\}$, and functions $\tilde{f}(x), \; \tilde{g}_i(x), \; i \in J_r$ are convex on the convex set $X \supset E$. Problem (23) is convex and its optimal value is a lower bound on the minimum of the original objective function $f(x)$ on E.

6 Convex Extensions in Combinatorial Optimization

Of particular interest is the theory of convex extensions in the study of combinatorial optimization. The solution to combinatorial optimization problems is connected with the analysis of the properties of combinatorial sets and the study of the behavior of functions defined on these sets. The definition of a combinatorial set is directly related to the concept of a combinatorial configuration.

Consider a abstract finite strictly ordered set $G = \{G_1, ..., G_m\}$. According to [28], a combinatorial configuration $\langle \upsilon, \ G, \ \Lambda \rangle$ is a mapping $\upsilon : J_n \to G$ that satisfies a given collection of constraints Λ. A set J_n is called an enumerating set, and G is a generating set. Note that the requirements for the finiteness and strict ordering of the generating set follow from the problems that arise in the enumeration and construction of various combinatorial configurations.

Consider the set P as a set of combinatorial configurations, which are ordered sets of elements of the generating set G, i.e. $\pi = \langle \pi_1, .., \pi_n \rangle$, where $\pi_i \in G, \ i \in J_n$. We call such a set a Euclidean combinatorial set. Taking into account the properties of the mapping $\upsilon : J_n \to G$ (bijectivity, injectivity, surjectivity), as well as additional requirements that form the collection of constraints Λ, it is possible to form various classes of Euclidean combinatorial sets, such as permutations (without repetitions and with repetitions), cyclic permutations, even permutations, signed permutations, partial permutations (without repetitions, with repetitions, with unbounded repetitions), permutation matrices, Boolean sets, their special subsets, etc.

The selection of Euclidean combinatorial sets into a special class is justified by a number of specific properties associated with the possibility of their bijective mapping ϕ onto some point subset E of an arithmetic Euclidean space of a certain dimension, i.e. $\phi : P \to E \subset \mathbb{R}^n$. Moreover, any combinatorial configuration $\pi \in P$ will correspond to a point $x \in E$ such that $x = \phi(\pi), \quad \pi = \phi^{-1}(x)$. Note that for finite set P, a bijective mapping $\phi : P \to E \subset \mathbb{R}^n$ exists for any $n \geq 1$. At the same time, mappings ϕ under which the set $E = \phi(P)$ has a special structure acquire special significance. For finite point set E, this is primarily due to the properties of convex hull (polytope) $\Pi = \ conv\ E$.

In order to emphasize that the points of a set $E = \phi(P)$ are images of combinatorial configurations $\pi \in P$ under a given bijective mapping $\phi : P \to E \subset \mathbb{R}^n$, the monograph [29] introduces the concept of a Euclidean combinatorial configuration, which is represented by a tuple $\langle \phi, \upsilon, \ G, \ \Lambda \rangle$ and defines the image of a combinatorial configuration $\langle \phi, \upsilon, \ G, \ \Lambda \rangle$ in $\mathbb{R}^n$. In this case, $E = \phi(P)$ is called the set of Euclidean combinatorial configurations.

We confine ourselves to consideration of Euclidean combinatorial sets P generated by combinatorial configurations $\langle \upsilon, \ G, \ \Lambda \rangle$ under the assumption that the generating set G is finite and is given by different sets of numerical parameters (vectors). We single out the class of combinatorial configurations $\langle \upsilon, G, \ \Lambda \rangle$, where $G = \{G_1, ..., G_m\}$, $G_i = g^i = (g_1^i, ..., g_{k_i}^i) \in \mathbb{R}^{k_i}, \ i \in J_m$. Then $\pi \in P$ represents an ordered set of vectors from the set G, i.e.

$$\pi = \langle \pi_1, ..., \pi_n \rangle = \langle G_{i_1}, ..., G_{i_n} \rangle, \ i_j \in J_m, \ j \in J_n. \tag{24}$$

We define a bijective mapping $\phi : P \to E$ by setting each element of $\pi \in P$ of the form (24) in matching point $x = (x_1, x_2, ..., x_N) \in R^N$ by the following rule.

Let $K = \max_{i \in J_m} k_i$. Assume $x = (x_1, x_2, ..., x_N) = (\tilde{g}^{i_1}, ..., \tilde{g}^{i_n})$, where $\tilde{g}^j = (g^j, \mathbf{0}) = (\mathbf{a^j_1}, ..., \mathbf{a^j_{k_j}}, 0, ..., 0) \in \mathbf{R^K}$, $j \in J_m$, $N = nK$.

At the same time, if $k_j = K$ for some $j \in J_m$, then $\tilde{g}^j = g^j$.

This approach to constructing the mapping $\phi : P \to E \subset \mathbb{R}^N$ is called the *immersion* of a combinatorial Euclidean set P to $\mathbb{R}^N$.

An important property of many Euclidean combinatorial sets is that they belong to the so-called well-described sets [30], for which a linear optimization problem is polynomially solvable. In particular, this applies to the sets of permutations, partial permutations, partial permutations, permutation matrices, Boolean set and their certain subsets etc. [1, 18, 31, 32]. These properties allowed us to specify the statements of Theorems 4–7 taking into account the solution of optimization problems on the right-hand sides of expressions. The corresponding results are described in [1, 19, 20].

In [29], properties of various classes of combinatorial sets are established. Most common, a set $G = \{g_1, ..., g_n\}$ of real numbers was considered as a generating set. To emphasize that we study an image in Euclidean space of the corresponding combinatorial set, further, we will add the refinement "Euclidean" to their title. Denote by $E_n(G)$ the Euclidean combinatorial set of permutations without repetition formed from G; a set of permutations with repetitions of $k < n$ various elements in G will be denoted as $E_{nk}(G)$. $B_n(m)$ is a partial case of $E_{nk}(G)$ with $k = 2$ and $G = \{0^{n-m}, 1^m\}$. In literature, less common is a set of partial permutations of size n (with repetitions if $\eta > k$) induced by a multiset $G = \{g_1, ..., g_\eta\}$, where $\eta > n$. There are two partial cases of the set. First corresponds to $\eta = k$ and called the Euclidean set of partial permutations without repetitions. Another case if $\eta = n \cdot k$, when the set is called the Euclidean set of partial permutations with unbounded repetitions. Cartesian products of two or more Euclidean combinatorial sets of permutations/partial permutations a called tuples of the Euclidean combinatorial sets [33]. Of particular importance in the study of various classes of sets of Euclidean combinatorial configurations is the study of the properties of combinatorial polytopes as convex hulls of point sets $E = \phi(P) \subset \mathbb{R}^n$. Analytic methods for describing polytopes of Euclidean sets of permutations and partial permutations, determining their dimension, establishing criteria for vertices, faces, adjacency of vertices and faces, and minimal irreducible representations are the subject of papers [31, 32, 34, 35].

We formulate the optimization problem on the Euclidean combinatorial set P in the following formulation. Let the functional $F : P \to \mathbb{R}^1$ be given. It is required to find

$$\pi^* = \arg \min_{\pi \in P'} F(\pi), \tag{25}$$

where $P' \subseteq P$ is a set of feasible solutions.

Suppose that $\phi : P \to E$ is a bijective mapping such that for some function $f : E \to \mathbb{R}^1$, for all $x \in E$, the condition

$$f(x) = F(\phi^{-1}(x)). \tag{26}$$

Then problem (25) can be formulated as: to find

$$x^* = \arg \min_{x \in E' \subseteq E} f(x), \tag{27}$$

where $E' = \phi(P')$ is the image of the set P' in $\mathbb{R}^n$.

Problem (27) is called the problem of Euclidean combinatorial optimization, and the model constructed as a result of specifying a bijective mapping $\phi : P \to E$ and forming a function $f : E \to \mathbb{R}^1$ that satisfies condition (26) is called the Euclidean combinatorial model.

The main results in the field of the theory of Euclidean combinatorial optimization are discussed in the article [36].

Thus, Euclidean combinatorial optimization problems can be considered as a special class of discrete optimization problems, which means that the results described in the previous section can be used for them. It is necessary to solve two interrelated problems: specify a bijective mapping $\phi : P \to E \subset \mathbb{R}^n$ and find an analytic expression consistent with it for a function $f : E \to \mathbb{R}^1$ that satisfies condition (26).

In [15, 22], a constructive iterative approach to the construction of convex/concave, strongly convex/concave extensions of monomials from $E_{nk}(G)$ with $G \geq 0$ onto the orthant $\mathbb{R}^n_+$. The complexity of constructing these extensions depends on the type of monomials, in particular, on its sign and the number of variables involved. As a result, a nonsmooth convex extension is constructed. Later this method was called the "Stoyan-Yakovlev method" [37, 38] (the Stoyan-Yakovlev Method, *SYM*) and was generalized in several directions. For example, in [37] it was noted that the number of arithmetic operations for constructing convex monomial extensions using a constructive method of the SYM type essentially depends on the method of splitting them into pairs of factors, and its modification (the Modified SYM, *MSYM*), using various such methods. The main disadvantage of SYM and MSYM is the disproportionate increase in the components in the resulting convex mono extension with a positive sign when the sign is reversed. Besides, in [37], MSYM was transferred to the image in $\mathbb{R}^n$ of a set of a tuple of permutation sets [38] and estimates of the complexity of SYM and different versions of MSYM are presented. Besides, in work [18], a method was proposed for constructing convex extensions of polynomials, the number of components of which for monomials with a negative sign linearly depends on the number of components in the corresponding monomials with a positive sign. In addition, three methods of generating extensions were presented, depending on the monomial partition into factors, and much better than SYM, MSYM estimates of the number of arithmetic operations were given, namely, quadratic and cubic. In the paper [19], a method was proposed for constructing convex extensions of monomials from the set $E_{nk}(G) \subset \mathbb{R}^n_+$ onto $\mathbb{R}^n_+$ in the forms of posynomials.

Analysis of the existing analytic methods for constructing convex extensions of quadratic functions from combinatorial sets inscribed in a hypersphere allows stating that, in many cases, they yield convex extensions in the form (10), (11), where a matrix of a quadratic form defines μ. Thus, the question arises of identifying the classes of FPCs to which this method is applicable, as well as comparing different convex extensions given in the form (10), (11) depending on a choice x^0.

7 Conclusions

In the paper, a comprehensive review of extension theory focusing on Convex Extension Theory is given, and the application of the results in Combinatorial Optimization is listed. An extended typology of function extensions is given. Several main problems in Convex Extension Theory are formulated, allowing the developing research areas more systematically. Close relation of the existence of function extension with vertex-located sets is established. Two ways to replace a general combinatorial optimization problem equivalently over a finite point configuration are outlined with the one or ones on vertex-located sets. Both these reformulation techniques rely on Euclidean Configuration Theory, studying the properties of finite point configurations as images of combinatorial configuration sets. This stimulates the development of these research areas further.

References

1. Yakovlev, S.: In: Optimization Methods and Applications: In Honor of Ivan V. Sergienko's 80th Birthday, Butenko, S., Pardalos, P.M., Shylo, V. (eds.), Springer Optimization and Its Applications (Springer International Publishing), pp. 567–584. https://doi.org/10.1007/978-3-319-68640-0_27
2. Christ, M.: Arkiv för Matematik **22**(1–2), 63 (1984). https://doi.org/10.1007/BF02384371
3. Christ, M., Lima: Proc. London Math. Soc. **s3–25**(1), 27 (1972). https://doi.org/10.1112/plms/s3-25.1.27
4. Peters, H.J.M., Wakker, P.P.: Econ. Lett. **22**(2), 251 (1986). https://doi.org/10.1016/0165-1765(86)90242-9
5. Bucicovschi, O., Lebl, J.: **1012** (2010). arXiv:1012.5796
6. Rzymowski, W.: J. Math. Anal. Appl. **212**(1), 30 (1997). https://doi.org/10.1006/jmaa.1997.5093
7. Laptin, Y.P.: Cybern. Syst. Anal. **52**(1), 85 (2016). https://doi.org/10.1007/s10559-016-9803-8
8. Crama, Y.: Math. Program. **61**(1–3), 53 (1993). https://doi.org/10.1007/BF01582138
9. Whitney, H.: Hassler Whitney Collected Papers, Contemporary Mathematicians, pp. 228–254. Birkhäuser Boston (1992). https://doi.org/10.1007/978-1-4612-2972-8_14
10. Tawarmalani, M., Sahinidis, N.V.: In: Convexification and Global Optimization in Continuous and Mixed-Integer Nonlinear Programming, no. 65 in Nonconvex Optimization and Its Applications, pp. 25–70. Springer, US (2002). https://doi.org/10.1007/978-1-4757-3532-1_2

11. Frölicher, A., Kriegl, A.: Diff. Geom. Appl. **3**(1), 71 (1993). https://doi.org/10.1016/0926-2245(93)90023-T
12. Jansson, C.: BIT Numer. Math. **40**(2), 291 (2000). https://doi.org/10.1023/A:1022343007844
13. Kirchheim, B., Kristensen, J.: Comptes Rendus de l'Académie des Sciences—Series I-Mathematics **333**(8), 725 (2001). https://doi.org/10.1016/S0764-4442(01)02117-6
14. Sherali, H.D.: In: Pardalos, P.M., Romeijn, H.E. (eds.), Handbook of Global Optimization, no. 62 in Nonconvex Optimization and Its Applications, pp. 1–63. Springer, US (2002)
15. Yakovlev, S.V.: Comput. Math. Math. Phys. **34**(7), 1112 (1994)
16. Yan, M.: J. Convex Anal. **21**(4), 965 (2014)
17. Alexandrov, A.: Leningrad State University. Ann. [Uchenye Zapiski] Math. Ser. (6), 3 (1939)
18. Pichugina, O., Yakovlev, S.: In: Shakhovska, N., Medykovskyy, M.O. (eds.), Advances in Intelligent Systems and Computing IV. Advances in Intelligent Systems and Computing, pp. 231–246. Springer International Publishing (2019). https://doi.org/10.1007/978-3-030-33695-0_17
19. Yakovlev, S., Pichugina, O., Koliechkina, L.: In: Lecture Notes in Computational Intelligence and Decision Making. Advances in Intelligent Systems and Computing, pp. 195–212. Springer International Publishing, Cham (2021). https://doi.org/10.1007/978-3-030-54215-3_13
20. Yakovlev, S., Pichugina, O.: In: Proceedings of the Second International Workshop on Computer Modeling and Intelligent Systems (CMIS-2019), pp. 570–580. CEUR Vol-2353 urn:nbn:de:0074-2353-0, Zaporizhzhia, Ukraine (2019)
21. Dragomirescu, F., Ivan, C.: Optimization **24**(3–4), 193 (1992). https://doi.org/10.1080/02331939208843789
22. Yakovlev, S.V.: Cybernetics **25**(3), 385 (1989). https://doi.org/10.1007/BF01069996
23. Hadjisavvas, N., Komlósi, S., Schaible, S.S. (eds.): Handbook of Generalized Convexity and Generalized Monotonicity, 2005th edn. Springer (2006)
24. Azagra, D., Mudarra, C.: Cal. Var. Part. Diff. Equ. **58**(3), 84 (2019). https://doi.org/10.1007/s00526-019-1542-z
25. Stoyan, Y.G., Yakovlev, S.V., Emets, O.A., Valuĭskaya, O.A.: Cybern. Syst. Anal. **34**(2), 27 (1998). https://doi.org/10.1007/BF02742066
26. Jiang, B., Liu, Y., Wen, Z.: SIAM J. Optim. **26**(4), 2284 (2016). https://doi.org/10.1137/15M1048021
27. Yakovlev, S.V., Pichugina, O.S.: Cybern. Syst. Anal. **54**(1), 99 (2018). https://doi.org/10.1007/s10559-018-0011-6
28. Berge, C.: Principles of Combinatorics. Academic Press (1971)
29. Stoyan, Y.G., Yakovlev, S.V., Pichugina, O.S.: The Euclidean Combinatorial Aonfigurations: a Monograph. Constanta, Kharkiv (2017)
30. Berstein, Y., Lee, J., Onn, S., Weismantel, R.: Math. Program. **124**(1–2), 233 (2010). https://doi.org/10.1007/s10107-010-0358-6
31. Stoyan, Y.G., Yakovlev, S.V.: Mathematical Models and Optimization Methods in Geometric Design. Naukova Dumka, Kiev (2020)
32. Yemelichev, V.A., Kovalëv, M.M., Kravtsov, M.K.: Polytopes, Graphs and Optimisation. Cambridge University Press, Cambridge (1984). http://www.ams.org/mathscinet-getitem?mr=744197
33. Grebennik, I.V., Kovalenko, A.A., Romanova, T.E., Urniaieva, I.A., Shekhovtsov, S.B.: Cybern. Syst. Anal. **54**(2), 221 (2018). https://doi.org/10.1007/s10559-018-0023-2
34. Iemets, O.O., Yemets', O.O., Polyakov, I.M.: **54**(5), 796. https://doi.org/10.1007/s10559-018-0081-5
35. Yemets, O.A., Yemets, A.O., Polyakov, I.M.: **49**(12). https://doi.org/10.1615/JAutomatInfScien.v49.i12.20
36. Stoyan, Y.G., Yakovlev, S.V.: Cybern. Syst. Anal. **56**(3), 366 (2020). https://doi.org/10.1007/s10559-020-00253-6
37. Valuiskaya, O.A., Pichugina, O.S., Yakovlev, S.V.: Radioelectron. Inf. J. (2(15)), 121 (2001)
38. Yemets', O., Romanova, N.G.: Optimization Over Polypermutations. Naukova Dumka, Kyiv (2010)

Action Encoding in Algorithms for Learning Controllable Environment

Andrii Tytarenko

Abstract The work considers the Learning Controllable Embedding (LCE) algorithmic framework which is tasked to learn a lower-dimensional latent state space steerable using simpler control algorithms like iLQR. However, most LCE papers consider raw action spaces, which limits their applicability to simpler low-dimensional continuous action spaces. This paper introduces the support of a larger set of action spaces by adding the action encoding to two LCE frameworks and provides empirical evidence on the algorithms' performance. Algorithms are rederived for this purpose, and the suboptimality analysis is extended. The experiments are discussed, including benchmark tasks, data collection policies, and a comparison of the considered algorithms. Moreover, the paper investigates the effect of the size of the action encoding on the performance and explores the possibilities of using LCE for more complex action spaces, such as discrete or high-dimensional ones. The results show that the proposed method can effectively learn controllable embeddings for a wider range of action spaces.

Keywords Learning controllable environment · Representation learning · Action encoding · Reinforcement learning · State-action embeddings

1 Introduction

Learning autonomous agents capable of control in complex state and action spaces in Markov decision processes (MDP) is a challenging problem. The difficulties can arise from factors such as large action spaces [1], limited interaction with the environment [2], partial observability (POMDP) [3]. To optimize a good policy, a lot of samples are required, along with online interactive learning and neural networks capable of handling higher dimensional observations with a large number of trainable parameters [4, 5]. There are various algorithms available to address the issue of

A. Tytarenko (✉)
Igor Sikorsky Kyiv Polytechnic Institute, Kyiv, Ukraine
e-mail: tytarenko.andrii@lll.kpi.ua

M. Zgurovsky and N. Pankratova (eds.), *System Analysis and Artificial Intelligence*, Studies in Computational Intelligence 1107,
https://doi.org/10.1007/978-3-031-37450-0_16

sample inefficiency, limited data, or high cost of data collection, such as model-based reinforcement learning [6] and offline reinforcement learning methods [2].

Learning Controllable Embedding (LCE) is an algorithmic framework that addresses the problem of controlling agents by learning a lower-dimensional latent state space and using simpler control algorithms like iLQR to perform control in this latent space. The challenge here is to ensure that the learned latent space has a simpler structure, i.e., the next states are easier to predict. Some specific instances of this framework are described in [7–10]. For example, E2C [7] proposes to learn a locally-linear latent space, allowing the use of algorithms like LQG for goal-reaching tasks. PCC [8] attempts to address some of the issues encountered in E2C by deriving losses that enable explicit minimization of the latent space's curvature. PC3 [9] improves upon PCC mainly by replacing the reconstruction loss required to ensure that the learned state space carries enough information to generate observations from latent states, using predictive coding instead. SOLAR [11] introduces a probabilistic framework for learning latent spaces specifically amenable to the guided policy search algorithm [12].

Most LCE papers consider raw action spaces, without any parametrized processing. This limits their applicability to simpler low-dimensional continuous action spaces. In this work, we study the ways to introduce action encoding to two LCE frameworks: PCC and PC3. We discuss a suboptimality of the algorithms and provide empirical evidence on the algorithms' performance.

2 General Results

In this section, the main results and contributions are presented.

In Sect. 2.1, the minimal preliminaries in Reinforcement Learning and Learning Controllable Environment are given. These are crucial for understanding the results. In Sect. 2.2, action encoding is described, which is the main subject of this study.

Sections. 2.3 and 2.4 present the proposed extensions to the two considered LCE frameworks. In each, an algorithm is derived along with the extension of the subobtimality analysis. Experiments are discussed in Sect. 2.5. The benchmark tasks are discussed along with the corresponding data collection policies, and the comparison of the considered algorithms is provided.

2.1 Preliminaries

In the following subsection, the background information is presented necessary for understanding the main results.

2.1.1 Reinforcement Learning

We denote a Markov Decision Process (MDP) $\mathcal{M}$ as a tuple $(S, A, c, \mathcal{T})$, where

S—state space

A—action space

$c : S \times A \rightarrow \mathbb{R}$—cost function. It's common for reinforcement learning to consider the reward function instead. However, these options are equivalent up to the sign, and we choose costs for consistency with the related work.

$\mathcal{T}$ – transition kernel. $\mathcal{T}(s_t, a_t, s_{t+1}) = p(s_{t+1}|s_t, a_t)$—probability of state s_{t+1} given the current state s_t and action taken a_t.

A state of an MDP is a sufficient statistic for a transition kernel, possessing a Markov property.

A reinforcement learning algorithm is tasked to find a policy π which maximized the expected return. This objective in the finite-horizon case of MDP can be written as:

$$\min_{\pi} \mathbb{E}_{\tau \sim p(\cdot)} \sum_{t=0}^{T-1} c(s_t, a_t) \tag{1}$$

where τ denotes a trajectory $(s_0, a_0, s_1, a_1, ...)$ obtained by sampling actions using a stochastic π.

In the current work, we are interested in controlling a non-linear dynamical system of the form $s_{t+1} = p(s_t, a_t) + \omega$. Here, ω is the Gaussian system noise, and P is a smooth function. In addition, we assume that the actual states of a system are not accessible to an algorithm. Instead, it operates using high-dimensional observations $x_t \in X \subseteq \mathbb{R}^N$, where $N >> |S|$. This leads to a visual-based stochastic control problem we work with in the present work.

$$\min_{\mathbf{a}} L(\mathbf{a}, p, c, x_0) = \mathbb{E}[\sum_{t=0}^{T-1} c(x_t, a_t)|p, x_0] \tag{2}$$

Note, however, that x_t is a sufficient statistic for recovering s_t. We leave the case of Partially Observed MDPs for future work.

In the current work, we assume that the cost functions are bounded by $c_{max} > 0$ and Lipschitz with $c_{lip} > 0$.

2.1.2 Learning Controllable Environment

One of the ways to solve the presented stochastic optimal control problem (2), is to employ Learning Controllable Environment (LCE) algorithms. In LCE framework, an algorithm is tasked to approximate a mapping from the high-dimensional observation space X to the lower-dimensional latent state space Z, along with a latent state space transition dynamics $p_F(z_{t+1}|z_t, a_t)$. This allows for solving the high-dimensional problem (2) by replacing it with a lower-dimensional stochastic optimal control problem:

$$\min_{\mathbf{a}} \mathbb{E}[L(\mathbf{a}, p_F, \hat{c}, z_0] \tag{3}$$

where z_0 corresponds to an "encoded" initial observation x_0 and is obtained by sampling $z_0 \sim p_E(z_0|x_0)$.

This latent problem can be solved with a large variety of reinforcement learning algorithms such as [13, 14]. However, in LCE we are interested in such latent spaces, which lead to problems solvable using locally-linear control algorithms, such as iLQR [15] or DDP [16]. Therefore, LCE algorithms are also tasked to enforce the local linearity in the latent state space.

Clearly, the quality of the solution $\hat{a}^*$ is determined by the quality of the latent state space. The formal analysis of the suboptimality is specific to each particular instance of the LCE framework and will be provided in Sects. 2.3 and 2.4.

The motivation for studying the LCE framework is as follows. Compared to most deep reinforcement learning algorithms, locally-linear control algorithms are incredibly fast. Moreover, there are no deep neural networks needed to be trained for each separate task in the same environment. Yet the current state-of-the-art LCE methods are not as powerful as other deep reinforcement learning algorithms, they possess great potential and properties inaccessible to other methods.

2.2 Action Encoding

Previous works [7–9, 11, 17] considered the problem of stochastic optimal control with continuous action spaces. The latent stochastic optimal control problem structure in those works considers raw actions, so this design choice imposes restrictions on the methods' applicability.

First, this implies that discrete action spaces are not supported. Discrete action spaces can be found in lots of practical problems: from Operations Research tasks like floor planning through self-driving [18], robotics [19], to games [4].

Second, LCE methods strive to "learn" a transformation of raw state spaces into latent state spaces with desired properties. By working directly with raw actions, the expressivity of the transformation may be reduced. Also, there are maybe many problems with complex action spaces, such as dexterous manipulation [20] or large-scale problems like [4, 21].

Lastly, this makes an algorithm highly dependent on the action space design choice, which is left to a possibly unaware algorithm user. A highly redundant action space may break the algorithm, even though the space is technically continuous and should be supported by it. By introducing the parametrized action encoding, we allow for further research on these useful applications.

In this work, we study the ways to introduce action encoding to the LCE method in the form of a stochastic mapping:

$$\hat{a}_t \sim p_E(\hat{a}_t|a_t) \tag{4}$$

or a deterministic function:

$$\hat{a}_t = E(a_t) \tag{5}$$

Two related methods are generalized with the action encoding: PCC and PC3. This generalization requires an extension of the corresponding frameworks, suboptimality analyses, and algorithm instances. These extensions are one of the key contributions of the present paper. Concrete design choices, implications, and evaluation results are discussed in further sections.

2.3 PCC: Prediction, Consistency, Curvature

In the scope of the Prediction, Consistency, Curvature (PCC) framework, the algorithm is supposed to learn three mappings: an observation encoder $p_E(z|x)$, a latent space dynamics $p_F(z'|z, a)$, and an observation decoder $p_D(x|z)$, where $x \in X$, $z \in Z$, and $a \in A$. For convenience, a composition of encoding, dynamics, and decoding is denoted $\hat{p}(x'|x, a)$.

We extend this approach by adding an action encoder $p_A(\hat{a}|a)$. Hence, a latent space dynamics is now $p_F(z'|z, \hat{a})$, where $\hat{a} \in \hat{A}$. The composition $\hat{p}$ now also includes the action encoding step, but the notation remains the same. See the diagram Fig. 1 for visualization.

As we discussed in Sect. 2.2, we are interested in solving a stochastic control problem in the latent space. However, we also want to reduce the suboptimality caused by the encoders, and the latent space dynamics. This is a way to optimize the parameters of our latent space transformation, as shown in [8].

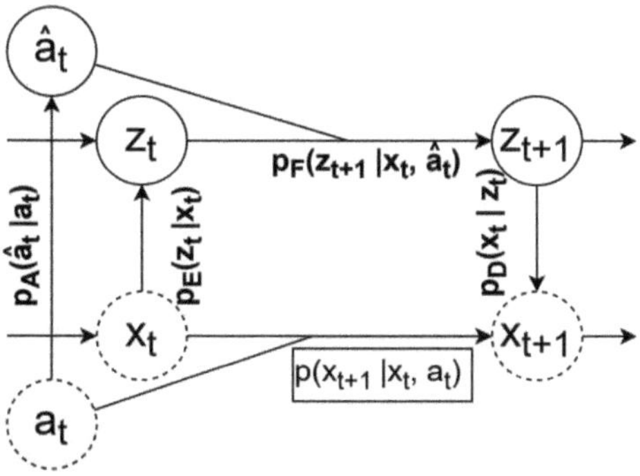

Fig. 1 The diagram depicting probabilistic models considered in this work. The mappings in bold compose $\hat{p}(x_{t+1}|x_t, a_t)$, while the direct transition mapping $p(x_{t+1}|x_t, a_t)$ is a natural model of the original MDP. The key idea is to match these two in a way, such that the obtained latent state space possesses certain desired properties (e.g. smoothness)

Namely, the joint problem of control and suboptimality reduction is as follows:

$$\min_{\mathbf{a},\hat{p}} \mathbb{E}_{z_0\sim p_E(\cdot|x_0),\hat{a}_t\sim p_A(\cdot|a_t))}[L(\hat{\mathbf{a}}, p_F, \hat{c}, z_0)|x_0] + \lambda\sqrt{R(\hat{p})} \tag{6}$$

where λ is a regularization coefficient, and $R\hat{p}$ is a regularization term, which characterizes suboptimality of the latent space transformation. The cost function $\hat{c}$ is called the latent cost function and is defined as:

$$\hat{c}(z, \hat{a}) = \int c(x, a) p_D(x|z) p_{A_D}(a|\hat{a}) da dx \tag{7}$$

Here, p_{A_D} is an action decoding mapping.

The regularization term is selected according to the suboptimality analysis provided in the following subsection.

2.3.1 Suboptimality

Reference [8] gives a characterization of control suboptimality. It is extended with an action encoding in the form of the following statement.

We are interested in the in an upper-bound of the form:

$$L(\mathbf{a}^*_{raw}, p, c, x_0) \geq \mathbb{E}_{\hat{a}^*_t\sim p_A(\cdot|a^*_t)}[L(\hat{\mathbf{a}}^*, p, c, x_0)] + \lambda\sqrt{R(\hat{p})} \tag{8}$$

where $\mathbf{a}$ denotes $a_{0:T-1}$, $\mathbf{a}^*_{raw}$ denotes a solution for the stochastic optimal control problem (2), and $\mathbf{a}^*$ denotes a solution for the latent problem (6).

Our reasoning builds on suboptimality analysis from [8]. First, let us assume $t > 0$, and $a_{0:T-1}$ are known.

Following the proof of Lemma 11 in [22] (technical result from a paper proposing policy improvement procedure with a baseline regret minimization), we derive the following decomposition.

$$\mathbb{E}[c(x_t, a_t)|x_t] = \int_{x_{0:t}} c(x_t, a_t) \prod_{k=1}^{t-1} p(x_k|x_{k-1}, a_{k-1}) \hat{p}(x_t|x_{t-1}, a_{t-1}) dx_{0:t} \tag{9}$$

$$+c_{max} \int_{x_{0:t}} D_{TV}[p_t||\hat{p}_t] \prod_{k=1}^{t-1} p(x_k|x_{k-1}, a_{k-1}) dx_{0:t-1} \tag{10}$$

Here, $D_{TV}[p_t||\hat{p}_t]$ denotes a total variation distance between probability densities $p_t = p(\cdot|x_{t-1}, a_{t-1})$ and $\hat{p}_t = \hat{p}(\cdot|x_{t-1}, a_{t-1})$.

We also denote $\prod_{k=1}^{t-1} p_k$ as Π_{t-1} for brevity. Marginalizing and factorizing the density $\hat{p}$ and using (7):

$$= c_{max} \int_{x_{0:t-1}} D_{TV}[p_t || \hat{p}_t] \Pi_{t-1} dx_{0:t-1} \tag{11}$$

$$+ \int_{x_{0:t-1}} \Pi_{t-1} \int_{x_t} \int_{z_{t-1:t}} \int_{\hat{a}_{t-1:t}} c(x_t, a_t) p_D(x_t|z_t) p_{A_D}(a_t|\hat{a}_t) \cdot \tag{12}$$

$$p_E(z_{t-1}|x_{t-1}) p_A(\hat{a}_{t-1}|a_{t-1}) p_F(z_t|z_{t-1}, \hat{a}_{t-1}) d\hat{a}_{t-1:t} dz_{t-1:t} dx_t dx_{0:t-1} \tag{13}$$

$$= c_{max} \int_{x_{0:t-1}} D_{TV}[p_t || \hat{p}_t] \Pi_{t-1} dx_{0:t-1} \tag{14}$$

$$+ \int_{x_{0:t-1}} \Pi_{t-1} \int_{z_t} \int_{\hat{a}_t} \hat{c}(z_t, \hat{a}_t) p(z_t|x_{t-1}, a_t) dz_t d\hat{a}_t dx_{0:t-1} \tag{15}$$

Here, $p(z_t|x_t, a_t)$ is density for a distribution composed of p_E, p_A, and p_F and marginalized over z_{t-1}, and $\hat{a}_{t-1}$.

Next, we follow a similar idea for the expectation over the sum of cost functions in two consecutive time steps. Let us denote the compositional factorized density $p_{FEA}(z_{t+1}|x_t, a_t) = \int_{z,\hat{a}} p_F(z_{t+1}|z, \hat{a}) p_E(z|x_t) p_A(\hat{a}|a_t) dz d\hat{a}$.

This time, we operate with expectations for brevity and clarity.

$$\mathbb{E}[c(x_{t-1}, a_{t-1}) + c(x_t, a_t)|a_{0:t}, x_0] = \tag{16}$$

$$\mathbb{E}[\mathbb{E}_{\substack{\hat{a}_{t-1} \sim p_A(\cdot|a_{t-1}) \\ z_{t-1} \sim p_{FEA}(\cdot|x_{t-2}, a_{t-2})}} [\hat{c}(z_{t-1}, \hat{a}_{t-1}) + \mathbb{E}_{\substack{\hat{a}_t \sim p_A(\cdot|a_t) \\ z_t \sim p_{FEA}(\cdot|x_{t-1}, a_{t-1})}} [\hat{c}(z_t, \hat{a}_t)]] \mid a_{0:t}, x_0]$$

$$+ c_{max} \mathbb{E}\left[D_{TV}[p_t || \hat{p}_t] \mid a_{0:t-1}, x_0\right]$$
$$+ c_{max} \mathbb{E}\left[D_{TV}[p_{t-1} || \hat{p}_{t-1}] \mid a_{0:t-2}, x_0\right]$$

$$\leq \mathbb{E}[\mathbb{E}_{\substack{\hat{a}_{t-1} \sim p_A(\cdot|a_{t-1}) \\ z_{t-1} \sim p_{FEA}(\cdot|x_{t-2}, a_{t-2})}} [\hat{c}(z_{t-1}, \hat{a}_{t-1}) + \mathbb{E}_{\substack{\hat{a}_t \sim p_A(\cdot|a_t) \\ z_t \sim p_F(\cdot|z_{t-1}, \hat{a}_{t-1})}} [\hat{c}(z_t, \hat{a}_t)]] \mid a_{0:t}, x_0]$$

$$+ c_{max} \mathbb{E}\left[D_{TV}[p_t || \hat{p}_t] \mid a_{0:t-1}, x_0\right]$$
$$+ c_{max} \mathbb{E}\left[D_{TV}[p_{t-1} || \hat{p}_{t-1}] \mid a_{0:t-2}, x_0\right]$$
$$+ c_{max} \mathbb{E}\left[D_{TV}\left[\int_{x'} \hat{p}(x', x_{t-2}, a_{t-2}) p_E(\cdot|x') dx' || p_{FEA}(\cdot|x_{t-2}, a_{t-2})\right] \mid a_{0:t-1}, x_0\right]$$

Expanding the total cost further (for three, four, five cost terms, and so on) we are able to derive the following bound. We skip the intermediate steps for brevity, as they are applying Pinsker's inequality (the one which connects TV distance and KL divergence), using the properties of the KL divergence, and Jensen's inequality. After these steps, the result is as follows.

$$\left|\mathbb{E}_{\hat{a}_t \sim p_A(\cdot|a_t)} L(\hat{\mathbf{a}}, F, \hat{c}, z_0)|x_0, \mathbf{a}]\right| \leq \tag{17}$$

$$2T^2 c_{max}(\mathbb{E}[\frac{1}{T}\sum_{t=0}^{T-1} 3D_{KL}[p(\cdot|x_t, a_t)||\hat{p}(\cdot|x_t, a_t)]+$$
$$2D_{KL}[p_E(\cdot|x_{t+1})||p_{FEA}(\cdot|x_t, a_t) \mid x_0, a_{0:T-1}]])^{\frac{1}{2}}$$

Let us consider $\mathbf{a}^*_{raw}$—a solution for the stochastic optimal control problem (2), and $(\mathbf{a}^*, \hat{p}^*)$—a solution for the latent problem (6). Following the reasoning from Lemma 1 of [8], we obtain the desired bound.

$$L(\mathbf{a}^*_{raw}, p, c, x_0) \geq \mathbb{E}_{\hat{a}^*_t \sim p_A(\cdot|a^*_t)}[L(\hat{\mathbf{a}}^*, p, c, x_0)] - \tag{18}$$
$$\lambda\sqrt{2R_{pred}(\hat{p}^*) + 3R_{consist}(\hat{p}^*)} \tag{19}$$

$$R_{pred}(\hat{p}) = \mathbb{E}[D_{KL}(p(\cdot|x, u)||\hat{p}(\cdot|x, u))] \tag{20}$$
$$R_{consis}(\hat{p}) = \mathbb{E}[D_{KL}[p_E(\cdot|x')||p_{FEA}(\cdot|x, a)] \tag{21}$$

Thus, optimizing the regularizing terms R_{pred} and $R_{consist}$ decreases the discrepancy between the latent and original MDPs. Interestingly enough, even in the presence of action encoding, the general interpretation of these losses as *prediction* and *consistency* suboptimality measures. Nevertheless, the terms differ for the original ones and the surrogate objectives for both must be rederived accordingly.

In the remaining part of the current subsection, we derive a tractable way to optimize both these terms and formulate a full algorithm for PCC with action encoding. We omit the local linearity term for brevity, as it is exactly identical to the original one [8, 9]. It will show up in the discussion of the algorithm.

Note, that in offline LCE methods, the control part of the problem is solved separately and is omitted during the model training. Therefore, we are not concerned with the control part of the objective, but only with suboptimality terms.

2.3.2 Prediction with Action Encoding

The suboptimality analysis suggests that one has to minimize regularization terms in order to learn a reasonable latent space transformation. First, we consider $R_{pred}(\hat{p})$ term, which basically boils down to the minimization of the negative log-likelihood $-\log(\hat{p}(x'|x, a)$ given the data triplets $(x_{t+1}|x_t, a_t)$. It is intractable for practical optimization. However, for a negative log-likelihood, a variational bound can be derived. It serves as a surrogate loss function, further used for optimization.

$$\log \hat{p}(x_{t+1}|x_t, a_t) \tag{22}$$

$$= \log \int_{z_t, z_{t+1}, \hat{a}_t} \hat{p}(x_{t+1}, z_t, z_{t+1}, \hat{a}_t|x_t, a_t) dz_t dz_{t+1} d\hat{a}_t \tag{23}$$

$$= \log \mathbb{E}_{z_t, z_{t+1}, \hat{a}_t \sim q(z_t, z_{t+1}, \hat{a}_t|x_t, x_{t+1}, a_t)} \left[\frac{\hat{p}(x_{t+1}, z_t, z_{t+1}, \hat{a}_t|x_t, a_t)}{q(z_t, z_{t+1}, \hat{a}_t|x_t, x_{t+1}, a_t)} \right] \tag{24}$$

$$\geq \mathbb{E}_{z_t, z_{t+1}, \hat{a}_t \sim q(z_t, z_{t+1}, \hat{a}_t|x_t, x_{t+1}, a_t)} \left[\log \frac{\hat{p}(x_{t+1}, z_t, z_{t+1}, \hat{a}_t|x_t, a_t)}{q(z_t, z_{t+1}, \hat{a}_t|x_t, x_{t+1}, a_t)} \right] \tag{25}$$

$$= \quad \mathbb{E}_{z_{t+1} \sim q(\cdot|x_{t+1})} \log p_D(x_{t+1}|z_{t+1}) \tag{26}$$

$$- \mathbb{E}_{z_{t+1} \sim q(\cdot|x_{t+1})} D_{KL}[q(z_t|z_{t+1}, x_t, \hat{a}_t||p_E(z_t|x_t)] \tag{27}$$

$$- \mathcal{H}(q(z_{t+1}|x_{t+1})) \tag{28}$$

$$+ \mathbb{E}_{\substack{z_{t+1} \sim q(\cdot|x_{t+1}) \\ \hat{a}_t \sim q(\cdot|a_t) \\ z_t \sim q(\cdot|z_{t+1}, x_t, \hat{a}_t)}} [\log p_F(z_{t+1}|z_t, \hat{a}_t)] \tag{29}$$

Here, $q(z_t|z_{t+1}, x_t, \hat{a}_t)$ is called backward dynamics and is modeled as a factorized normal density parametrized with a deep learning model. Encoders $q(z|x) = p(z|x)$ are parametrized by a single network. The decoder $p_D(x|z) = q(x|z)$is also parametrized with a deep learning model, but a particular form of distribution is dependent on the task. The novel action encoder is also a network $q(\hat{a}|a) = p_A(\hat{a}|a)$ parametrizing an action-space dependent distribution.

2.3.3 Consistency with Action Encoding

Now, we consider $R_{consist}(\hat{p})$ term. It is also intractable for practical optimization. Nevertheless, a variational bound can be derived as well. It serves as a surrogate loss function, further used for optimization. The derivation follows the same reasoning as in the case of R_{pred}. Essentially, we decompose the KL divergence, introduce a variation family $q(z_t, \hat{a}_t|z_{t+1}, x_t, a_t)$, use Jensen's inequality to propagate logarithm, and factorize the expression into tractable terms.

$$
\begin{aligned}
D_{KL}[p_E(\cdot|x_{t+1}) \;||\; p_{FEA}(\cdot|x_t, a_t)] \leq \qquad & \qquad (30)\\
- \mathcal{H}(q(z_{t+1}|x_{t+1})) \qquad & \qquad (31)\\
+ \mathbb{E}_{z_{t+1}\sim q(\cdot|x_{t+1}} D_{KL}(q(z_t|z_{t+1}, x_t, \hat{a}_t)||p_E(z_t|x_t)) \qquad & \qquad (32)\\
- \mathbb{E}_{\substack{z_{t+1}\sim q(\cdot|x_{t+1}) \\ \hat{a}_t\sim q(\cdot|a_t) \\ z_t\sim q(\cdot|z_{t+1}, x_t, \hat{a}_t)}} [\log p_F(z_{t+1}|z_t, \hat{a}_t)] \qquad & \qquad (33)
\end{aligned}
$$

This bound does not differ much from the original [8], except for the last term, where the action encoding step is used for sampling.

2.3.4 Action Decoding

Although the suboptimality analysis does not include action decoders $p_{A_D}(a|\hat{a})$, we still need access to them in practice. The main reason is that as soon as we find an optimal plan in the latent space, we need to decode the raw actions which the controlled agent has to take.

The action decoder is trained with a separate loss function based on the variational auto-encoder [23] (here $q(\hat{a}|a)$ essentially is equivalent to p_A):

$$
L_{act} = -\mathbb{E}_{\hat{a}\sim q(\cdot|a)} \log p_{A_D}(a|\hat{a}) + D_{KL}(q(\hat{a}|a)||p(\hat{a})) \qquad (34)
$$

In practice, the marginal distribution $p(\hat{a})$ is set to be $\mathcal{N}(0, 1)$, and the expectation is also estimated using one-point estimates. Note, that the encoder is also trained as a part of other loss functions, this is exactly the same action encoder.

2.3.5 Algorithm

Finally, we describe a full algorithm for LCE offline pretraining.

Algorithm (parameter update)

Learning Controllable Environment with action encoding (based on PCC framework).

State encoder $q_{\theta_1}(z|x) = \mathcal{N}(\mu_{\theta_1}(x))$
Action encoder $q_{\theta_2}(\hat{a}|a) = \mathcal{N}(\mu_{\theta_2}(a))$.
Backward dynamics $q_{\theta_3}(z|z', x, \hat{a}) = \mathcal{N}(\mu_{\theta_3}(x, z', \hat{a}))$
Forward dynamics $q_{\theta_4}(z'|z, \hat{a}) = \mathcal{N}(\mu_{\theta_4}(z, \hat{a}))$
State decoder $q_{\theta_5}(x|z)$. Distribution is task specific. E.g. Ber$(\mu_{\theta_5}(z))$
Action decoder $q_{\theta_6}(a|\hat{a})$ Distribution is task specific. E.g. categorical or Gaussian.
μ_{θ_i} $(i = 1..5)$ - deep learning models (neural networks).
$\lambda_1, \lambda_2, \lambda_3, \lambda_4, \gamma$ - hyperparameters. Loss weights and a learning rate.

1. Sample a triplet of data $(x, a, x') \sim$ Dataset[1]
2. Sample next latent state from encoder $z' \sim q_{\theta_1}(z'|x')$
3. Sample latent action from encoder $\hat{a} \sim q_{\theta_2}(\hat{a}|a)$
4. Sample latent state from backward dynamics $z \sim q_{\theta_3}(z|z', x, \hat{a})$
5. Compute loss L_{pred} with (26). [2]
6. Compute loss $L_{consist}$ with (30)
7. Compute loss L_{act} with (34)
8. Compute the smoothness LCE penalty L_{LLC} (See [8, 9])
9. Updating parameters $\theta_i = \theta_i - \gamma \nabla_{\theta_i}(\lambda_1 L_{pred} + \lambda_2 L_{consist} + \lambda_3 L_{act} + \lambda_4 L_{LLC})$, $i = 1..6$

2.4 PC3: Predictive Coding, Consistency, Curvature

PC3 [9] takes a similar approach to the problem of LCE. The authors also propose to learn separate models for the encoder $p_E(z|x)$ and the dynamics $p_F(z'|z, a)$, but not the backward dynamics nor the decoder $p_D(x|z)$.

Instead of taking the same variational approach to suboptimality minimization, they build on the idea of predictive coding [24] to maximize the mutual information between the state-action pair and the next state.

We adopt this approach and incorporate the action encoding procedure. In this subsection, $E(x)$ and $E_A(a)$ are the deterministic state and action encoder, respectively.

we consider the predictive model

$$q(x'|x, \hat{a}) \propto \phi_0(x')\phi_1(E(x')|E(x), E_A(a)) \tag{35}$$

With $\phi_0 \geq 0$ and $\phi_1 \geq 0$.

Following the [9] and the reasoning from the analysis in the previous subsection, the predictive suboptimality minimization problem is as follows.

$$\min_q \mathbb{E}_{\hat{a}_t \sim p_A(\cdot|a_t)}[L(\hat{\mathbf{a}}, p, c, x_0)] - \lambda\sqrt{R_{pred}(q)} \tag{36}$$

$$R_{pred}(q) = \mathbb{E}[D_{KL}(p(\cdot|x, a)||q(\cdot|x, a))] \tag{37}$$

From Lemma 1 [9] it follows immediately, that given a reasonable choice of $e(x)$ and $f(y)$,

---

[1] In practice, batches are used.

[2] Expectations are estimated using one-point estimation with previously sampled latent states and action.

$$\mathbb{E}[D_{KL}(p(y|x)||\phi_0(y)\phi_1(e(x), f(y))] \leq I(X; Y) - I(e(X); f(Y)) \quad (38)$$

where X and Y are the random variables corresponding to the data distribution $p(x, y)$

Using this result, we may set:

$$x = (x_t, a_t) \quad (39)$$

$$y = x_{t+1} \quad (40)$$

$$e(x) = (E(x_t), E_A(a_t)) \quad (41)$$

$$f(y) = E(x_{t+1}) \quad (42)$$

and obtain the following information gap bound for R_{pred}:

$$R_{pred} \leq I(x_{t+1}; x_t, a_t) - I(E(x_{t+1}); E(x_t), E_A(a)) \quad (43)$$

Here, we swapped X and Y for readability, which is allowed due to the symmetry of I.

Thus, instead of resorting to variational inference, we now have a mutual-information-based surrogate objective. And because the first term on the right-hand side of (43) is constant, maximizing the mutual information between the encoded state-action pair and the encoded next state effectively minimizes suboptimality term R_{pred}.

2.4.1 Predictive Coding

The maximization of mutual information is a widely used procedure in modern reinforcement learning [25–27]. Interestingly enough, in [28] it is stated that the objective we demonstrated in this subsection—so-called forward information—is unbiased and leads to empirically better representation performance than concurrent formulations such as backward information.

One of the ways to estimate the mutual information is to employ contrastive predictive coding from [24]. The idea is to construct a critic-function $f : Z \times Z \times \hat{A} \to \mathbb{R}$ and the lower bound

$$I(E(x_{t+1});\ E(x_t), E_A(a_t)) \geq \quad (44)$$

$$\mathbb{E}\frac{1}{K}\sum_{i=0}^{K} \log \frac{\exp f(E(x^i_{t+1}), E(x^i_t), E_A(a^i_t))}{\frac{1}{K}\sum_{j\neq i} \exp f(E(x^i_{t+1}), E(x^j_t), E_A(a^j_t))} \quad (45)$$

The f is chosen to be equal to $\log p_F(z_{t+1}|z_t, \hat{a}_t)$, in order to enable parameter sharing and tune the p_F for the lower bound. In practice, $(x_{t+1}, x_t, a_t)^K_{i=1}$.

Except for the contrastive predictive coding loss, it is important to keep the consistency of p_F, analogously to the method described in the previous Sect. 2.4. Hence, the maximum likelihood objective is added:

$$L_{cons} = \log p_F(E(x_{t+1})|E(x_t), E_A(a_t)) \tag{46}$$

In practice, however, the usage of contrastive predictive coding is shown to be prone to collapsing the latent space [29]. Shu et al. [9] addresses this problem by adding fixed-variance Gaussian noise to the result of encoding of the next state.

In our findings, in the case of the discrete raw action spaces, it is important to add noise to the encoded action as well. The intuition behind this conclusion is to distribute the image of each particular discrete action to some region of latent space, thus making sure the resulting mapping is non-injective. It is important, as the locally-linear control algorithm that is used after the pretraining on the latent space is iterative w.r.t. to actions. Thus, it is important to reassure, that the regions surrounding the $E_A(a)$ for each $a \in A$ are mapped to a.

2.4.2 Algorithm

Finally, we summarize the algorithm for LCE offline pretraining with predictive codes.

Algorithm (parameter update)

Learning Controllable Environment with action encoding (based on PCC framework).

State encoder $q_{\theta_1}(z|x) = \mathcal{N}(\mu_{\theta_1}(x), \sigma_E^2)$[3]

Action encoder $q_{\theta_2}(\hat{a}|a) = \mathcal{N}(\mu_{\theta_2}(a), \sigma_A^2)$.

Forward dynamics $q_{\theta_3}(z'|z, \hat{a}) = \mathcal{N}(\mu_{\theta_3}(z, \hat{a}))$

Action decoder $q_{\theta_4}(a|\hat{a})$ Distribution is task specific. E.g. categorical or Gaussian.

μ_{θ_i} $(i = 1..4)$ - deep learning models (neural networks).

$\lambda_1, \lambda_2, \lambda_3, \lambda_4, \gamma, \sigma_E^2, \sigma_A^2$ - hyperparameters. Loss weights, a learning rate, and fixed variances.

1. Sample a triplet of data $(x, a, x') \sim$ Dataset[4]
2. Sample next latent state from encoder $z' \sim q_{\theta_1}(z'|x')$
3. Sample latent action from encoder $\hat{a} \sim q_{\theta_2}(\hat{a}|a)$
4. Compute loss L_{pred} with (44). [5]

---

[3] Note, neural net μ predicts only the mean vector. σ is a hyperparameter. Unlike the first algorithm, where the net predicts both.

[4] In practice, batches are used.

[5] Expectations are estimated using one-point estimation with previously sampled latent states and action.

5. Compute loss $L_{consist}$ with (46)
6. Compute loss L_{act} with (34)
7. Compute the smoothness LCE penalty L_{LLC} (See [9])
8. Updating parameters $\theta_i = \theta_i - \gamma \nabla_{\theta_i}(\lambda_1 L_{pred} + \lambda_2 L_{consist} + \lambda_3 L_{act} + \lambda_4 L_{LLC})$, $i = 1..4$

One may immediately observe, that the total model contains only 4 networks instead of 6. This reduction significantly increases the speed of training. Other performance characteristics are discussed in Sect. 2.5.

2.5 Experiments

The algorithms are paired with iLQR with the receding horizon method. The choice of this particular controller is made for consistency with prior work. SOLAR, however, uses a model-based guided policy search algorithm LQR-FLM [12]. However, we do not use it as we do not include the SOLAR algorithm in our comparison and because it is not directly applicable to our approach.

Because all the tasks are of a goal-reaching type, the cost function $c(z_t, u_t)$ is a quadratic cost of a form:

$$c(z_t, u_t) = (z_t - z_{goal})^T Q(z_t - z_{goal}) + u_t^T R u_t \ , \tag{47}$$

where z_{goal} is an encoding of a goal state x_{goal}, and $\hat{u}$ is an encoding of a raw action u. In our experiments, $Q = \alpha I$ and $R = \beta I$. Parameters α and β are specified for each task separately.

Following [7, 8, 11] the experiments are performed on image-based control benchmark domains. Namely, Planar System, Inverted Pendulum, and Cartpole. Different options for the action spaces are included. In this section, these tasks, along with action spaces, and the data collection policies are discussed. A comparison of the results of both proposed algorithms and the original PCC/PC3 is provided for each task.

2.5.1 Tasks

We use three task domains for benchmarking the proposed algorithms, namely, Planar system, Inverted Pendulum, and Cartpole. The state spaces and observation spaces are shared between prior LCE papers [7, 9, 17]. Each domain is tested in discrete action space settings.

For the Planar system, the action spaces include A_{simple} = {up, bottom, left, right}, A_{rot} = {rotate 90 degrees clockwise, rotate 90 degrees counterclockwise, forward} (the state space[6] is modified to reflect the orientation of the agent), and $A_{red} = \{\text{up, bottom, left, right}\}^2$. A_{rot} (rotational) and A_{red} (redundant) are considered for comparison with A_{simple} in terms of resulting performance. Ideally, A_{red} must perform as good as A_{simple} and slightly better than A_{rot} (due to the rotation steps).

Lastly, we also test the algorithm with A_{nonlin}:

$$A_{nonlin} = \{\phi_0, \phi_1, \phi_2, r_1, r_2\} \tag{48}$$

$$u_x = \cos(\phi_0 + \phi_1)r_1 + \cos(\phi_2)r_2 \tag{49}$$

$$u_y = \sin(\phi_0 + \phi_1)r_1 + \sin(\phi_2)r_2 \tag{50}$$

$$-\pi/3 \leq \phi_i \leq \pi/3, i = 0..2 \tag{51}$$

$$1.5 \leq r_j \leq 1.5, j = 0, 1 \tag{52}$$

This action space is continuous and designed to test action encoder capabilities. It is expected to match PC3's performance.

For Inverted Pendulum and Cartpole domains, the action space is just a discretized version of the original continuous action space (A_{discr}). Again, the results are better the closer they match the results of continuous space evaluations (denoted A_{cont}).

2.5.2 Data Collection

The current work does not provide a solution for the exploration problem, since the considered and proposed methods are fully offline.

Nevertheless, for the experiments, a data collection policy must be specified. We sample the date in the form of triplets (x_t, u_t, x_{t+1}). Each triplet is sampled using the following procedure. First, a state s_t is sampled from a uniform distribution over the state space S, and corresponding image observation x_t is generated. Next, a random action u_t is sampled from a uniform distribution over the action space U and applied to the environment in the state, s_t thus giving the next state s_{t+1}. Finally, the obtained state is used to generate an image observation x_{t+1}.

A reported evaluation metric is the percentage of time steps a controlled agent spent in the nearest region around the goal.

2.5.3 Comparison

The comparison of the results of the evaluation is summarized in Table 1.

The metrics for different action spaces are related as expected. A_{rot} (rotational) and A_{red} (redundant) are considered for comparison with A_{simple} in terms of resulting performance. Ideally, A_{red} performs on par A_{simple} and is slightly better than A_{rot} (due to the rotation steps). One of the biggest problems in Planar with bare one-hot

[6] "up", "bottom", "left", "right" for A_{simple} come in two step-size options: 1.0 and 2.5. So the number of actions is actually doubled.

Table 1 Evaluation of proposed algorithms

Task	Action space	AE-PCC*	AE-PC3*	PC3**
Planar	A_{simple}	71.5 ± 0.8	75.5 ± 0.3	45.5 ± 1.3
Planar	A_{rot}	68.4 ± 0.3	67.8 ± 0.2	N/A
Planar	A_{red}	75.5 ± 0.9	75.5 ± 0.3	N/A
Planar	A_{nonlin}	75.0 ± 1.0	73.5 ± 1.1	67.8 ± 2.2
Inverted Pendulum	A_{discr}	61.8 ± 4.9	61.8 ± 4.9	N/A
Inverted Pendulum	A_{cont}	31.6 ± 1.4	62.1 ± 2.8	61.8 ± 4.9
Cartpole	A_{discr}	95.8 ± 1.0	96.1 ± 0.4	N/A
Cartpole	A_{cont}	94.4 ± 1.2	95.8 ± 0.4	95.4 ± 0.9

*A reported evaluation metric is the percentage of time steps a controlled agent spent in the nearest region around the goal. The results are reported for the average model
**PC3 is only applicable in the continuous case. In order to support the simplest Planar discrete space, we simply use one-hot encoding

encoding or without a decent amount of additional noise in actions is stucking on the boundaries of the obstacles. In the case of AE-PC3 the effect is mitigated due to the VAE action encoder's variance.

The comparison of algorithms against each other is more tricky. The AE-PC3 (action-encoded PC3) performs better as a rule, but worse on the action spaces which we consider difficult. We suspect, that this discrepancy is caused by the degraded action decoder performance.

Another important note is that AE-PC3 performs mildly better on continuous action spaces of the Inverted Pendulum (Swing Up) and Cartpole than PC3. This is evidence that the encoding strategy does not degrade the performance even if the action space is simple and continuous.

3 Conclusion

In this work, we propose a way to integrate action encoding in the Learning Controllable Environment framework. Particularly, we build on two popular algorithms: PC3 and PCC. We provide a suboptimality analysis, rederive the algorithms, and conduct experiments to collect empirical proof of the concept. In future work, it is worth doing a more thorough study of the action encoding practices applied to naturally complex Markov Decision Processes. Also, an interesting direction is trying to use control algorithms that originally support discrete action spaces.

References

1. Dulac-Arnold, G., Evans, R., van Hasselt, H., Sunehag, P., Lillicrap, T., Hunt, J., Mann, T., Weber, T., Degris, T., Coppin, B.: (2015). arXiv:1512.07679
2. Levine, S., Kumar, A., Tucker, G., Fu, J.: (2020). arXiv:2005.01643
3. Feinberg, E.A., Kasyanov, P.O., Zgurovsky, M.Z.: Math. Oper. Res. **41**(2), 656 (2016)
4. Vinyals, O., Babuschkin, I., Czarnecki, W.M., Mathieu, M., Dudzik, A., Chung, J., Choi, D.H., Powell, R., Ewalds, T., Georgiev, P., et al.: Nature **575**(7782), 350 (2019)
5. Reed, S., Zolna, K., Parisotto, E., Colmenarejo, S.G., Novikov, A., Barth-Maron, G., Gimenez, M., Sulsky, Y., Kay, J., Springenberg, J.T. et al.: (2022).arXiv:2205.06175
6. Moerland, T.M., Broekens, J., Jonker, C.M.: (2020). arXiv:2006.16712
7. Watter, M., Springenberg, J., Boedecker, J., Riedmiller, M.: Adv. Neural Inf. Proc. Syst. **28** (2015)
8. Levine, N., Chow, Y., Shu, R., Li, A., Ghavamzadeh, M., Bui, H.: (2019). arXiv:1909.01506 (2019)
9. Shu, R., Nguyen, T., Chow, Y., Pham, T., Than, K., Ghavamzadeh, Ermon, S., Bui, H.: In: International Conference on Machine Learning. PMLR (2020)
10. Hafner, D., Lillicrap, T., Fischer, I., Villegas, R., Ha, D., Lee, H., Davidson, J.: In: International Conference on Machine Learning, pp. 2555–2565. PMLR (2019)
11. Zhang, M., Vikram, S., Smith, L., Abbeel, P., Johnson, M., Levine, S.: In: International Conference on Machine Learning, pp. 7444–7453. PMLR (2019)
12. Levine, S., Abbeel, P.: Adv. Neural Inf. Proc. Syst. **27** (2014)
13. Haarnoja, T., Zhou, A., Hartikainen, K., Tucker, G., Ha, S., Tan, J., Kumar, V., Zhu, H., Gupta, A., Abbeel, P. et al.: (2018). arXiv:1812.05905
14. Schulman, J., Wolski, F., Dhariwal, P., Radford, A., Klimov, O.: (2017). arXiv:1707.06347
15. Li, W., Todorov, E.: In: ICINCO (1), pp. 222–229. Citeseer (2004)
16. Theodorou, E., Tassa, Y., Todorov, E.: In: Proceedings of the 2010 American Control Conference, pp. 1125–1132. IEEE (2010)
17. Banijamali, E., Shu, R., Bui, H., Ghodsi, A. et al.: In: International Conference on Artificial Intelligence and Statistics, pp. 1751–1759. PMLR (2018)
18. Wang, S., Jia, D., Weng, X.: (2018). arXiv:1811.11329
19. Savva, M., Kadian, A., Maksymets, O., Zhao, Y., Wijmans, E., Jain, B., Straub, J., Liu, J., Koltun, V., Malik, J. et al.: In: Proceedings of the IEEE/CVF International Conference on Computer Vision, pp. 9339–9347 (2019)
20. Nagabandi, A., Konolige, K., Levine, S., Kumar, V.: In: Conference on Robot Learning, pp. 1101–1112. PMLR (2020)
21. Baker, B., Kanitscheider, I., Markov, T., Wu, Y., Powell, G., McGrew, B., Mordatch, I.: Machine Learning. Cornell University (2019)
22. Ghavamzadeh, M., Petrik, M., Chow, Y.: Adv. Neural Inf. Proc. Syst. **29** (2016)
23. Kingma, D.P., Welling, M.: (2013). arXiv:1312.6114
24. Oord, A.V.D., Li, Y., Vinyals, O.: (2018). arXiv:1807.03748
25. Anand, A., Racah, E., Ozair, S., Bengio, Y., Côté, M.A., Hjelm, R.D.: Adv. Neural Inf. Proc. Syst. **32** (2019)
26. Karl, M., Soelch, M., Bayer, J., Van der Smagt, P.: (2016). arXiv:1605.06432
27. Mazoure, B., des Combes, R.T., Doan, T.L., Bachman, P., Hjelm, R.D.: Adv. Neural Inf. Proc. Syst. **33**, 3686 (2020)
28. Rakelly, K., Gupta, A., Florensa, C., Levine, S.: Adv. Neural Inf. Proc. Syst. **34**, 26345 (2021)
29. Rezende, D., Mohamed, S.: In: International Conference on Machine Learning, pp. 1530–1538. PMLR (2015)

Comparison of Constrained Bayesian and Classical Methods of Testing Statistical Hypotheses in Sequential Experiments

Kartlos Kachiashvili, Vakhtang Kvaratskhelia, and Archil Prangishvili

Abstract The article focuses on the discussion of basic approaches to hypotheses testing in sequential experiments, which are Wald and Berger sequential tests and the test based on Constrained Bayesian Method (CBM). The positive and negative aspects of these approaches are considered and demonstrated on the basis of computed examples.

Keywords Hypotheses testing · p-value · Likelihood ratio · Frequentist approaches · Bayesian approach · Constrained Bayesian method · Wald's test

1 Introduction

One of the basic branches of statistical science is the theory of hypotheses testing which involves deciding on the plausibility of two or more hypothetical models based on some data. The modern theory of hypotheses testing began with Student's discovery of the t test in 1908 [22]. This was followed by Fisher [7], who created a new paradigm for hypothesis testing. The Fisher's criteria for the observation result

K. Kachiashvili (✉) · V. Kvaratskhelia · A. Prangishvili
Faculty of Informatics and Control Systems, Georgian Technical University, 77 Kostava Str., Tbilisi 0175, Georgia
e-mail: k.kachiashvili@gtu.ge; kkachiashvili@gmail.com

V. Kvaratskhelia
e-mail: v.kvaratskhelia@gtu.ge

A. Prangishvili
e-mail: a.prangishvili@gtu.ge

K. Kachiashvili · V. Kvaratskhelia
Muskhelishvili Institute of Computational Mathematics of the Georgian Technical University, 4 Grigol Peradze Str., Tbilisi 0159, Georgia

K. Kachiashvili
Ilia Vekua Institute of Applied Mathematics of Ivane Javakhishvili Tbilisi State University, 2 University Str., Tbilisi 0186, Georgia

M. Zgurovsky and N. Pankratova (eds.), *System Analysis and Artificial Intelligence*, Studies in Computational Intelligence 1107,
https://doi.org/10.1007/978-3-031-37450-0_17

x is based on $p\text{-}value = \sum_i P(X = x_i \mid H) \le \alpha$ (for discrete random variable) or $p\text{-}value = \int_G p(x \mid H)\,dx \le \alpha$ (for continuous random variable) $G \in R$, R is the observation space, where X is the suitable random variable and $p(x \mid H)$ is the probability distribution density of X at hypothesis H [5]. For appropriate values x_i from sub-space G, hypothesis H is rejected. In particular, if the *p-value* is less than or equal to α, the null hypothesis is rejected. The α level rejection region is defined as a set of all data points that have a *p-value* less than or equal to α. The philosophical basis of the Fisherian test consists in the examination of the extent to which the data contradict the model corresponding to the test hypothesis.

Neyman and Pearson [23, 24] involved not only the hypothesis but also a class of possible alternatives and the probabilities of two kinds of errors: false rejection (Error I) and false acceptance (Error II) [22]. The "best" test was the one that minimized the probability of an alternative at validity of the basic hypothesis (Error II) subject to a bound on probability of the basic hypothesis at validity of the alternative (Error I). The latter is the significance level of the test.

A prominent alternative approach to testing is the Bayesian approach introduced by Jeffreys [9]. The essence of the Bayesian approach is [1]: to define the likelihood ratio $B(x) = \frac{f_A(x)}{f_H(x)}$; to accept the alternative hypothesis if $B(x) > 1$; to report the posterior probabilities of the hypotheses $P(H \mid x) = \frac{B(x)}{1+B(x)}$ or $P(A \mid x) = \frac{1}{1+B(x)}$ which have been obtained on the basis of assigning the equal prior probabilities of $1/2$ to the two hypotheses and applying the Bayes theorem.

The considered methods of Fisher, Neyman–Pearson and Bayes are called parallel methods as they make decisions on the basis of an existing set of observations i.e. on the basis of existing information about investigating phenomenon despite this information is sufficient for making decisions with given reliability, or is not. For making a decision with the given reliability the sequential test was developed by Wald [27, 28] that makes a decision on the basis of sequentially obtained observation results only when required reliability of the decision is guaranteed. For this purpose, on each stage of making a decision, the observation space is divided into three not mutually intersecting regions in two of which a decision is made but in the third one to continue observation is decided.

By the influence of the Wald's work a new conditional test which gave the region of acceptance of null hypothesis, the region of acceptance of alternative hypothesis and the region of making no decision were offered in [1–3]. In [2] it was mentioned that "In practice the no-decision region is typically innocuous, corresponding to a region in which virtually no statistician would feel that the evidence is strong enough for a conclusive decision... In some settings, even unconditional frequentists should probably introduce a no-decision region to avoid paradoxical behavior". The article [1] was focused on the discussion of the conditional frequentist approach to testing, which was argued to provide the basis for methodological unification of Fisher, Jeffreys and Neyman approaches. In that paper, the author considers the positive and negative points of three different philosophies of hypotheses testing. An attempt to reconcile these different points of view was realized and as a result there was offered a new, compromise T^C method of testing which used Fisher's *p-value* criterion for making a decision, Neyman–Pearson's statement (using basic and alternative hypotheses) and Jeffrey's formulae for computing the Type I and Type II conditional error probabilities for every observation result x on the basis of which the decision is made. Despite such a noble aim, the scientific community met this attempt not identically, as it is seen from the comments attached to the paper. In our opinion, despite some inconveniences in some specific cases, the offered method is interesting and deserves attention, and can be usefully used for solving many practical problems.

For the same purpose that has the Wald's approach, new methods of hypotheses testing called *constrained Bayesian methods* (CBM) were offered in [10, 11, 17, 19–21]. They incorporate different aspects of above-considered classical approaches. In particular, they use the Neyman–Pearson constrained optimization statement for Bayesian formulation and get data-dependent measures of evidence with regard to the level of restriction. They are optimum in the sense of the chosen criterion and convenient for testing any number of different type of hypotheses.

CBM differs from the traditional Bayesian approach with a risk function split into two parts, reflecting risks for incorrect rejection and incorrect acceptance of hypotheses and stating the risk minimization problem as a constrained optimization problem when one of the risk components is restricted and the another one is minimized [11, 17, 20]. Application of this method to different types of hypotheses (two and many simple, composite and multiple hypotheses) with parallel and sequential experiments showed the advantage and uniqueness of the method in comparison with existing ones [13–17]. The uniqueness of the method consists in the emergence of the regions of impossibility of making a simple or any decision alongside with the regions of acceptance of tested hypotheses (like the sequential analysis method), which allows us based on this approach to develop both parallel and sequential method without any additional efforts. The advantage of the method is the optimality of made decisions with guaranteed reliability and minimality of necessary observations for given reliability (see, for example [13–17]. CBM uses not only loss functions and a priori probabilities for making decisions as the classical Bayesian rule does, but also a significance level as the frequentist method does. The combination of these opportunities improves the quality of made decisions in CBM in comparison with other methods.

The results of investigation and comparison of three noted sequential methods the Wald's, the Berger's and CBM are given below.

2 Description of the Investigated Methods of Hypotheses Testing

The essence of the Wald's sequential test consists in the following [27]: compute the likelihood ratio

$$B(x) = \frac{p(x_1, x_2, \ldots, x_n \mid H_0)}{p(x_1, x_2, \ldots, x_n \mid H_A)}$$

for n sequentially obtained observation results, and, if

$$B < B(x) < A,$$

do not make the decision and continue the observation of the random variable. If

$$B(x) \geq A,$$

accept the hypothesis H_0 on the basis of n observation results. If

$$B(x) \leq B,$$

accept the hypothesis H_A on the basis of n observation results.

The thresholds A and B are chosen so that

$$A = \frac{1-\beta}{\alpha} \text{ and } B = \frac{\beta}{1-\alpha}.$$

Here α and β are the desirable values of the error probabilities of Types I and II, respectively.

It is proved [27] that in this case the real values of the error probabilities of Types I and II are close enough to the desired values, but still are distinguished from them.

Since Wald's pioneer works, a lot of different investigations were dedicated to the sequential analysis problems (see, for example [8, 17, 25]) and efforts to the development of this approach constantly increase as it has many important advantages in comparison with the parallel methods (see, for example [26]).

In spite of absolutely different motivations of introduction of T^C and CBM, they lead to the hypotheses acceptance regions with identical properties in principle. Namely, in despite of the classical cases when the observation space is divided into two complementary sub-spaces for acceptance and rejection of tested hypotheses, similarly to the Wald's method, here the observation space contains the regions for making the decision and the regions for no-making the decision. Though, for CBM, the situation is more differentiated than for T^C. For CBM the regions for no-making the decision are divided into the regions of impossibility of making the decision and the regions of impossibility of making unique decision. In the first case, the impossibility of making the decision is equivalent to the impossibility of making the decision with given probability of the error for a given observation result, and it becomes possible when the probability of the error decreases. In the second case, it is impossible to make a unique decision when the probability of the error is required to be small, and it is unattainable for the given observation result. By increasing the error probability, it becomes possible to make a decision.

In our opinion these properties of T^C and CBM are very interesting and useful. They bring the statistical hypotheses testing rule much close to the everyday decision-making rule when, at shortage of necessary information, acceptance of one of made suppositions is not compulsory.

As was mentioned above, our aim is to compare the listed methods for elucidation of their positive and negative points and revealing the best one if such exists. For this reason, let us introduce a brief formal description of these methods. T^C method is described below in accordance with [1, 6]. CBM is presented in accordance with [17, 20, 21].

Let us suppose that the observation result $X \sim f(x \mid \theta)$, and it is necessary to test basic hypothesis $H_0 : \theta = \theta_0$ versus alternative one $H_A : \theta = \theta_A, \theta_A > \Theta_0$. Let us choose the test statistic $T = t(X)$ such that small values of T reflects evidence against H_0.

The Conditional Test T^C (The Berger's Method)

The considered test has the following form

$$T^C = \begin{cases} \text{if } B(x) \le c_0, & \text{reject } H_0 \text{ and report conditional error probability} \\ & \qquad \text{(CEP)} \ \alpha(x) = \dfrac{B(x)}{1+B(x)}, \\ \text{if } B(x) > c_0, & \text{accept } H_0 \text{ and report CEP } \beta(x) = \dfrac{1}{1+B(x)}, \end{cases}$$

where $B(x) = \frac{f(x|\theta_0)}{f(x|\theta_A)}$ is the likelihood ratio, and c_0 is the minimax critical value defined as

$$P(B(x) < c \mid H_0) = 1 - P(B(x) < c \mid H_1). \tag{1}$$

The Modified Conditional Test T^*

The test consists in the following

$$T^* = \begin{cases} \text{if } B(x) \le r, & \text{reject } H_0 \text{ and report conditional error probability} \\ & \qquad \text{(CEP)} \ \alpha(B(x)) = \dfrac{B(x)}{1+B(x)}, \\ \text{if } r < B(x) < a, & \text{make no decision,} \\ \text{if } B(x) \ge a, & \text{accept } H_0 \text{ and report CEP } \beta(x) = \dfrac{1}{1+B(x)}, \end{cases}$$

where a and r are defined as follows

$$\begin{aligned} r = 1 \ \text{and} \ a = F_0^{-1}(1 - F_A(1)) \ \text{if} \ F_0(1) \le 1 - F_A(1), \\ r = F_A^{-1}(1 - F_0(1)) \ \text{and} \ a = 1 \ \text{if} \ F_0(1) > 1 - F_A(1), \end{aligned} \tag{2}$$

where F_0 and F_A are the c.d.f. of $B(X)$ under $p(x \mid H_0)$ and $p(x \mid H_A)$, respectively.

Constrained Bayesian Method (CBM)

Let us consider a set of hypotheses H_i, $i = 1, \ldots, S$ ($S \geq 2$), involving that the random vector X is distributed by the law $p(x, \theta_i)$, i.e. $H_i : X \sim p(x, \theta_i) \equiv p(x \mid H_i)$; $p(H_i)$ is a priori probability of hypothesis H_i; Γ_i is the region of acceptance of H_i (Γ_i belongs to the observation space of random variable X, i.e. $\Gamma_i \in R^n$, where n is the dimension of the observation vector). The decision is made on the basis of $\mathbf{x}^T = (x_1, \ldots, x_n)$, the measured value of the random vector $\mathbf{X}$. It is possible to formulate different constrained tasks of testing the considered hypotheses [17, 20]. Here we consider only one of them, namely the task with restrictions on the averaged probability of rejection of true hypotheses for stepwise loss function with two possible values 0 and 1. The essence of this method is the minimization of the averaged probability of incorrect acceptance of hypotheses at restriction of the averaged probability of rejection of true hypotheses, i.e.

$$1 - \sum_{i=1}^{S} p(H_i) P\big(\mathbf{X} \in \Gamma_i \mid H_i\big) \Longrightarrow \min_{\{\Gamma_i\}}, \tag{3}$$

subject to

$$\sum_{i=1}^{S} p(H_i) \sum_{i=1,\ j \neq i}^{S} P\big(\mathbf{X} \in \Gamma_j \mid H_i\big) \leq \gamma. \tag{4}$$

Solution of task (3) and (4) is [11, 20]

$$\Gamma_j = \Big\{\mathbf{x} :\ p(H_j) p(\mathbf{x} \mid H_j) > \lambda \sum_{i=1,\ j \neq i}^{S} p(H_i) P(\mathbf{x} \mid H_i)\Big\}, \quad j = 1, \ldots, S. \tag{5}$$

Coefficient λ is the same for all regions of acceptance of hypotheses, and it is determined so that in (4) the equality takes place.

When the number of hypotheses is equal to two and their a priori probabilities are equal to 1/2, solution (5) can be rewritten using the Bayes factor: the hypothesis H_0 rejection region is defined as $B(\mathbf{x}) \leq \lambda$, and the alternative hypothesis rejection region is $B(\mathbf{x}) \geq 1/\lambda$.

Probabilities (3) and (4) take the forms

$$\min_{\{\Gamma_0, \Gamma_A\}} \left\{1 - \frac{P(B(\mathbf{x}) > \lambda \mid H_0) + P(B(\mathbf{x}) < \frac{1}{\lambda} \mid H_A)}{2}\right\} \tag{6}$$

and

$$\frac{P(B(\mathbf{x}) > \lambda \mid H_A) + P\big(B(\mathbf{x}) < \frac{1}{\lambda} \mid H_0\big)}{2} \leq \gamma, \tag{7}$$

respectively.

The posterior probabilities of the hypotheses are calculated similarly to the above given Bayes method.

The probabilities of incorrect rejection of basic and alternative hypotheses when they are true are

$$\alpha_0 = P\big(B(\mathbf{x}) \le \lambda \mid H_0\big) = 1 - P\big(B(\mathbf{x}) > \lambda \mid H_0\big)$$

and

$$\alpha_A = P\left(B(\mathbf{x}) \ge \frac{1}{\lambda} \mid H_A\right) = 1 - P\left(B(\mathbf{x}) < \frac{1}{\lambda} \mid H_A\right),$$

respectively, and the probabilities of incorrect acceptance of hypotheses when they are erroneous are

$$\beta_0 = P\big(B(\mathbf{x}) > \lambda \mid H_A\big) \text{ and } \beta_A = P\left(B(\mathbf{x}) < \frac{1}{\lambda} \mid H_0\right),$$

respectively.

It is clear that, at $\lambda = 1$, CBM completely coincides with the Bayes method but, at $\lambda \neq 1$ it has new properties [20, 21]. Namely, when $\lambda < 1$ hypotheses acceptance regions intersect and for the data from this intersecting area it is impossible to make an unambiguous decision; when $\lambda > 1$ in the observation space here arise sub-region which does not belong to any hypothesis acceptance region and it is impossible to make a simple decision [17, 19, 20]. Therefore, probabilities of errors of Types I and II are computed by the following ratios:

at $\lambda = 1$,

$$\alpha_0 = P\big(B(\mathbf{x}) \le \lambda \mid H_0\big), \quad \alpha_A = P\left(B(\mathbf{x}) \ge \frac{1}{\lambda} \mid H_A\right),$$
$$\beta_0 = P\big(B(\mathbf{x}) > \lambda \mid H_A\big), \quad \beta_A = P\left(B(\mathbf{x}) < \frac{1}{\lambda} \mid H_0\right);$$

at $\lambda > 1$,

$$\alpha_0 = P\left(B(\mathbf{x}) < \frac{1}{\lambda} \mid H_0\right), \quad \alpha_A = P\big(B(\mathbf{x}) > \lambda \mid H_A\big),$$
$$\beta_0 = P\big(B(\mathbf{x}) > \lambda \mid H_A\big), \quad \beta_A = P\left(B(\mathbf{x}) < \frac{1}{\lambda} \mid H_0\right);$$

at $\lambda < 1$,

$$\alpha_0 = P\big(B(\mathbf{x}) \le \lambda \mid H_0\big), \quad \alpha_A = P\left(B(\mathbf{x}) \ge \frac{1}{\lambda} \mid H_A\right),$$
$$\beta_0 = P\left(B(\mathbf{x}) \ge \frac{1}{\lambda} \mid H_A\right), \quad \beta_A = P\big(B(\mathbf{x}) < \lambda \mid H_0\big).$$

While the probabilities of making no decision are

$$P\Big(\frac{1}{\lambda} \le B(\mathbf{x}) \le \lambda \mid H_0\Big) \text{ and } P\Big(\frac{1}{\lambda} \le B(\mathbf{x}) \le \lambda \mid H_A\Big) \text{ at } \lambda > 1,$$

and

$$P\Big(\lambda < B(\mathbf{x}) < \frac{1}{\lambda} \mid H_0\Big) \text{ and } P\big(\lambda < B(\mathbf{x}) < \lambda \mid H_A\big) \text{ at } \lambda < 1,$$

respectively.

Let us denote by α' and β' the probability of no accepting true hypothesis and the probability of accepting false hypothesis, respectively. Then, it is obvious that,

when $\lambda = 1$,

$$\alpha_0' = \alpha_0, \quad \alpha_A' = \alpha_A, \quad \beta_0' = \beta_0, \quad \beta_A' = \beta_A;$$

when $\lambda > 1$,

$$\alpha_0' = \alpha_0 + P\Big(\tfrac{1}{\lambda} \le B(\mathbf{x}) \le \lambda \mid H_0\Big), \quad \alpha_A' = \alpha_A + P\Big(\tfrac{1}{\lambda} \le B(\mathbf{x}) \le \lambda \mid H_A\Big),$$
$$\beta_0' = \beta_0, \quad \beta_A' = \beta_A;$$

when $\lambda < 1$,

$$\alpha_0' = \alpha_0 + P\Big(\lambda \le B(\mathbf{x}) \le \tfrac{1}{\lambda} \mid H_0\Big), \quad \alpha_A' = \alpha_A + P\Big(\lambda \le B(\mathbf{x}) \le \tfrac{1}{\lambda} \mid H_A\Big),$$
$$\beta_0' = \beta_0, \quad \beta_A' = \beta_A.$$

As was mentioned in [6, p. 196], "T^* is an actual frequentist test; the reported CEPs, $\alpha(B(\mathbf{x}))$ and $\beta(B(\mathbf{x}))$, are conditional frequentist Type I and Type II error probabilities, conditional on the statistic we use to measure strength of evidence in the data. Furthermore, $\alpha(B(\mathbf{x}))$ and $\beta(B(\mathbf{x}))$ will be seen to have the Bayesian interpretation of being (objective) posterior probabilities of H_0 and H_A, respectively. Thus, T^* is simultaneously a conditional frequentist and a Bayesian test". It is not difficult to be convinced that the same is true for the considered CBM. Generalization of the T^* test for any number of hypotheses seems quite problematic. For the general case, it is possible only by simulation because the definition of exact distribution of $B(\mathbf{x})$ likelihood ratio for arbitrary hypothetical distributions is very difficult if not impossible. Generalization of CBM for any number of hypotheses does not represent any problem. It is stated and solved namely for the arbitrary number of hypotheses [11, 17, 19–21]. The properties of the decision rules are common and do not depend on the number of hypotheses.

In [6] it is also noted that, because T^* is a Bayesian test, it inherits many of the positive features of Bayesian tests; as the sample size grows, the test chooses the right model. If the data actually arise from the third model, the test chooses the hypothesis which is the closest to the true model in Kullback–Leibler divergence [4]. CBM has the same positive features that the T^* test has and chooses the right model with greater reliability at the increasing sample size (see [17, 21] and examples given below). If the data arise from the model which is not included in the hypothetical set of tested hypotheses and γ is quite small in restriction (4), the CBM does not choose any tested hypotheses.

3 Comparison of Sequential Hypotheses Testing Methods

The specific features of hypotheses testing regions of the Berger's T^* test and CBM (see previous sections), namely, the existence of the no-decision region in the T^* test and the existence of regions of impossibility of making a unique or any decision in CBM give the opportunities to develop the sequential tests on their basis. Using the concrete example taken from [3], below these tests are compared among themselves and with the Wald sequential test [27]. For clarity, let us briefly describe these tests.

The sequential test developed on the basis of T^* test is as follows [3]:

- if the likelihood ratio $B(x) \leq r$, reject H_0 and report the conditional error probability $\alpha(B(x)) = \frac{B(x)}{1+B(x)}$;
- if $r < B(x) < a$, make no decision;
- if $B(x) \geq a$, accept H_0 and report the conditional error probability $\beta(B(x)) = \frac{1}{1+B(x)}$.

Here r and a are determined by ratios (2).

The sequential test developed on the basis of CBM consists in the following [12, 18, 21]. Let Γ_i^n be the H_i hypothesis acceptance region (5) on the basis of n sequentially obtained repeated observation results; R_n^m is the decision-making space in the sequential method; m is the dimensionality of the observation vector; I_i^n is the population of sub-regions of intersections of hypotheses H_i acceptance regions Γ_i^n $(i = 1, \ldots, S)$ with the regions of acceptance of other hypotheses H_j $(j = 1, \ldots, S)$, $j \neq i$; $E_n^m = R_n^m = \bigcup_{i=1}^{S} \Gamma_i^n$ is the population of regions of space R_n^m which do not belong to any of hypotheses acceptance regions.

The H_i hypotheses acceptance regions for n sequentially obtained observation results in the sequential method are:

$$R_{n,i}^m = \frac{\Gamma_i^n}{I_i^n}, \quad i = 1, \ldots, S; \tag{8}$$

the no-decision region is:

$$R_{n,S+1}^m = \left(\bigcup_{i=1}^{S} I_i^n \right) \cup E_n^m, \tag{9}$$

where

$$\Gamma_i^n = \left\{ \mathbf{x} : \ p(\mathbf{x} \mid H_i) > \sum_{\ell=1,\ \ell \neq i}^{S} \lambda_\ell^i p(\mathbf{x} \mid H_\ell) \right\}, \quad 0 \leq \lambda_\ell^i < +\infty, \quad \ell = 1, \ldots, S. \quad (10)$$

Coefficients $\lambda_\ell^i = \lambda \frac{p(H_\ell)}{p(H_i)}$ are defined from the equality in the suitable restriction (4).

This test is called the sequential test of Bayesian type [18]. Such tests could be considered for all constrained Bayesian methods offered in [17, 19, 20] and differing from each other in restrictions.

Example 1 ([3]) Suppose that $X_1, X_2, \ldots, X_n$ are i.i.d. $N(\theta, 1)$ and that it is desired to test $H_0 : \theta = -1$ versus $H_A : \theta = 1$. Then

$$B = \prod_{i=1}^{n} \frac{(2\pi)^{-1/2} \exp\{-\frac{1}{2}(x_i + 1)^2\}}{(2\pi)^{-1/2} \exp\{-\frac{1}{2}(x_i - 1)^2\}} = \exp\{-2n\overline{x}\}.$$

Let us suppose the data are observed sequentially and the hypotheses are identically probable.

The sequential test developed on the basis of T^* test for this concrete example is as follows [3]:

- if $\overline{x}_n \geq g(n)$, where n is the number of sequentially obtained observations, stop experimentation, reject H_0 and report the conditional error probability

$$\alpha(B_n) = \frac{1}{1 + \exp(2n\overline{x}_n)};$$

- if $\overline{x}_n < -g(n)$, stop experimentation, accept H_0 and report the conditional error probability

$$\beta(B_n) = \frac{1}{1 + \exp(-2n\overline{x}_n)};$$

The choice

$$g(n) = \frac{1}{2n} \ln\left(\frac{1}{\alpha} - 1\right) \quad (11)$$

guarantees that the reported error probability will not exceed α [3].

The sequential test developed on the basis of CBM in this case is as follows:

- if

$$\overline{x} < \min\left\{ \left(\frac{\Phi^{-1}(\gamma)}{\sqrt{n}} + 1\right), -\left(\frac{\Phi^{-1}(\gamma)}{\sqrt{n}} + 1\right) \right\},$$

stop experimentation, accept H_0 and report the conditional error probability

$$\beta_{\text{CBM}}(\gamma, n) = P(\overline{x} < B_{\text{CBM}} \mid H_A) = \Phi\left(\sqrt{n}\,(B_{\text{CBM}} - 1)\right);$$

– if

$$\overline{x} > \max\left\{-\left(\frac{\Phi^{-1}(\gamma)}{\sqrt{n}}+1\right), \left(\frac{\Phi^{-1}(\gamma)}{\sqrt{n}}+1\right)\right\},$$

accept H_A and report the conditional error probability

$$\alpha_{\text{CBM}}(\gamma, n) = P(\overline{x} > A_{\text{CBM}} \mid H_0) = 1 - \Phi\left(\sqrt{n}\,(A_{\text{CBM}}+1)\right).$$

Otherwise do not make the decision and continue the observation of the random variable.

Here γ is the desired value of restriction in (4), Φ is the standard normal c.d.f. and

$$A_{\text{CBM}} = \max\left\{-\left(\frac{\Phi^{-1}(\gamma)}{\sqrt{n}}+1\right), \left(\frac{\Phi^{-1}(\gamma)}{\sqrt{n}}+1\right)\right\},$$
$$B_{\text{CBM}} = \min\left\{\left(\frac{\Phi^{-1}(\gamma)}{\sqrt{n}}+1\right), -\left(\frac{\Phi^{-1}(\gamma)}{\sqrt{n}}+1\right)\right\}.$$

The Wald's sequential test for this concrete example is as follows:

– if $\overline{x} < -\frac{1}{2n}\ln\frac{1-\beta}{\alpha}$, stop experimentation, accept H_0;
– if $\overline{x} > -\frac{1}{2n}\ln\frac{\beta}{1-\alpha}$, stop experimentation, accept H_A;

otherwise do not make the decision and continue the observation of the random variable. The error probabilities of Types I and II computed similarly to the previous case are:

$$\alpha_W(\alpha, \beta, n) = P(\overline{x} > A_W \mid H_0) = 1 - \Phi\left(\sqrt{n}\,(A_W+1)\right)$$

and

$$\beta_W(\alpha, \beta, n) = P(\overline{x} < B_W \mid H_A) = \Phi\left(\sqrt{n}\,(B_W-1)\right),$$

respectively. Here

$$A_W = -\frac{1}{2n}\ln\frac{1-\beta}{\alpha} \quad \text{and} \quad B_W = -\frac{1}{2n}\ln\frac{\beta}{1-\alpha}.$$

It is obvious that, when $\alpha = \beta$ in the Wald's test and α of the Berger's test from (11) are equal, the hypotheses acceptance thresholds in both these tests are the same. That means that these tests become identical.

Let us consider the case when, for the Wald's test, $\alpha = \beta = 0.05$, for the Berger's test, $\alpha = 0.05$, and, for the sequential test of Bayesian type, $\gamma = 0.05$. The dependences of the thresholds on the number of observations in the considered tests for chosen error probabilities are shown in Fig. 1. The computed values are given in Table 1. The dependence of error probabilities on the number of observations in the

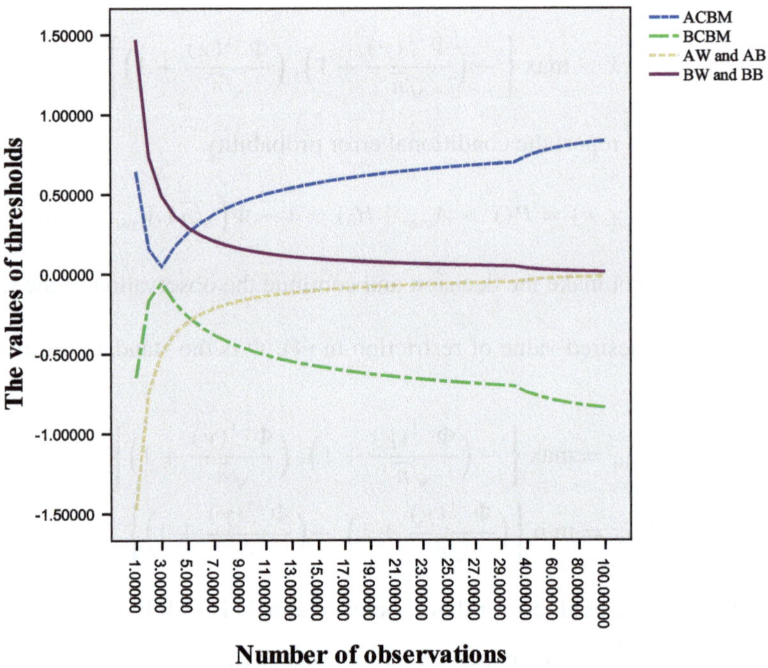

Fig. 1 Dependence of the thresholds on the number of observations in the considered tests (Kullback's divergence between the considered hypotheses $J(1:2) = 2$). ACBM and BCBM—the upper and lower thresholds of the sequential test of Bayesian type; AW and AB—the upper thresholds of the Wald and Berger's sequential tests, respectively; BW and BB—the lower thresholds of the Wald and Berger's sequential tests, respectively

sequential test of Bayesian type and in the Wald's test (that is the same, in the Berger's test) is shown in Fig. 2, and the computed values are given in Table 2. From these data, it is seen that, in the sequential test of Bayesian type, the probability of incorrect acceptance of a hypothesis when other hypothesis is true at increasing n decreases more significantly than in Wald's test, but the probability of no acceptance of a true hypothesis in the Wald's test decreases more significantly at increasing n than in the sequential test of Bayesian type. Though, it should be noted that Berger computed the error probabilities in the similar manner as Fisher had for the given value of the statistics [3]. These probabilities given in Table 2 were computed as the averaged possibilities of occurrence of such events in the manner similar to the Neyman's principle.

The computation results of the sequentially processed sample generated by $N(1, 1)$ with 17 observations are given in Table 3, where the arithmetic mean of the observations $x_k, \ldots, x_m$ is denoted by $\overline{x}_{k,m}$. From here it is seen that the Wald and Berger's tests yield absolutely the same results, though the reported error probabilities in the Berger's test are a little less than in the Wald's test for the reason mentioned above (Berger computed the error probabilities for the given value of the

Table 1 The computed values of the thresholds depending on the number of observations in the considered tests

n	A_{CBM}	B_{CBM}	A_W and A_B	B_W and B_B
1	0.64485	−0.64485	1.47222	−1.47222
2	0.16309	−0.16309	0.73611	−0.73611
3	0.05034	−0.05034	0.49074	−0.49074
4	0.17757	−0.17757	0.36805	−0.36805
5	0.2644	−0.2644	0.29444	−0.29444
6	0.32849	−0.32849	0.24537	−0.24537
7	0.3783	−0.3783	0.21032	−0.21032
8	0.41846	−0.41846	0.18403	−0.18403
9	0.45172	−0.45172	0.16358	−0.163585
10	0.47985	−0.47985	0.14722	−0.14722
11	0.50406	−0.50406	0.13384	−0.13384
12	0.52517	−0.52517	0.12268	−0.12268
13	0.5438	−0.5438	0.11325	−0.11325
14	0.56039	−0.56039	0.10516	−0.10516
15	0.5753	−0.5753	0.09815	−0.09815
16	0.58879	−0.58879	0.09201	−0.09201
17	0.60106	−0.60106	0.0866	−0.0866
18	0.6123	−0.6123	0.08179	−0.08179
19	0.62264	−0.62264	0.07749	−0.07749
20	0.6322	−0.6322	0.07361	−0.07361
21	0.64106	−0.64106	0.07011	−0.07011
22	0.64932	−0.64932	0.06692	−0.06692
23	0.65702	−0.65702	0.06401	−0.06401
24	0.66425	−0.66425	0.06134	−0.06134
25	0.67103	−0.67103	0.05889	−0.05889
26	0.67742	−0.67742	0.05662	−0.05662
27	0.68345	−0.68345	0.05453	−0.05453
28	0.68915	−0.68915	0.05258	−0.05258
29	0.69456	−0.69456	0.05077	−0.05077
30	0.69969	−0.69969	0.04907	−0.04907
40	0.73993	−0.73993	0.03681	−0.03681
50	0.76738	−0.76738	0.02944	−0.02944
60	0.78765	−0.78765	0.02454	−0.02454
70	0.8034	−0.8034	0.02103	−0.02103
80	0.8161	−0.8161	0.0184	−0.0184
90	0.82662	−0.82662	0.01636	−0.01636
100	0.83551	−0.83551	0.01472	−0.01472

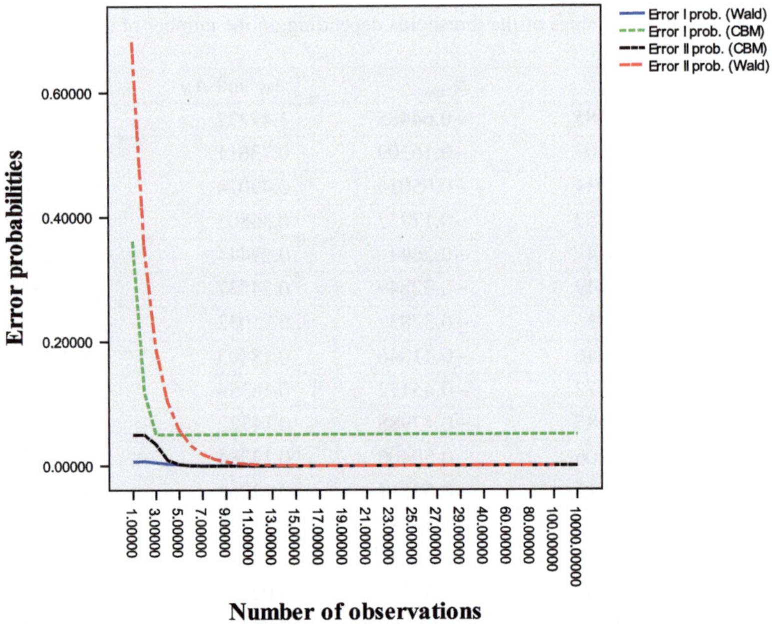

Fig. 2 Dependence of the error probabilities on the number of observations in the sequential test of Bayesian type

statistics). Out of 17 observations, correct decisions were taken 7 times on the basis of 3, 3, 5, 1, 1, 3 and 1 observations in both tests. The average value of observations for making the decision is equal to 2.43. In the sequential test of Bayesian type for the same sample correct decisions were taken 10 times on the basis of 1, 2, 2, 1, 3, 2, 1, 1, 3 and 1 observations. The average value of observations for making the decision is equal to 1.7.

The reported error probabilities in the sequential test of Bayesian type and the Wald's test decrease depending on the number of observations used for making the decision (see Table 2). By the Type II error probability it strongly surpasses the Wald's test. While these characteristics for the Berger's test have no monotonous dependence on the number of observations (for the reason mentioned above). They basically are defined by the value of the likelihood ratio. For example, the value of the Type I error probability for 5 observations $(x_7, \ldots, x_{11})$ surpasses the analogous value for 3 observations x_{14}, x_{15}, x_{16} and both of them surpass the same value for 1 observation x_{17}.

Example 2 Let us briefly consider Example 7 from [3]. The sequential experiment is conducted involving i.i.d. $N(\theta, 1)$ data for testing $H_0 : \theta = 0$ versus $H_A : \theta = 1$ under a symmetric stopping rule (or at least a rule for which $\alpha = \beta$). Suppose the report states that sampling stopped after 20 observations, with $\overline{x}_{20} = 0.7$. In this case, the likelihood ratio

Table 2 The values of error probabilities depending on the number of observations

n	$P(\overline{x} > A_{\text{CBM}} \mid H_0)$ $= P(\overline{x} < B_{\text{CBM}} \mid H_A)$ Error II probability in CBM	$P(\overline{x} < A_{\text{CBM}} \mid H_A)$ $= P(\overline{x} > B_{\text{CBM}} \mid H_0)$ Error I probability in CBM	$P(\overline{x} > A_W \mid H_0)$ $= P(\overline{x} < B_W \mid H_A)$ Error II probability for Wald's test	$P(\overline{x} < A_W \mid H_A)$ $= P(\overline{x} > B_W \mid H_0)$ Error I probability for Wald's test
1	0.05	0.36124	0.68161	0.00671
2	0.05	0.11829	0.3545	0.00704
3	0.03444	0.05	0.18887	0.00491
4	0.00926	0.05	0.10313	0.00311
5	0.00235	0.05	0.05732	0.0019
6	0.00057	0.05	0.03227	0.00114
7	0.00013	0.05	0.01834	0.00068
8	0.00003	0.05	0.0105	0.00041
9	0.00001	0.05	0.00605	0.00024
10	0	0.05	0.0035	0.00014
11	0	0.05	0.00203	0.00008
12	0	0.05	0.00119	0.00005
13	0	0.05	0.00069	0.00003
14	0	0.05	0.00041	0.00002
15	0	0.05	0.00024	0.00001
16	0	0.05	0.00014	0.00001
17	0	0.05	0.00008	0
18	0	0.05	0.00005	0
19	0	0.05	0.00003	0
20	0	0.05	0.00002	0
21	0	0.05	0.00001	0
22	0	0.05	0.00001	0
≥23	0	0.05	0	0

$$B_{20} = \prod_{i=1}^{20} \frac{f(x_i \mid 0)}{f(x_i \mid 1)} = \exp\left\{ -20(\overline{x}_{20} - 0.5)\right\} = 0.018.$$

T^* Test

Compute $F_0(1) = 0.8413$. Therefore, $a = 1$ and $r = 0.3174$. Because $B_{20} = 0.018 < r = 0.3174$, the basic hypothesis H_0 is rejected and the associated conditional error probability $\alpha(B_{20}) = \frac{B_{20}}{1+B_{20}} = 0.01768$ is reported.

Table 3 The results of testing of a normal sample

n	Observation results x_i	$\overline{x}_{k,m}$	The Berger's test	The Wald's test	$\overline{x}_{k,m}$ for sequential test of Bayesian type	The sequential teat of Bayesian type
1	1.201596	$\overline{x}_{1,3}$=0.6347	H_A	H_A	$\overline{x}_1$=1.2016	H_A
2	0.043484		$\alpha = 0.0217$			
3	0.658932		$\beta = 0.9783$		$\overline{x}_{2,3}$=0.3512	H_A
4	0.039022	$\overline{x}_{4,6}$=0.8942	H_A	H_A	$\overline{x}_{4,5}$=0.3280	H_A
5	0.616960		$\alpha = 0.0047$			
6	2.026540		$\beta = 0.9953$		$\overline{x}_6$=2.02654	H_A
7	−0.422764	$\overline{x}_{7,11}$=0.2992	H_A	H_A	$\overline{x}_{7,9}$=0.1643	H_A
8	0.562569		$\alpha = 0.0047$			
9	0.353047		$\beta = 0.9522$			
10	−0.123311					
11	1.126263				$\overline{x}_{10,11}$=0.5015	H_A
12	1.521061	$\overline{x}_{12}$=1.521061	H_A $\alpha = 0.0456$ $\beta = 0.9544$	H_A	$\overline{x}_{12}$=1.521061	H_A
13	1.486411	$\overline{x}_{13}$=1.4864	H_A $\alpha = 0.0487$ $\beta = 0.9513$	H_A	$\overline{x}_{13}$=1.486411	H_A
14	−0.578935	$\overline{x}_{14,16}$=0.5536	H_A	H_A	$\overline{x}_{14,16}$= 0.55362	H_A
15	0.043484		$\alpha = 0.0348$			
16	0.658932		$\beta = 0.9652$			
17	1.754413	$\overline{x}_{17}$=1.7544	H_A $\alpha = 0.0291$ $\beta = 0.9709$	H_A	$\overline{x}_{17}$=1.754413	H_A
$\overline{n}$			2.43	2.43		1.7

Wald Test

Choosing $\alpha = 0.05$ and $\beta = 0.05$ the thresholds are computed $A = 19$ and $B = 0.0526$. Because $B_{20} = 0.018 < B = 0.0526$, the alternative hypothesis H_A is accepted. Error probabilities are $\alpha = P(B_{20} < 0.0526 \mid H_0) = 0.001899$ and $\beta = P(B_{20} > 19 \mid H_A) = 0.001899$.

CBM Test

The results of computation obtained by CBM for the data $\overline{x}_{20} = 0.7$, $\sigma^2(\overline{x}_{20}) = \frac{1}{20} = 0.05$ and $\gamma = 0.05$ are the following: $\lambda = 3.141981$ and $\frac{1}{\lambda} = 0.31827$. Because $B_{20} = 0.018 < B_{\text{CBM}} = \frac{1}{\lambda} = 0.31827$, the alternative hypothesis H_A is accept and error probabilities

$$\alpha = P(B_{20} < 0.31827 \mid H_0) = 0.00635$$

and

$$\beta = P(B_{20} > 3.141981 \mid H_A) = 0.00635.$$

If $\gamma = 0.01$ is chosen the computation results are the following: accept the alternative hypothesis H_A with error probabilities $\alpha = 0.01$ and $\beta = 0.015945$.

It is obvious that, for this example, by error probabilities CBM *surpasses* the T^* and the Wald's method surpasses the CBM. Though, for the sake of justice, it is necessary to note that the error probabilities of CBM are also quite small.

4 Conclusion

The offered CBM method is a more general method of hypotheses testing than the existing classical Fisher's, Jeffreys', Neyman's, Berger's and Wald's methods. It has all positive properties of the mentioned methods. Namely, it is a data-dependent measure like Fisher's test, for making the decision it uses posteriori probabilities like Jeffreys' test and computes Type I and Type II error probabilities like Neyman–Pearson's approach. Like the Berger's methods, it has no-decision-making regions. Moreover, the regions of making decisions have new, more general properties than the same regions in the considered methods. These properties allow us to make more well-founded and reliable decisions. Particularly, do not accept a unique hypothesis or do not accept any hypothesis when the information on the basis of which the decision must be made is not enough for distinguishing the informationally close hypotheses or for choosing a hypothesis among informationally distant ones.

Very interesting peculiarity of CBM is the possibility of its use in parallel and sequential experiments without any changes and when it is necessary smoothly transit from parallel to sequential methodology. In despite of Berger's and Wald's methods, the sequential test of Bayesian type is universal and without modification can be used for any number of hypotheses and any dimensionality of observation vector. It is simple and very convenient for use and methodologically practically does not depend on the number of tested hypotheses and dimensionality of the observation space. The computed results, presented in the paper, clearly demonstrate high quality of the sequential test of Bayesian type.

References

1. Berger, J.O.: Could Fisher, Jeffreys and Neyman have agreed on testing? With comments and a rejoinder by the author. Statist. Sci. **18**(1), 1–32 (2003)
2. Berger, J.O., Boukai, B., Wang, Y.: Unified frequentist and Bayesian testing of a precise hypothesis. Statist. Sci. **12**(3), 133–160 (1997)

3. Berger, J.O., Brown, L.D., Wolpert, R.L.: A unified conditional frequentist and Bayesian test for fixed and sequential simple hypothesis testing. Ann. Statist. **22**(4), 1787–1807 (1994)
4. Berk, R.H.: Limiting behavior of posterior distributions when the model is incorrect. Ann. Math. Statist. **37**, 51–58 (1966); Correction, ibid. 745–746
5. Christensen, R.: Testing Fisher, Neyman, Pearson, and Bayes. Amer. Statist. **59**(2), 121–126 (2005)
6. Dass, S.C., Berger, J.O.: Unified conditional frequentist and Bayesian testing of composite hypotheses. Scand. J. Statist. **30**(1), 193–210 (2003)
7. Fisher, R.A.: Statistical Methods for Research Workers. Oliver and Boyd, London (1925)
8. Ghosh, B.K.: Sequential Tests of Statistical Hypotheses. Addison–Wesley Publishing Co., Reading, Mass.–London–Don Mills, Ont. (1970)
9. Jeffreys, H.: Theory of Probability. Oxford University Press, Oxford (1939)
10. Kachiashvili, K.J.: Generalization Of Bayesian rule of many simple hypotheses testing. Int. J. Inf. Technol. Decis. Mak. **02**(01), 41–70 (2003)
11. Kachiashvili, K.J.: Investigation and computation of unconditional and conditional Bayesian problems of hypothesis testing. ARPN J. Syst. Softw. **1**(2), 47–59 (2011)
12. Kachiashvili, K.J.: The methods of sequential analysis of Bayesian type for the multiple testing problem. Sequ. Anal. **33**(1), 23–38 (2014)
13. Kachiashvili, K.J.: Comparison of some methods of testing statistical hypotheses, Part I. Parallel methods. Int. J. Stat. Med. Res. **3**, 174–197 (2014)
14. Kachiashvili, K.J.: Comparison of some methods of testing statistical hypotheses, Part II. Sequential methods. Int. J. Stat. Med. Res. **3**, 189–197 (2014)
15. Kachiashvili, K.J.: Constrained Bayesian method for testing multiple hypotheses in sequential experiments. Sequ. Anal. **34**(2), 171–186 (2015)
16. Kachiashvili, K.J.: Constrained Bayesian method of composite hypotheses testing: singularities and capabilities. Int. J. Stat. Med. Res. **5**, 135–167 (2016)
17. Kachiashvili, K.: Constrained Bayesian Methods of Hypotheses Testing: A New Philosophy of Hypotheses Testing in Parallel and Sequential Experiments. Nova Science Publishers Inc, New York (2018)
18. Kachiashvili, K., Hashmi, M.A.: About using sequential analysis approach for testing many hypotheses. Bull. Georgian Natl. Acad. Sci. (N.S.) **4**(2), 20–26 (2010)
19. Kachiashvili, G.K., Kachiashvili, K.J., Mueed, A.: Specific features of regions of acceptance of hypotheses in conditional Bayesian problems of statistical hypotheses testing. Sankhya A **74**(1), 112–125 (2012)
20. Kachiashvili, K.J., Hashmi, M.A., Mueed, A.: Sensitivity analysis of classical and conditional Bayesian problems of many hypotheses testing. Comm. Statist. Theory Methods **41**(4), 591–605 (2012)
21. Kachiashvili, K.J., Mueed, A.: Conditional Bayesian task of testing many hypotheses. Statistics **47**(2), 274–293 (2013)
22. Lehmann, E.L.: The Fisher, Neyman-Pearson theories of testing hypotheses: one theory or two? J. Am. Stat. Assoc. **88**(424), 1242–1249 (1993)
23. Neyman, J., Pearson, E.S.: On the use and interpretation of certain test criteria for purposes of statistical inference. Part I. Biometrika **20A**(1/2), 175–240 (1928)
24. Neyman, J., Pearson, E.S.: On the problem of the most efficient tests of statistical hypotheses. Philos. Trans. R. Soc. Lond. Ser. A Contain. Pap. Math. Phys. Character **231**, 289–337 (1933)
25. Siegmund, D.: Sequential analysis. Tests and Confidence Intervals. Springer Series in Statistics. Springer, New York (1985)
26. Tartakovsky, A., Nikiforov, I., Basseville, M.: Sequential analysis. Hypothesis testing and changepoint detection. Monographs on Statistics and Applied Probability, vol. 136. CRC Press, Boca Raton, FL (2015)
27. Wald, A.: Sequential Analysis. Wiley, New York; Chapman & Hall, Ltd., London (1947)
28. Wald, A.: Foundations of a general theory of sequential decision functions. Econometrica **15**, 279–313 (1947)

Investigation of Artificial Intelligence Methods in the Short-Term and Middle-Term Forecasting in Financial Sphere

Yuriy Zaychenko, Helen Zaichenko, and Oleksii Kuzmenko

Abstract In this paper the problems of short- and middle-term forecasting at the financial sphere are considered. For this problem intelligent forecasting methods: GMDH and hybrid deep learning networks based on self-organization are suggested. As alternative method was used ARIMA The optimal parameters of hybrid networks and ARIMA were found. Optimal structures of hybrid networks were constructed for short-term and middle-term forecasting. The experimental investigations of GMDH and hybrid DL networks in forecasting problems in finance are carried out and compared with ARIMA thereafter the best forecasting spheres for GMDH and hybrid DL networks were determined.

Keywords Short-term · Middle-term financial forecasting · GMDH · Hybrid DL network · Optimization

1 Introduction. Analysis of Previous Works

Problems of forecasting share prices and market indexes at stock exchanges pay great attention of investors and various money funds. For its solution were developed and for a long time applied powerful statistical methods, first of all ARIMA [1, 2]. Last years different intelligent methods and technologies were also suggested and widely used for forecasting in financial sphere, in particular among them neural networks and fuzzy logic systems.

Y. Zaychenko (✉) · H. Zaichenko · O. Kuzmenko
Institute for Applied System Analysis, Igor Sikorsky Kyiv Polytechnic Institute, Peremogy avenue 37, Kyiv 03056, Ukraine
e-mail: zaychenkoyuri@ukr.net

H. Zaichenko
e-mail: syncmaster@bigmir.net

O. Kuzmenko
e-mail: oleksii.kuzmenko@ukr.net

M. Zgurovsky and N. Pankratova (eds.), *System Analysis and Artificial Intelligence*, Studies in Computational Intelligence 1107,
https://doi.org/10.1007/978-3-031-37450-0_18

The efficient tools of modeling and forecasting of non-stationary time series is Group Method of Data Handling (GMDH) suggested and developed by acad. Alexey Ivakhnenko [3, 4]. This method is based on self-organization and enables to construct optimal structure of forecasting model automatically in the process of algorithm run. Method GMDH and fuzzy GMDH were successfully applied for forecasting at stock exchanges for long time.

As alternative approach for forecasting in finance is application of various types of neural network: MLP [5], fuzzy neural networks [6, 7], neo-fuzzy networks [8] and Deep learning (DL) networks [9]. New trend in sphere DL networks is new class of neural networks-hybrid DL networks based on GMDH method [10]. The application self-organization in these networks enables to train not only neuron weights but to construct optimal structure of a network. Due to method of training in these networks weights are adjusted not simultaneously but layer after layer. That prevents the phenomenon of vanishing or explosion of gradient. It's very important for networks with many layers.

The first works in this field used as nodes of hybrid network Wang-Mendel neurons with two inputs [10]. But drawback of such neurons is the demand to train not only neural weights but the parameters of fuzzy sets in antecedents of rules as well. That needs a lot of calculational expenses and large training time as well. Therefore later DL neo-fuzzy networks were developed in which as nodes are used neo-fuzzy neurons by Yamakawa [11, 12]. The main property of such neurons is that it's necessary to train only neuron weights not fuzzy sets. That demands less computations as compared with Wang-Mendel neurons and significantly cuts training time as a whole. The investigation of both classes of hybrid DL networks were performed and their efficiency at forecasting in financial sphere was compared in [13].

Therefore it presents great interest to compare the efficiency of intelligent methods: hybrid DL networks, GMDH and conventional statistical method ARIMA at the problem of forecasting at financial sphere.

The goal of this paper is to investigate the accuracy of hybrid DL networks based on self-organization, GMDH and ARIMA at the problem of forecasting share prices and market indices at the stock exchange, to compare their efficiency at the different intervals and to determine the classes of forecasting problems for which the application of corresponding computational intelligence technologies is the most perspective.

2 The Description of the Evolving Hybrid GMDH-Neo-Fuzzy Network

The evolving hybrid DL-network architecture is presented in Fig. 1. To the system's input layer a $(n \times 1)$-dimensional vector of input signals is fed. After that this signal is transferred to the first hidden layer. This layer contains $n_1 = c_n^2$ nodes, and each of these neurons has only two inputs.

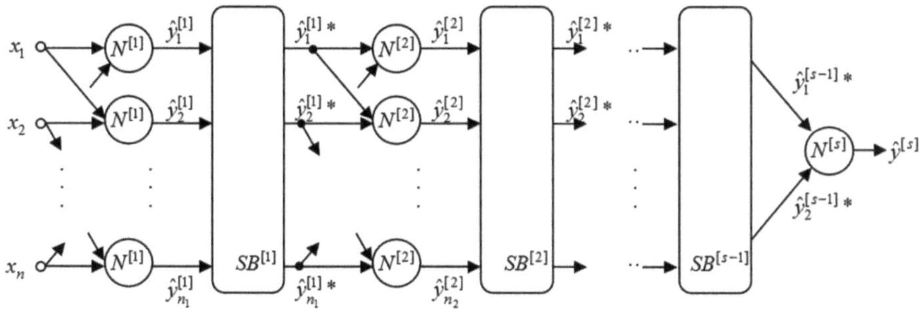

Fig. 1 Evolving GMDH-network

At the outputs $N^{[1]}$ of the first hidden layer the output signals are formed. Then these signals are fed to the selection block of the first hidden layer.

It selects among the output signals $\hat{y}_l^{[1]} n_1^*$ (where $n_1^* = F$ is so-called freedom of choice) most precise signals by some chosen criterion (mostly by the mean squared error $\sigma^2_{y_l^{[1]}}$). Among these n_1^* best outputs of the first hidden layer $\hat{y}_l^{[1]*} n_2$ pairwise combinations $\hat{y}_l^{[1]*}$, $\hat{y}_p^{[1]*}$ are formed. These signals are fed to the second hidden layer, that is formed by neurons $N^{[2]}$. After training these neurons output signals of this layer $\hat{y}_l^{[2]}$ are transferred to the selection block $SB^{[2]}$ which chooses F best neurons by accuracy (e.g. by the value of $\sigma^2_{y_l^{[2]}}$) if the best signal of the second layer is better than the best signal of the first hidden layer $\hat{y}_l^{[1]*}$. Other hidden layers work similarly. The system evolution process continues until the best signal of the selection block $SB^{[s+1]}$ appears to be worse than the best signal of the previous s layer. Then it's necessary to return to the previous layer and choose its best node neuron $N^{[s]}$ with output signal $\hat{y}^{[s]}$. And moving from this neuron (node) along its connections backwards and sequentially passing all previous layers the final structure of the GMDH-neo-fuzzy network is constructed.

It should be noted that in such a way not only the optimal structure of the network may be constructed but also well-trained network due to the GMDH algorithm. Besides, since the training is performed sequentially layer by layer the problems of high dimensionality as well as vanishing or exploding gradient are avoided.

2.1 *Neo-Fuzzy Neuron as a Node of Hybrid GMDH-System*

Let's consider the architecture of the node that is presented in Fig. 2 and is suggested as a neuron of the proposed GMDH-system. As a node of this structure a neo-fuzzy neuron (NFN) developed by Takeshi Yamakawa and co-authors in [9] is used. The neo-fuzzy neuron is a nonlinear multi-input single-output system shown in Fig. 2.

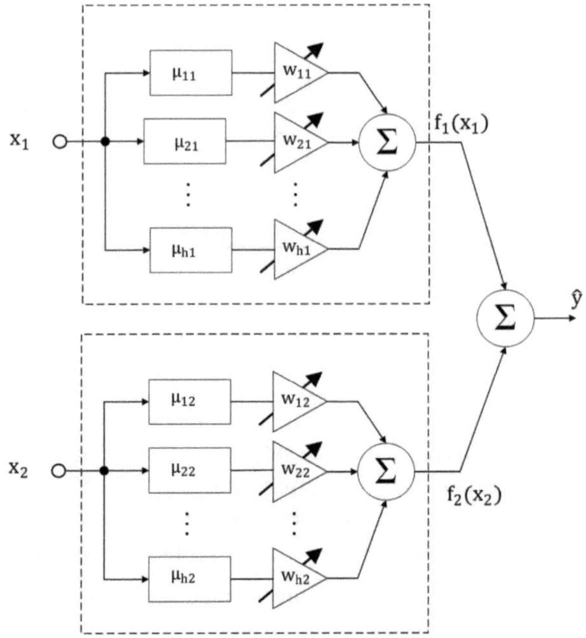

Fig. 2 Architecture of neo-fuzzy neuron with two inputs

The main difference of this node from the general neo-fuzzy neuron structure is that each node uses only two inputs.

It realizes the following mapping:

$$\hat{y} = \sum_{i=1}^{2} f_i(x_i) \tag{1}$$

where x_i is the input i $(i = 1, 2, \ldots, n)$, $\hat{y}$ is a system output. Structural blocks of neo-fuzzy neuron are nonlinear synapses NS_i which perform transformation of input signal in the form

$$f_i(x_i) = \sum_{j=1}^{h} w_{ji}\mu_{ji}(x_i) \tag{2}$$

and realize fuzzy inference: *if* x_i *is* x_{ji} *then the output is* w_{ji}, where x_{ji} is a fuzzy set which membership function is μ_{ji}, w_{ji} is a synaptic weight in consequent [14].

2.2 The Neo-Fuzzy Neuron Learning Algorithm

The learning criterion (goal function) is the standard local quadratic error function:

$$E(k) = \frac{1}{2}(y(k) - \hat{y}(k))^2 = \frac{1}{2}e(k)^2 = \frac{1}{2}(y(k) - \sum_{i=1}^{2}\sum_{j=1}^{h} w_{ji}\mu_{ji}(x_i(k)))^2 \quad (3)$$

It is minimized via the conventional stochastic gradient descent algorithm.

In case we have a priori defined data set the training process can be performed in a batch mode at one epoch using conventional least squares method [11]

$$w^{[1]}(N) = \left(\sum_{k=1}^{N} \mu^{[1]}(k)\mu^{[1]T}(k)\right)^{+} \sum_{k=1}^{N} \mu^{[1]}(k)y(k) \quad (4)$$

$$= P^{[1]}(N)\sum_{k=1}^{N} \mu^{[1]}(k)y(k) \quad (5)$$

where $(\bullet)^+$ means pseudo inverse of Moore–Penrose (here $y(k)$ denotes external reference signal (real value).

If training observations are fed sequentially in on-line mode, the recurrent form of the LSM can be used in the form (6)

$$\begin{cases} w_l^{ij}(k) = w_l^{ij}(k-1) + \frac{P^{ij}(k-1)(y(k)-(w_l^{ij}(k-1))^T\varphi^{ij}(x(k)))\varphi^{ij}(x(k))}{1+(\varphi^{ij}(x(k)))^T P^{ij}(k-1)\varphi^{ij}(x(k))}, \\ P^{ij}(k) = P^{ij}(k-1) - \frac{P^{ij}(k-1)\varphi^{ij}(x(k))(\varphi^{ij}(x(k)))^T P^{ij}(k-1)}{1+(\varphi^{ij}(x(k)))^T P^{ij}(k-1)\varphi^{ij}(x(k))}. \end{cases} \quad (6)$$

3 Experimental Investigation and Results Analysis

3.1 Data Set

As input data was taken share prices of APPLE corp. at NYSE stock exchange at the period since 3 January till 8 of November 2022. The sample consisted of 215 instances which were divided into training and test subsamples.

Flow chart of Apple close prices is presented at the Fig. 3.

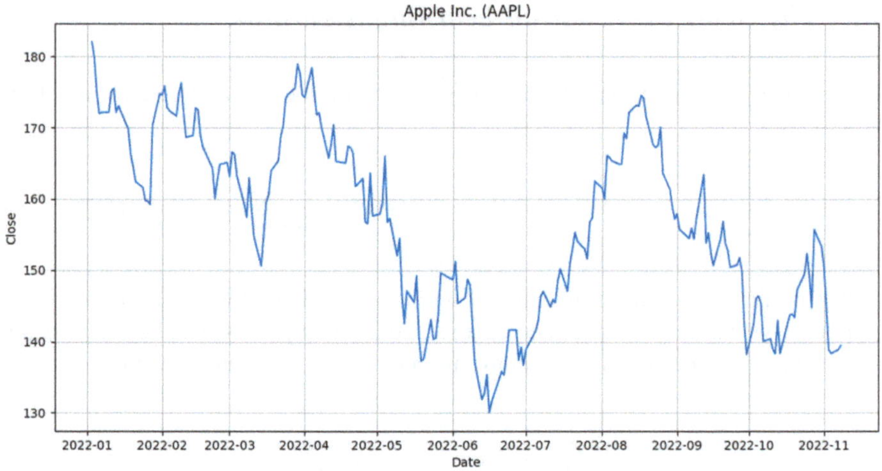

Fig. 3 Dynamics of the index Apple close

Table 1 Experimental parameters

Parameter	Value
Membership functions	Gaussian
Number of inputs	3; 4; 5
Number of linguistic variables	3; 4
Ratio (percentage of the training sample)	0,6 (60%); 0,7 (70%); 0,8 (80%)
Criterion	MSE; MAPE
Forecast interval	1; 3; 5; 7; 20

3.2 Experimental Investigations of Hybrid DL Networks

The first series of experiments were performed with hybrid Deep Learning network with neo-fuzzy neurons as nodes. At experiments the following parameters were varied: ratio training/test sample, number of inputs 3–5, number of fuzzy sets per variable. The membership functions were taken Gaussian form. The goal of experiments was to find optimal parameters values (Table 1).

At the first experiment forecast period was taken 1 day, as the accuracy metrics were taken MSE and MAPE.

After experiments optimal parameters of hybrid DL network were found: number of inputs—3, number of rules—4, ratio training/test—0.6. The process of optimal structure generation is presented in the Table 2.

Flow chart of the best forecast of hybrid neo-fuzzy network for 1 day is shown in the Fig. 4 (Table 3).

Table 2 The process of structure generation for the best forecast (inputs: 3; variables: 4; ratio: 0,6)

NFN	SB 1	SB 2	SB 3
(0, 1)	2,062709732		
(0, 2)	4,643638129		
(1, 2)	3,142933032		
((0, 1), (0, 2))		0,0135017552	
((0, 1), (1, 2))		0,01339908479	
((0, 2), (1, 2))		0,02584785481	
(((0, 1), (0, 2)), ((0, 1), (1, 2)))			0,1419107228
(((0, 1), (0, 2)), ((0, 2), (1, 2)))			0,2327715495
(((0, 1), (1, 2)), ((0, 2), (1, 2)))			0,1409591637

Fig. 4 Flow chart of the best forecast of hybrid DL network (inputs:3; variables: 4; ratio: 0,6)

The next experiments were performed with forecasting interval of 3 days. The flow chart of the best forecast of hybrid DL network is presented in the Fig. 5. Values of criteria MSE and MAPE for this experiment are shown in the Table 4.

In the succeeding experiments the forecasting accuracy of hybrid DL network was investigated in middle-term forecasting at forecasting interval of 5, 7 and 20 days.

The values of MSE and MAPE at the forecasting intervals 7 and 20 days are presented in the Tables 5 and 6 correspondingly.

The flow chart of the best forecast of hybrid DL network for interval 7 days is presented in the Fig. 6 and for 20 days in the Fig. 7.

In the next series of experiments the forecasting accuracy of GMDH method was explored in the problem of short and middle-term forecasting of APPLE share prices. During experiments the following parameters were varied: inputs number—3, 4, 5; ratio train/test—60%, 70%, 80%; forecast interval—1, 3, 5, 7, 20 (days); linear partial

Table 3 The best forecast at the interval 1 day (inputs: 3; variables: 4; ratio: 0,6)

Date	Real	Forecast	MSE	MAPE
30.08.2022	158,91	160,0752	1,357698	0,733247
31.08.2022	157,22	155,9666	1,571053	0,797237
01.09.2022	157,96	153,2552	22,13497	2,978464
02.09.2022	155,81	151,5822	17,87408	2,713417
06.09.2022	154,53	151,577	8,72047	1,910984
07.09.2022	155,96	150,1485	33,77412	3,726308
27.10.2022	144,8	146,5022	2,897403	1,175536
...	...	...	...	...
31.10.2022	153,34	143,8247	90,54012	6,205333
01.11.2022	150,65	148,1426	6,28697	1,664376
02.11.2022	145,03	148,0086	8,872028	2,053779
03.11.2022	138,88	145,9827	50,4488	5,114294
04.11.2022	138,38	143,2844	24,05327	3,544163
07.11.2022	138,92	141,2582	5,467025	1,683103
08.11.2022	139,5	140,7952	1,677528	0,928455
Min:			**0,000235**	**0,010755**
Avg:			**25,74452**	**2,767978**
Max:			**202,3936**	**8,704958**

Fig. 5 Flow chart of the best forecast for 3 days interval

Table 4 Forecasting accuracy of hybrid DL network for 3 days interval

	MSE	MAPE
Min	0,03003	0,11122
Avg	31,32012	3,150427
Max	212,328	8,916037

Table 5 Forecasting accuracy of hybrid DL network for 7 days interval

	MSE	MAPE
Min	0,007047	0,057459
Avg	43,06783	3,628598
Max	216,1277	9,251327

Table 6 Forecasting accuracy of hybrid DL network for 20 days interval

	MSE	MAPE
Min	0,016875	0,090367
Avg	51,35114	3,647911
Max	298,9944	11,47409

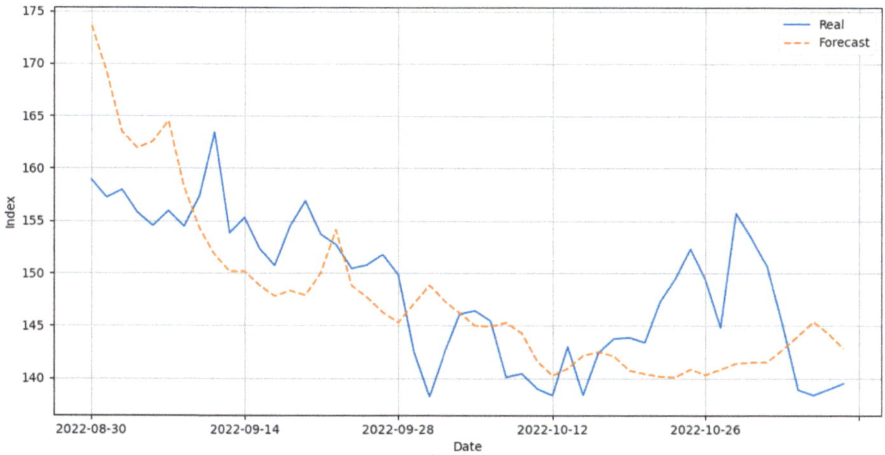

Fig. 6 Flow chart of the best forecast for 7 days interval

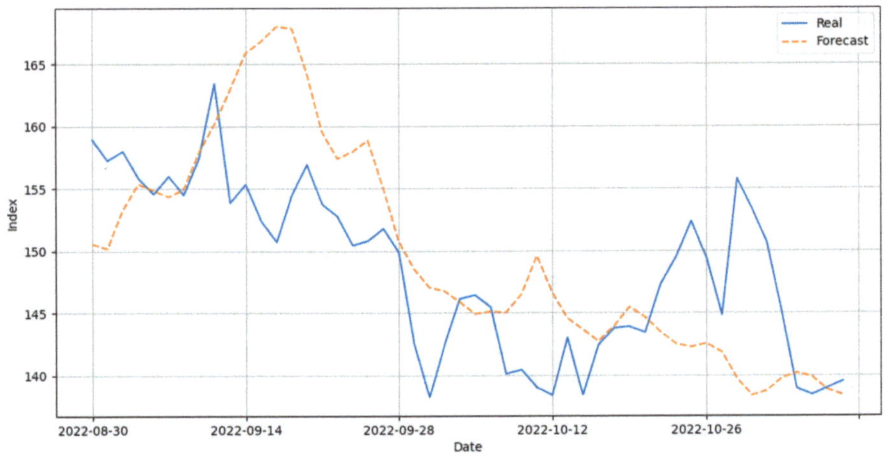

Fig. 7 Flow chart of the best forecast for 20 days interval

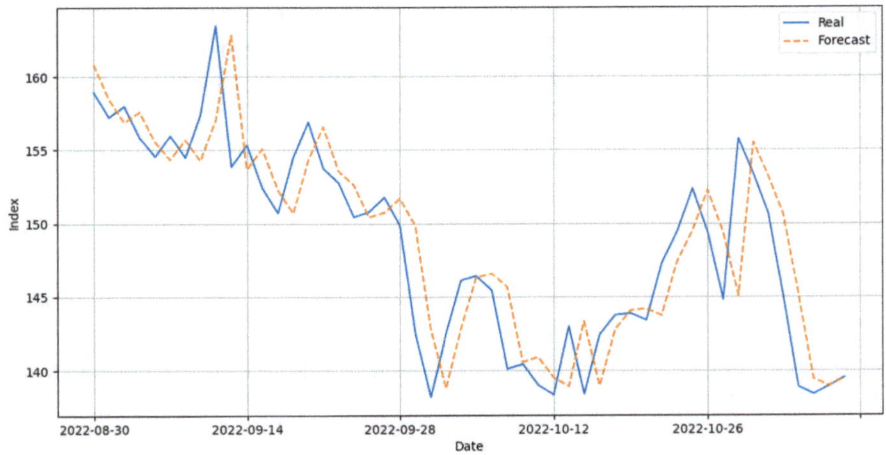

Fig. 8 The flow chart of the best forecast by GMDH (function: linear; inputs: 5; ratio: 0,7)

description; freedom choice was taken 3, 4. Flow chart of the best forecast for 1 day is shown in the Fig. 8.

In the next series of experiments the method ARIMA was investigated in the problem of forecasting APPLE share prices. First the autocorrelation function (ACF) of time series was calculated and its plot is presented in the Fig. 9 and partial ACF in the Fig. 10.

After that the time series was transformed to a stationary one using differencing. Fuller test was applied to transformed time series and $P - value$ was calculated: $P - value = 2.952140431702318e - 24 < 0.05$ which confirmed its stationarity. The parameters of the **ARIMA(p, d, q)** model were defined as follows:

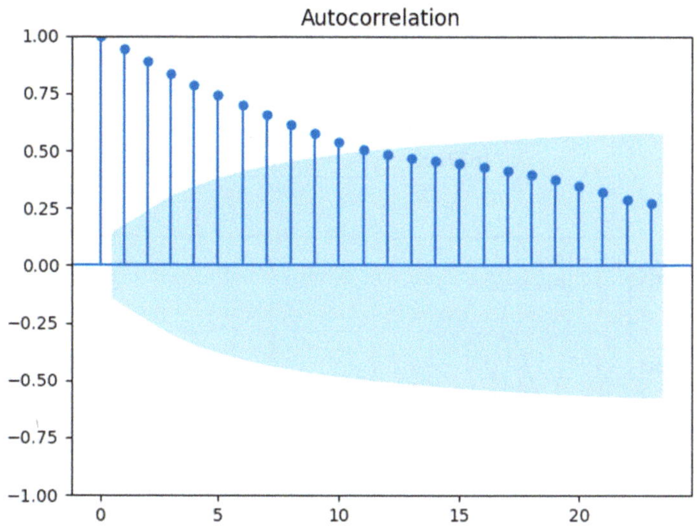

Fig. 9 Plot of ACF APPLE prices time series

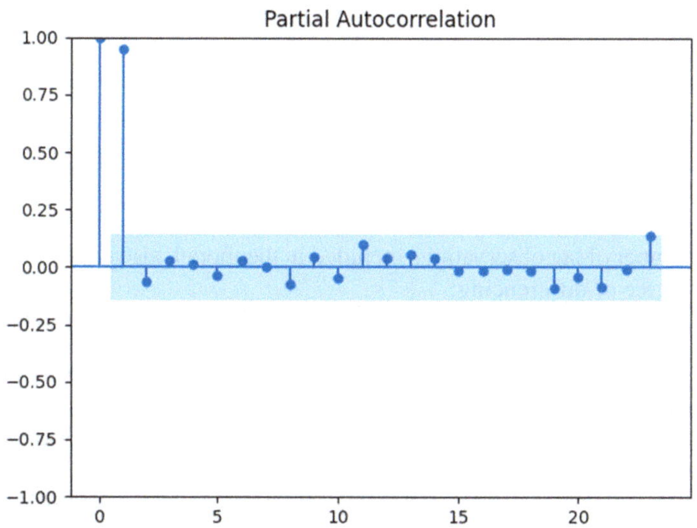

Fig. 10 PACF original of APPLE prices time series

Table 7 Forecasting accuracy of ARIMA for 1 day interval

	MSE	MAPE
Min	5,6169	0,0074
Avg	187832,5	1,125262
Max	1629120	4,1034

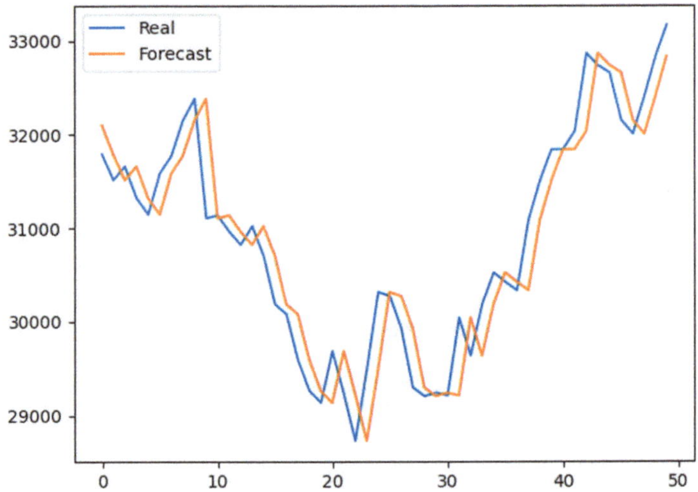

Fig. 11 Comparison of forecast data with real data by ARIMA model for one day interval

p: The number of lag observations included in the model, also called the lag order;
d: The degree of differencing;
q: The size of the moving average window, also called the order of moving average.

The experiments were performed to find best parameters of ARIMA model for 1 day forecasting interval. Best model: ARIMA(0, 1, 1)(0, 0, 0)[0] intercept.

After that the experimental investigations with the optimal model were performed and forecasting accuracy was evaluated at test sample presented in the Table 7.

Flow charts of the real and forecasting share prices using ARIMA model are presented in the Fig. 11.

3.3 General Comparison of All Methods: Hybrid DL Networks, GMDH and ARIMA

At the final series of experiments all methods: GMDH, hybrid DL-neo-fuzzy networks and ARIMA were investigated at short-term and middle-term forecasting at forecasting intervals 1, 3, 5, 7 and 20 days. The accuracy of forecasting of investi-

Table 8 Average MSE values of the best models for different intervals

	GMDH-neo-fuzzy	GMDH	ARIMA
Interval 1	25,74451785	12,84965	13,28682
Interval 3	31,32012269	32,77652	44,86191
Interval 5	32,9170912	37,66422	63,42466
Interval 7	43,06782659	47,36439	78,85156
Interval 20	51,35113807	79,49455	117,8323

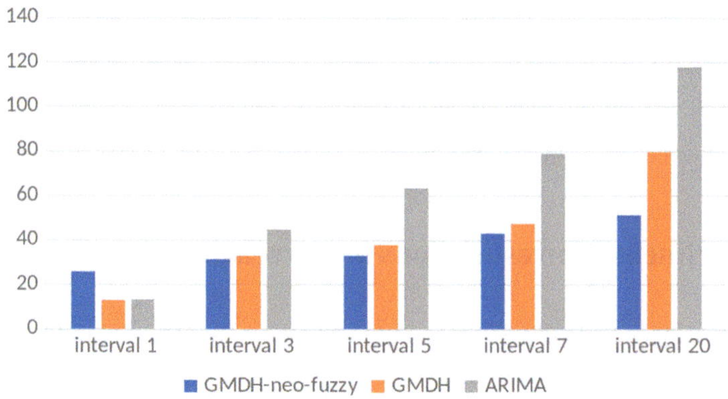

Fig. 12 Comparison of the average MSE values of the best models for different intervals

Table 9 Average MAPE values of the best models for different intervals

	GMDH-neo-fuzzy	GMDH	ARIMA
Interval 1	2,767977744	1,853726	1,962852
Interval 3	3,150426983	3,055711	3,391568
Interval 5	3,178736648	3,291786	4,150802
Interval 7	3,628598328	3,821228	4,839673
Interval 20	3,647911268	5,06835	6,280025

gated methods for different intervals by criterion MSE is presented in the Table 8. The average MAPE values for all intervals are shown in the Table 9. Comparison of the average MSE values of the best models for different forecasting intervals are presented in the Fig. 12 while MAPE values are shown in the Fig. 13.

After analysis of the presented results it was detected that method GMDH has the best accuracy for short-term forecasting 1 and 3 days while hybrid GMDH-neo-fuzzy network has the best accuracy for middle-term forecasting 5, 7 and 20 days. As for statistical method ARIMA it's takes second place after GMDH at short-term forecasting for 1 day interval and is the worst for middle-term forecasting.

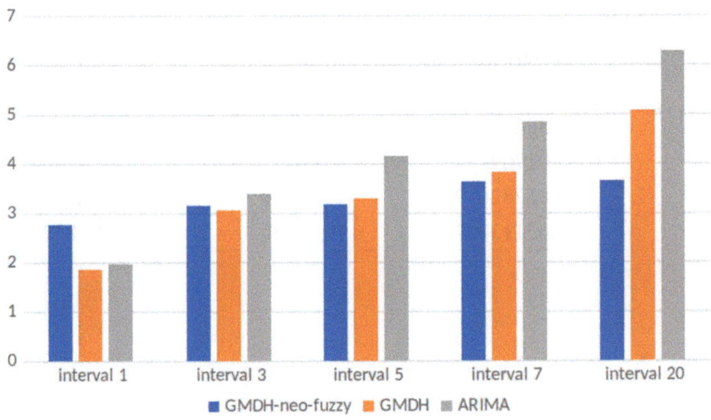

Fig. 13 Comparison of the average MAPE values of the best models for different intervals

4 Experimental Investigations at Forecasting Dow-Jones

In the next experiments the forecasting models were investigated in the problem of forecasting Dow-Jones index. The flow chart of this index dynamics in 2022 year is presented in the Fig. 14.

The whole sample included values from 03.01.22 to 08.11.22. It was divided into train and test samples. During experiments forecasting interval varied so: 1, 3, 5, 7 and 20 days.

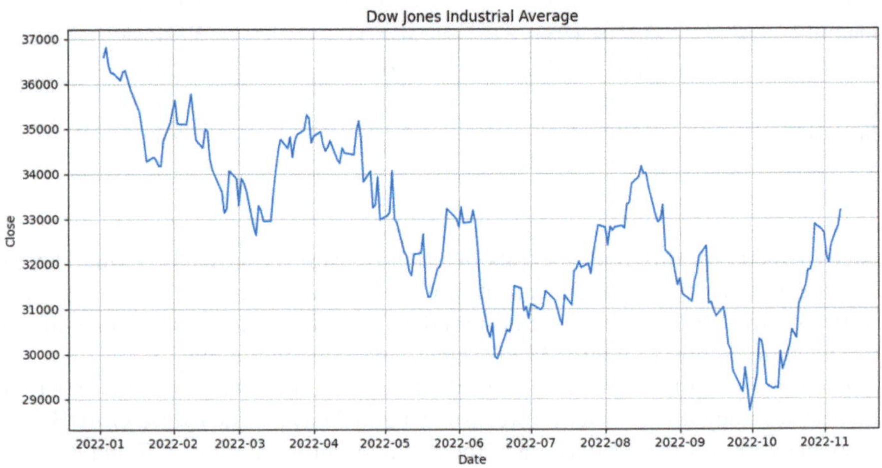

Fig. 14 Flow chart of Dow-Jones index dynamics

Table 10 Average MAPE values of the best models for different intervals of Dow-Jones index

	GMDH-neo-fuzzy	GMDH	ARIMA
Interval 1	2,199804	1,179718	1,125262
Interval 3	2,420808	2,281228	2,364068
Interval 5	2,724774	2,722114	2,80689
Interval 7	3,057794	3,373748	3,379318
Interval 20	4,212836	6,142261	6,167008

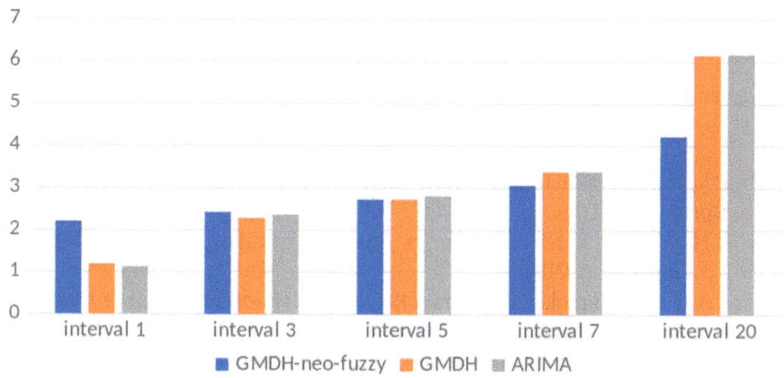

Fig. 15 Comparison of the average MAPE values of the best models for Dow-Jones index

After experiments the forecasting accuracy of models GMDH, hybrid DL network and ARIMA was estimated. The values of MAPE criterion for different forecasting intervals are presented in the Table 10 and are shown in the Fig. 15.

Analysis of these results in general confirms the conclusions of experiments of forecasting APPLE share prices: **Method GMDH and ARIMA are the best for short-term forecasting problem (1, 3 days) while hybrid GMDH DL network is the best tools for middle-term forecasting (7, 20 days).**

5 Conclusions

1. In this paper the problem of forecasting at financial market by different forecasting intervals was considered (short-term and middle-term forecasting). For its solution it was suggested to apply GMDH, hybrid deep learning (DL) networks based on GMDH and ARIMA.
2. The experimental investigations were performed an the problem of forecasting APPLE share prices at period since January till August 2022 and Dow-Jones Index Aver.
3. Optimization of parameters of hybrid DL networks and ARIMA was performed during the experiments. The optimal structure of hybrid DL network was constructed using GMDH method.

4. The experimental investigations of optimized ARIMA, GMDH and hybrid DL networks were carried out at different forecasting intervals and their accuracy was compared.

In result it was established that application of GMDH has the best forecasting accuracy for short-term forecasting at the interval 1, 3 days while hybrid DL neo-fuzzy network is the best for middle-term forecasting—7, 20 days and in a whole the intelligent methods based on GMDH outperform statistical method ARIMA at the problems of short-term and middle-term forecasting at stock exchanges.

References

1. Brockwell, P.J.: Introduction to Time Series and Forecasting. Brockwell, P.J., Davis, R.A., 2nd edn, 429 p. Springer (2002)
2. Shumway, R.H., Stoffer, D.S.: Time Series Analysis and Its Applications With R Examples, 4th edn, 562 p. Springer (2017)
3. Ivakhnenko, A.G., Ivakhnenko, G.A., Mueller, J.A.: Self-organization of the neural networks with active neurons. Pattern Recognit. Image Anal. **4**(2), 177–188 (1994)
4. Ivakhnenko, A.G., Wuensch, D., Ivakhnenko, G.A.: Inductive sorting-out GMDH algorithms with polynomial complexity for active neurons of neural networks. Neural Netw. **2**, 1169–1173 (1999)
5. Haykin, S.S.: Neural Networks: A Comprehensive Foundation, 2nd edn. Prentice Hall, Upper Saddle River (1999)
6. Ossovsky, S.: Neural networks for information processing. Translate from Polish, M. Finance Stat. 344 (2002)
7. Wang, F.: Neural networks genetic algorithms and fuzzy logic for forecasting. In: Proceeding of International Conference on Advanced Trading Technologies, New York, pp. 504–532 (1992)
8. Yamakawa, T., Uchino, E., Miki, T., Kusanagi, H.: A neo-fuzzy neuron and its applications to system identification and prediction of the system behavior. In: Proceedings 2nd International Conference on Fuzzy Logic and Neural Networks "LIZUKA-92", pp. 477–483. Lizuka (1992)
9. Goodfellow, I., Bengio, Y., Courville, A.: Deep Learning, MIT Press (2016). http://www.deeplearningbook.org
10. Zaychenko, Y., Bodyanskiy, Y., Tyshchenko, O., Boiko, O., Hamidov, G.: Hybrid GMDH-neuro-fuzzy system and its training scheme. Int. J. Inf. Theories Appl. **24**(2), 156–172 (2018)
11. Zaychenko, Yu., Hamidov, G.: The hybrid deep learning GMDH-neo-fuzzy neural network and its applications. In: Proceedings of 13th IEEE International Conference on Application of Information and Communication Technologies – AICT2019, pp. 72–77, 23–25 October 2019, Baku
12. Bodyanskiy, E., Zaychenko, Y., Boiko, O., Hamidov, G., Zelikman, A.: Structure optimization and investigations of hybrid GMDH-Neo-fuzzy neural networks in forecasting problems. In: Zgurovsky, M., Pankratova, N. (eds.) System Analysis & Intelligent Computing. Book Studies in Computational Intelligence, SCI, vol. 1022, pp. 209–228. Springer (2022)
13. Zaychenko, Y., Zaichenko, H., Hamidov, G: Hybrid GMDH deep learning networks - analysis, optimization and applications in forecasting at financial sphere. Syst. Res. Inf. Technol. (1), 73–86 (2022). https://doi.org/10.20535/SRIT.2308-8893.2022.1.06
14. Bodyanskiy, Eu., Zaychenko, Yu., Hamidov, G., Kuleshova, N.: Multilayer GMDH-neuro-fuzzy network based on extended neo-fuzzy neurons and its application in online facial expression recognition. Syst. Res. Inf. Technol. (3), 67–78 (2020)

Neo-Fuzzy Radial-Basis Function Neural Network and Its Combined Learning

Yevgeniy Bodyanskiy, Olha Chala, Valentin Filatov, and Iryna Pliss

Abstract The Neo-Fuzzy Radial Basis Function Stacking Neural Network (NF-RBFNN) is a proposed hybrid neural network system that combines traditional RBFNN and neo-fuzzy neuron. The combination aims to create an efficient system that performs well in scenarios with limited data sets or when fast learning is required due to online data acquisition. NF-RBFNN consists of two independent subsystems, which facilitate quick parameter setting and implementation. With the ability to approximate nonlinear functions and handle uncertain information, the neo-fuzzy neuron and traditional radial basis function neural network are known for their respective strengths. By merging these two approaches, the NF-RBFNN is an effective hybrid neural network that can improve approximation properties while handling uncertainties. Ultimately, this simple yet efficient system excels in situations where quick learning and uncertain information handling come into play. With its impressive performance and straightforward implementation, the system holds great potential for a multitude of practical applications in various fields.

Y. Bodyanskiy (✉) · O. Chala · V. Filatov
Artificial Intelligence Department, Kharkiv National University of Radio Electronics, Nauky Ave. 14, Kharkiv, Ukraine
e-mail: yevgeniy.bodyanskiy@nure.ua

O. Chala
e-mail: olha.chala@nure.ua

V. Filatov
e-mail: valentin.filatov@nure.ua

I. Pliss
Control Systems Research Laboratory, Kharkiv National University of Radio Electronics, Nauky Ave. 14, Kharkiv, Ukraine
e-mail: iryna.pliss@nure.ua

M. Zgurovsky and N. Pankratova (eds.), *System Analysis and Artificial Intelligence*, Studies in Computational Intelligence 1107,
https://doi.org/10.1007/978-3-031-37450-0_19

1 Introduction

Today, methods and systems of computational intelligence [1–3] are widely used to solve various information processing problems, specifically Data Mining problems. The most common and effective systems of computational intelligence include artificial neural networks, neuro-fuzzy systems, and hybrid systems, due to their universal approximating properties. In recent years, there has been a special focus on deep neural networks (DNNs), because they provide great classification accuracy, and also they are based on multilayer perceptions (MLPs). These networks are highly efficient in solving image, natural language, and multidimensional time series prediction problems, among others. The main element of deep neural networks is the Rosenblatt perceptron [1], which has activation functions that do not satisfy the conditions of the approximation theorem of Cybenko [4]. As a result, the number of neurons and adjustable synaptic weights can become very large, requiring extremely large volumes of training data sets that are not always available when solving real problems. In addition, multi-epoch learning based on classic error back-propagation may require a significant amount of time, as DNNs are relatively slow systems and not suitable for solving data stream mining problems.

An alternative to the quite big family of multilayer perceptrons can be neural networks with kernel activation functions [5], where radial-basis function neural networks [6–9] are the most widespread. Based on RBFNN, a convolutional deep neural network [10] has already been built, which has many advantages over perceptron deep neural networks. The main advantage of radial basis function neural networks over standard perceptron networks is that the input signal of the network depends linearly on adjustable synaptic weights, enabling the use of fast Gaussian-Newtonian optimization algorithms and, most importantly, the least squares method, both in batch and recurrent forms (for data stream mining problems). However, RBFNs may suffer from the curse of dimensionality effect when the number of R-neurons (kernel activation functions) in the hidden layer increases exponentially with the increasing dimensionality of the input signal. This problem can be addressed by using the previous convolution of the input signal [10], formulating a simplified architecture based on evolving systems methods [11–14], or using combined learning to adjust both the number of activation functions and their parameters [15].

We propose to enhance the approximation properties of RBFN by replacing the layer of linear synaptic weights with a layer of nonlinear synapses [16, 17] that implement a nonlinear F-transform [18] with high approximation properties. However, the output signals should remain linearly dependent on the adjusted parameters to maintain a high speed of learning the synaptic weights of the network.

2 Architecture of Neo-Fuzzy Radial Basis Function Stacking Neural Network (NF-RBFNN)

Figure 1 shows the architecture of the proposed NF-RBFNN, which, similar to the traditional NF-RBFNN, has two layers of information processing.

n-dimensional vector-observation is received at the input of the zero input layer

$$x(k) = (x_1(k), \ldots, x_i(k), \ldots, x_i(k))^T \in R^n, \tag{1}$$

where k - observation number, $k = 1, 2, \ldots, N$, in the training data set if the network is trained in batch mode, or the index of the current discrete time, $k = 1, 2, \ldots, N, N+1 \ldots$, if the training takes place online. Also, previously the components of the input vectors $x(k)$ were coded for a certain interval, which often comes across as:

$$0 \leq x_i(k) \leq 1 \text{ or } -1 \leq x_i(k) \leq 1, \tag{2}$$

The first hidden layer is formed by the so-called R-neurons, that implement nonlinear transformation using kernel activation functions, among which the most common are Gaussians.

$$\varphi_l\left(x(k)\right) = u_l(k) = \varphi_l\left(\left\|x(k) - c_l\right\|^2, \sigma^2\right) = exp\left(-\frac{\left\|x(k) - c_l\right\|^2}{2\sigma^2}\right) \tag{3}$$

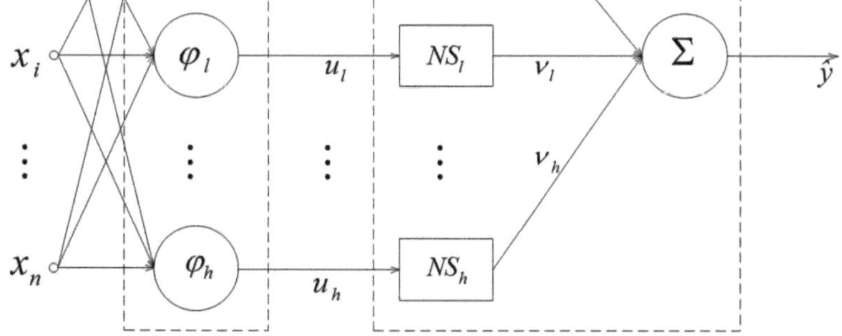

Fig. 1 Neo-fuzzy radial basis function stacking neural network architecture

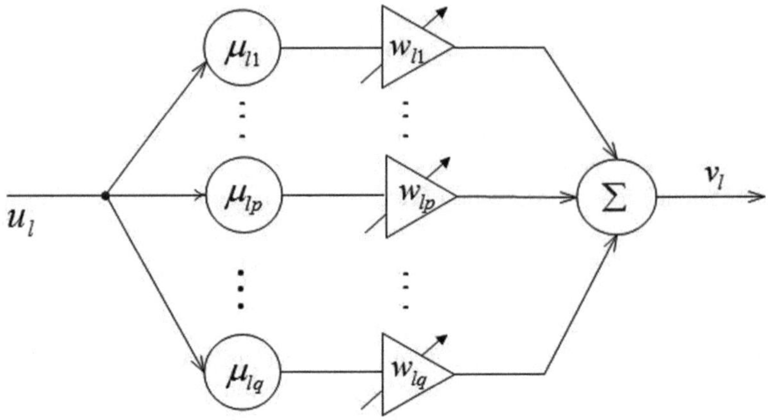

Fig. 2 Nonlinear synapse

where c_l is a *(n+1)* vector which determines the coordinates of the kernel functions' center φ_l and σ^2 is a scalar spread-width parameter of this function.

Note that other kernel bell-shaped functions can be used as activation ones. For example, Cauchians [15] or a multidimensional version of Yepanechnikov's kernel functions. A vector signal is formed at the output of the entire layer that completely coincides with the corresponding layer of the standard RBFNN, forming a vector signal $\varphi(x(k)) = u(k) = \left(\varphi_1(x(k)), \ldots, \varphi_l(x(k)), \ldots, \varphi_h(x(k))\right)^T = \left(u_1(k), .., u_l(k), \ldots, u_h(k)\right)^T$ which is subsequently fed to the output layer of the network, which in the RBFNN is formed by adjustable synaptic weights $w_l, l = 0, 1, \ldots, h$ and adders. In this way, RBFNN implements the transformation:

$$\hat{y}(k) = \sum_{l=1}^{h} w_l \varphi_l(x(k)) = \sum_{l=1}^{h} w_l u_l(x(k)) = w^T \varphi(x(k)) = w^T u(k) \tag{4}$$

where w^T is a vector of adjustable synaptic weights.

In order to provide the necessary approximation functions h can be large enough, which leads to the effect of the so-called "Curse of dimensionality".

To ensure approximation properties of the proposed system, instead of an elementary set of synaptic weights, we will use nonlinear synapses $NS_l, l = 1, 2, \ldots, h$ [16] which provide high-accuracy recovery of one-dimensional nonlinear functions using the F-transformation [18]. The scheme of the NS_l nonlinear synapse is shown in Fig. 2.

The output signals of R-neurons of the hidden layer $u_l(k)$ of the formula are received at the input of nonlinear synapses, which are fuzzified using a system of membership functions $\mu_{lp}, p = 1, 2, \ldots, q$ of the formula that satisfies the Ruspini unity partitioning conditions. Figure 3 shows a system of triangular membership functions that implement the signals $u_l(k)$ in fuzzification procedure formula.

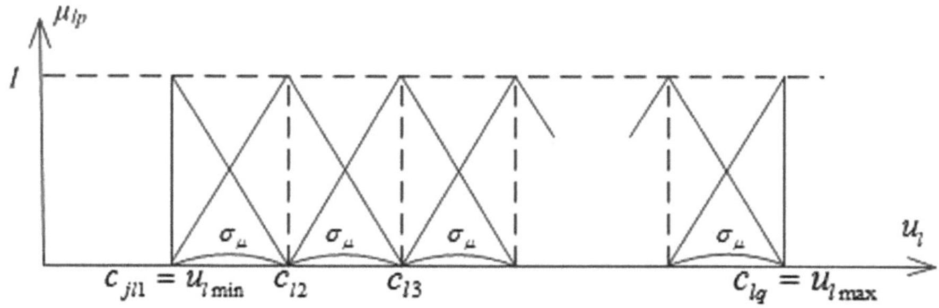

Fig. 3 Triangular membership functions of the lth nonlinear synapse

In principle, any others satisfying the Ruspini partition conditions, such as B-splines, can be used as membership functions. The advantage of triangular functions is that at each moment k only two adjacent functions are fired, that is, only two synaptic weights are adjusted. Thus, a system is proposed that adjusts only twice the number of synaptic weights at each moment compared to the standard RBFNN. The use of B-splines usually improves the approximation capabilities, but it will be necessary to adjust the parameter h_q every moment k.

If, as utility functions, those are used that do not satisfy the Ruspini unity partition conditions then an additional layer of defuzzification must be introduced into the system, which makes this system more cumbersome. The fuzzified, by membership functions, signals are then sent to the synaptic weights layer of the nonlinear synapse $w_l = (w_{l1}, \dots w_{lp}, \dots w_{lq})^T$, on output of which appears a value $v_l(k) = \sum_{p=1} q w_{lp} \mu_{lp}(u_l(k))$.

A set of parallel-connected nonlinear signals, $NS_1, \dots NS_l, \dots NS_h$ of which forms a neo-fuzzy neuron that forms an output layer-stack NF-RBFNN, the input of which is the output of the system as a whole

$$\hat{y}(k) = \sum_{l=1}^{h} v_l(k) = \sum_{l=1}^{h} \sum_{p=1}^{q} w_{lp} \mu_{lp}(u_l(k)). \tag{5}$$

Through introducing the $(hq \times 1)$ of the vectors of synaptic weights, $w = (w_{11}, w_{12}, \dots w_{1p}, \dots w_{1q}, \dots w_{21}, \dots w_{lp}, \dots w_{hq})^T$ and the membership functions $\mu(k) = \left(\mu_{11}(u_1(k)), \mu_{12}(u_1(k)), \dots \mu_{1p}(u_1(k)), \mu_{21}(u_2(k)), \dots \mu_{lp}(u_l(k)), \dots, \mu_{hq}(u_h(k))\right)^T$, it is easy to rewrite the transformation implemented by a neo-fuzzy neuron in a compact form

$$\hat{y}(k) = w^T \mu(k). \tag{6}$$

3 Combined Learning of NF-RBFNN

The proposed system configuration is a combination of supervised learning, self-learning, and lazy learning using the principle of "Neurons in data points" [19, 20]. At the same time, the learning of the R-neurons layer and the stack formed by the neo-fuzzy neuron occur independently of each other. Learning of the R-neurons layer is implemented based on the methods of evolving systems of computational intelligence [11–14] in the form of a sequence of steps [15].

In the beginning, the maximum possible number of R-neurons in the network H and some threshold of inseparability between centers c_l and c_{l+1} are set. At the input of the system of the first observation from the training data set, $x(1)$ is formed as the first activation function $\varphi_1(x(1))$ with the center $c_1 = x(1)$ with the principle "Neurons at data points" [21]. Then, upon receipt of the observation $x(2)$ condition is checked

$$\left\| x(2) - c_1 \right\| \leq \delta \tag{7}$$

and if it is fulfilled, then a new center c_2 is not formed on its basis. If the inequality is fulfilled:

$$\delta \leq \left\| x(2) - c_1 \right\| \leq 2\delta \tag{8}$$

then the correction of the centers of the formula takes place according to Kohonen's self-learning rule [22], "Winner takes all"

$$c_1(2) = c_1 + \nu_1(2)(x(2) - c_1(1)) \tag{9}$$

where $\nu(2) < 1$ self learning rate parameter.

If the inequality is true

$$\left\| x(2) - c_1 \right\| \leq 2\delta \tag{10}$$

then a second R-neuron is formed with the activation function $\varphi_2(x(2))$ with the center $c_2 = x(2)$. At the kth step of self-learning, when an input signal $x(N)$ is fed, the neuron-winner is determined for which the distance is calculated with Eq. 11

$$\left\| x(N) - c_l(N-1) \right\| \forall l \tag{11}$$

is minimal among all already formed R-neurons. Then inequalities are checked similarly to the previous one:

$$\begin{gathered} \left\| x(N) - c^*(N-1) \right\| \leq \delta, \\ \delta \leq \left\| x(N) - c^*(N-1) \right\| \leq 2\delta, \\ 2\delta \leq \left\| x(N) - c^*(N-1) \right\| \end{gathered} \tag{12}$$

after which either the observation $x(N)$ is ignored and a new activation function is not formed, or the center of $c * (N-1)$ is adjusted according to their "Winner takes all" rule or a new activation function $\varphi_n(x(N))$ is formed. The process of forming the first hidden layer occurs until the total number of neurons in this layer reaches the value of H.

After the layer of R-neurons is formed and the parameters of the centers of all these neurons are determined, it is possible to proceed to learn the output stack formed by the neo-fuzzy neuron. Since the output signal of the system depends linearly on the adjustable synaptic weights, the ordinary least squares method can be used to determine them

$$w(N) = \left(\sum_{k=1}^{N} \mu(k)\mu^T(k)\right)^{+} \sum_{k=1}^{N} \mu(k)y(k) \tag{13}$$

where $y(k)$ is reference signal, and $(\bullet)^+$ is a symbol pf matrix pseudo inversion

At the same time, all hq parameters of the neo-fuzzy neuron are simultaneously activated. If the system is configured online, it is more convenient to use a recursive adaptive algorithm [23].

$$\begin{cases} w(N\ 1) = w(N) \quad r^{-1}(N\ \ 1)(y(N\ 1) - w^T(N)\mu(N\ \ 1))\mu(N\ \ 1) \\ r(N\ \ 1) = \qquad \alpha\, r(N)\ \ \|\ \mu(N\ \ 1)\|^2 \\ 0 \leq \qquad \alpha\ \leq\ 1 \end{cases}$$

where α is a forgetting factor that determines the smoothing properties of this algorithm. It is not difficult to see that with $\alpha = 1$ we see the stochastic approximation algorithm [24], designed for the adaptive identification of stochastic objects, and with $\alpha = 0$ we get the three-way optimal speed algorithm of the Katchmarz algorithm [25], the most effective in cases of real-time operation when the speed of learning comes to the fore. In addition, in this case, at each moment of discrete time, all of $2h$ synaptic weights are simultaneously adjusted, which is $0.5q$ time less than in the case of training using a traditional LSM.

4 Computer Modelling

4.1 Actuality of the Problematic in the Cybersecurity Field

The increasing frequency and sophistication of cyberattacks on organizations of all sizes rises their concern about the ways of how to protect themselves. This is why, over the past few years interest in artificial intelligence approaches to solve appeared problematic increased. The reason behind it is the ability of these approaches carefully scrutinize all necessary details, which a person cannot always grasp to create a full picture of the potential threat, in particular, detect and respond to threats, predict future attacks, personalize defenses, and automate tasks.

One of the primary benefits of AI in cybersecurity is automation. With AI algorithms, organizations can automate tasks such as threat detection and response in real time, reducing the need for human intervention and allowing for faster response times. This can help organizations detect and respond to threats much more quickly than traditional methods, leading to improved security outcomes.

Another significant advantage of AI in cybersecurity is its ability to detect threats that would be difficult or impossible for humans to identify. By analyzing massive amounts of data, AI algorithms can identify patterns and anomalies that may indicate a cyberattack is underway. This can help organizations identify threats before they cause significant damage, improving their overall cybersecurity posture.

In addition to threat detection, AI can also help predict future attacks. By analyzing past attacks and identifying commonalities, AI can help organizations anticipate and prepare for new threats. This predictive capability can be particularly useful in the constantly evolving threat landscape, where new threats can emerge quickly and unexpectedly.

Finally, AI can be used to personalize cybersecurity defenses to better protect against targeted attacks. By creating individualized threat models for specific users or devices, organizations can tailor their defenses to the specific needs of their network. This personalized approach can be particularly effective in defending against increasingly sophisticated and targeted cyberattacks.

In conclusion, AI has the potential to revolutionize cybersecurity by automating tasks, detecting and responding to threats, predicting future attacks, and personalizing defenses. As cyberattacks continue to become more common and more sophisticated, organizations that fail to incorporate AI into their cybersecurity strategies may find themselves at a significant disadvantage. Also, considering all these factors, we can conclude that it is important to have such an approach that can quickly and efficiently adapt to new malicious developments and detect them.

4.2 Models and Methods

Generally speaking, there are a few classes of approaches to solving the problem and they are signature-based detection, heuristic-based detection, anomaly-based detection, machine learning-based detection, and finally hybrid approaches [26, 27]. The signature-based detection method involves using a known set of signatures to identify malware. If the signature of a file matches a known signature, it is flagged as malware.

The heuristic-based detection technique involves using a set of rules to identify potential malware. The rules are based on known characteristics of malware, such as file size or the presence of certain strings.

Anomaly-based detection involves looking for deviations from normal behavior to identify malware. The system is trained to recognize normal behavior, and any deviations from it are flagged as potential malware.

As it was previously mentioned, computational intelligence approaches (machine learning ones) are able to analyze files and identify patterns that are characteristic of viruses. These algorithms can be trained on large datasets of known viruses and non-viruses, in other words, binary training data sets, but also it is possible to train on the labeled data with certain types of viruses, in order to learn to recognize the characteristics of different types of viruses. Once the algorithm is trained, it can be used to scan new files and determine whether they are likely to contain a virus.

Hybrid approaches are considered the most advanced and this is why many modern malware detection systems use a combination of different techniques, such as signature-based and behavior-based detection or heuristic-based and machine learning-based detection, to improve overall detection rates and reduce false positives. Hybrid approaches can provide a more comprehensive and robust solution, but they can also be more complex and resource-intensive to implement.

The most common approach is to use deep learning models, rather than typical machine learning algorithms such as K-nearest neighbor (KNN), Support Vector Machines (SVMs), and others due to the complexity of the data. This is why models such as convolutional neural networks (CNNs) and recurrent neural networks (RNNs), became extremely to analyze network traffic and identify patterns that are associated with malicious activity. These models can be trained on large datasets of network traffic and can learn to recognize patterns that are indicative of a virus or other type of malware.

4.2.1 Machine Learning Approaches: SVM and KNN

SVMs are highly accurate in detecting malware, achieving accuracy above 95% [28]. They perform well even with small data sets, making them a suitable option for detecting new, previously unseen malware. However, overfitting can occur if not properly tuned, reducing performance on new data. At the same time, such techniques as feature selection and ensemble methods improve SVMs' performance. Ensemble methods combine multiple SVM models to create a more robust and accurate model.

Computational complexity is a major challenge for SVM when dealing with large datasets, and high-dimensional data can result in poor performance. Sensitivity to parameter tuning is another issue, and poor choices can lead to under or overfitting, which negatively impacts performance. Adversarial attacks can modify malicious code to evade detection by the classifier, highlighting the need for continuous updating and monitoring of the SVM model. Finally, interpreting the decision-making process of SVMs can be difficult, especially for non-experts, limiting its interpretability.

The K-Nearest Neighbor (KNN)calculates the distance between a test data set and its k-nearest neighbors from the training data set and then test data is classified based on the majority class of its k-nearest neighbors. Thus, the performance of the algorithm largely depends on the choice of distance metric and the value of k, which can be tricky leading to underfitting or overfitting if the k value was chosen without proper analysis.

KNN has some disadvantages that can affect its performance in malware detection. One significant disadvantage is its sensitivity to the choice of a distance metric, as choosing the wrong metric can lead to poor classification accuracy. Another disadvantage is that KNN requires a large amount of memory to store the training dataset, which can be a challenge for systems with limited memory. Additionally, KNN can be computationally expensive during the testing phase, as it requires distance calculations between the test data set and all training data sets. This can lead to slow prediction times and limit its effectiveness in real-time detection scenarios. Finally, KNN can be vulnerable to noise and outliers in the training dataset, which can negatively impact its classification accuracy.

4.2.2 RNN, CNN, and LSTM in Malware Detection

With new malware strains being constantly created and evolving led to the exploration of machine learning techniques, particularly deep neural networks, for malware detection. Deep neural networks can automatically learn and extract relevant features from large datasets, making them a promising option for detecting new and unknown malware. However, while deep neural networks have shown impressive performance in detecting malware, they also present some challenges, such as the potential for overfitting and the need for large amounts of training data.

Recurrent Neural Networks (RNNs) have shown promising results in detecting malware [29], particularly in the field of behavior-based malware detection. RNNs are capable of analyzing sequences of events, making them suitable for detecting patterns of malicious behavior in software. The ability to learn from sequential data is especially useful for detecting advanced persistent threats and other types of sophisticated malware.

RNNs have also demonstrated high accuracy rates in detecting malware, often outperforming other machine-learning approaches and providing outstanding results. However, there are also significant disadvantages to using RNNs for malware detection. RNNs can suffer from the vanishing gradient problem, where the gradient used in the backpropagation process becomes too small, making it difficult to update the weights of the network. This can lead to slower convergence and decreased accuracy.

Another significant disadvantage of RNNs is their interpretability. The decision-making process of RNNs can be difficult to understand, making it challenging to identify which features or behaviors contribute to the classification decision. This can be a critical issue in security applications, where explainability and interpretability are crucial for detecting and responding to new and emerging threats.

The next model that we can take into consideration is Long Short-Term Memory (LSTM) which is a type of recurrent neural network that has been widely used for malware detection. One advantage of LSTMs is their ability to model sequential data and capture long-term dependencies, making them well-suited for analyzing malware behavior over time. LSTMs can learn to detect subtle changes in the behavior of malware, such as changes in system calls or network traffic patterns, that may not be detectable by traditional signature-based methods.

Several studies have reported high accuracy rates for malware detection using LSTMs in Android apps and Windows PE files [30, 31]. LSTMs can be prone to overfitting if not properly regularized, which can lead to poor generalization performance on new data. Finally, like other deep learning models, LSTMs can be difficult to interpret, making it challenging to understand how the model is making decisions or which features are most important for classification.

Both recurrent neural networks (RNNs) and long short-term memory (LSTM) are types of neural networks that have been used for malware classification. While RNNs are designed to handle sequential data, still LSTMs are a more advanced version of RNNs that can better handle long-term dependencies in sequential data.

In terms of performance, both RNNs and LSTMs have been shown to achieve high accuracy rates in detecting malware. However, LSTMs have been found to outperform traditional RNNs in many cases, particularly when dealing with longer sequences of data which can appear quite often in real-world tasks.

Another class of deep neural networks that are able to solve malware detection tasks is convolutional neural networks (CNNs). CNNs have shown promise in the field of malware detection due to their ability to automatically learn features from raw data such as byte sequences or opcode sequences [32]. CNNs have been used for various tasks in malware detection such as binary classification, family classification, and malware behavior analysis. One significant advantage of CNNs is their ability to capture local patterns and features that are relevant to the task, leading to improved classification accuracy. CNNs have also been shown to perform well with small to medium-sized datasets.

However, CNNs also have some limitations. One significant disadvantage is their inability to handle variable-length inputs, which can be a challenge in malware detection tasks where the input data can vary in size. Tuning the hyperparameters of a CNN model can be time-consuming and require significant computational resources.

The problematic aspect that unites all these DNNs is the computational complexity caused by not just the chosen architecture of the neural network (the amount of computational units can directly influence the processing time and cost), but also their nature. This type of network requires large datasets since they use linear activation functions, hence, piece-wise approximation leads to an increase of activation functions to do proper approximation to the goal function leading to an increase of synaptic weights. Usually, their number can be enormously large leading to an increase in training data for their tunning.

All these factors create a major challenge which is the difficulty in training and optimizing DNNs since they require significant computational resources for both these processes, making them time-consuming.

4.3 Dataset

In the field of cybersecurity, having a sufficient amount of labeled training data is crucial for building effective machine learning models, including neural networks. However, obtaining labeled data in the cybersecurity domain can be challenging for several reasons.

Malware creators often modify their code to evade detection by security tools, resulting in constantly evolving and dynamic threats. This means that security

Table 1 Statistics on the "PE Malware Machine Learning Dataset"

Characteristic	Value
Number of observations	201 549
Software	86 812
Malware	114 737
Size	117GB
File types	.exe

researchers and analysts must constantly update their datasets to ensure that their models remain effective.

There is a lack of standardized and widely accepted datasets for cybersecurity research. While some publicly available datasets exist, they may not be representative of the wide range of threats that organizations face. Additionally, organizations may be hesitant to share their own datasets due to concerns about data privacy and confidentiality.

Labeling data in the cybersecurity domain can be time-consuming and expensive. It requires domain expertise to accurately label data, and the process can be slowed down by the need to verify and validate each sample.

As a result of these challenges, many machine learning models in the cybersecurity domain may not have access to enough training data, which can result in poor performance and accuracy. Researchers have explored various techniques to address this problem, such as transfer learning and synthetic data generation, but these techniques may also have limitations and trade-offs. Overall, the issue of limited training data in the cybersecurity field remains a significant challenge for building effective machine learning models.

The biggest dataset available online with open access at the moment is "PE Malware Machine Learning Dataset" [33] which contains 200 000 files labeled as "Whitelist" for regular software or "Blacklist" for the malware. The detailed statistic on the dataset is shon in the Table 1.

The files in the "samples" directory are named after their corresponding entry in the ID field of the samples.csv file, which contains the labels for each of the samples in the directory. It is worth noting that the file extensions have been removed from all files in the samples directory to prevent unintentional execution. To run the malware properly, the correct extension can be determined by parsing the PE header and manually renaming the file.

The description of the dataset and examples are shown in the Table 2.

The majority of the malware samples were obtained from easily accessible sources and consist of similar families of adware, spyware, and ransomware, which dominate the dataset. Although the dataset also contains more complex malware from Advanced Persistent Threat actors, these samples are significantly less common. As a result, the dataset may not accurately reflect the malware used in real-world attacks, but may still generalize to more advanced malware.

Table 2 The description of the "PE Malware Machine Learning Dataset"

Column	Description	Example
Id	The identifier for each sample corresponds to the name of its corresponding file in the samples directory	2
md5	The MD5 hash of the file	ad27f1a72dda61d1659...
sha1	The file's SHA1 hash value	f8fd630c880257c...
sha256	The file's SHA256 hash value	984d732c9f321972...
Total	The count of antivirus systems that have scanned the file at the time of the query	67
Positives	The count of antivirus engines that detect this file as malicious..	0
List	The file is labeled as either malicious or legitimate, denoted by either blacklist or whitelist respectively	Whitelist
Filetype	For this dataset, the value of this field will always be "exe"	exe
Submitted	The date on which the sample was added to the dataset owner's database	6/24/2018 4:18:38 PM
Length	The file size in bytes	211 456
Entropy	The entropy of the file using the Shannon entropy formula. The resulting values will fall within the range of 0–8	2.231824

The majority of the legitimate files were obtained from instances of Windows 7 and above with various software installed and downloaded. The dataset also includes false positives from the sources mentioned above. This may result in a bias towards Microsoft-produced software, as these binaries are prevalent in the legitimate file dataset.

Also, during explanatory analysis, we excluded unnecessary columns, leaving a valuable subset of data, which would be id, list, and file type. Additionally, from Fig. 4 we can see that the dataset has a bias toward the class "Malware", and due to hardware limitations, the dataset was narrowed down to a balanced subset with 10 000 observations. Since all the PE files were stripped from the filetype extension we did an extra step of data preparation. Takes place further transformation that was common for all the systems that were considered for the comparative analysis, and that is the transformation into the string of binaries.

4.4 Computational Simulation

To evaluate the effectiveness of the proposed system, we have selected two deep neural network architectures CNN [34] and LSTM [35] for comparative analysis in this article. The main aim of this comparative analysis is to identify which architecture

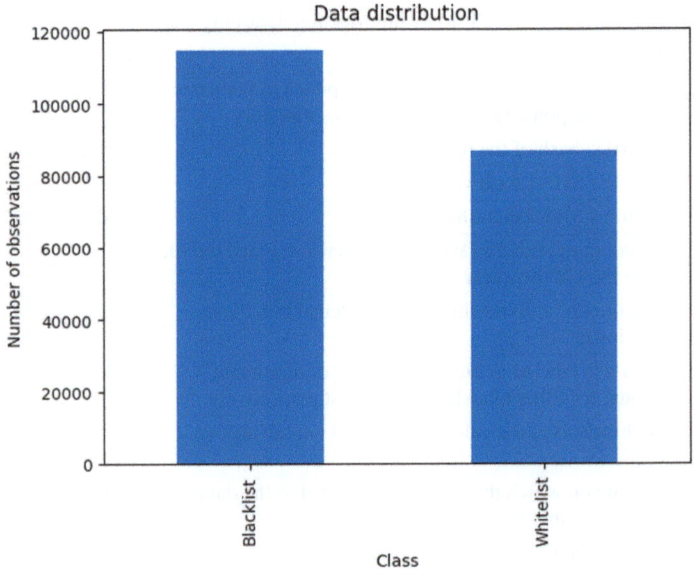

Fig. 4 Data distribution

performs better in terms of malware classification accuracy, taking into consideration the training time consumption and the time spent on the classification. The comparative analysis involves training both network architectures on the same dataset and evaluating their performance using standard evaluation metrics. This approach allows us to make an informed decision about which architecture to use in our proposed system and provides insights into the strengths and limitations of different deep-learning models for malware detection.

The architecture of the first Convolutional model was empirically changed in comparison to the original article, to reach higher quality respectfully to the dataset we chose, so the architecture we achieved Conv2D-Conv2D-MaxPooling2D-Conv2D-MaxPooling2D-Conv2D-MaxPooling2D-Flatten-Dence-Dence-Dropout-Dence. Also, we have to take into account preprocessing for CNN, in particular, we transformed the string of binaries to the grayscale image Fig. 5 that later was fed to the CNN.

The dropout rate was set at 0.25, the number of epochs 15, and the size of the batch 50. All these parameters were obtained empirically, reaching the highest accuracy.

As for the Using hexdump the data was preprocessed for the LSTM approach as for the architecture, it remained untouched: Input-LSTM-Droput-Dence, however, the dropout rate increased to 0.3, the number of epochs reached 25 and batch size 50.

The initial experiment was designed to assess the performance of the proposed system, in comparison to deep neural networks, across varying sizes of the dataset. The training set consisted of 8000 samples, and the evaluation set included 2000

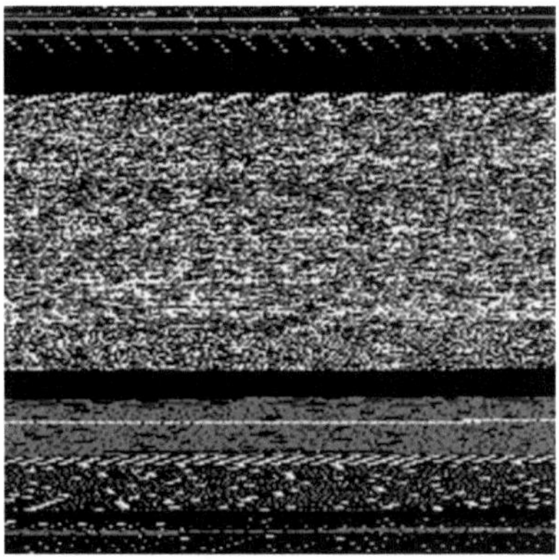

Fig. 5 Prepossessed grayscale image

Table 3 Time consumption on the training process on full dataset (8000 samples)

Model	Time, min	Epochs	Batch
CNN	189, 3	15	50
LSTM	144, 8	25	50
NF-RBFNN	19, 1	1	1

observations. However, to accurately measure the performance of the systems, the full dataset was partitioned into subsets and the accuracy of each system was assessed.

From this graphic, we can see that the NF-RBFNN outperforms other advanced and more complex deep neural network architectures on small datasets. Specifically, when the number of training samples is limited, traditional deep learning models such as LSTM and CNN perform poorly compared to the NF-RBFNN. The reason for this poor performance can be attributed to the high complexity and large number of parameters in these models, which require a considerable amount of training data to learn effectively. On the other hand, the NF-RBFNN is less complex and requires fewer parameters, making it more suitable for datasets with a limited number of training samples. Therefore, the results suggest that the NF-RBFNN can be a viable option for tasks that involve small datasets, where more complex deep learning models may not perform well.

When evaluating the performance of different models, it is important to also consider the time consumption on the training process. For this purpose, we used a specific hardware configuration during the simulation. The processor used was an Intel(R) Core(TM) i5-7300U CPU @ 2.60GHz with a clock speed of 2.70 GHz, and the system had 8.00 GB of RAM. The processor was x64-based.

To develop the models, we used Python programming language along with popular deep learning frameworks such as PyTorch and Keras. We also used supporting libraries such as NumPy and Pandas to manipulate and analyze the data. It is important to note that the performance of the models can also be influenced by the hardware and software used for their development (Table 3).

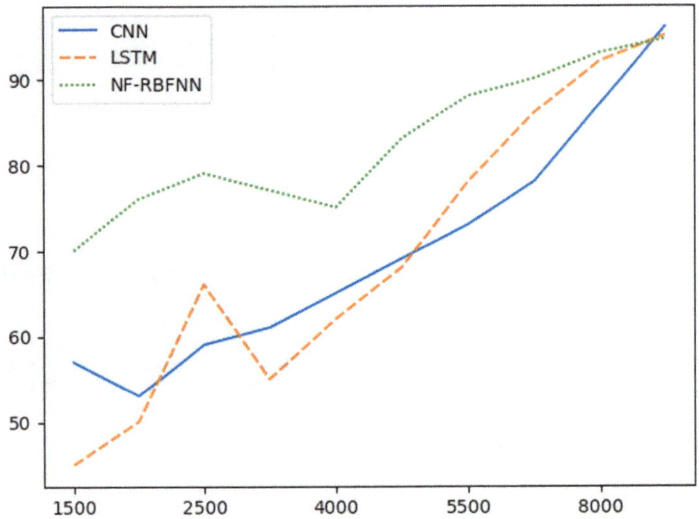

Fig. 6 Dependency between the size of the data set and the accuracy of the system

The presented table demonstrates that the proposed system outperforms both deep neural networks in terms of speed while maintaining high accuracy on a small dataset. This suggests that the proposed system can be effectively used for classification tasks with limited data. It should be noted that the deep neural networks show great results on the 8000 training sample, but their training process is considerably slower compared to the proposed system.

Moreover, it is important to highlight that the proposed system performs comparably well on the full dataset, taking into account its higher speed. This can be observed in Fig. 6, where the proposed system achieves a similar level of accuracy as the deep neural networks while requiring significantly less time for training. Therefore, the proposed system can be a suitable choice for classification tasks that require both high accuracy and efficiency, especially when dealing with large datasets.

In conclusion, while the dataset used in this article has a large number of observations that may be sufficient for deep neural networks to train on, it is important to note that the dataset is from 2021 and the field of cybersecurity threats is rapidly evolving. This means that the dataset may only provide outdated information to the DNNs. While the system is still effective at identifying known threats, it is limited in its ability to identify new potential threats.

Unfortunately, it is extremely challenging to find high-quality datasets that can adequately train DNNs. However, small datasets can be found to train the NF-RBFNN. In order to improve the system's performance, it would be ideal to create a dataset with different types of viruses, with evenly spread classes of malware so that the system could work with multiclass classification, which it is capable of doing. This would allow for more comprehensive training of the system and would provide better accuracy in identifying potential threats.

5 Conclusion

The proposed neo-fuzzy radial basis function stacking neural network is essentially a hybrid of the traditional radial basis function neural network and the neo-fuzzy neuron. While the two subsystems are learned independently of each other, this ensures a high speed of parameter setting for the system as a whole. The system is designed for use under conditions where the training data set is small in size, or when data is received online and learning speed is a priority. The NF-RBFNN is simple to implement numerically and is characterized by high performance.

References

1. Mumford, C., Jain, L.: Computational Intelligence, Collaboration, Fuzzy and Emergence. Springer, Berlin (2009)
2. Kruse, R., Borgelt, C., Klawonn, F., et al.: Computational Intelligence. A Methodological Introduction. Springer, Berlin (2013)
3. Kacprzyk, J., Pedrycz, W.: Springer Handbook of Computational Intelligence. Springer, Berlin, Heidelberg (2015)
4. Cybenko, G.: Approximation by superpositions of a sigmoidal function. Math. Control Signal Syst. **2**, 303–314 (1989). https://doi.org/10.1007/BF02551274
5. Kung, S.Y.: Kernel Methods and Machine Learning. University, Cambridge (2014)
6. Moody, J.: Cjd: Fast learning in networks of locally tuned processing units. Neural Comput. **1**, 281–294 (1989)
7. Poggio, T., Girosi, F.: Networks for approximation and learning. Proc. IEEE **78**, 1481–1497 (1990)
8. Park, J., Sandberg, I.W.: Universal approximation using radial-basis-function networks. Neural Comput. **3**, 246–257 (1991)
9. Leonard, J.A., Kramer, M.A., Ungar, L.H.: Using radial basis functions to approximate a function and its error bounds. IEEE Trans. Neural Netw. **3**, 624–627 (1992)
10. Amirian, M., Schwenker, F.: Radial basis function networks for convolutional neural networks to learn similarity distance metric and improve interpretability. IEEE Access **8**, 123087–123097 (2020)
11. Angelov P.P.: Evolving Rule-Based Models: A Tool for Design of Flexible Adaptive Systems. Physica Verlag HD: Imprint: Physica, Heidelberg (2002)
12. Kasabov, N.K.: Evolving Connectionist Systems: The Knowledge Engineering Approach, 2nd edn. Springer, New York, London (2007)
13. Angelov, P.P., Kasabov, N.K.: Evolving computational intelligence systems. In: Proceedings of the 1st International Workshop on Genetic Fuzzy Systems, Granada, pp. 76–82 (2005)
14. Lughofer, E.: Evolving Fuzzy Systems - Methodologies. Advanced Concepts and Applications. Springer, Berlin, Heidelberg (2011)
15. Bodyanskiy, Y., Deineko, A., Tyshchenko, A.: An evolving radial basis neural network with adaptive learning of its parameters and architecture. Autom. Control Comput. Sci. **49**, 255–260 (2015)
16. Miki, T.: Analog Implementation of Neo-Fuzzy Neuron and Its On-board Learning (1999)
17. Yamakawa, J., Uchino, E., Miki, J., Kusanagi, H.: A neo-fuzzy neuron and its application to system identification and prediction of the system behavior. In: Proceedings of the 2nd International Conference on Fuzzy Logic & Neural Networks, Iizuka, Japan (1992)
18. Perfilieva, I.: Fuzzy transforms: theory and applications. Fuzzy Sets Syst. **157**, 993–1023 (2006). https://doi.org/10.1016/j.fss.2005.11.012

19. Epanechnikov, V.A.: Non-parametric estimation of a multivariate probability density. Theory Probab. Appl. **14**, 153–158 (1969). https://doi.org/10.1137/1114019
20. Nelles, O.: Nonlinear Systems Identification. Springer, Berlin (2001)
21. Zahirniak, D.R., Chapman, R., Rogers, S.K., et al: Pattern recognition using radial basis function network. Aerospace Appl. Artif. Intell. 249–260 (1990)
22. Kohonen, T.: Self-Organizing Maps. Springer, Berlin, Heidelberg (2001)
23. Bodyanskiy, Y.V., Kokshenev. I., Kolodyazhniy. V.: An adaptive learning algorithm for a neo fuzzy neuron. In: Wagenknecht, M., Hampel, R. (eds.) Proceedings of the 3rd Conference of the European Society for Fuzzy Logic and Technology, Zittau, Germany, September 10–12, 2003, pp. 375–379. University of Applied Sciences at Zittau/Görlitz, Germany (2003)
24. Goodwin G.C., Ramadge P.J., Caines P.E.: Discrete time stochastic adaptive control. SIAM J. Control Optim. **19**, 829–853 (1981) https://doi.org/10.1137/0319052
25. Kaczmarz, S.: Approximate solution of systems of linear equations. Int. J. Control **57**, 1269–1271 (1993). https://doi.org/10.1080/00207179308934446
26. Arora, S., Bansal, R., Singhal, V.: Malware detection techniques: a brief survey. In: International Conference on Inventive Communication and Computational Technologies (ICICCT) (2018)
27. Gharakheili, V., van Beest, E.P.C., Havinga, P.J.M., Hancke, G.P.: A survey of intrusion detection systems using machine learning: techniques, applications, and challenges. J. Inf. Secur. Appl. **50**, 102390 (2020)
28. Islam, S.M., Islam, R., Islam, M.N., Hossain, M.A.: Detection of computer viruses using machine learning: a review. Inte. J. Comput. Sci. Inf. Secur. (IJCSIS) **17**(3), 97–105 (2019)
29. Zhang, X., Yang, J., Zhang, Y., Huang, K.: Malware detection using recurrent neural networks with long short-term memory. IEEE Access **8**, 128642–128651 (2020)
30. Kwon, O., Ryu, W.J., Lee, K.H., Kim, D.: Malware classification using long short-term memory recurrent neural network. J. Intell. & Fuzzy Syst. **36**(1), 677–684 (2019)
31. Lin, Y., Zhu, X., Xie, Y.: Malware detection using deep learning with long short-term memory. J. Ambient Intell. Humaniz. Comput. **10**(8), 3033–3043 (2019)
32. Alhusain, A., Singh, P., Sivakumar, P.: Convolutional neural networks for malware detection: a survey. IEEE Access **9**, 53137–53150 (2021)
33. https://practicalsecurityanalytics.com/pe-malware-machine-learning-dataset/
34. Bensaoud, A., Abudawaood, N., Kalita, J.: Classifying Malware Images with Convolutional Neural Network Models. Int. J. Netw. Secur. **22** (2020)
35. Andrade, E.O., Viterbo, J., Vasconcelos, C.N., Guérin, J., Bernardini, F.C.: A model based on LSTM neural networks to identify five different types of malware. Procedia Comput. Sci. **159**, 182–191 (2019)

Investigations of Different Classes Hybrid Deep Learning Networks and Analysis of Their Efficiency in Forecasting

Galib Hamidov and Yuriy Zaychenko

Abstract Investigations of hybrid GMDH deep learning neuro-fuzzy networks and neo-fuzzy networks in the problem of forecasting in the financial sphere were carried out. The problem of forecasting NASDAQ stock prices with an interval of 1 day, 1 week and month was considered. In the process of forecasting the hyperparameters of hybrid fuzzy neural networks were optimized, namely the number of inputs, the number of rules and the size of the sliding window. Optimal structures of hybrid fuzzy neural networks for different prediction intervals were synthesized. Experimental studies of the efficiency of forecasting hybrid neo-fuzzy and neuro-fuzzy networks were carried out and compared with conventional DL networks according to MAPE criterion. Training time of hybrid DL networks was estimated and compared.

Keywords Hybrid neuro-fuzzy networks · Self-organization · Forecasting · Structure and parameters optimization

1 Introduction

Nowadays third-generation neural networks—deep learning (DL) networks are increasingly used to solve various problems of computational intelligence—forecasting, pattern recognition and classification, and so on [1–4]. Their main features are high flexibility, the ability to solve complex problems of recognition of 2xD and 3xD images etc.

G. Hamidov
Information Technologies Department, Azershig, Baku, Azerbaijan
e-mail: galib.hamidov@gmail.com

Y. Zaychenko (✉)
Institute for Applied System Analysis, Igor Sikorsky Kyiv Polytechnic Institute, Kyiv, Ukraine
e-mail: zaychenkoyuri@ukr.net

M. Zgurovsky and N. Pankratova (eds.), *System Analysis and Artificial Intelligence*, Studies in Computational Intelligence 1107,
https://doi.org/10.1007/978-3-031-37450-0_20

However, in classical deep learning networks the problem of finding the optimal or at least suboptimal network structure has not been solved yet. In addition, there are disadvantages of gradient descent algorithms such as the disappearance or explosion of the gradient.

These shortcomings have led to the search for new approaches and structures of deep neural networks capable of solving these problems.

One such new approach is the development of hybrid deep learning networks based on the Group Method of Data Handling (GMDH) [7–10].

Using the idea of self-organization makes it possible to synthesize the suboptimal structure of the deep learning network in the process of run GMDH algorithm itself, step by step, until the stop condition is met.

Therefore in recent years, in our previous works hybrid neural networks with different types of nodes: GMDH-neuro-fuzzy network with fuzzy Wang–Mendel nodes with two inputs [12, 13] and GMDH-neo-fuzzy networks with nodes-neo-fuzzy neurons have been developed and studied.

In Wang–Mendel network we must train parameters of Membership functions and output weights while in neo-fuzzy neuron we need train only output weights. That demands less computational expenses and training time.

Therefore it's important problem of comparing the properties and efficiency of these classes of hybrid neural networks in computational intelligence problems.

The aim of this work is study of efficiency and comparative analysis of these classes of hybrid deep learning networks in forecasting problems in the financial sphere.

2 Synthesis of Optimal Structure of Hybrid GMDH-Neo-Fuzzy Network

The GMDH method was used to synthesize the structure of the hybrid network based on the principle of self-organization. The structure of evolving DL network is shown at the Fig. 1.

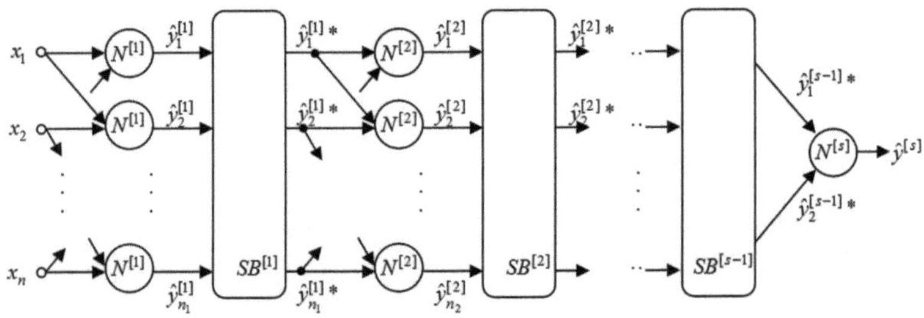

Fig. 1 Evolving DL network generated by GMDH

The successive increase in the number of layers is carried out until the value of the external criterion of optimality MSE begins to increase for the best model of the current layer. In this case it is necessary to return to the previous layer, to find there the best model with the minimum value of criterion. Then we move backward, go through its connections, find the corresponding neurons of the previous layer.

This process continues until we reach the first layer and the corresponding structure is automatically determined.

3 Neuro-Fuzzy Network with Small Number of Tuning Parameters as a Node of DL Network

Let us consider the node architecture, shown in Fig. 2 and proposed as a neuron of the suggested GMDH-system. This architecture is in fact a Wang–Mendel neuro-fuzzy system with only two inputs x_i and x_j, and one output $\hat{y}_l$ [11, 13].

To the node input a two-dimensional vector of signals $x(k) = (x_i(k), x_j(k))^T$ is fed, where $k = 1, 2, ..., N$ is either the observation number in training set or the current discrete time.

The first layer of a node contains $2h$ membership functions $\mu_{pi}(x_i(k)), \mu_{pj}(x_j(k))$, $p = 1, 2, ..., h$ and provides fuzzification of input variables. The bell-shaped constructions with non-strictly local receptive support are usually used as membership functions. It allows avoiding appearing of "gaps" in the fuzzified space while using scatter partitioning of input space. Usually the Gaussians are used as membership functions of the first layer

$$\begin{cases} \mu_{pi}(x_i(k)) = exp\Big(-\frac{(x_i(k)-c_{pi})^2}{2\sigma_i^2}\Big), \\ \mu_{pj}(x_j(k)) = exp\Big(-\frac{(x_j(k)-c_{pj})^2}{2\sigma_j^2}\Big). \end{cases} \quad (1)$$

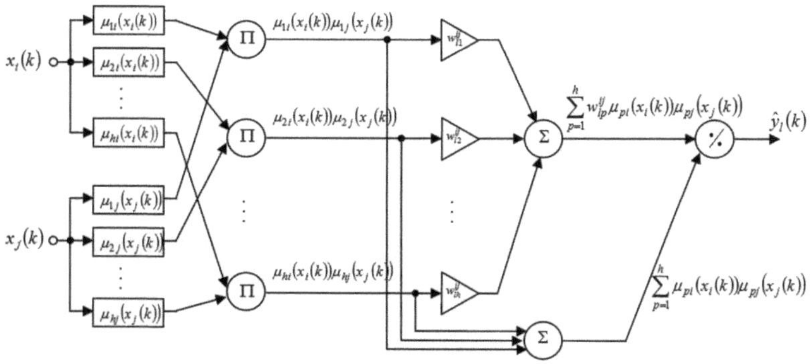

Fig. 2 GMDH-neuro-fuzzy system node

where c_{pi}, c_{pj} are parameters, that define the centers of the membership functions, σ_i, σ_j are width parameters of these functions. The second layer provides aggregation of the membership levels. It consists of h multiplication units and forms two-dimensional radial basis activation functions

$$\bar{x}_p(k) = \mu_{pi}(x_i(k))\mu_{pj}(x_j(k)) \tag{2}$$

and for Gaussians with the same values $\sigma_i = \sigma_j = \sigma$ we can write

$$\bar{x}_p(k) = exp\left(-\frac{\left\|x(k) - c_p\right\|^2}{2\sigma^2}\right) \tag{3}$$

(here $c_p = (c_{pi}, c_{pj})^T$), i.e. the elements of the first and the second layers process the input signal similarly to the R-neurons of the radial basis function neural networks.

The third layer is one of synaptic weights that are adjusted during learning process. The outputs of this layer are values

$$w_{lp}^{ij}\mu_{pi}(x_i(k))\mu_{pj}(x_j(k)) = w_{lp}^{ij}\bar{x}_p(k) \tag{4}$$

The fourth layer is formed by two summation units and computes the sums of output signals of the second and the third hidden layers

$$\begin{cases} \sum\limits_{p=l}^{h} w_{lp}^{ij}\mu_{pi}(x_i(k))\mu_{pj}(x_j(k)) = \sum\limits_{p=l}^{h} w_{lp}^{ij}\bar{x}_p(k), \\ \sum\limits_{p=l}^{h} \mu_{pi}(x_i(k))\mu_{pj}(x_j(k)) = \sum\limits_{p=l}^{h} \bar{x}_p(k). \end{cases} \tag{5}$$

And finally in the fifth layer of the neuron normalization is realized, as a result the node output signal $\hat{y}_l$ is formed [13]:

$$\hat{y}_l(k) = \frac{\sum\limits_{p=l}^{h} w_{lp}^{ij}\mu_{pi}(x_i(k))\mu_{pj}(x_j(k))}{\sum\limits_{p=l}^{h} \mu_{pi}(x_i(k))\mu_{pj}(x_j(k))} = \frac{\sum\limits_{p=l}^{h} w_{lp}^{ij}\bar{x}_p(k)}{\sum\limits_{p=l}^{h} \bar{x}_p(k)} = \\ = \sum_{p=l}^{h} w_{lp}^{ij}\varphi_p^{ij}(x(k)) = (w_l^{ij})^T\varphi^{ij}(x(k)) \tag{6}$$

where $\varphi^{ij}(x(k)) = \left(\varphi_1^{ij}(x(k)), \ldots, \varphi_p^{ij}(x(k)), \ldots, \varphi_p^{ij}(x(k))\right)^T$,
$\varphi_p^{ij}(x(k)) = \mu_{pi}(x_i(k))\mu_{pj}(x_j(k))\left(\sum_{p=1}^{h} \mu_{pi}(x_i(k))\mu_{pj}(x_j(k))\right)^{-1}$,
$w_l^{ij} = \left(w_{l1}^{ij}, \ldots, w_{lp}^{ij}, \ldots, w_{lh}^{ij}\right)^T$.

It is easy to see that the node implements nonlinear mapping of input signals to output signal like normalized radial basis function neural network, however the NFS contains significantly lower number h of adjusted parameters comparing with the neural network.

Using introduced notation and choosing transformations in every node of the standard GMDH in the form

$$\hat{y}_l(k) = w_{l0}^{ij} + w_{l1}^{ij}x_i(k) + w_{l2}^{ij}x_j(k) \quad (7)$$

that contains three unknown parameters, it is easy to see that with three membership functions being on the each input of the proposed node we get the same three synaptic weights that should be adjusted.

In the simplest case the estimation of these synaptic weights can be realized with the conventional least squares method (LSM), traditionally used in the GMDH. If the entire training set is presented, we can use the LSM in its batch form

$$w_l^{ij}(N) = \left(\sum_{k=1}^{N} \varphi^{ij}\left(x(k)\right)\left(\varphi_{ij}(x(k))\right)\right)^{+} \sum_{k=1}^{N} \varphi^{ij}\left(x(k)\right)y(k) \quad (8)$$

(here $y(k)$ is external reference signal). If training samples are fed sequentially in on-line mode, the recurrent form of the LSM is used in the form

$$\begin{cases} w_l^{ij}(k) = w_l^{ij}(k-1) + \dfrac{P^{ij}(k-1)\left(y(k) - (w_l^{ij}(k-1))^T\varphi^{ij}(x(k))\right)\varphi^{ij}(x(k))}{1+\left(\varphi^{ij}(x(k))\right)^T P^{ij}(k-1)\varphi^{ij}(x(k))}, \\ P^{ij}(k) = P^{ij}(k-1) - \dfrac{P^{ij}(k-1)\varphi^{ij}\left(x(k)\right)\left(\varphi^{ij}(x(k))\right)^T P^{ij}(k-1)}{1+\left(\varphi^{ij}(x(k))\right)^T P^{ij}(k-1)\varphi^{ij}(x(k))}. \end{cases} \quad (9)$$

4 Experimental Investigations and Analysis

The implementation of the neo-fuzzy network model was performed using the Python language, adding two classes to the previously developed library.

The learning process of each neo-fuzzy network module is performed using a stochastic gradient descent with momentum.

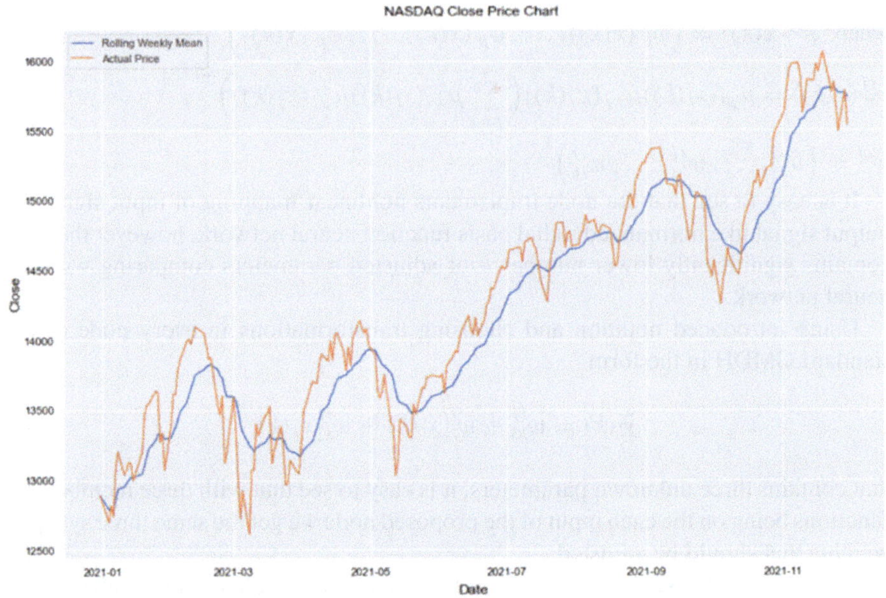

Fig. 3 Flow chart of NASDAQ index

The experiments consist of short- and medium-term forecasting of the NASDAQ stock exchange index. For the experiment, the daily data in the period from 1.12.2020 to 1.12.2021 were taken. There are several *input data in the data set*:

1. Price at the time of opening of trading.
2. The lowest price per day.
3. The highest price per day.
4. Price at the close of trading.
5. Trading volume per day.

Figure 3 shows the initial price flow chart of the NASDAQ index.

As it's well-known, financial time series are non-stationary, so that to determine the order of autoregression, it is necessary to bring them as close as possible to stationary one.

In this paper, it was decided to use logarithm of the time series of closing prices, and then apply the discrete differentiation operator, i.e., taking the difference between the closing prices of the exchange for two consecutive days. In this case, the value:

$$\Delta Y\_Close(t) = Y\_close(t) - Y\_close(t-1), \text{ where } t \in (1, T) \qquad (10)$$

is stationary, which confirms the Dickie–Fuller test −4.55, p = 0.001.

After achieving stationarity, a study was conducted to establish the order of the autoregression model. As a result of the application of the partial autocorrelation function (PACF) and its analysis it was determined that for the task of short-term forecasting it is enough to know the indicators of the series for the previous day.

The data set itself is non-normalized, so it is necessary to bring it to the range [0, 1].

In this work it was decided to implement the procedure of min-max transformation:

$$\overline{X} = \frac{X - X_{min}}{X_{max} - X_{min}} \tag{11}$$

The ratio of 2:1 was chosen as the division of the sample into training and test. Also, as a membership function of belonging of the input x_i to the rule R_j the bell-shaped membership function was used:

$$\mu_j(x_i) = \left(1 + \frac{(x_i - c_j)^2}{a_j^2}\right)^{-1} \tag{12}$$

The opening, closing, highest and lowest prices for the day and trading volume are used as indicators.

The following hyperparameters were investigated that to find the optimal model:

1. Forecasting period—1 day, 1 week or 1 month.
2. The size of the sliding window—from 1 to 6.
3. The number of rules for each neo-fuzzy layer—from 2 to 9.

Mean Squared Error (MSE) was used as a model quality, while RMSE metrics and MAPE metrics were used as forecasting quality. As a level of freedom of choice 10 modules were selected.

Table 1 lists the top 10 MAPE models for the one-day forecasting task.

Table 1 Results of cross-validation for one day

Number of rules	Sliding window size	MAPE for test sample (%)
8	4	0,185145
7	4	0,248567
6	2	0,330886
5	3	0,336418
9	4	0,338967
4	4	0,355478
9	3	0,391252
5	4	0,406969
8	3	0,428600
7	3	0,431992

Table 2 Results of cross-validation for forecasting for 1 week

Number of rules	Sliding window size	MAPE for test sample (%)
9	5	0,298815
8	5	0,793335
7	5	1,307635
4	4	1,295359
5	4	1,754165
4	1	1,917575
7	1	1,970467
2	5	2,092332
4	5	2,130194

Table 3 Results of cross-validation 1-month forecasting

Number of rules	Sliding window size	MAPE for test sample (%)	RMSE for test sample (%)
8	1	1,258013	0,39683876
8	2	1,450802	0,458064084
8	3	1,482047	0,458321843
8	4	1,560433	0,466505255
9	3	1,644714	0,484253573
8	5	1,883745	0,499209484
9	4	2,502631	0,537447431
9	2	2,369392	0,590771646

Depending on the forecast interval, we obtained different sets of hyperparameters of the models. Thus, for one week forecasting task the corresponding results according to the MAPE criterion are given in Table 2.

For the task of forecasting for the month, the corresponding results according to the criteria of RMSE and MAPE are given in the Table 3.

From presented data it follows that the model with the number of rules equal to 7, the level of freedom of choice 10, and the sliding window 5 is able to make quality forecasts for one week by both criteria RMSE and MAPE, while for a month forecast good models with the number of rules equal to 8, the level of freedom of choice 10 and the sliding window 1, which coincides with the analysis of the series according to the criterion of PACF.

We compared the obtained metrics at the synthesized models with hybrid neuro-fuzzy networks with similar hyperparameters. Table 4 shows a comparison of such models for the one-week forecasting task.

From these results it follows that forecasting accuracy that hybrid neo-fuzzy network is better than neuro-fuzzy one.

Table 4 The results of comparisons according to the MAPE criterion for forecasting for 1 week

Number of rules	Sliding window size	MAPE of hybrid neo-fuzzy networks (%)	MAPE of hybrid neuro-fuzzy networks (%)
9	5	0,298815	1,735491
8	5	0,793335	0,939993
7	5	1,295359	1,462190
4	4	1,307635	3,300800
5	4	1,754165	9,282890
4	1	1,917575	1,342019
7	1	1,970467	1,251306
2	5	2,062085	3,032662
4	5	2,092332	2,338723
9	1	2,130194	3,888323

Table 5 Results of comparisons for 1 month forecasting

Number of rules	Sliding window size	MAPE hybrid neo-fuzzy networks (%)	MAPE hybrid neuro-fuzzy networks (%)
8	1	1,258013	2,165929
8	2	1,450802	2,433154
8	3	1,482047	2,460703
8	4	1,560433	3,702811
9	3	1,644714	1,206005
8	5	1,883745	4,135492
9	2	2,369392	1,106095
9	4	2,502631	1,250217
9	1	2,517481	0,755839
9	5	2,620769	1,267467

In the Table 5 the experimental results of comparison of hybrid deep learning models for the one-month forecasting task are presented.

In the Table 6 the results of comparisons by this criterion for the problem of forecasting for one day are shown.

From these comparisons it can be concluded that hybrid neo-fuzzy networks make better forecasts for 1 week and 1 month, while hybrid neuro-fuzzy networks give slightly better forecasts for one day by MAPE metric, although this difference is not significant.

Figure 4 shows a chart of the dependence MAPE on the length of the forecast interval and the number of rules.

Table 6 The results of comparisons for 1 day forecasting

Number of rules	Sliding window size	MAPE hybrid neo-fuzzy networks (%)	MAPE hybrid neuro-fuzzy networks (%)
8	4	0,185145	0,246549
7	4	0,248567	0,225886
6	2	0,330886	0,336275
5	3	0,336418	0,315224
9	4	0,338967	0,141076
4	4	0,355478	0,250930
9	3	0,391252	0,476951
5	4	0,406969	0,193799
8	3	0,428600	0,320434
7	3	0,431992	0,411948

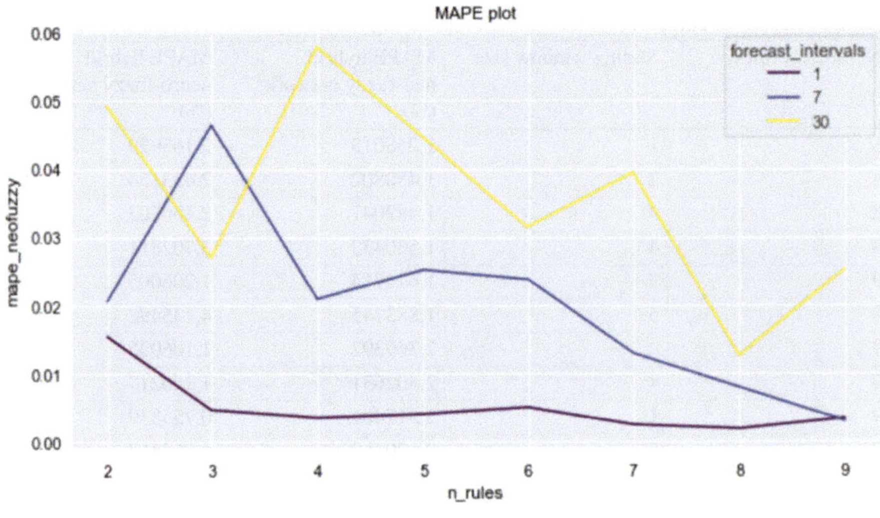

Fig. 4 MAPE dependence versus forecasting interval

From this figure one may conclude that for the short-term forecasts it is reasonable to set a small number of rules, as the percentage error does not change much with increasing number of rules, while for medium-term forecasts there is a greater oscillation of this metric.

5 Structure Optimization of Hybrid Neo-Fuzzy Network

In the next experiments neo-fuzzy network structure optimization was performed with application of GMDH method for different forecasting intervals. The optimal synthesized models for forecasting for a period of 1 day, week and month are presented in the Table 7.

Analyzing this data we may conclude that with increase of forecasting interval the structure of network becomes more complex.

In the Table 8 number of rules for hybrid neo-fuzzy network in dependence of interval length are presented.

The results of forecasting by optimal model for one month interval are presented in the Fig. 5.

Table 7 Optimal structures of hybrid neo-fuzzy networks

Forecasting interval/ structure	One day	One week	One month
Number of layers	3	4	4
1 layer number of nodes	4	4	4
2 layer number of nodes	2	3	4
3 layer number of nodes	1	2	2
4 layer number of nodes	–	1	1

Table 8 Number of rules in dependence of interval length

Forecasting interval	One day	One week	One month
Number of rules	3	7	9

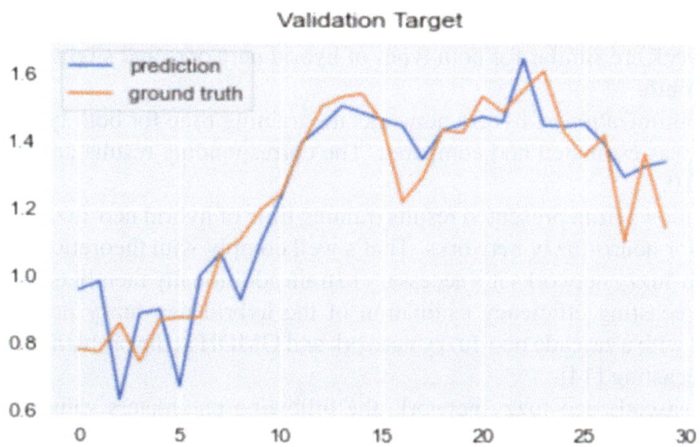

Fig. 5 Flow chart of forecasting NASDAQ for 1 month

Table 9 Optimal structures of hybrid neuro-fuzzy networks versus forecasting interval

Forecasting interval/structure	One day	One week	One month
Number of layers	3	4	3
1 layer-number of nodes	3	4	4
2 layer-number of nodes	2	3	2
3 layer-number of nodes	1	2	1
4 layer-number of nodes	–	1	–

Table 10 Training time for hybrid networks versus interval length (in sec)

Type of hybrid network	One day (s)	One week (s)	One month (s)
Hybrid neo-fuzzy network	36.3	60.3	70.2
Hybrid neuro-fuzzy network	70	86.2	90.5

Table 11 MAPE values for different forecasting methods

Inputs number/method	Hybrid GMDH-neo-fuzzy network	GMDH	Cascade neo-fuzzy neural network
4 inputs	4,31	4,19	6,04
5 inputs	3,91	4,11	6,09
6 inputs	4,36	5,53	8,01
7 inputs	4,77	6,26	8,68

In the next experiments structure optimization of neuro-fuzzy networks was performed using GMDH. The structure of the best synthesized hybrid neuro-fuzzy network for one day, week and month are presented in the Table 9.

Analyzing these results in Tables 7 and 9 we conclude that structures for one day and one week are similar for both types of hybrid networks and slightly differs only for one month.

After optimization of hybrid networks the training time for both types of hybrid networks was estimated and compared. The corresponding results are presented in the Table 10.

As it follows from presented results training time of hybrid neo-fuzzy networks is less than for neuro-fuzzy networks. That's well comply with theoretical estimations as in neuro-fuzzy networks it's necessary to train additionally membership functions.

For forecasting efficiency estimation of the hybrid neo-fuzzy network, it was compared with a cascade neo-fuzzy network and GMDH at the problem of NASDAQ index forecasting [14].

In the cascade neo-fuzzy network, the following parameters values were used: number of inputs 9, number of rules 9, cascades number is 3. The comparative forecasting results are presented in the Table 11, training sample—70%

Analyzing these results one can easily conclude the suggested hybrid neo-fuzzy network has the best accuracy, the second one is GMDH method, and the worst is the cascade neo-fuzzy network.

6 Conclusions

Investigations of hybrid GMDH neuro-fuzzy networks and neo-fuzzy networks in the problem of forecasting in the financial sphere were carried out. The problem of forecasting NASDAQ stock prices with intervals of one day, one week and month was considered.

In the process of forecasting the hyperparameters of hybrid fuzzy neural networks were optimized, namely the number of inputs, the number of rules and the size of the sliding window.

Optimal structures of hybrid fuzzy neural networks for different prediction intervals were synthesized with application of GMDH.

Experimental explorations of the efficiency of forecasting by hybrid neo-fuzzy and neuro-fuzzy networks according to MAPE criterion were performed.

As a result of experiments, it was found that the error of MAPE is lower in models based on neo-fuzzy networks for one week and one month forecasting, while for one day hybrid GMDH neuro-fuzzy networks appeared to be slightly better.

In the process of experiments, it was found that the training time of hybrid neo-fuzzy networks was shorter as compared to alternative hybrid neuro-fuzzy networks.

References

1. Goodfellow, I., Bengio, Y., Courville, A.: Deep Learning Book (2016). https://www.deeplearningbook.org/
2. Hinton, G.: A Practical Guide to Training Restricted Boltzmann Machines (2010). https://www.cs.toronto.edu/~hinton/absps/guideTR.pdf
3. Gan, Z., Henao, R., Carlson, D., Carin, L.: Learning Deep Sigmoid Belief Networks with Data Augmentation (2015). https://www.researchgate.net/publication/278244188_Learning_Deep_Sigmoid_Belief_Networks_with_Data_Augmentation
4. Hinton, G.: Deep Belief Networks (2009). http://www.scholarpedia.org/article/Deep_belief_networks
5. Erhan, D., Bengio, Y., Courville A., et al.: Why Does Unsupervised Pre-training Help Deep Learning? (2010). http://www.jmlr.org/papers/volume11/erhan10a/erhan10a.pdf
6. Hinton, G., Osindero, S., Teh, Y.: A Fast Learning Algorithm for Deep Belief Nets (2006). http://www.jmlr.org/papers/volume11/erhan10a/erhan10a.pdf
7. Ivakhnenko, A.G.: Self-learning systems of recognition and automatic control. Technica, Kiev (1969)
8. Ivakhnenko, A.G., Stepashko, V.S.: Disturbance tolerance of modeling. Naukova dumka, Kiev (1985)
9. Zaychenko, Yu.: The fuzzy group method of data handling and its application for economical processes forecasting. Sci. Inquiry **7**(1), 83–96 (2006)

10. Ivakhnenko, G.A.: Self-organization of neuronet with active neurons for effects of nuclear test explosions forecasting. Syst. Anal. Model. Simul. **20**, 107–116 (1995)
11. Jang, R.J.-S.: ANFIS: adaptive-network-based fuzzy inference systems. IEEE Trans. Syst. Man Cybernet. **23**, 665–685 (1993)
12. Bodyanskiy, Y., Boiko, O., Zaychenko, Y., Hamidov G.: Evolving hybrid GMDH-neuro-fuzzy network and its applications. In: Proceedings of the International Conference, SAIC 2018, Kiev, 8.10–10.10.18
13. Zaychenko, Y., Bodyanskiy, Y., Tyshchenko, O., Boiko, O., Hamidov, G.: Hybrid GMDH-neuro-fuzzy system and its training scheme. Int. J. Inf. Theories Appl. **24**(2), 156–172 (2018)
14. Bodyanskiy, E., Zaychenko, Y., Boiko, O., Hamidov, G., Zelikman, A.: Structure optimization and investigations of hybrid GMDH-Neo-fuzzy neural networks in forecasting problems. In: Zgurovsky, M., Pankratova, N. (eds.) System Analysis & Intelligent Computing. Book Studies in Computational Intelligence, SCI, vol. 1022, pp. 209–228. Springer (2022)

Generalized Models of Logistics Problems and Approaches to Their Solution Based on the Synthesis of the Theory of Optimal Partitioning and Neuro-Fuzzy Technologies

Anatolii Bulat, Elena Kiseleva, Liudmyla Hart, and Olga Prytomanova

Abstract The most common logistics problems are considered: the transportation problems and the optimal partitioning-allocation ones. Mathematical models and approaches to solving two-stage optimal partitioning-allocation problems are presented. Such problems generalize, on the one hand, the classical finite-dimensional transportation problems to the case when the volumes of production (storage, processing) at given points are not known in advance. In this case, the volumes are found as a solution to the corresponding continuous problem of optimal partitioning of the set of consumers (suppliers of a continuously distributed resource) into the regions of their service by these points. On the other hand, two-stage optimal partitioning-allocation problems generalize discrete two-stage production and transportation problems to the case of a continuously distributed resource. Statements of two new two-stage problems of optimal partitioning-allocation under uncertainty and approaches to their solution based on the synthesis of the theory of optimal partitioning of sets and neuro-fuzzy technologies are formulated.

Keywords Logistics · Infinite-dimensional transportation problem · Optimal partitioning-allocation problem · Fuzzy parameters · Neurolinguistic approximation · Neuro-fuzzy technologies

1 Introduction

Currently, the term *logistics* is widely used in business. It defines the theory and practice of moving raw materials, materials, production, labor and financial resources, finished products from their source to the consumer. The main purpose of logistics

A. Bulat
M.S. Polyakov Institute of Geotechnical Mechanics of the National Academy of Science of Ukraine, Simferopolska Str., 2a, Dnipro 49005, Ukraine

E. Kiseleva (✉) · L. Hart · O. Prytomanova
Oles Honchar Dnipro National University, Gagarin Avenue, 72, Dnipro 49010, Ukraine
e-mail: kiseleva47@ukr.net

M. Zgurovsky and N. Pankratova (eds.), *System Analysis and Artificial Intelligence*, Studies in Computational Intelligence 1107,
https://doi.org/10.1007/978-3-031-37450-0_21

is to ensure the availability of the necessary product in the required quantity, in the required condition, in the required place, at the required time and at a price acceptable to the consumer and at the lowest cost for the enterprise. In some cases, the problem of reducing costs in a particular domain can be formalized by reducing it to one of the known logistic models, followed by the use of mathematical methods for solving problems of this class to obtain the optimal solution.

The most commonly used logistics models in practice include transportation models (problems). The purpose of solving the transportation problem is to find such a plan for the transportation of products, in which the total transportation costs would be the lowest. F. Hitchcock first formulated such a problem and its mathematical model in 1941 in his paper "The Distribution of a Product from Several Sources to Numerous Localities".

Over time, the scope of the transportation model is expanding, and the model itself is being improved and integrated with other models, primarily with models of the production sector. We are talking, for example, about production-transportation problems, such as problems of planning the development and location of enterprises [1]; about multi-stage transportation problems, when products are sent from suppliers to consumers through some intermediate points [2]; about infinite-dimensional transportation problems, etc.

Note that the vast majority of logistics problems were studied under conditions of certainty. However, real situations for which logistics models are created are most often characterized by a certain degree of uncertainty. In these cases, the quality of the decisions made in the optimization logistics models is directly dependent on the completeness of consideration of all the uncertain factors that are significant for the consequences of the decisions made.

In the global energy sector, there is a steady trend to increase production and consumption of energy. Even with significant structural changes in industry and the transition to energy-saving technologies, energy needs will increase in the coming decades, especially the need for electricity. When determining the ways of energy development, it is necessary to take into account both the availability of resources sufficient for long-term development, and the impact of each of these ways on the environment. It is hardly possible to assume that all energy will be only coal, nuclear or solar, wind, geothermal. Each of these approaches to the development of the energy sector combines both positive and negative aspects [3]. Unfortunately, there are still no balanced comprehensive integrated approaches to energy development, especially taking into account the specifics of individual regions.

Energy, by its physical nature, can be considered as a flow process, and thus, the classical logistics approach can be applied to analyze the problems that arise in the energy sector. The scheme of the logistics series in the energy complex is shown in Fig. 1.

The main manufacturers of electricity and heat are currently large thermal, nuclear and hydraulic power plants based on power units of 50 MW and above. It is considered that these power plants produce electricity and heat more economically. As you know, in logistics, the object of management is economic flows, an important place among which is occupied by financial flows. This provision defines the limits of the complete

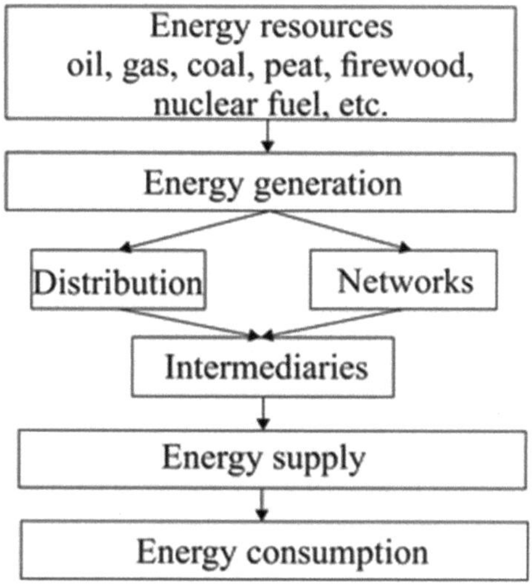

Fig. 1 Scheme of the main elements of the logistics system in the energy complex

logistics system. It is necessary to note that the existing high-capacity energy complexes allocate only 50% of the price of electricity to production. Other components of tariff pricing are system operator services, electricity losses on long-distance high-voltage transmission lines, electricity losses on interregional low-voltage networks, transportation by local utility networks and distribution to consumers of electricity. Thus, the efficiency of power supply to consumers is significantly reduced, despite the higher efficiency of power production plants. In addition, powerful thermal power plants are characterized by a high degree of equipment wear and tear due to its long-term use (on average more than 40 years).

One of the main tasks that energy logistics solves is load distribution between the elements of the energy supply system. Therefore, small distributed energy is currently becoming increasingly important. When implementing small-scale energy projects, alternative types of local fuel balances can be used—peat, coal, shale, gas [4], even wind. In addition, unlike large-scale energy, which hardly increases its capacity and requires significant investments, small-scale energy is able to increase capacity for direct consumers in a matter of months. Thus, the small-scale energy closes some of the problems and allows large-scale energy to redirect the released capacities to another area, and the qualitative differentiation of the logistics system of the electric power complex depends on the reliability of all parts of the system.

The introduction of small generating facilities will allow:

- to disperse the country's generating capacities, which are mainly concentrated in the central regions and to increase the reliability of energy supply to distant areas;
- to reduce losses in low-voltage networks;
- to cover the lack of power during peak loads;
- to get a cheaper product in the end [5].

The feasibility of their implementation is an important addition to the overall energy system and an essential factor in the country's energy security. In the context of increasing consumption, small-scale energy will provide an opportunity to predict the sustainable development of energy supply in the regions, and the economic efficiency of small distributed generation will depend on the potential of the region, that is, its resource base and the determination of the needs of the end consumer [6].

Small-scale energy can act as a catalyst for the transition from the traditional organization of energy systems to new technologies and practices. This transition should be based on decentralization, digitalization, intellectualization of energy supply systems, with the active involvement of consumers themselves and all types of energy resources and be characterized by increased energy efficiency and reduced greenhouse gas emissions. To achieve the set goals, energy logistics should solve the problems of managing and optimizing energy flows and related information and financial flows in the energy supply system, both for individual generating capacities and for the entire energy supply system as a whole [7].

Therefore, it is relevant to develop logistics problems not only in conditions of certainty, but also in cases where either the individual parameters included in the model description are fuzzy, inaccurate, undetermined, or there is an unreliable mathematical description of some dependencies in the model, etc.

The purpose of the study is to analyze the statements of logistics transportation problems, location-allocation problems and, based on the analysis, to formulate new two-stage partitioning-allocation problems under uncertainty and approaches to their solution.

2 Materials and Methods

2.1 Classical Transportation Problem

Transportation models often describe the movement (transportation) of any product from a point of departure (starting point, e.g. production site) to a point of destination (warehouse, store, storage). The purpose of the transportation problems is to determine the volume of transportation from points of departure to points of destination with the minimum total cost of transportation. In addition, the restrictions imposed on the volume of cargo available at the points of departure (supply) and the restrictions that take into account the need for cargo at the points of destination (demand) should be taken into account. The transportation model assumes that the cost of transportation on any route is directly proportional to the volume of cargo transported on this route. In general, the transportation model can be used to describe situations related to inventory management, capital flow management, scheduling, staff assignment, etc.

Figure 2 shows a general representation of the transportation problem in the form of a network with m points of departure and n points of destinations, which are

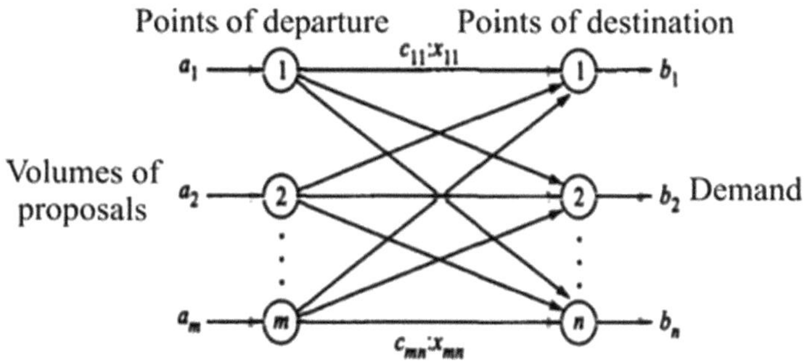

Fig. 2 Scheme of the classical transportation problem

shown as network nodes. The arcs connecting the network nodes correspond to the routes connecting the points of departure and destination. Two types of data are related to the arc (i, j) connecting the point of departure i and the point of destination j: the cost c_{ij} of transporting a unit of cargo from point i to point j and the quantity x_{ij} of cargo transported. The volume of cargo at the point of departure i is equal to a_i, and the volume of cargo at the point of destination j is equal to b_j $(i = 1, 2, \ldots, m; j = 1, 2, \ldots, n)$.

The problem is to determine the unknown values x_{ij} that minimize the total transportation costs $\sum_{i=1}^{m}\sum_{j=1}^{n} c_{ij}x_{ij}$ and satisfy the constraints imposed on the cargo volumes at the points of departure (proposals) and destination (demand).

Let us give a formal statement of the classical transportation problem.

Probelm 1 *Classical transportation problem* [8]. It is required to find such values $x_{ij}, i = 1, 2, \ldots, m;\ j = 1, 2, \ldots, n$, which provide

$$\sum_{i=1}^{m}\sum_{j=1}^{n} c_{ij}x_{ij} \to min \tag{1}$$

subject to

$$\sum_{j=1}^{n} x_{ij} \le a_i, \quad i = 1, 2, \ldots, m; \tag{2}$$

$$\sum_{i=1}^{m} x_{ij} \ge b_j, \quad j = 1, 2, \ldots, n; \tag{3}$$

$$x_{ij} \ge 0, \quad i = 1, 2, \ldots, m; \quad j = 1, 2, \ldots, n, \tag{4}$$

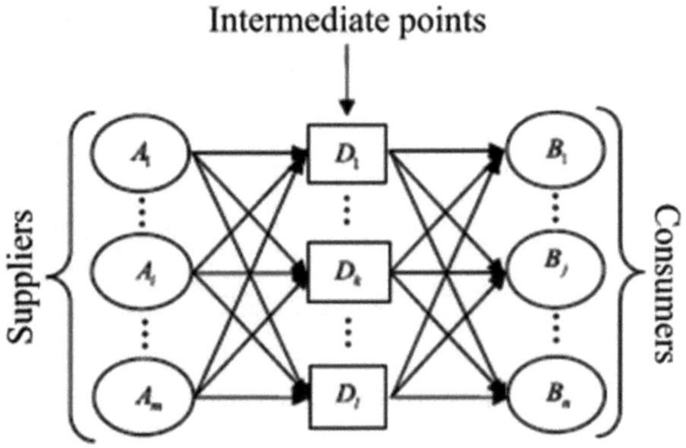

Fig. 3 The system "suppliers—intermediate points—consumers" in the two-stage transportation problem

where the above formulas have the following meaning: the objective function of the transportation problem tends to the minimum (1); the total volume of cargo exported from each supplier should not exceed its supply (2); the volume of cargo received by each consumer should not be less than its demand (3). In addition, cargo volumes on each of the routes should be non-negative (4).

The system of relations (1)–(4) is a *mathematical model of the transportation problem.*

Obviously, the transportation problem is a linear programming problem with $m \times n$ variables and $m + n$ indirect constraints. Let us first consider the ideal case, when the sum of possible supplies is exactly equal to the sum of demands

$$\sum_{i=1}^{m} a_i = \sum_{j=1}^{n} b_j = d. \quad (5)$$

This is the so-called *closed transportation problem.* In this case, the transportation problem (1)–(5) is admissible and solvable. Although the transportation problem can be solved as a conventional linear programming problem, its special structure allows the development of an algorithm with simplified calculations based on simplex duality relations, for example, the potential method [9].

If the transportation of products is carried out not specifically from the supplier to the consumer, but through some intermediate points (Fig. 3), then the so-called two-stage transportation problem arises.

2.2 *Two-Stage Finite-Dimensional Transportation Problem*

The formulation of the finite-dimensional two-stage transportation problem [2] assumes that the cargo is transported from suppliers to consumers only through intermediate points. The scheme of cargo transportation functioning is shown in Fig. 3. Intermediary firms and various storage facilities (warehouses) can be intermediate points.

Assume that, as in Problem 1, at the m supply points $A_1, \ldots, A_m$, there are, respectively, $a_1, \ldots, a_m$ units of products that need to be transported to n consumers $B_1, \ldots, B_n$, in order to satisfy their demands $b_1, \ldots, b_n$, but at the same time, p intermediate points $D_1, \ldots, D_p$ can be used to transport products from suppliers to consumers. Denote by $x_{ik}^{(1)}$ $(i = 1, 2, \ldots, m;\ k = 1, 2, \ldots, p)$ the quantity of products transported from the ith supply point A_i to the kth intermediate point D_k, and by $c_{ik}^{(1)}$—the cost of transporting a unit of this product. Similarly, we denote by $x_{kj}^{(2)}$ $(k = 1, 2, \ldots, p;\ j = 1, 2, \ldots, n)$ the quantity of products transported from the kth intermediate point D_k to the jth consumer B_j, and by $c_{ik}^{(2)}$—the cost of transporting a unit of this product. Let us write down the mathematical model of the problem.

Probelm 2 *Two-stage transportation problem* [2]. It is required to find such values $x_{ik}^{(1)}, x_{kj}^{(2)}, i = 1, 2, \ldots, m;\ k = 1, 2, \ldots, p;\ j = 1, 2, \ldots, n$, which provide

$$\sum_{i=1}^{m}\sum_{k=1}^{p} c_{ik}^{(1)} x_{ik}^{(1)} + \sum_{k=1}^{p}\sum_{j=1}^{n} c_{kj}^{(2)} x_{kj}^{(2)} \to min \tag{6}$$

subject to

$$\sum_{k=1}^{p} x_{ik}^{(1)} = a_i, \quad i = 1, 2, \ldots, m; \tag{7}$$

$$\sum_{k=1}^{p} x_{kj}^{(2)} = b_j, \quad j = 1, 2, \ldots, n; \tag{8}$$

$$\sum_{i=1}^{m} x_{ik}^{(1)} - \sum_{j=1}^{n} x_{kj}^{(2)} = 0, \quad k = 1, 2, \ldots, p; \tag{9}$$

$$x_{ik}^{(1)} \geq 0, x_{kj}^{(2)} \geq 0, i = 1, 2, \ldots, m;\ k = 1, 2, \ldots, p;\ j = 1, 2, \ldots, n. \tag{10}$$

Problem (6)–(10) is a linear programming problem containing $m \times p + p \times n$ variables $x_{ik}^{(1)}$, $x_{kj}^{(2)}$ and $m + n + p$ indirect constraints. The objective function (6) specifies the total costs of transporting products from suppliers to consumers through intermediate points. Constraints (7) mean the transportation of all products $a_1, \ldots, a_m$ from supply points to intermediate points, and constraints (8) mean the transportation of products $b_1, \ldots, b_n$ to be delivered from intermediate points to consumers. Constraints (9) specify the conditions that all products coming from suppliers to each intermediate point must be sent to consumers. This determines the compatibility conditions for the system of constraints in the form of linear equalities and linear inequalities (7)–(10).

To solve a two-stage transportation problem, various specialized algorithms are used [2, 8].

It should be noted that the above Problems 1 and 2 are finite-dimensional. The need to develop infinite-dimensional transportation problems arises when there are "too many" consumers and the formulation of the transportation problem as a discrete mathematical model becomes inappropriate due to the difficulties associated with solving problems of excessively high dimensionality. There are also problems in which the set that is partitioned into subsets is already initially continual in its structure.

2.3 Infinite-Dimensional Transportation Problem and Continuous Optimal Allocation-Partitioning Problem

The first theoretical results and methods for solving the infinite-dimensional transportation problem were published by L. V. Kantorovich in 1942 in connection with the solution of the classical G. Monge problem (the problem of ditches and embankments), formulated by him in 1784. Further, L. V. Kantorovich and G. Sh. Rubinshtein, using the functional-analytical method developed by them [10], studied modifications and generalizations of the problem on mass movement.

More general than infinite-dimensional transportation problems are infinite-dimensional problems of location of enterprises with simultaneous partitioning of a given region (set), continuously filled with consumers, into consumer areas (subsets), each of which is serviced by one enterprise, in order to minimize transportation and production costs. Optimization problems of production and transportation planning are the most common type of logistics problems that arise in the analysis of both long-term and current planning issues [1]. Such "location-allocation problems" are considered both in continuous and discrete formulations. Telephone subscribers, schoolchildren, voters, points of irrigated territory, patients for diagnosis of diseases, etc. can also act as consumers here. Of particular interest are continuous infinite-dimensional allocation-partitioning problems, in which the consumer demand in the field of allocation is given by a continuous density function.

The development of the mathematical theory of optimal set partitioning (OSP) began with the solution of a simplified model of the infinite-dimensional production location-allocation problem [11]. Let us give a formal statement of this problem (Problem 3).

Let the consumer of some homogeneous product be uniformly distributed in the domain $\Omega \subset E_2$. A finite number N of manufacturers of this product are located in isolated points $\tau_i = (\tau_i^{(1)}, \tau_i^{(2)})$, $i = 1, \ldots, N$ of domain Ω. The following are assumed to be given: demand $\rho(x)$ for the product of the consumer with coordinates $x = (x^{(1)}, x^{(2)})$; cost $c(x, \tau_i)$ of transporting a unit of production from the manufacturer $\tau_i = (\tau_i^{(1)}, \tau_i^{(2)})$ to the consumer with coordinates $x = (x^{(1)}, x^{(2)})$. It is assumed that the manufacturer's profit depends only on his costs, which are the sum of production and transportation costs.

For each ith manufacturer, a function $\varphi_i(Y_i)$ is given that describes the dependence of the cost of production on its capacity Y_i, determined by the formula $Y_i = \int_{\Omega_i} \rho(x)dx$, and the adjusted capital expenditures for the reconstruction of the ith manufacturer to increase its capacity from the existing one to the projected one Y_i.

The set of consumers Ω can be partitioned into areas Ω_i of consumer service by the ith manufacturer so that

$$\bigcup_{i=1}^{N} \Omega_i = \Omega;\ mes\left(\Omega_i \cap \Omega_j\right) = 0, \forall i, j = 1, \ldots, N\ (i \neq j), \tag{11}$$

where $mes(\cdot)$ means the Lebesgue measure; it is not excluded that some of the subsets $\Omega_i,\ i = 1, \ldots, N$ will be empty.

At the same time, the capacity of the ith manufacturer is determined by the total demand of consumers belonging to Ω_i, and does not exceed the given volumes:

$$\int_{\Omega_i} \rho(x)dx \leq b_i, i = 1, \ldots, N. \tag{12}$$

Figure 4 shows the partitioning of the set $\Omega \subset E_2$ into three subsets $\Omega_1,\ \Omega_2,\ \Omega_3$ with the centers $\tau_1,\ \tau_2,\ \tau_3$ of these subsets, respectively.

Probelm 3 *Infinite-dimensional production location-allocation problem.* It is required to partition the set of consumers Ω into areas of their service by N manufacturers, i.e., into subsets $\Omega_i,\ i = 1, \ldots, N$, and to place these manufacturers in Ω in such a way as to minimize the functional of the total costs for the production of product and its delivery to consumers:

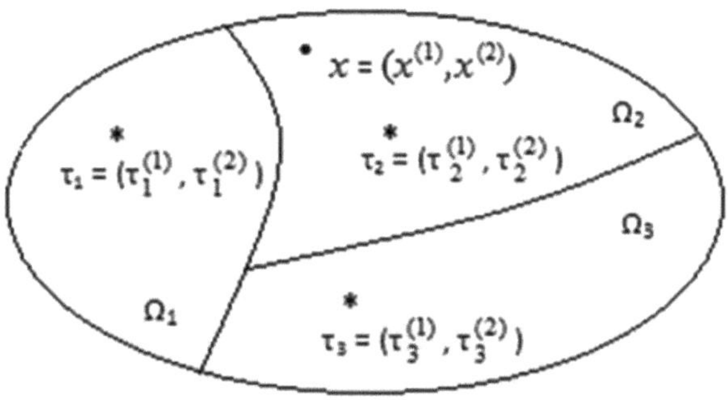

Fig. 4 Partitioning the set $\Omega \subset E_2$ into three subsets

$$F\left(\{\Omega_1, \ldots, \Omega_N\}, \{\tau_1, \ldots, \tau_N\}\right) = \sum_{i=1}^{N} \left\{ \int_{\Omega_i} c\left(x, \ \tau_i\right) \rho(x) dx + \varphi_i \left(\int_{\Omega_i} \rho(x) dx \right) \right\} \tag{13}$$

subject to (11), (12).

Problem (11)–(13) is *an infinite-dimensional location-allocation problem.*

In most practical problems, the cost of manufacturing a product at an industrial enterprise with capacity Y_i is equal to the product of the cost of this product and its quantity. Thus, we have

$$\varphi_i\left(Y_i\right) = d_i + a_i Y_i, i = 1, \ldots, N. \tag{14}$$

Substituting expression (14) into (13), we obtain

$$F\left(\{\Omega_1, \ldots, \Omega_N\}, \{\tau_1, \ldots, \tau_N\}\right) = \sum_{i=1}^{N} \int_{\Omega_i} \left(c\left(x, \ \tau_i\right) + a_i\right) \rho(x) dx. \tag{15}$$

Problem (11)–(15) with $a_i = 0, \quad i = 1, \ldots, N$, is *an infinite-dimensional transportation problem.*

The founder of the theory of optimal set partitioning, E. Kiseleva, obtained the necessary optimality conditions for problem (11)–(15) and developed a method and algorithm for its solution [12].

The main results of the mathematical theory of continuous OSP problems in n-dimensional Euclidean space, which are non-classical problems of infinite-dimensional mathematical programming with Boolean variables, developed over the past fifty years by the author and her students, are presented in more than 400 scientific papers, including six monographs. The structure of the theory of optimal set partitioning that has developed to date can be represented in the form of a block diagram (Fig. 5).

Let us give the mathematical formulation of one of the problems of this block diagram, namely, the linear multiproduct problem of optimal set partitioning under constraints in the form of equalities and inequalities with finding the coordinates of the centers of subsets (Problem 4). It should be noted that this problem is a generalization of problem (11), (12), (15) for the case when each ith manufacturer with the coordinate τ_i ($i = 1, \ldots, \ N$) manufactures several types of products.

Probelm 4 The linear multiproduct problem of optimal partitioning of a set Ω from n-dimensional Euclidean space E_n into its disjoint subsets $\Omega_1, \ldots, \Omega_N$ (among which there may be empty ones) under constraints in the form of equalities and inequalities with finding the coordinates of the centers $\tau_1, \ldots, \tau_N$ of these subsets, respectively. It is required to find

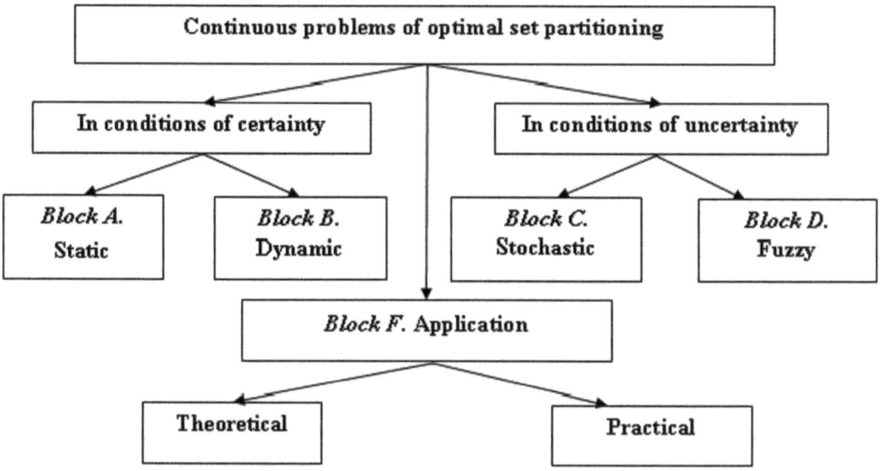

Fig. 5 Structure of the theory of optimal set partitioning [13]

$$\min_{\{\Omega_1^1,\ldots,\Omega_i^j,\ldots,\ \Omega_N^M\},\ \{\tau_1,\ldots,\ \tau_N\}} \sum_{j=1}^{M}\sum_{i=1}^{N}\int_{\Omega_i^j}(c^j\,(x,\ \tau_i)+a_i^j)\rho^j(x)dx$$

subject to

$$\sum_{j=1}^{M}\int_{\Omega_i^j}\rho^j(x)dx=b_i,\ \ i=1,\ldots,p;\sum_{j=1}^{M}\int_{\Omega_i^j}\rho^j(x)dx\le b_i,\ \ i=p+1,\ldots,N;$$

$$\{\Omega_1^1,\ldots,\Omega_i^j,\ldots,\ \Omega_N^M\}\in\Sigma_\Omega^{\mathrm{NM}},\ \ \{\tau_1,\ldots,\tau_N\}\in\underbrace{\Omega\times\cdots\times\Omega}_{N}=\Omega^N,$$

where $\sum_\Omega^{\mathrm{NM}}=\{\{\Omega_1^1,\ldots,\Omega_i^j,\ldots,\ \Omega_N^M\}:\ \bigcup_{i=1}^N\Omega_i^j=\Omega,\ \ mes(\Omega_i^j\cap\Omega_k^j)=0,$ $i,k=1,\ldots,N\ \ (i\neq k);\ \ j=1,\ldots,M\}$ is the class of all possible partitions of the set Ω into N disjoint subsets according to M products; $\tau_i=(\tau_i^{(1)},\ldots,\tau_i^{(n)})\in\Omega$ is the common center of the subsets $\Omega_i^1,\ldots,\Omega_i^M$ $(i=1,\ldots,N); x=(x^{(1)},\ldots,x^{(n)})\in\Omega$. Here the functions $c^j(x,\ \tau_i)$ are given, real, bounded on $\Omega\times\Omega$, measurable by x for any fixed $\tau_i\in\Omega$ for all $i=1,\ldots,N;\ \ j=1,\ldots,M;$ the functions $\rho^j(x)$ are given, non-negative, bounded, measurable on Ω for all $j=1,\ldots,M;$ $a_1^1,\ldots,a_i^j,\ldots,\ a_N^M$ and $b_1,\ldots,b_N$ are given non-negative numbers such that

$$S=\int_\Omega\sum_{j=1}^{M}\rho^j(x)dx\le\sum_{i=1}^{N}b_i,0\le b_i\le S,i=1,\ldots,N.$$

Special cases of Problem 4 are the problems of optimal partitioning of a set both with given coordinates of the centers of subsets and with those unknown in advance; both without constraints and with constraints. A classic example of such problems is the problem of locating enterprises with simultaneous partitioning of the territory into areas served by one enterprise. In this problem, as a rule, the criterion for the quality of location and partitioning is the total cost of production and delivery of products to the consumer. In the case when each enterprise manufactures several types of products, we get a multiproduct problem of optimal set partitioning.

The created theory of optimal set partitioning is based on a unified approach, which consists in reducing the original infinite-dimensional optimization problems in a certain way (for example, using the Lagrange functional) to non-smooth, usually finite-dimensional optimization problems. For the numerical solution of the latter, modern effective methods of non-differential optimization are used, namely, various versions of the r-algorithm developed at the V.M. Glushkov Institute of Cybernetics of the National Academy of Sciences of Ukraine under the direction of N.Z. Shor [14].

The theory of optimal set partitioning is effectively used to solve a wide range of theoretical and practical classes of optimization problems of different nature, which are reduced in the mathematical formulation to continuous models of optimal set partitioning.

3 Results and Discussion

3.1 Two-Stage Continuous-Discrete Optimal Allocation-Partitioning Problem with Given Centers of Subsets in the Set to be Partitioned

The considered two-stage continuous-discrete optimal allocation-partitioning problem, on the one hand, generalizes the classical finite-dimensional transportation Problem 1 to the case when the volumes of production (storage, processing) at given points are unknown in advance. These volumes are found as a solution to the corresponding continuous problem of optimal partitioning of the set of consumers (suppliers of a continuously allocated resource) into the areas of their service by these points. On the other hand, the two-stage continuous-discrete optimal allocation-partitioning problem generalizes the discrete two-stage production-transportation Problem 2 to the case of a continuously allocated resource.

Real-world problems, reducible to two-stage continuous-discrete optimal partitioning-allocation problems, are characterized by the presence of two stages and consist in determining the regions of collection of a continuously allocated resource by enterprises (points of the first stage) and the volume of transportation of the processed product from these enterprises to consumers (points of the second stage) in order to minimize the total cost of transportation of the resource transferred from suppliers to processing points (collection, storage) and then to consumers. Note that such problems can be easily found in real-life [1, 15]: these are ones where, for example, natural raw materials (oil, gas, ore) or the harvesting crops can be represented as a continuously allocated resource; the problems of organizing wood waste collection for fuel production with its subsequent allocation between points of heat energy production while minimizing the total transportation cost; problems of optimizing the deposit and credit activity of bank branches aiming to attract deposits from individuals with a subsequent allocation of the funds between borrowers, and many others [16–18].

Let us give a formal statement of the problem.

Let Ω be a bounded, closed, Lebesgue measurable set in n-dimensional Euclidean space E_n. A set $\Omega_1, \ldots, \Omega_N$ of Lebesgue measurable subsets of the set $\Omega \subset E_n$ is called a possible partition of the set Ω into its disjoint subsets $\Omega_1, \ldots, \Omega_N$ if

$$\bigcup_{i=1}^{N} \Omega_i = \Omega, mes(\Omega_i \cap \Omega_j) = 0, \forall i, j = 1, \ldots, N \ (i \neq j),$$

where $mes(\cdot)$ means the Lebesgue measure. Denote the class of all possible partitions of the set Ω into disjoint subsets $\Omega_1, \ldots, \Omega_N$ by $\sum_{\Omega}^{N}$, i.e.

$$\sum_{\Omega}^{N} = \left\{ (\Omega_1, \ldots, \Omega_N) : \bigcup_{i=1}^{N} \Omega_i = \Omega, \ \ mes(\Omega_i \cap \Omega_j) = 0, \ \ i, j = 1, \ldots, N(i \neq j) \right\}.$$

Let us introduce the functional

$$F(\{\Omega_1, \ldots, \Omega_i, \ldots, \Omega_N\}, \ \{v_{11}, \ldots, v_{ij}, \ldots, v_{\text{NM}}\}) =$$

$$= \sum_{i=1}^{N} \int_{\Omega_i} c_i^I(x, \ \tau_i^I) \rho(x) dx + \sum_{i=1}^{N} \sum_{j=1}^{M} c_{ij}^{\text{II}}(\tau_i^I, \tau_j^{\text{II}}) v_{ij},$$

where $\tau_i^I = (\tau_i^{I(1)}, \ldots, \tau_i^{I(n)})$ is some given reference point for the subset Ω_i, called the center of this subset ($i = 1, \ldots, N$); $\tau_j^{\text{II}} = (\tau_j^{II(1)}, \ldots, \tau_j^{II(n)})$ is some given point of the set Ω (j = 1,...,M)); v_{ij} ($i = 1, \ldots, N; \ j = 1, \ldots, M$) are weight parameters defining the connection between the subsets centers $\tau_i^I \in \Omega_i, \ \ i = 1, \ldots, N$ and the points $\tau_j^{\text{II}} \in \Omega, \ \ j = 1, \ldots, M$. The functions $c_i^I \left(x, \ \tau_i^I\right), \ i = 1, \ldots, N$ are given real bounded on $\Omega \times \Omega$, measurable by the argument $x = (x^{(1)}, \ldots, x^{(n)})$ for any fixed $\tau_i^I \in \Omega_i$; the functions $c_{ij}^{\text{II}} \left(\tau_i^I, \tau_j^{\text{II}}\right), \ \ i = 1, \ldots, N; \ j = 1, \ldots, M$ are given bounded on $\Omega \times \Omega$ and have the meaning of the "distance" functions between the points $\tau_i^I \in \Omega_i$ and $\tau_j^{\text{II}} \in \Omega$ in the corresponding metric of E_n; the function $\rho(x)$ is given non-negative one, bounded and measurable on Ω.

Then, under the two-stage continuous-discrete linear single-product optimal allocation-partitioning problem with a given location of subsets' centers under constraints in the form of equalities, we mean the following problem.

Probelm 5 *The two-stage continuous-discrete optimal allocation-partitioning problem with a given location of centers of subsets in a given set to be partitioned.* It is required to find such a partition of the set Ω into N Lebesgue measurable subsets $\Omega_1^*, \ldots, \Omega_i^*, \ldots, \Omega_N^*$ and such a non-negative vector $v^* = (v_{11}^*, \ldots, v_{ij}^*, \ldots, v_{\mathrm{NM}}^*) \in E_{\mathrm{NM}}$, which provide

$$\min_{\{\Omega_1,\ldots,\Omega_N\},\ \{v_{11},\ldots,v_{\mathrm{NM}}\}} F(\{\Omega_1, \ldots, \Omega_N\},\ \{v_{11}, \ldots, v_{\mathrm{NM}}\})$$

subject to

$$\sum_{j=1}^{M} v_{ij} = \int_{\Omega_i} \rho(x)dx, \quad i = 1, \ldots, N; \sum_{i=1}^{N} v_{ij} = b_j^{\mathrm{II}}, \quad j = 1, \ldots, M;$$

$$\{\Omega_1, \ldots, \Omega_N\} \in \sum_{\Omega}^{N}; \; v_{ij} \geq 0, \quad i = 1, \ldots, N; \quad j = 1, \ldots, M;$$

$$x = (x^{(1)}, \ldots, x^{(n)}) \in \Omega; \quad \tau^I = \left(\tau_1^I, \ldots, \tau_N^I\right) \in \Omega^N, \quad \tau^{\mathrm{II}} = \left(\tau_1^{\mathrm{II}}, \ldots, \tau_M^{\mathrm{II}}\right) \in \Omega^M.$$

Here $b_j^{\mathrm{II}}, \ j = 1, \ldots, M$ are given non-negative numbers such that the solvability conditions for the problem hold true:

$$S = \int_{\Omega} \rho(x)dx = \sum_{i=1}^{N} \int_{\Omega_i} \rho(x)dx = \sum_{i=1}^{N} \sum_{j=1}^{M} v_{ij} = \sum_{j=1}^{M} b_j^{\mathrm{II}};$$

$$0 \leq b_j^{\mathrm{II}} \leq S, \quad j = 1, \ldots, M.$$

Note that in terms of the classical transportation problem, the vector $v = (v_{11}, \ldots, v_{\mathrm{NM}})$ has the meaning of the volume of transportation of products from the points $\tau_1^I, \ldots, \tau_N^I$ of the first stage to the points $\tau_1^{\mathrm{II}}, \ldots, \tau_M^{\mathrm{II}}$ of final consumption (of the second stage).

Here and below, we consider integrals in the sense of Lebesgue and assume that the measure of the set of boundary points of the subsets $\Omega_1, \ldots, \Omega_N$ is equal to zero. In [19], a method and an algorithm for solving Problem 5 were developed and substantiated. A theorem was proved that determines the form of the optimal solution of the two-stage continuous-discrete optimal partitioning-allocation problem with a given location of the subsets centers under constraints in the form of equalities.

3.2 Two-Stage Continuous-Discrete Optimal Partitioning-Allocation Problem with Allocation (Finding Coordinates) of the Subsets Centers in a Given Set to be Partitioned

This problem is a generalization of Problem 5 to the case when the coordinates of the points of production (collection, storage, processing) of a homogeneous, continuously distributed in a given set of products and the production volumes at these points are not known in advance. They are found as a solution to the corresponding continuous problem of optimal partitioning of the set from E_n into subsets with the allocation (finding the optimal coordinates) of the centers of these subsets.

To formulate this problem, we introduce the functional

$$F(\{\Omega_1,\ldots,\Omega_N\},\{\tau_1^I,\ldots,\tau_N^I\},\ \{\upsilon_{11},\ldots,\upsilon_{\mathrm{NM}}\}) =$$

$$= \sum_{i=1}^{N} \int\limits_{\Omega_i} c_i^I(x,\ \tau_i^I)\rho(x)dx + \sum_{i=1}^{N}\sum_{j=1}^{M} c_{ij}^{\mathrm{II}}(\tau_i^I,\tau_j^{\mathrm{II}})\upsilon_{ij},$$

where $\tau_1^I,\ldots,\tau_N^I$ is a set of some reference points for the subsets $\Omega_1,\ldots,\Omega_N$, respectively, called the centers of these subsets ($\tau_i^I \in \Omega_i,\quad i=1,\ldots,N$), moreover the coordinates of the centers $\tau_i^I = (\tau_i^{I(1)},\ldots,\tau_i^{I(n)}),\quad i=1,\ldots,N$ are unknown and need to be determined; $\tau_1^{\mathrm{II}},\ldots,\tau_M^{\mathrm{II}}$ are some given points of the set Ω; υ_{ij} $(i=1,\ldots,N;\ j=1,\ldots,M)$ are weight parameters defining the connection between the subsets centers $\tau_i^I \in \Omega_i,\quad i=1,\ldots,N$ and given points $\tau_j^{\mathrm{II}} \in \Omega,\quad j=1,\ldots,M$. The function $\rho(x)\geq 0$ is a given bounded one, measurable on Ω; $c_i^I\left(x,\ \tau_i^I\right),\quad i=1,\ldots,N$ are given real functions bounded on $\Omega\times\Omega$ and measurable by the argument $x=(x^{(1)},\ldots,x^{(n)})$ for any fixed $\tau_i^I\in\Omega_i$; $c_{\mathrm{ij}}^{\mathrm{II}}\left(\tau_i^I,\tau_j^{\mathrm{II}}\right),\quad i=1,\ldots,N;\ j=1,\ldots,M$ are given functions bounded on $\Omega\times\Omega$ that have the meaning of "distance" between points $\tau_i^I\in\Omega_i$ and $\tau_j^{\mathrm{II}}\in\Omega$ in the corresponding metric of E_n.

Probelm 6 *Two-stage continuous-discrete optimal partitioning-allocation problem with location (finding coordinates) of the subsets centers.* It is required to find such a partition of the set $\Omega\subset E_n$ into N of Lebesgue measurable subsets $\Omega_1^*,\ldots,\Omega_i^*,\ldots,\Omega_N^*$, such coordinates of the centers $\tau_1^{I*},\ldots,\tau_i^{I*},\ldots,\tau_N^{I*}$ of these subsets in the domain Ω, and such a non-negative vector $\upsilon^*=(\upsilon_{11}^*,\ldots,\upsilon_{\mathrm{ij}}^*,\ldots,\upsilon_{\mathrm{NM}}^*)$ $\in E_{\mathrm{NM}}$, which provide

$$\min_{\{\Omega_1,\ldots,\Omega_N\},\{\tau_1^I,\ldots,\tau_N^I\},\{\upsilon_{11},\ldots,\upsilon_{\mathrm{NM}}\}} F(\{\Omega_1,\ldots,\Omega_N\},\{\tau_1^I,\ldots,\tau_N^I\},\{\upsilon_{11},\ldots,\upsilon_{\mathrm{NM}}\})$$

subject to

$$\sum_{j=1}^{M} v_{ij} = \int_{\Omega_i} \rho(x)dx, \ \ i = 1, \ldots, N; \sum_{i=1}^{N} v_{ij} = b_j^{\text{II}}, \ \ j = 1, \ldots, M;$$

$$\{\Omega_1, \ldots, \Omega_N\} \in \sum_{\Omega}^{N}; \ v_{ij} \geq 0, \ \ i = 1, \ldots, N; \ \ j = 1, \ldots, M;$$

$$x = (x^{(1)}, \ldots, x^{(n)}) \in \Omega; \quad \tau^I = \left(\tau_1^I, \ldots, \tau_N^I\right) \in \Omega^N, \quad \tau^{\text{II}} = \left(\tau_1^{\text{II}}, \ldots, \tau_M^{\text{II}}\right) \in \Omega^M.$$

Here $b_j^{\text{II}}, \ \ j = 1, \ldots, M$ are given non-negative numbers such that the solvability conditions for the problem hold true:

$$S = \int_{\Omega} \rho(x)dx = \sum_{i=1}^{N} \int_{\Omega_i} \rho(x)dx = \sum_{i=1}^{N} \sum_{j=1}^{M} v_{\text{ij}} = \sum_{j=1}^{M} b_j^{\text{II}};$$

$$0 \leq b_j^{\text{II}} \leq S, \ \ j = 1, \ldots, M.$$

Note that in terms of the classical transportation problem, the vector $v = (v_{11}, \ldots, v_{\text{NM}})$ has the meaning of the volume of transportation of products from the points $\tau_i^I, i = 1, \ldots, N$ of the first stage to the points $\tau_j^{\text{II}}, \ \ j = 1, \ldots, M$ of final consumption (of the second stage).

In [16], to solve Problem 6, an algorithm is proposed, based on the transition from the original problem to the dual one, for the numerical solution of which Shor's r-algorithm is used [14]. We present the dual problem in the form

$$G_1\left(\psi, \tau^I\right) \to \max_{\psi \in E_N} \min_{\tau^I \in \Omega^N}, \tag{16}$$

where

$$G_1\left(\psi, \tau^I\right) = \int_{\Omega} \min_{k=\overline{1,N}} \left(c_k^I\left(x, \tau_k^I\right) + \psi_k\right) \rho\left(x\right) \text{dx} +$$

$$+ \sum_{j=1}^{M} b_j^{\text{II}} \min_{k=\overline{1,N}} (c_{kj}^{\text{II}}(\tau_k^I, \tau_j^{\text{II}}) - \psi_k), \psi \in E_N, \ \ \tau^I \in \Omega^N. \tag{17}$$

To apply the r-algorithm to the solution of the dual problem (16), (17), we write the subgradient $g_{G_1}\left(\psi, \tau^I\right) = (-g_{G_1}^{\psi}; \ g_{G_1}^{\tau^I}) = (-g_{G_1}^{\psi_1}, \ldots, -g_{G_1}^{\psi_N}; \ g_{G_1}^{\tau_1^I}, \ldots, g_{G_1}^{\tau_N^I})$ of the objective function (17) in the following form:

$$g_{G_1}^{\psi_i}\left(\psi, \tau_i^I\right) = \int_{\Omega} \rho\left(x\right) \lambda_i\left(x\right) \text{dx} + \sum_{j=1}^{M} b_j^{\text{II}} q_{ij}, i = 1, \ldots, N; \ \ j = 1, \ldots, M,$$

where

$$q_{ij} = \begin{cases} -1, \; c_{\mathrm{ij}}^{\mathrm{II}}\left(\tau_i^I, \tau_j^{\mathrm{II}}\right) - \psi_i = \min_{k=\overline{1,N}}\left(c_{\mathrm{kj}}^{\mathrm{II}}\left(\tau_k^I, \tau_j^{\mathrm{II}}\right) - \psi_k\right), \\ 0, \qquad\qquad\qquad\qquad \text{in other cases;} \end{cases}$$

$$g_{G_1}^{\tau_i^I}\left(\psi, \tau_i^I\right) = \int_{\Omega} \rho(x)\, g_{c^I}^{\tau_i^I}(x, \tau^I)\lambda_i(x)\,\mathrm{dx} + \sum_{j=1}^{M} b_j^{\mathrm{II}} r_{ij}, i = 1, \ldots, N; \; j = 1, \ldots, M,$$

where

$$r_{ij} = \begin{cases} g_{c^{\mathrm{II}}}^{\tau_i^I}(\tau_i^I, \tau_j^{\mathrm{II}}), \; c_{ij}^{\mathrm{II}}\left(\tau_i^I, \tau_j^{\mathrm{II}}\right) - \psi_i = \min_{k=\overline{1,N}}\left(c_{kj}^{\mathrm{II}}\left(\tau_k^I, \tau_j^{\mathrm{II}}\right) - \psi_k\right), \\ \qquad 0, \qquad\qquad\qquad\qquad \text{in other cases,} \end{cases}$$

where $g_{c^I}^{\tau_i^I}(x, \tau^I)$ $(i = 1, \ldots, N)$ is the ith component of the N-dimensional vector of the generalized gradient $g_{c^I}^{\tau^I}(x, \tau^I)$ of the function $c_i^I\left(x, \; \tau_i^I\right)$ at the point $\tau^I = \left(\tau_1^I, \tau_2^I, \ldots, \tau_N^I\right)$; $g_{c^{\mathrm{II}}}^{\tau_i^I}(\tau_i^I, \tau_j^{\mathrm{II}})$ $(i = 1, \ldots, N; \; j = 1, \ldots, M)$ is the ith component of the N-dimensional vector of the generalized gradient $g_{c^{\mathrm{II}}}^{\tau^I}(\tau_i^I, \tau_j^{\mathrm{II}})$ of the function $c_{ij}^{\mathrm{II}}\left(\tau_i^I, \tau_j^{\mathrm{II}}\right)$ at the point $\tau^I = \left(\tau_1^I, \tau_2^I, \ldots, \tau_N^I\right)$;

$$\lambda_i(x) = \begin{cases} 1, \; c_i^I\left(x, \tau_i^I\right) + \psi_i = \min_{k=\overline{1,N}}\left(c_k^I\left(x, \tau_k^I\right) + \psi_k\right), \\ 0, \qquad\qquad\qquad \text{in other cases} \end{cases}$$

is the characteristic function of the subset $\Omega_i \subset \Omega$ $(i = 1, \ldots, N)$.

To solve the problem, a heuristic pseudo-gradient algorithm close to the r-algorithm is used. As a result of its application, we obtain $\lambda^*(x) = (\lambda_1^*(x), \ldots, \lambda_N^*(x))$, $x \in \Omega$; $\tau^{I*} = (\tau_1^{I*}, \ldots, \tau_N^{I*}) \in \Omega^N$; $\psi^* = (\psi_1^*, \ldots, \psi_N^*) \in E_N$, while we find the vector $\upsilon^* = (\upsilon_{11}^*, \ldots, \upsilon_{\mathrm{NM}}^*) \in E_{\mathrm{NM}}$ as a solution to the following transportation problem:

$$\sum_{i=1}^{N}\sum_{j=1}^{M} c_{ij}^{\mathrm{II}}\left(\tau_i^I, \tau_j^{\mathrm{II}}\right) \upsilon_{ij} \to \min_{\upsilon \in E_{\mathrm{NM}}},$$

$$\sum_{j=1}^{M} \upsilon_{ij} = \int_{\Omega} \rho(x)\lambda_i^*(x)dx, \quad i = 1, \ldots, N; \sum_{i=1}^{N} \upsilon_{ij} = b_j^{\mathrm{II}}, \quad j = 1, \ldots, M;$$

$$\upsilon_{ij} \geq 0, \quad i = 1, \ldots, N; \; j = 1, \ldots, M.$$

Note that two-stage continuous-discrete optimal partitioning-allocation problems under uncertainty of initial data are very common in practice. Consider the formulation of such problems for the case when some input parameters can be specified incompletely, inaccurately, unreliably.

3.3 Two-Stage Continuous-Discrete Optimal Partitioning-Allocation Problem with Fuzzy Parameters in the Objective Functional

Consider, for example, in the objective functional of Problem 4, functions $c_i^I\left(x, \tau_i^I\right)$, $i = 1, \ldots, N$, which are functions of the distance between points $x \in \Omega$ and $\tau_i^I \in \Omega_i$ for the first stage and are defined as one of the following metrics [12]:

$$c(x, \tau_i) = \|x - \tau_i\|_2 = \sqrt{\sum_{k=1}^{n} (x^{(k)} - \tau_i^{(k)})^2} \text{ (Euclidean distance);} \tag{18}$$

$$c(x, \tau_i) = \|x - \tau_i\|_1 = \sum_{k=1}^{n} |x^{(k)} - \tau_i^{(k)}| \text{ (Manhattan distance);} \tag{19}$$

$$c(x, \tau_i) = \|x - \tau_i\|_0 = \max_{1 \le k \le n} |x^{(k)} - \tau_i^{(k)}| \text{ (Chebyshev distance).} \tag{20}$$

Let the functions $c_{ij}^{\text{II}}\left(\tau_i^I, \tau_j^{\text{II}}\right)$, $i = 1, \ldots, N$, $j = 1, \ldots, M$ be functions of the distance between points $\tau_i^I \in \Omega_i$ and $\tau_j^{\text{II}} \in \Omega$ for the second stage and be defined similarly to the distance functions of the first stage.

For practical problems, the distances between consumers and enterprises of the first stage, as well as the distances between enterprises of the first and second stages may differ significantly from the distances calculated using "formal" metrics (18)–(20). These differences can be specified fuzzily using multipliers—a vector of fuzzy parameters $\widetilde{a} = (\widetilde{a}_1, \ldots, \widetilde{a}_N)$ for each function $c_i^I\left(x, \tau_i^I\right)$, $i = 1, \ldots, N$ and a vector of fuzzy parameters $\widetilde{w} = (\widetilde{w}_{11}, \ldots, \widetilde{w}_{\text{NM}})$ for each function $c_{ij}^{\text{II}}\left(\tau_i^I, \tau_j^{\text{II}}\right)$, $i = 1, \ldots, N$, $j = 1, \ldots, M$.

Then the functional can be written in the form

$$F(\{\Omega_1, \ldots, \Omega_i, \ldots, \Omega_N\}, \{v_{11}, \ldots, v_{ij}, \ldots, v_{\text{NM}}\}, \widetilde{a}, \widetilde{w}) =$$

$$= \sum_{i=1}^{N} \int_{\Omega_i} \widetilde{a}_i c_i^I(x, \tau_i^I) \rho(x) dx + \sum_{i=1}^{N} \sum_{j=1}^{M} \widetilde{w}_{ij} c_{ij}^{\text{II}}\left(\tau_i^I, \tau_j^{\text{II}}\right) v_{ij}, \tag{21}$$

where the parameters $\widetilde{a} = (\widetilde{a}_1, \ldots, \widetilde{a}_N)$ and $\widetilde{w} = (\widetilde{w}_{11}, \ldots, \widetilde{w}_{\text{NM}})$ can be considered as linguistic variables, which, in turn, may depend on the factors affecting them. To restore the exact values of these parameters, the method of neurolinguistic identification of unknown complex nonlinear dependencies can be used [20].

In [21], a method for solving the formulated problem is proposed, based on the use of the method of neurolinguistic identification of unknown dependencies (to restore the exact values of those parameters of the problem that are not clearly specified), methods of the theory of optimal set partitioning and the potential method for solving the transportation problem.

3.4 Two-Stage Continuous-Discrete Optimal Partitioning-Allocation Problem with Neurolinguistic Identification of Functions Included in the Objective Functional, the Explicit Analytical Form of Which is Unknown

Note that in the mathematical formulation of Problem 3 and in various generalizations of this problem [22], the objective functional includes the functions $c(x, \tau_i), \quad i = 1, \ldots, N$ and $\rho(x)$, which, for example, in terms of infinite-dimensional transportation and location problems, have the following meaning. $c(x, \tau_i)$ is the cost of transporting a unit of products from the production point τ_i $(i = 1, \ldots, N)$ to the consumption point $x = (x^{(1)}, \ldots, x^{(n)})$; $\rho(x)$ is the demand function of the consumer x for products manufactured by the production points $\tau_i, \quad i = 1, \ldots, N$.

Also, in infinite-dimensional transportation and location problems, it is assumed that the demand is either uniformly allocated in a given domain or allocated in this domain with a given density $\rho(x)$ (because of this, the function $\rho(x)$ is often called demand or density).

In Problems 4 and 5, it was assumed for the functions $c\,(x, \tau_i)\,, \quad i = 1, \ldots, N$ and $\rho(x)$ that the explicit analytical dependence on their arguments is known. However, in practice, this dependence (usually a complex nonlinear one) is unknown. In addition, it is often impossible to take into account the affecting of some real factors in an explicit analytical form, either due to lack of information about the modeled dependence, or due to the difficulties of taking into account the variety of factors that affect the nature of this dependence. For the infinite-dimensional transportation problem, for example, the demand $\rho(x)$ can be affected by factors such as changes in consumer income, exchange rate fluctuations or political instability, gasoline price fluctuations, and many others. In cases when such dependencies are not probabilistic, but fuzzy-set in nature, it is proposed to use neuro-fuzzy technologies to identify the dependence of the function $\rho(x)$ on its arguments [23].

In [24], a method was developed and substantiated for identifying the dependence of the function $\rho(x)$, included in the objective functional of Problems 4 and 5, on its arguments using neuro-fuzzy technology.

4 Conclusions

The paper analyzes the most common logistics transportation problems, including two-stage continuous-discrete optimal partitioning-allocation problems.

Two-stage continuous-discrete optimal partitioning-allocation problems are characterized by the presence of two stages and consist in determining the regions of collection of a continuously distributed resource (raw materials) by the enterprises of the first stage and the volumes of transportation of the processed product from the enterprises of the first stage to consumers (points of the second stage). At the same time, the total costs of transporting the resource from suppliers to consumers through processing (collection, storage) points should be minimal.

Two-stage continuous-discrete optimal partitioning-allocation problems generalize, on the one hand, infinite-dimensional transportation problems to the case when the volumes of production (storage, processing) at given points are unknown in advance. These volumes are found as a solution to the corresponding continuous problem of optimal partitioning of the set of consumers (suppliers of a continuously distributed resource) into the regions of their service by these points. On the other hand, two-stage continuous-discrete optimal partitioning-allocation problems generalize finite-dimensional two-stage production-transportation problems to the case of a continuously distributed resource.

Based on the analysis, two new two-stage partitioning-allocation problems under uncertainty are formulated: with fuzzy parameters in the objective functional and with neurolinguistic identification of functions included in the objective functional, the explicit analytical form of which is unknown.

An approach is proposed for solving these new two-stage partitioning-allocation problems under uncertainty using the synthesis of the theory of optimal set partitioning and neuro-fuzzy technologies.

The solution of the two-stage continuous-discrete optimal partitioning-allocation problem is based on a unified approach, which consists in reducing the original infinite-dimensional optimal partitioning-allocation problems to non-smooth, usually finite-dimensional optimization problems. For the numerical solution of the latter, effective methods of nondifferentiable optimization are used, namely, different versions of Shor's r-algorithm.

The method and algorithm for solving a two-stage continuous-discrete optimal partitioning-allocation problem with fuzziness in the objective functional is based on the principle that initially, to restore the exact values of fuzzy parameters in the objective functional, the neurolinguistic identification method is used. Then the optimal partition is found using the methods of the theory of optimal set partitioning and the potential method for solving the transportation problem.

References

1. Mikhalevich, V.S., Trubin, V.A., Shor, N.Z.: Optimization Problems of Production-Transportation Planning: Models, Methods, and Algorithms, 264 p. Nauka, Moscow (1986). [in Russian]
2. Stetsyuk, P.I., Lyashko, V.I., Mazyutinets, G.V.: Two-stage transportation problem and its AMPL-implementation. Naukovi Zapiski NaUKMA. Comput. Sci. **1**, 14–20 (2018)
3. Energy Transition Investment Trends 2022, 21 p. Bloomberg NEF, New York, USA (2022)
4. Bulat, A.F., Voziyanov, V.S., Slobodyannikova, I.L., Vitushko, O.V.: The method of complex processing of minerals. Patent of Ukraine 151317 IPC (2022.01) E21C 41/00; Dec. 11/15/2021, Publ. 07/07/2022, Bull. No. 27. [in Ukrainian]
5. Bulat, A.F., Dyakun, I.L.: Substantiation of the effectiveness of creating heat and power complexes with steam turbine cogeneration at coal mining enterprises. In: Miners' Forum-2017: Proceedings of the International Conference, October 4–7 (2017), pp. 368–374. National Mining University, Dnipro (2017). [in Russian]
6. Eid, C., et al.: Managing electric flexibility from distributed energy resources: a review of incentives for market design. Renew. Sustain. Energy Rev. **64**, 237–247 (2016). https://doi.org/10.1016/j.rser.2016.06.008
7. Kulińska, E., Dendera-Gruszka, M.: New perspectives for logistics processes in the energy sector. Energies **15**, 5708 (2022). https://doi.org/10.3390/en15155708
8. Taha, H.A.: Operations Research: An Introduction, 10th edn. (Global edn.), 848 p. Pearson Education Ltd. (2017)
9. Holstein, E.G., Yudin, D.B.: Transport-type Linear Programming Problems, 382 p. Nauka, Moscow (1969). [in Russian]
10. Kantorovich, L.V., Rubinstein, GSh.: On a functional space and some extremal problems. Doklady Akademii Nauk SSSR **115**(6), 1058–1061 (1957). [in Russian]
11. Kiseleva, E.M.: The emergence and formation of the theory of optimal set partitioning for sets of the n-dimensional Euclidean space. Theory and application. J. Aut. Inf. Sci. **50**(9), 1–24 (2018). https://doi.org/10.1615/JAutomatInfScien.v50.i9.10
12. Kiseleva, E.M., Shor, N.Z.: Continuous Problems of Optimal Set Partitioning: Theory, Algorithms, Applications, 564 p. Naukova Dumka, Kyiv (2005). [in Russian]
13. Kiseleva, E.M., Hart, L.L., Prytomanova, O.M., Baleiko, N.V.: Fuzzy Problems of Optimal Set Partitioning: Theoretical Foundations, Algorithms, Applications, 400 p. Lyra, Dnipro (2020). [in Ukrainian]
14. Shor, N.Z.: Minimization Methods for Non-differentiable Functions. Springer Series in Computational Mathematics, 162 p. Springer, Berlin (1985). https://doi.org/10.1007/978-3-642-82118-9
15. Us, S.A, Stanina, O.D.: On some mathematical models of facility location problems of mining and concentration industry. In: Theoretical and Practical Solutions of Mineral Resources Mining, pp. 419–424. CRC Press, Balkema – Taylor & Francis Group, London (2015). https://doi.org/10.1201/b19901
16. Kiseleva, E., Prytomanova, O., Hart, L.: Solving a two-stage continuous-discrete problem of optimal partitioning-allocation with the subsets centers placement. Open Comput. Sci. DeGruyter **10**, 124–136 (2020). https://doi.org/10.1515/comp-2020-0142
17. Yakovlev, S.V.: The concept of modeling packing and covering problems using modern computational geometry software. Cybern. Syst. Anal. **59**(1), 108–119 (2023). https://doi.org/10.1007/s10559-023-00547-5
18. Yakovlev, S., Kartashov, O., Podzeha, D.: Mathematical models and nonlinear optimization in continuous maximum coverage location problem. Computation **10**(7), 119 (2022). https://doi.org/10.3390/computation10070119
19. Kiseleva, E.M., Prytomanova, O.M., Us, S.A.: Solving a two-stage continuous-discrete optimal partitioning-distribution problem with a given position of the subsets centers. Cybernetics and Systems Analysis. **56**(1), 3–15 (2020). https://doi.org/10.1007/s10559-020-00215-y

20. Kiseleva, E., Prytomanova, O., Zhuravel, S.: An algorithm for solving location-allocation problem with fuzzy parameters. Sci. Discuss. Praha, Czech Republic **40**(1), 11–18 (2020)
21. Kiseleva, E.M., Prytomanova, O.M., Dzyuba, S.V., Padalko, V.G.: Solving a two-stage continuous-discrete optimal partitioning-allocation problem with fuzzy parameters. Probl. Appl. Math. Math. Model. **19**, 106–116 (2019). https://doi.org/10.15421/321911 [in Ukrainian]
22. Kiseleva, E.M., Shor, N.Z.: Analysis of algorithms for a class of continuous partition problems. Cybern. Syst. Anal. **30**(1), 64–74 (1994). https://doi.org/10.1007/BF02366365
23. Kiseleva, E.M., Prytomanova, O.M., Zhuravel, S.V.: Mathematical aspects of neuro-fuzzy technology application in project management. Eur. Coop. Warsaw, Poland **12**(31), 61–70 (2017)
24. Kiseleva, E., Prytomanova, O., Zhuravel, S.: Algorithm for solving a continuous problem of optimal partitioning with neurolinguistic identification of functions in target functional. J. Autom. Inf. Sci. **50**(3), 1–20 (2018). https://doi.org/10.1615/JAutomatInfScien.v50.i3.10

Intelligent Data Analysis for Complex Systems and Processes

Technological Principles of Using Media Content for Evaluating Social Opinion

Michael Zgurovsky, Dmytro Lande, Oleh Dmytrenko, Kostiantyn Yefremov, Andriy Boldak, and Artem Soboliev

Abstract The technological principles of using content from Internet media and social networks to evaluate social phenomena, socially significant events, and social opinion is presented. These principles include new methods of identifying and evaluating information sources, presenting the semantics of documents as a Directed Weighted Network of Terms, allowing implementation search procedures using signs of closeness to the semantics of text messages. The above technological tools are integrated based on microservice architecture for the implementation of a system for evaluating the effectiveness of public opinion. The developed system is part of a single analytical and expert environment based on the concept of the Information and Analytical Situation Center (IASC) of the World Data Center "Geoinformatics and Sustainable Development", and it is used to solve tasks of intelligent data analysis.

Keywords Internet-media · Social networks · Semantic networks · DWNT · Information retrieval · User's information needs · Intent document

---

M. Zgurovsky
National Technical University of Ukraine "Igor Sikorsky Kyiv Polytechnic Institute", Kyiv 03056, Ukraine

D. Lande
Department of Information Security, National Technical University of Ukraine "Igor Sikorsky Kyiv Polytechnic Institute", Kyiv 03056, Ukraine

O. Dmytrenko · K. Yefremov (✉) · A. Soboliev
World Data Center for Geoinformatics and Sustainable Development, National Technical University of Ukraine "Igor Sikorsky Kyiv Polytechnic Institute", Kyiv 03056, Ukraine
e-mail: k.yefremov@gmail.com

A. Boldak
Department of Computer Engineering, National Technical University of Ukraine "Igor Sikorsky Kyiv Polytechnic Institute", Kyiv 03056, Ukraine

M. Zgurovsky and N. Pankratova (eds.), *System Analysis and Artificial Intelligence*, Studies in Computational Intelligence 1107,
https://doi.org/10.1007/978-3-031-37450-0_22

1 Introduction

The mass distribution and accessibility of this information and communication environment of the Internet, in particular, social networks, which contain numerous user ratings, allows considering the media content of this environment as an information reflection of social phenomena, socially significant events, social opinion, etc.

The so-called Open Source INTelligence (OSINT) is one of the most important tools for researching media content. It includes the retrieval, collection and analysis of intelligence information obtained from publicly available sources [1]. The task of evaluating social opinion based on the application of the open information environment of the Internet can also be considered as a special case of OSINT. The effectiveness of OSINT usage in analytical work is determined, first of all, by the characteristics of technical means that ensure the promptness of receiving information, the ability to manage large information arrays, the ability to process and intellectually analyze data, and the ease of their further and reusability, etc. [2].

In works [3, 4], an approach to the quantitative assessment of the effectiveness of social transformations was proposed. The assessment is based on the assessment of the proximity to expectations vector and social activity of society, as well as the vector of government actions, considering social synergy, which is understood as additional social activity created by the interaction of the country's citizens. The interaction can be both positive, when society realizes that successes generate new successes, and negative, when failures lead to people's loss of hope for positive changes (in the other words, apathy) and, as a result, generate new failures.

Vectors of society's expectations and social activity can be formed due to the analysis of media content. In the work [3], the authors proposed a method for evaluating the effectiveness of social transformations, based on the emotional tone analysis of information messages obtained through online monitoring of media resources and social networks. The method is based on the hypothesis of adequate reflection of the emotional tone of messages on the components of the social expectations vector. The authors also proposed a method of evaluating the consistency of the results obtained by both methods.

In the works [3, 4] there is a link to a web application [5] in which these methods are implemented and used to solve the task of analyzing the attitude of the public to the government actions related to the strengthening of quarantine measures due to the spread of the pandemic COVID-19 in Ukraine. A test set of data collected over 200 days from the most popular 11 social media and 100 Ukrainian news websites was used (the sample size was more than 2 million messages).

In particular, the experience of using the "PRO ET CONTRA" toolkit [5] of the integrated online platform "Advanced Analytics" of the World Data Center "Geoinformatics and Sustainable Development" to solve other tasks related to the analysis of the emotional tone of informational messages showed that there is a certain gap between the expected and actual correspondence of the results obtained from the system to the actual information needs [6–8]. The reasons for this inconsistency are the insufficient precision and accuracy of the results, caused on the one hand by the lack of semantic search/collection of information, and on the other hand by the inability of the system to adapt to constant changes in the information environment, to the appearance of new information sources and disappearing existing ones, and to their features changes.

Therefore, the purpose of this work is to increase the effectiveness of the public opinion assessment system based on the analysis of messages from websites and social networks using the new methods and algorithms for identifying and evaluating the properties of information sources, developing semantic search procedures target at increasing the correspondence of requests to information tasks within a complex information system.

2 Materials and Methods

2.1 Identifying and Evaluating Information Sources

To solve a wide range of information and analytical issues information published on news sites is increasingly used. The task of identifying new information sources is quite simply solved by processing links contained in messages.

At the same time, there are many sources of informational influence aimed at spreading fake information, the consumption of which can negatively affect the reliability of the results. Therefore, information sources must be evaluated for reliability, that is, the properties of the information source to consistently produce reliable information. That is, there are two aspects related to the reliability of the information source: stability of information production and its reliability.

Assessing the reliability of the information contained in messages is a complex, poorly formalized task, the solution of which is beyond the scope of this work. However, there is a certain correlation between the disruption of news traffic stability and manipulative informational influences through the production of fake news. An example of a high level of news traffic stability at the sources can be large news agencies that regularly provide users with approximately the same amount of information over a long period.

Therefore, we will consider the level of news traffic stability as an indirect measure of the reliability of an information source, that is, the maintenance of conditions that ensure the continuity of the creation and transmission of data (news) by this source.

To identify the level of news traffic stability at an information source, it is necessary to research the distribution of the number of publications over a certain period of time by calculating the ratio of the normalized scope of each source.

The Hurst index (H) is related to the coefficient of normalized swing [9]:

$$\frac{R}{S} = \left(\frac{N}{2}\right), H \geq 1, \tag{1}$$

where N are the days during which the observations take place; S is the root mean square deviation of a series of observations on the information source; R is the scale of the amount of published information. H is the Hurst index.

$$S = \sqrt{\frac{1}{N}\sum_{i=1}^{n}(x_i - \overline{x})^2}, \tag{2}$$

where $\overline{x}$ is the average of an observations series x for N days; N is the number of observation days; x_i is the number of publications from the researched information source.

$$R = \max_{1 \leq i \leq N} W(i, N) - \min_{1 \leq i \leq N} W(i, N), \tag{3}$$

where $W(i, N)$ is an accumulated deviation of the x series from the mean value $\overline{x}$.

$$W(i, N) = \sum_{j=1}^{i}(x_j - \overline{x}), \tag{4}$$

where $\overline{x}$ is the arithmetic mean of a series of observations for a day; x_i is the number of publications from the researched information source.

Based on Eq. 1, we obtain the following expression for calculating the Hurst index:

$$H = \frac{ln(\frac{R}{S})}{ln(\frac{N}{2})}. \tag{5}$$

It is known that the Hurst index is a measure of persistence—the propensity of the process to trend (in contrast to the usual Brownian motion). A value of $H > 0.5$ means that the dynamic of the process directed in the past will most likely cause the continuation of movement in the same direction in the future. If $H < 0.5$, it is predicted that the process will change direction. $H = 0.5$ means uncertainty—Brownian motion.

To make a decision based on the uncertainty factor, the method of determining the root mean square deviation Eq. 2 is used—the degree of deviation of all values of the characteristic from its average.

The application of a modified method of multi-criteria Pareto optimization using two criteria: the value of the Hurst index and the root mean square deviation allows obtaining the distribution of information sources by groups with a similar level of news traffic stability [10].

2.2 Semantic Search Engine Model

The main idea of a search engine based on a semantic network is to represent search images of queries and documents by network structures that model the content of these queries or documents. As the basis for the construction of these network formations, the Directed Weighted Network of Terms (DWNT) [17] is considered. DWNT is a semantic model for the presentation of textual data, where the nodes are key terms (words and phrases) of the text of the document, which are used as the names of concepts of a specific subject domain, and edges are semantic and syntactic connections between these terms. Comparing DWNT as search images obtained for different texts makes it possible to determine the semantic closeness of the corresponding texts.

A procedure consisting of several steps is used to construct a DWNT:

1. text pre-processing;
2. extraction of key terms;
3. construction of DWNT.

Some of the most common techniques for the pre-processing of text data, including automatic segmentation into separate sentences and further tokenization of the text—segmentation of the input text into elementary units (tokens, lexemes) are used [11]. After tokenization, within each sentence parts of speech are marked. Part-of-Speech tagging or PoS tagging [12] is assigning a word in the text to a certain part of speech and the corresponding tag. In addition, the lemmatization of individually marked lexemes is carried out to obtain their canonical, dictionary forms—lemmas. This step allows you to further group different forms of the same word so that they can be analyzed as a single element.

To implement pre-processing of text data, software functions of Python programming language libraries are used. For example, StanfordNLP [13] is used for processing Ukrainian text documents, pymorphy2 [14]—for Russian text documents, NLTK (Natural Language Toolkit open-source library) [15]—for English tests.

It is assumed that key terms can be individual words that are nouns, as well as phrases formed as following patterns:

- unigram "$ADJ \sim NOUN$";
- threegrams "$NOUN \sim CCONJ \sim NOUN$", and "$ADJ \sim ADJ \sim NOUN$";
- fourgrams "$ADJ \sim NOUN \sim CCONJ \sim NOUN$", and "$ADJ \sim CCONJ \sim ADJ \sim NOUN$".

As you can see here, words that are nouns (NOUN tag), adjectives (ADJ tag) and conjunctions (CCONJ tag) were used. Also, for ease of further use, the proper noun (PROPN tag) has been renamed to the NOUN tag.

In the next step, a sequence of phrases is formed. In this sequence, the phrases with more words are placed before phrases and words that are part of them, considering the initial order of occurrence of words in the sentence.

For each formed term, a tuple of three elements is built:

- a term (a word or phrase formed according to the presented templates);
- a tag that is assigned to a word depending on its belonging to a certain part of the language, or a collective tag for the corresponding template;
- numerical value of GTF.

Global Term Frequency (GTF) is a global indicator of the importance of a term [16]:

$$GTF = \frac{n_i}{\sum_k n_k} \tag{6}$$

where n_i is a number of the appearance of term i in the text; the sum $\sum_k n_k$ is a general (global) number of formed terms in the whole text.

In contrast to the conventional statistical measure $TF - IDF$, GTF allows more efficient finding information-important text elements when working with a text corpus of a predefined subject domain, when an information-important term occurs in almost every document of the corpus.

The construction of DWNT is carried out in three stages:

1. a construction of an undirected network;
2. a determination of connection directions;
3. a determination of connection weights.

To build an undirected network of key terms, the horizontal visibility graph algorithm for time series (Horizontal Visibility Graph algorithm—HVG) is used [18, 19]. This approach also allows building of network structures based on texts in which individual words or word combinations are specially matched with numerical weight values.

Horizontal visibility graphs are built within each individual sentence, in which each term corresponds to a pre-calculated GTF score as a weight value. First, nodes are marked on the horizontal axis, each of which corresponds to a term and weighted numerical estimates—GTF are marked along the vertical axis. Next, a traditional graph of horizontal visibility is constructed. Next, a standard graph of horizontal visibility is constructed. It is considered, two nodes t_i and t_j corresponding to the elements of the time series x_i and x_j, are connected in a HVG if and only if,

$$x_k \leq min(x_i; x_j) \tag{7}$$

for all t_k ($t_i < t_k < t_j$), where $i < k < j$ are the nodes of graph.

As a result, an undirected network of key terms will be obtained, in which the directions of connections are established according to the principle of the entry of a shorter term into a term that is its extension [18]. The direction of all other remaining undirected connections is set from left to right.

The weight values of connections are determined as a result of combining nodes corresponding to the same terms (the number of equally directed connections between the corresponding nodes determines the weight value of the connection between these nodes) [18].

In order to remove loops that may arise at the stage of applying the HVG algorithm from the resulting graph, the diagonal elements of the adjacency matrix are zeroed.

2.3 *Method of Information Intent Mapping*

If the query language allows defining search based on the phrase, we can withdraw the use of operations on DWNT to implement semantic search, replacing them by building a query to a full-text search system.

In this case, after reduction with a certain threshold and indication of DWNT connections, it is necessary to map it in the pairs of terms form.

Some primary document that satisfies the user's informational need can be considered informational intention. If the user does not have such a document at his disposal, he can simulate such a document by writing down the desired information in natural language.

Based on the primary document, a DWNT is built and pairs of terms map. This semantic model is used to build a query. For example, for the information intent expressed by the document:

> Ukraine extends validity of 'green' COVID-19 certificate to 365 days. The Cabinet of Ministers has extended the validity of green COVID certificates from 180 to 365 days. The corresponding decree was published on the government website. According to the text of the document, different COVID certificates will be valid in Ukraine—with information on vaccination, a negative test result (valid for 72 h) or recovery from COVID-19. Ukrainian Journal 2021.09.16 19:54.

Figure 1 shows the constructed DWNT obtained after the determination and reduction of connections.

This network is used to automatically construct a query to the Manticore Search system [20]:

> "COVID_validity COVID_Ukraine COVID_green COVID_vaccination COVID _certificate COVID_green COVID_day Ukraine_validity Ukraine_COVID Ukraine_vaccination validity_COVID validity_Ukraine validity_day vaccination_COVID vaccination_Ukraine vaccination_certificate certificate_COVID certificate_vaccination certificate_day certificate_green day_COVID day_ certificate day_ validity green_ certificate green_COVID"/5.

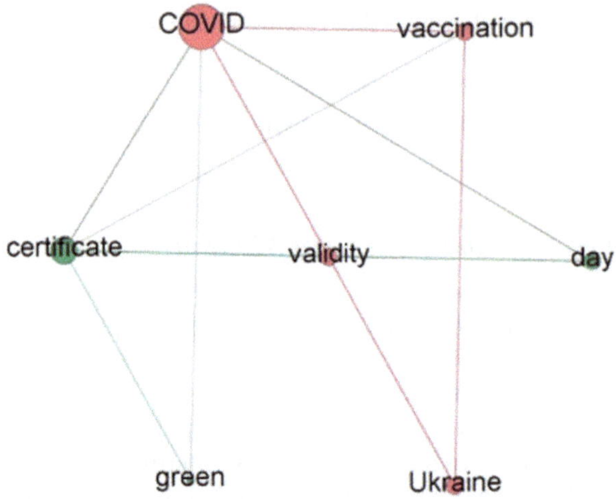

Fig. 1 DWNT of the primary document

1. 13:27 16.09.2021 Education Ministry explains work of educational institutions from Sept 20, 2021
Interfax-Ukraine 2021.09.16 14:17
The Ministry of Education and Science told how educational institutions will work from September 20, 2021. "In the event of the introduction of a yellow level of epidemic danger, education seekers are allowed to educational institutions provided that at least 80% of employees of the educational institution have a document confirming receipt of a full course of vaccination or an international, domestic or foreign certificate," the press service of the Ministry of

2. UKRAINE EXTENDS VALIDITY OF GREEN COVID CERTIFICATES FROM 180 TO 365 DAYS
Ukraine open for business 2021.09.15 14:54
15 September, 2021 The Cabinet of Ministers of Ukraine has extended the validity of green COVID certificates from 180 to 365 days. The corresponding decree was published on the government website

3. 15.09.21 12:26 Ukraine extending validity of COVID certificates to 365 days
Censor.NET 2021.09.15 13:09
The Cabinet of Ministers has issued a decree extending the validity of the international and domestic COVID certificate from 180 days to 365 days since the last jab. As reported by Censor.NET. The COVID certificate will have two options (yellow or green) and contain information about vaccination, a negative test result (valid for 72 hours) or recovery from COVID-19

4. Ukraine extending validity of COVID certificates to 365 days
Ukrinform 2021.09.15 12:30
15.09.2021 11:40 The Cabinet of Ministers has issued a decree extending the validity of the international and domestic COVID certificate from 180 days to 365 days since the last jab. According to Ukrinform, the relevant document was posted on the Government portal. The COVID certificate will have two options (yellow or green) and contain information about vaccination, a negative test result (valid for 72 hours)

5. 11:11 15.09.2021 Ukraine extends validity of 2-dose COVID certificates from 180 to 365 days
Interfax-Ukraine 2021.09.15 12:17
The Cabinet of Ministers of Ukraine has extended the validity of green COVID certificates from 180 to 365 days. The corresponding decree was published on the government website. According to the text of the document, different COVID certificates will be valid in Ukraine - with information on vaccination, a negative test result (valid for 72 hours) or recovery from COVID-19. Internal certificates with vaccination information will be "yellow" and

Fig. 2 The result of a search engine query

In this query, the quorum operator ($/N$) is used, which means that only those documents that contain at least $N = 5$ pairs of terms from the given list should be displayed. Thus, processing a query with a search engine allows getting a list of documents, the relevance of which is determined by the proximity of the corresponding semantic networks. It can be verified by looking at the result of this query, shown in Fig. 2.

The proposed method has two significant advantages over traditional search models. First, the search is not based on an artificially generated query, but on the basis of the text of the document provided by the user. The described approach is practically impossible to apply in other search models. Second, all resulting documents are relevant, that is, they satisfy the users' information needs. It is also unattainable in other models that provide a high level of relevance of the result to the query. These advantages are achieved due to a possible decrease in search precision (for example, if synonyms are not used while building DWNT networks), as well as due to an increase in the indexing time at which it is necessary to build DWNT networks for all documents in a database.

In the presented system, a number of types of microservices have been developed and the main workflows of data processing have been placed and configured. Since the amount of data that is accumulated and stored in the system can reach more than several terabytes of information (BigData), there is a need to use a data store that will allow you to get the result of a full-text search in an adequate time delay. We used Manticore Search [20]—an open-source database with high-performance full-text search.

3 Results

The proposed system for collecting, managing and analytical processing data from informational Internet sources is part of a single analytical and expert environment based on the concept of the Information and Analytical Situation Center (IASC) of the World Data Center "Geoinformatics and Sustainable Development" [21], which is used to solve tasks related to the preparation of strategic management decisions in the field of economy, public relations and national security in the conditions of multi-criteria problems to be solved, multifactorial risks, incompleteness and uncertainty of data, the presence of a large number of restrictions, etc. In particular, this analytical and expert environment was used to evaluate public protests against the implementation of quarantine restrictions due to the spread of the COVID-19 epidemic.

As part of this research, the stability of news traffic was analyzed for 1,355 information sources registered in the system, which are periodical daily publications, based on the analysis of their publications accumulated over three months. Figure 3 shows the resulting graph of the ratio of the Hurst index to the root mean square deviation, which is based on data published at the information sources.

Figure 3 shows that a significant number of the sources have Hurst index $H > 0.5$. This result indicates the stability of news traffic obtained from such sources and the fact that information from them can be used. All other sources with the Hurst index $H \leq 0.5$ change direction in the process of producing and distributing news and the behaviour of such sources are unpredicted.

Reduction of the set of information sources corresponding to a certain topic was also carried out due to the ranking by the similarity of the corresponding semantic networks.

At first, a simple general query containing only the basic concepts of the subject domain does not consider various aspects. It leads to obtaining a data array containing, in particular, elements of information noise. After processing this query by means of a traditional information-search system, an array of information formally relevant to the request was obtained. Next, the DWNT based on this array was

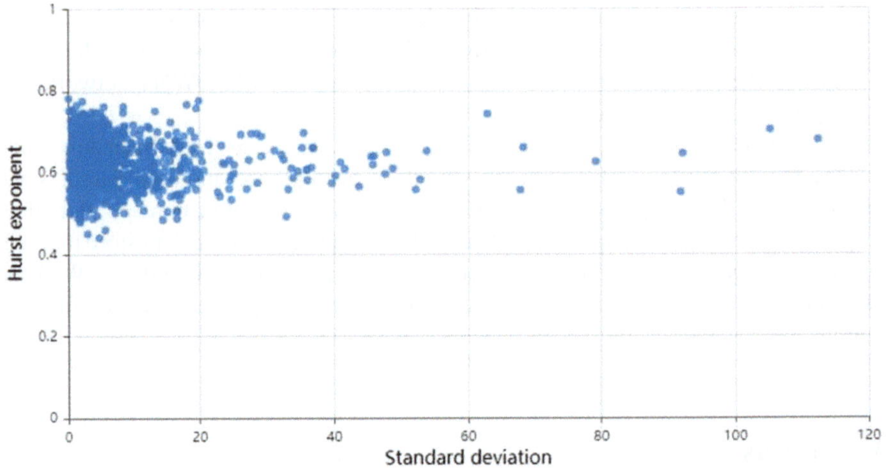

Fig. 3 Chart of the distribution of the Hurst index (y-axis) relative to the standard deviation of the investigated information sources (x-axis)

formed. It is assumed that this network contains the most significant terms or steady phrases (correspond to the nodes) and connections between them (correspond to the edges), which will reduce the influence of noisy documents in the array obtained as a result of processing the initial query. Within the framework of the model, the obtained network was considered a standard. The DWNT networks formed both by individual documents and by collections of documents related to individual information sources were compared with this standard network. The DWNT networks formed both by individual documents and by collections of documents related to individual information sources were compared with this against. In order to identify public protests against the implementation of quarantine restrictions in connection due to the spread of the COVID-19 epidemic, an initial query was formed:

"anti~lockdown protest covid"

During July–September 2021, 779 documents from more than 100 information sources were received under this query (Fig. 4).

Moreover, most information sources are represented by one document. For the 10 sources that contained the largest number of documents, namely from 20 to 76, the usefulness in relation to the subject domain was evaluated. For document arrays obtained from these sources, the corresponding semantic networks were built. Figure 5 shows a fragment of the semantic network corresponding to the "Independent.co.uk" source as an example.

Table 1 shows the values of discrepancy and usefulness, which were calculated in accordance with formulas Eqs. 3 and 4.

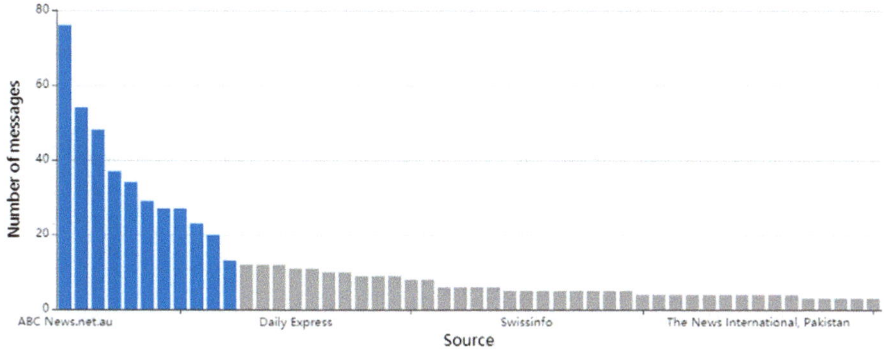

Fig. 4 Number of messages distribution by information sources

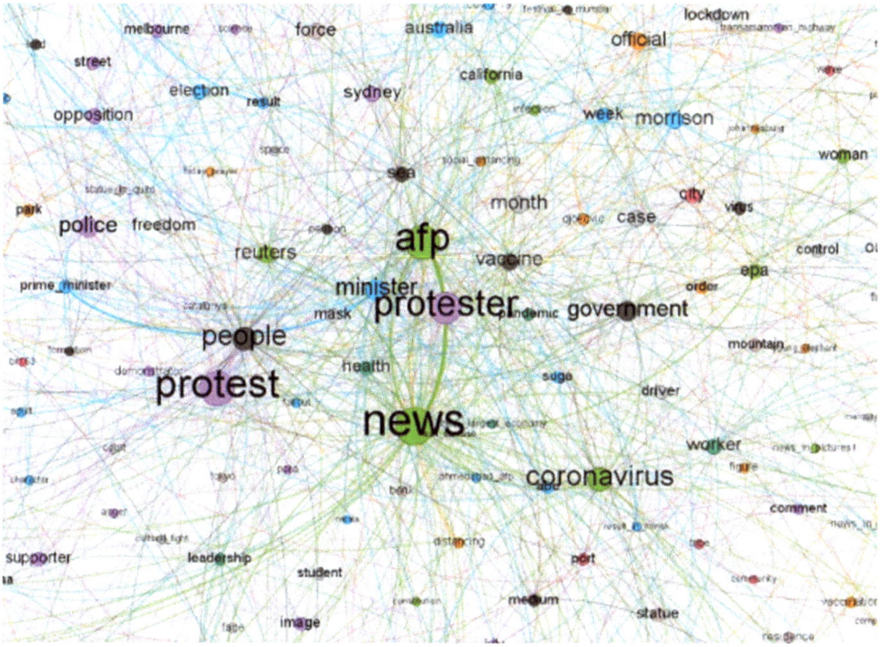

Fig. 5 Semantic web fragment corresponding to the "Independent.co.uk" source

The example shows the order of ranking according to the proposed semantic criterion of usefulness is slightly different from the order of ranking according to the number of found documents, but more corresponds to the selectivity of information sources in relation to the user's information needs. To identify events related to public protests against the implementation of quarantine restrictions during the last 6 months, the web application "PRO ET CONTRA" [5] was used. The application collected messages from two search engines: a traditional full-text search engine with

Table 1 Properties of arrays of documents corresponding to the selected information sources

No	Information source	Number of found documents	Discrepancy with the standard	Usefulness
1	ABC News.net.au http://www.abc.net.au/	76	0.165	0.835
2	Irish Times http://www.irishtimes.com/	54	0.11	0.89
3	News.com.au http://www.news.com.au/	48	0.143	0.857
4	Yahoo News https://news.yahoo.com/	37	0.147	0.853
5	The Epoch Times https://www.theepochtimes.com/	34	0.2	0.8
6	Nikkei Asian Review https://asia.nikkei.com/	29	0.114	0.886
7	Daily Mail https://www.dailymail.co.uk/	27	0.173	0.827
8	The Guardian https://amp.theguardian.com/	27	0.204	0.796
9	Deutsche Welle https://www.dw.com/en/	23	0.236	0.764
10	Independent.co.uk https://www.independent.co.uk/	20	0.123	0.877

query language support and an experimental system in which the proposed method of semantic search based on the use of DWNT is implemented.

As a result of the query `"anti~lockdown protest covid"`, 24153 messages were received to the first system. The corresponding news traffic is shown in Fig. 6a.

The document [22] reflects user's information needs was chosen for the experimental system. The DWNT (Fig. 7) build for the intent document was used to search for messages. As a result, 15698 messages were received. The relevant news traffic is shown in Fig. 6b.

Figure 6 shows news traffic data. Let's compare them with the data on protests published in the Global Protest Tracker [23]. According to the authors of the Global Protest Tracker, protests related with the outbreak of COVID-19 have a potential impact on governance and politics at the local, national and international levels.

The Global Protest Tracker uses expert analysis of reports from news sources such as Al Jazeera, the Atlantic, Balkan Insight, BBC, Bloomberg, CNN, DW News, the Economist, Euronews, Financial Times, Foreign Affairs, Foreign Policy, France24, the Guardian, the Nation, NBC News, New York Times, NPR, Reuters, Radio Free Europe/Radio Liberty, Vox, Wall Street Journal, Washington Post and World Politics Review.

In the last 6 months, the Global Protest Tracker recorded two powerful protests:

- In France in July 2021, there were protests (more than 150,000 participants) for 1 day. They were motivated by concerns about the government's adoption of excessive restrictions and the implementation of vaccination passports;
- Long-term protests (more than 4,000 participants) took place in Australia in July 2021 against the introduction of quarantine restrictions in many regions of the country.

Figure 6a shows the news traffic selected by the query does not have an increase in the specified period of intensification of protests (during July 2021). However, for

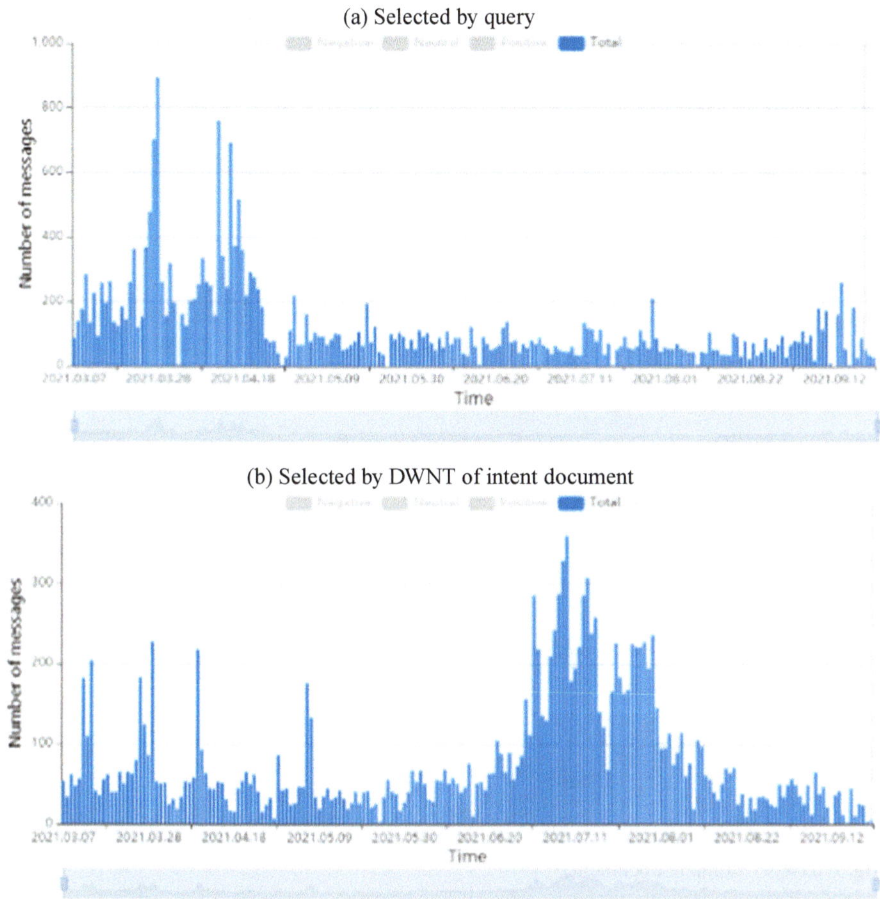

Fig. 6 Comparison of selected news traffic

the news traffic selected by the proposed method (Fig. 6b), there is an increase in traffic during this period.

An additional analysis of the text messages confirmed their relevance to the protests against the quarantine restrictions related to the implementation of vaccination passports topic. The messages related to the protest actions in Malaysia in August 2021, about public dissatisfaction with the ineffective government's actions, which led to an increase in cases of the coronavirus were also included in this document sample.

Let's evaluate the usefulness of the sample relative to the user's information needs using weighted entropy:

$$U = \frac{-p * log(p) - (1-p) * log(1-p)}{log(2)}. \tag{8}$$

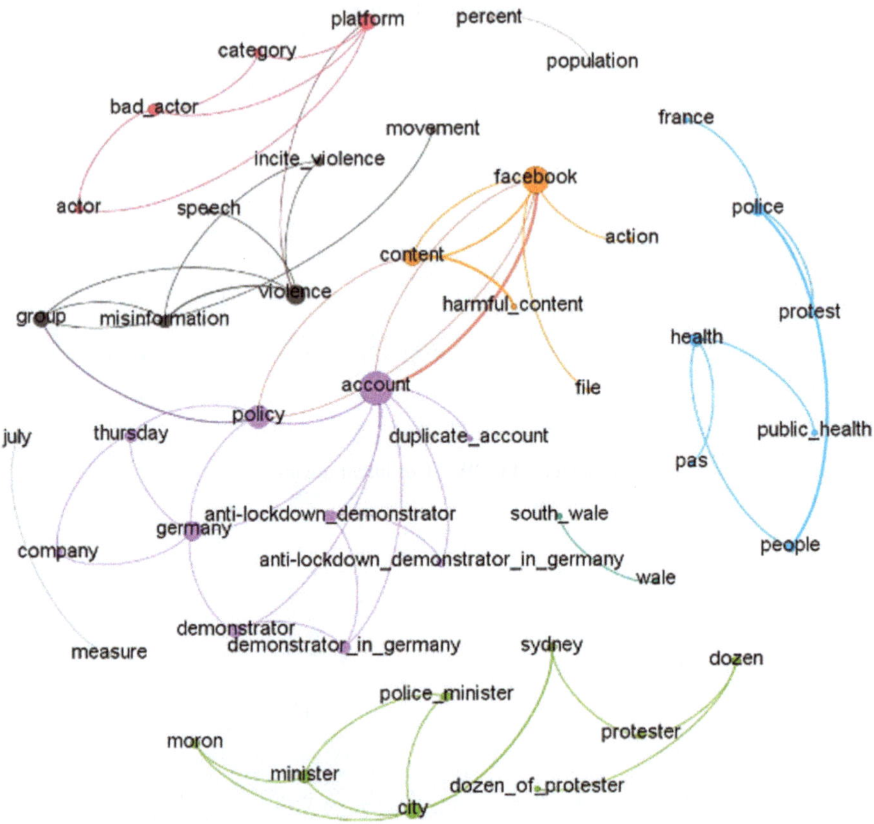

Fig. 7 DWNT of intent document

where U is the weighted entropy (usefulness) of messages in the range [0, 1] (the value 1 corresponds to the case all messages satisfy the information needs); p is part of messages that satisfy information needs.

If we consider that the messages recorded during July and August 2021 satisfy the user's information needs (3,861 such messages for the sample based on the query, and 8,300 for the sample based on the intent document), then we have for the sample based on the request $U = 0.634$ and for sampling based on intent document $U = 0.998$.

That is, the use of the proposed approach for information retrieval increases the usefulness of the data sample, estimated by the weighted entropy, by more than 57%. But we are cautious about such too approximate estimates!

4 Discussion

The use of information search based on the semantic closeness of messages feature allows automation of this process due to the automatic revealing of semantically similar messages provided to the expert for cross-analysis.

Therefore, the DWNT semantic model and the proposed mathematical apparatus for its manipulation create prerequisites for future scientific research on the formalization and development of methods for solving the problem related to the analysis of the reliability of the information that messages contain. Metadata of messages (source, date of publication, link) and results of analytical processing of the text messages (estimates of emotional tone, named entities in the text, DWNT) can serve as initial data for the research.

The proposed DWNT semantic model and based on its methods and algorithms are universal and can be applied to solve a variety of tasks, including identifying and comparing the semantics of both individual messages and sets of messages united by certain features (for example, a single information source or topic).

The proposed method provides a search not based on an artificially generated query, but on the text of the document provided by the user. This approach is practically impossible to apply in other methods. It allows unskilled users to express their information needs in natural language and receive search results corresponding to them. Moreover, the iterative DWNT construction algorithm will be applied in the future to implement step-by-step setting of the user's information needs using the inclusion/exclusion semantics of these needs in the form of partial DWNTs built on the basis of a number of documents that the user provides to the system as desired and/or unwanted samples in relation to the search topic. Such configured and stored information needs (search topics) can be reused to perform various applied tasks.

The implementation of the above possibilities is due to the properties of the proposed system architecture, which provides:

1. the ability to define pipelines composed of software components (microservices) registered in the system;
2. a dynamic control of properties' life cycle;
3. a control of system performance and fault tolerance;
4. an easy integration with other components and external systems, which is achieved through the use of standardized APIs.

The implementation of such architecture provides the infrastructure for research on the effectiveness of new data processing methods. Also, the microservice architecture provides an evolutionary improvement of the developed system to increase its usefulness by implementing new microservices and including them in existing or new data processing workflows.

5 Conclusions

The application of the proposed integrated approach allows for increasing the correspondence between the results of the intelligent processing of messages from websites and social networks. Also, the information tasks described in this paper relate to the analysis of the impact of the information and communication environment on society.

As part of the considered approach, a method of automatic recognition and evaluation of information sources based on the criterion of stability of news traffic is proposed. This approach allows for expanding an analyzed information environment, while simultaneously limiting fake information sources.

A semantic model of natural language text representation—DWNT—was proposed, and the corresponding mathematical apparatus was developed. Also, the information search methods and algorithms based on determining the semantic closeness of texts were also proposed. This allows using one or more intent documents the texts of which contain a description of the user's information needs in natural language to parameterize search operations, instead of queries traditionally used in full-text search systems.

Based on the results of using the developed system to solve an applied task related to the detection of public protests against the implementation of quarantine restrictions during the last 6 months and due to a more complete description of the semantics of the user's information needs compared to traditional query languages, the proposed method of information search was chosen. The proposed approach provides an increase in the usefulness of the system by almost 57%.

The application of this method of information search using DWNT allows determining the subject domain of messages based on the description of the semantics.

The information system for the revealing and analytical processing of data from Internet media resources and social networks, in which the proposed approach for increasing the usefulness of information search is implemented, has a microservice architecture. It creates prerequisites for experimental research of new methods of data revealing, storage and analytical processing, information retrieval, etc. This system is part of a single analytical and expert environment based on the concept of the Information and Analytical Situation Center (IASC) of the World Data Center "Geoinformatics and Sustainable Development".

Acknowledgements This research was partially supported by the National Research Foundation of Ukraine (2020.01/0283) and the Ministry of Education and Science of Ukraine (0121U109764). We thank our colleagues from the ISC WDS World Data Center for Geoinformatics and Sustainable Development, National Technical University of Ukraine "Igor Sikorsky Kyiv Polytechnic Institute", which provided insight and expertise that greatly assisted the research.

References

1. Army, U.S.: Open source intelligence In: Army Techniques Publication No. 2-22.9. US Government, Washington, DC (2012) Available via Google Scholar. www.fas.org/irp/doddir/army/atp2-22-9.pdf. Accessed 27 April 2023
2. Lande, D., Shnurko-Tabakova, E.: OSINT as a part of cyber defense system. Theor. Appl. Cybersecur. **1**, 103–108 (2019). https://doi.org/10.20535/tacs.2664-29132019.1.169091
3. Zgurovsky, M., Lande, D., Boldak, A., Yefremov, K., Perestyuk, M.: Linguistic analysis of internet media and social network data in the problems of social transformation assessment. Cybern. Syst. Anal. **57**, 228–237 (2021)
4. Zgurovsky, M., Boldak, A., Lande, D., Yefremov, K., Perestyuk, M.: Predictive online analysis of social transformations based on the assessment of dissimilarities between government actions and society's expectations. In: 2020 IEEE 2nd International Conference on System Analysis and Intelligent Computing (SAIC). IEEE (2020). https://doi.org/10.1109/SAIC51296.2020.9239186
5. PRO ET CONTRA v.2.0 Internet media analytics. http://wdc.org.ua/services/proEtContra/. Accessed 28 April 2023
6. Broder, A.: A taxonomy of web search. ACM SIGIR Forum. **36**, 3–10 (2002). https://doi.org/10.1145/792550.792552
7. Donato, D., Donmez, P., Noronha, S.: Toward a deeper understanding of user intent and query expressiveness. In: ACM SIGIR, Query Representation and Understanding Workshop (2011)
8. Jansen, B., Booth, D., Spink, A.: Determining the informational, navigational and transactional intent of Web queries. Inf. Proc. Manag. **44**, 1251–1266 (2008)
9. Feder, J.: Fractals. Plenum Press, New York (1988)
10. Soboliev, A.M.: Detection of information sources that spread unreliable information in the global Internet network. Regist. Storage Data Process. **21**, 56–68 (2019). https://doi.org/10.35681/1560-9189.2019.21.3.183717
11. Webster, J.J., Kit, C.: Tokenization as the initial phase in NLP. In: Proceedings of the 14th Conference on Computational Linguistics, COLING 1992 April, pp. 1106–1110 (1992)
12. Marcus, M., Santorini, B., Marcinkiewicz, M.A.: Building a large annotated corpus of english: the penn treebank. Comput. Linguist. (Special Issue on Using Large Corpora) **II**(19), 313–330 (1993)
13. The Stanford Natural Language Processing Group, Available via Google Scholar. https://nlp.stanford.edu/. Accessed 27 April 2023
14. Pymorphy2 morphological analyzer. https://pymorphy2.readthedocs.io/en/stable. Accessed 27 April 2023
15. NLTK 3.6.3 documentation. https://www.nltk.org. Accessed 27 April 2023
16. Lande, D., Dmytrenko, O.: Using part-of-speech tagging for building networks of terms in legal sphere. In: Proceedings of the 5th International Conference on Computational Linguistics & Intelligent Systems (COLINS 2021). Volume I: Main Conference Kharkiv, Ukraine, April 22–23, 2021. CEUR Workshop Proceedings (ceur-ws.org), vol. 2870, pp. 87–97 (2021)
17. Lande, D., Dmytrenko, O.: Creating directed weighted network of terms based on analysis of text corpora. In: 2020 IEEE 2nd International Conference on System Analysis & Intelligent Computing (SAIC), pp. 1–4. IEEE (2020). https://doi.org/10.1109/SAIC51296.2020.9239182
18. Luque, B., Lacasa, L., Ballesteros, F., Luque, J.: Horizontal visibility graphs: Exact results for random time series. Phys. Rev. E. **80** (2009)
19. Gutin, G., Mansour, T., Severini, S.: A characterization of horizontal visibility graphs and combinatoris on words. Phys. A **390**, 2421–2428 (2011)
20. Manticore Search. Available via Manticore Search. https://manticoresearch.com. Accessed 27 April 2023
21. World Data Center for Geoinformatics and Sustainable Development. http://wdc.org.ua/. Accessed 27 April 2023

22. Covid restrictions over Delta variant trigger protests in Europe, Australia. https://www.hindustantimes.com/world-news/covid-restrictions-over-delta-variant-trigger-protests-in-europe-australia-101627152365258.html. Accessed 27 April 2023
23. Global Protest Tracker. https://carnegieendowment.org/publications/interactive/protest-tracke. Accessed 27 April 2023

Scenario Modelling in the Context of Foresight Studies

Serhiy Nayev, Iryna Dzhygyrey, Kostiantyn Yefremov, Ivan Pyshnograiev, Andriy Boldak, and Sergii Gapon

Abstract Foresight projects can have different goals and applications, and take various forms and scales. Within the European Foresight Platform framework, monitoring covers thousands of studies. The prevailing number of foresight use scenario approach, and the higher the horizon the higher the share of scenarios in FL studies. Types of scenarios and development procedures are considered in the context of three schools of scenario development, Shell approach, PMT school and La Prospective. Foresight studies of the World Data Center for Geoinformatics and Sustainable Development are presented as cases of FL research's results implementation in policy-and decision-making. It's noted the efforts of the COVID-19 Foresight project team of the Center in the area of coronavirus pandemic scenario development. Also, artificial intelligence application prospects for forward-looking studies are highlighted in the context of The Kyiv School of mathematicians and system analytics activity.

S. Nayev · I. Dzhygyrey · K. Yefremov (✉) · I. Pyshnograiev · A. Boldak · S. Gapon
National Technical University of Ukraine "Igor Sikorsky Kyiv Polytechnic Institute", Kyiv, Ukraine
e-mail: k.yefremov@wdc.org.ua

S. Nayev
e-mail: mail@wdc.org.ua

I. Dzhygyrey
e-mail: dzhygyrey@wdc.org.ua

I. Pyshnograiev
e-mail: pyshnograiev@wdc.org.ua

A. Boldak
e-mail: boldak@wdc.org.ua

S. Gapon
e-mail: gapon@wdc.org.ua

M. Zgurovsky and N. Pankratova (eds.), *System Analysis and Artificial Intelligence*, Studies in Computational Intelligence 1107,
https://doi.org/10.1007/978-3-031-37450-0_23

1 Introduction

The Information and Analytical Situation Center of the World Data Center for Geoinformatics and Sustainable Development operates in Igor Sikorsky Kyiv Polytechnic Institute to study the behaviour of complex socio-economic systems. One of the main components of its software and methodological complex is scenario modelling technologies in the mode of the situation centre, which makes it possible to build the future states of the mentioned systems, choose the preferred one and determine the necessary actions to achieve it.

Currently, the scenario approach is a powerful tool used in various spheres of analytical, expert and scientific activities for various scale tasks, from planning the activities of enterprises [1, 2] to predicting the future of humanity [3]. Prospective, so-called forward-looking (FL) research can significantly strengthen the concept of sustainable development, raise to a new level a holistic, systemic perception, and not interpret only from the point of view of the three-pillar model "Society-Economy-Environment" [4]. Conversely, the sustainable development concept is the driving force for changing FL research itself that ensures the inclusiveness, participation and representativeness of such projects in particular [5].

This review aims to show the place of scenario modelling in the implementation of modern FL research, foresight in particular.

2 Foresight

Foresight study enables process participants to get clearer ideas about the research object, its future changes, and the consequences of the decisions made [6]. The measures taken will affect the object, will create new challenges that will require a reassessment of the situation, that is, an iterative foresight process. Thus, repetitive foresight studies of a complex object with high uncertainty are not interpreted as "predictions" but rather as a cyclical process of gradually acquiring new knowledge and adapting to improve judgments about the best actions to take regarding future risks and opportunities. Foresight research is conducted under TUNA or TUUNA conditions (abbreviation from turbulence, unpredictable uncertainty, novelty, and ambiguity) [7].

As part of the international initiative "Foresight4Food" [8], an integrated scheme of the foresight process is proposed. The scheme covers the identification of participants ("actors"); establishing the purpose of the foresight; determination of boundaries and relationships of the analyzed system; determination of key driving forces (drivers), trends, and uncertainties, which becomes the basis for scenario modelling; understanding interested parties' visions for the future and, thus, determining the preferred strategies of influence and adaptation to change. The process scheme is based on the so-called adaptive or exploratory approach to foresight, so it reflects an iterative process designed to change the future, requiring flexibility and constant

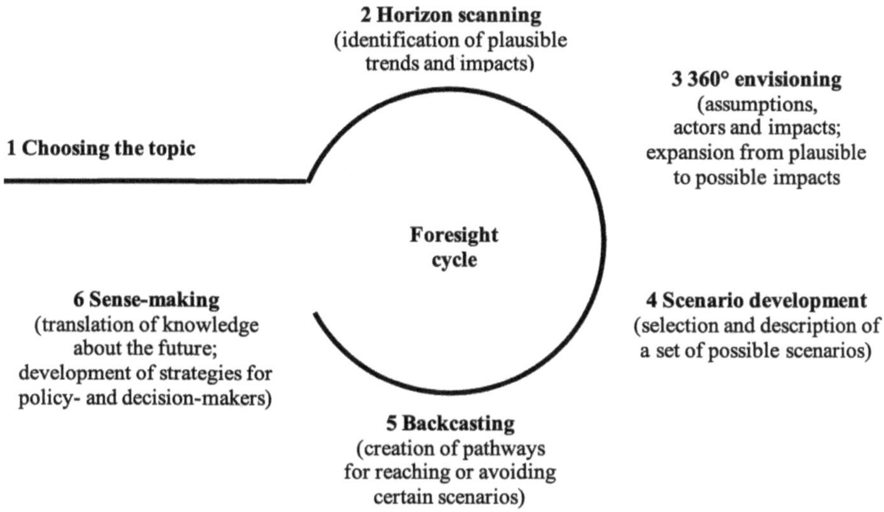

Fig. 1 Foresight cycle (based on [12])

refinement. There is another approach to foresight studies—a strategic foresight based on planning that allows one to determine measures for the preferred future implementation [9, 10]. A strategic foresight project can include such stages as "Framing" (result is project plan), "Scanning" (result is information), "Forecasting" (result is a baseline and its alternatives, scenarios), "Visioning" (result is preferred future, goals), "Planning" (result is a strategic plan, strategies), "Acting" (result is an action plan, initiatives) [11].

The Scientific Foresight (Science and Technology Options Assessment or STOA) Unit of the European Parliament proposed six phases of scientific foresight (Fig. 1), one of the goals of which is to provide European parliamentarians with reviews of the possible impact of technical and scientific trends regularly in the form of a 'What-if?' series of publications [12]. The development of scenarios in the foresight cycle is of great importance. This stage aims at the development of several exploratory scenarios, which provide not a prediction of the future, but rather an exploration of a range of possible ways of development of events.

Foresight projects can take different forms and scales in terms of stakeholder involvement, from very "closed" scientific studies to widely involved processes. It is also worth noting that there is significant interpenetration and layering, a deep integration of approaches in the field of foresight research, in particular, such concepts as strategic foresight, future studies, scenario analysis. Thus, the approaches of scenario thinking cover many aspects of foresight in its general sense on the other hand, many foresight approaches contain elements of scenario analysis [6].

The results of the review [13] show that FL research can be divided conventionally into several large groups, to which the author of the bibliographic study [13] gave the names "Corporate foresight" (1760 publications), "Past and futures" (1934 publications), “Humanity at the limen” (955 publications), “Environmental futures” (593 publications), “Post-normality and complexity” (70 publications) and “Technological trends” (69 publications), with the largest share of research after 2004 belonging to the cluster of corporate foresight.

3 European Foresight Platform

Developments in the field of foresight are so diverse, multi-scale and multi-sectoral that, with the financial support of the EU Sixth Framework Programme, the European Foresight Monitoring Network (EFMN) initiative was created in 2004 and the ForLearn project [14] aimed at consolidating and improving the availability of knowledge in the field of foresight in Europe. Over time and with the support of the EU Seventh Framework Programme these projects acquired a new quality and scale. In 2009, the EFMN was reorganized into the European Foresight Platform (EFP) [15]. The EFP consortium was formed by four partners—the Austrian Institute of Technology (AIT), the Institute for Prospective Technological Studies (IPTS), the Netherlands Organisation for Applied Scientific Research (TNO) and the Manchester Institute of Innovation Research (MIoIR) of the University of Manchester.

The objects of EFP monitoring are projects, tools, methods and organizations, and scientific and expert groups are involved in FL activities (FLA), which include foresight studies, forecasting, horizon scanning, and impact assessment [16]. Here, *foresight* is understood as a systemic, participatory, prospective and policy-oriented process, which is aimed at actively involving key stakeholders in a wide range of activities for predicting the future, preparing recommendations and implementing transformations in technological, economic, environmental, political, social and ethical (TEEPSE) spheres. *Horizon scanning* is a structured, ongoing activity for monitoring, analyzing, and positioning frontier issues important for strategy development, policymaking, and research planning. Horizon scanning covers the identification of new and emerging trends, policies, practices, stakeholders, services and products, technologies, behaviours, surprises (“jokers”, that is, hard-to-predict factors that can significantly change the state of the object under study) and weak signals [17], etc. *Forecasting* is considered as an activity based on subjective and statistical sources of information aimed at forming ideas about the future. *Impact assessment* is the identification and analysis of short-and long-term TEEPSE consequences of political initiatives, programs, legislation implementation or the application of new technology [16].

If within the framework of the EFMN, the monitoring of FLA projects covered hundreds of studies [18], among which Ukraine was also represented by the study “Forecast of scientific, technological and innovative development of Ukraine” [19], now within the framework of the EFP, monitoring covers thousands of studies, which makes it possible to effectively use monitoring results in practice, improve the quality of FL research projects and stimulate new foresight studies [16]. Due to the systematization of knowledge about completed projects, in particular the methods

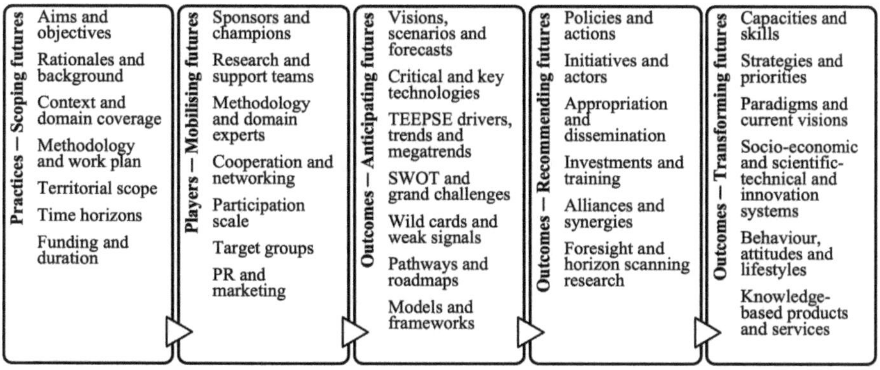

Fig. 2 The EFP mapping concept (based on [20])

and approaches used, the European foresight platform itself will eventually become a super-tool for future FL projects.

EFP monitoring is based on three upper-level elements, namely, FLA practices, players and outcomes, and 33 lower-level elements (Fig. 2), which are divided by phases of FL processes into scoping, mobilising, anticipating, recommending and transforming (SMART) futures [16, 20]. The characteristics of 263 FL programmes, projects, methods, and other case studies are presented on the EFP web resource [21]. The European foresight platform evaluates the context of FL research from the point of view of belonging to the EU Framework Programme, the thematic direction of the Framework Programme and the level of project implementation (global, national, corporate, etc.).

4 The Place of Scenario Modeling in Foresight Studies

The European Foresight Platform used R. Popper's foresight diamond to assess the contribution of 44 methods to each project. In the analysis [20], it is indicated that most FL studies use at least one method of each "vertice" of the diamond, which have the names "creativity", "expertise", "interaction", "evidence" (Fig. 3). Foresight researchers use a wide variety of foresight diamond methods, from the analysis of information flows to expert predictions, separated into four groups mentioned as the vertices of the foresight or futures diamond, namely exploratory, advisory, participatory and explainatory methods. Popper emphasizes that each of the approaches has its advantages and limitations: (1) the use of creativity-based tools requires original thinking and an imagination; (2) expertise-based methods are based on knowledge and competence in particular subject areas (expert panels, Delphi, etc.); (3) interaction-based methods are used to obtain additional information from non-experts (scenario seminars, voting, polling, stakeholder analysis, etc.); (4) evidence-based methods are tools for understanding and forecasting phenomena, assessing the actual

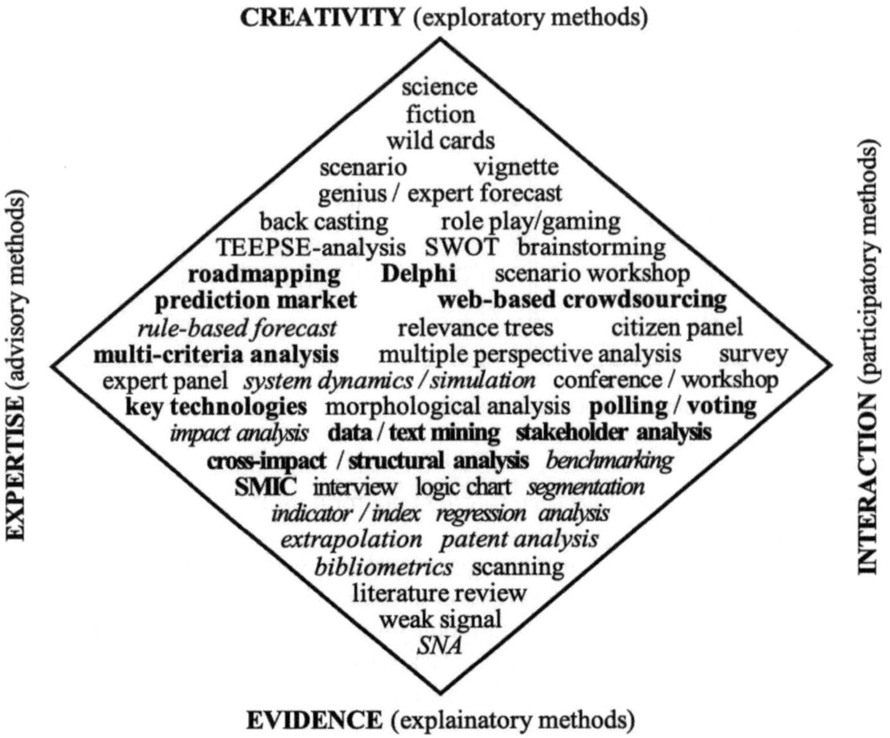

Fig. 3 Foresight diamond [20]: qualitative, quantitative and semi-quantitative methods; SMIC is cross-impact matrices and systems methodology, SNA is social network analysis

state of problems, effects of technologies (benchmarking, bibliometrics, data mining and indicators work, etc.).

The filling of the foresight diamond expands and supplements over time. Currently, the set of used methods includes [22, 23]

- qualitative methods: back-casting; brainstorming; citizens' panels or focus groups; expert panels; futures wheel [24]; idea networking; genius forecasting; horizon scanning; mind mapping; morphological analysis; relevance tree; scenario art; scenario development; serious gaming; SWOT analysis; wild cards and weak signals,
- semi-quantitative methods: causal layered analysis [25]; cross-impact analysis; Delphi survey; driving force-pressure-state-impact-response frameworks; futures triangle [26]; roadmapping; STEEP analysis [27]; stakeholder analysis; structural analysis; utility maximization and choice modeling,
- quantitative methods: bibliometrics; trend extrapolation; trend impact analysis; system dynamics modelling; global value chain analysis.

It is clear the presented list of tools used in current foresight studies, including foresight diamond methods and approaches, is far from complete. For example, in the UNDP "Foresight" guide [27], such methods as agent-based modelling, wind-tunnelling, and others are mentioned. With the expansion of the goals and scope of foresight, the number, quality and variability of the methods involved in such research also increases.

As part of the "The Millennium Project" [28], which was founded in 1996 by the American Council of the United Nations University and which has now turned into an independent, non-profit scientific centre for FL research at the world level with 63 centres, encyclopaedic dictionary [29] and electronic edition 'Futures Research Methodology Version 3.0' have been created [30], which is one of the largest, most comprehensive peer-reviewed collections of methods and tools for FL studies. Particular sections of this publication are devoted to the development of scenarios and interactive scenarios. The Millennium Project annually publishes the State of the Future [31] survey, which covers an overview of 15 global challenges, and launched the Global Futures Intelligence System (GFIS) web project, which integrates all available information, groups and software and enables updating both the study "The State of the Future" and the publication 'Futures Research Methodology', not periodically, but continuously [32].

A significant part of the publication 'Recent Developments in Foresight Methodologies' [33] and components of other monographs, collections, and individual publications are also devoted to scenario modelling use in foresight studies. The European Environment Agency understands FL studies as some platform for supporting long-term decision-making based on cooperation and six components [34, 35]: drivers and trends, indicators, scenarios, methods and tools, networking, capacity building and governance and use of information about the future, one of which, as we can see, is the development of scenarios.

Thus, multivariate event scenarios design and roadmapping with practical measures are key features of a significant amount of foresight studies. The results of the analysis [36] based on 1296 FL studies show that the frequency of scenarios use increases with the increase of the time horizon, and for studies with a horizon of more than 50 years it is 100% (Table 1). Scenarios are among the top five widely used foresight methods in the global context, among which (in order of decreasing frequency of use, without taking into account the literature review) expert panels, scenarios, trend extrapolation, futures workshops, brainstorming [36, 37], and there are one of the three tools that are often used in Northern Europe, and is the most used method in Eastern Europe [37]. Scenarios were used in one way or another in almost half of the thousand evaluated foresight projects [37]. The use of scenarios, along with expert panels and literature reviews, can be attributed to "geographically" independent research methods, as opposed to, for example, roadmapping and futures workshops, which are more often used in countries with significant expenditures on scientific and scientific and technical activities [36].

However, the above list of the most used methods should not limit future foresight researchers. As shown in paper [38], methods choice for a project is often unsystematic, impulsive, inexperienced etc. It bases on the nature of the methods themselves and their interconnectedness and interdependence in most cases, the so-called "mixture of methods". The nature of the methods dictates that quantitative methods are

Table 1 Widely used FLA methods (based on [36])

Method	Frequency of use, %		
		Time horizon up to 10 years	Time horizon 51–100 years
Literature review	54	50	56
Expert panels	50	49	47
Scenarios	42	33	100
Trend extrapolation	25	24	53
Futures workshops	24	22	
Brainstorming	19	12	28
Interviews	17	14	12
Delphi	16	12	1

preferred over qualitative methods. And the "mixture of methods" means that some methods "go side by side" in practice, as, for example, brainstorming results are input data for the Delphi method [38]. The depth of the forecast significantly influences the choice of particular methods. Scenarios are often combined with trend extrapolation, modelling, relevance trees, and various types of analysis (cross-impact, structural, or morphological) [39]. According to the results of the analysis of more than two thousand foresights at the regional, national and subnational level [36], in 25% of cases, scenarios are combined with futures workshops.

A significant part of the mentioned methods is difficult to implement without special software tools use. In general, ICT use both to obtain data for further analysis and to process this data is necessary. Foresight studies are now supported by a variety of software applications, including databases, analytical applications, scenario packages, and more. Here it is worth mentioning the package of strategic foresight tools of the "La Prospective" school [39] by Michel Godet, which contains applications for structural analysis, scenarios building based on morphological analysis, etc. Components of the package were turned into web applications in 2013 and are currently maintained by the Futuribles foresight centre [40].

ICT tools for foresight studies are diversifying, becoming more qualitative and may soon move into a new integrated quality—foresight support systems (FSSs), as they not only allow for the involvement of a greater number of diverse experts in projects but also more efficiently collect and analyze and interpret large arrays data [41]. The idea of implementing FSSs in the form of web platforms deserves special attention. Platforms such as Google Trends, Twitter Sentiment Analysis or Palantir applications are already actively used. Their integration in the future with web-oriented platforms of FL-projects, more complex and bigger than the current "Futurescaper", "SenseMaker", "Wikistrat", or others, can form large-scale trend monitoring systems that will work in real-time [42].

5 Classification of Scenarios and Scenario Development

The scenario component commonly used in foresight studies enables the development of multiple scenarios based on quantitative approaches to represent the consequences of various likely changes of the driving forces and components of the analysed object. Scenarios are used both for planning the activities of enterprises and for the purpose of formulating strategies for social, ecological, and economic systems development [43]. Scenarios are not an attempt to predict the future, but they provide decision-makers with a deeper understanding of the potential consequences of their decisions and explore new opportunities for responding to future changes [44]. Or, as one of the working documents of the IGLO-MP 2020 project [45] eloquently noted, the fundamental aspect of scenarios is that they "handle" uncertainty.

Among the scenario approaches, the following types of scenarios can be distinguished [11, 44, 46, 47]: baseline, reference or predictive scenarios, which try to investigate what is expected to happen within the given context without external changes (the "Forecasts" and "What-if" subtypes are distinguished); explorative scenarios, which try to find out what might happen, in other words, investigate how the future state of the object or system will be affected by internal or external drivers and build scenarios from the present to the future state (external and strategic research scenarios are distinguished); normative scenarios illustrate how a target might be reached and may include a backcasting to evaluate conditions, options and ways to achieve particular goals (divided into preserving and transforming subtypes), form the specified six subtypes. Since the scenarios differ from the point of view of the quantitative and qualitative application, it is possible to highlight qualitative scenarios, which mainly represent the development of the future in the form of phrases, storylines, and images; quantitative scenarios, mainly presented in the form of tables, graphs and maps, which are the result of simulation modelling [44, 47]. Qualitative scenarios are usually built based on stakeholder and expert knowledge, while quantitative scenarios make greater use of mathematical models and other tools under certain assumptions about driver changes and interactions [47]. However, in many foresight projects, both approaches are combined, which gives the advantage of flexibility next to the application of proven scientific knowledge and the validity of assumptions. In contrast to prediction and projection, which require a better understanding of the studied system, scenarios make it possible to delve into the issues of uncertainty and complexity (Fig. 4).

Philip van Notten proposed a classification of scenario characteristics based on the goals of scenario studies, design of the scenario process and content of scenarios [48]. These three macro scenario characteristics involved ten micro characteristics (Table 2). There are other classifications of scenario types in the field of foresight. In particular, a group of scientists from the Finland Futures Research Centre classified foresight research frames based on existing typologies of scenarios and as a result identified predictive, planning, scenaric, visionary, critical and transformative foresight frames [49].

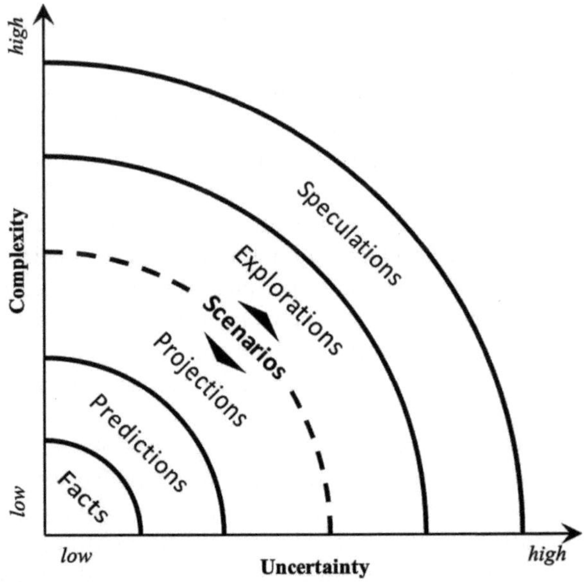

Fig. 4 The role of scenario modelling in projections and predictions [47]

Table 2 Van Notten's typology of scenarios [48]

Macro characteristics	Micro characteristics
The goal of scenario studies Exploration—Pre-policy research	*The function of the scenario exercise* Process—Product *The role of values in the scenario process* Descriptive—Normative *The subject area covered* Issue-based—Area based—Institutional based *The nature of change addressed* Evolutionary—Discontinuity
Design of the scenario process Intuitive—Analytical	*Inputs into the scenario process* Qualitative—Quantitative *Methods employed in the scenario process* Participatory—Model-based *Groups involved in the scenario process* Inclusive—Exclusive
Content of the scenarios Complex—Simple	*The role of time in the scenario* Chain—Snapshot *Issues covered by the scenario* Heterogeneous—Homogeneous *Level of integration* Integration—Fragmented

Scenario planning covers the entire foresight study, while scenario development is only an element of scenario planning [11]. Scenarios are parallel "stories" or models of expected images of the future which include certain assumed events and are built on the basis of the synthesis of data obtained using various foresight methodologies [50]. Different types of scenarios are developed using different types of scenario procedures. Each of these procedures has its advantages and disadvantages and requires different support tools, including special software applications. Peter Bishop and Andy Hines in their study [11] attempted to systematize the procedures of scenario development but it should be noticed that this field of knowledge has expanded considerably nowadays. These authors identified and evaluated eight categories of such procedures:

1. 'Judgment' is a category that covers genius forecasting, visualization, role playing, and the Coates and Jarratt domain procedure.
2. 'Baseline' (expected) is a category that contains trend extrapolation, the Manoa technique, systems scenarios, and trend impact analysis.
3. 'Elaboration of fixed scenarios' is a category whose procedures make it possible to obtain several scenarios, as opposed to the first two categories. This category includes incasting and SRI-matrix.
4. 'Event sequences' is a category that includes probability trees and modification, the sociovision procedure, as well as Harman divergence mapping.
5. 'Backcasting'. This category includes John Anderson's horizon mission methodology, IBM Corporation's Impact of Future Technologies, and David Mason's future mapping.
6. 'Dimensions of uncertainty' is a category that contains Schwartz's GBN matrix, morphological analysis and FAR (field anomaly relaxation), Option Development and Option Evaluation (OS/OE) applications by the Parmenides Foundation (Germany), and MORPHOL computer program by Michel Godet.
7. 'Cross-impact analysis' was implemented in the FL sphere by Michel Godet (SMIC and PROB-EXPERT applications) and the Battelle Memorial Institute (IFS application or Interactive future simulation using Monte Carlo simulation).
8. 'Modelling' includes trend impact analysis (TIA method by Ted Gordon), sensitivity analysis and dynamic scenarios, which are a combination of scenario development and system analysis using causal models.

Separately, it is possible to note the integrated approach to scenario planning of Inayatullah, the developer of causal layered analysis (CLA), which covers 'the used future'; 'the disowned future'; 'alternative futures'; 'alignment'; 'models of social change'; and 'uses of the future' concepts of futures thinking. [26]. He also suggested ways to combine several FL procedures, namely macro-history, scenarios, futures wheels, integral futures, and emerging issues analysis, to improve the efficiency of conducting and the quality of scenario planning results. It is worth mentioning Schwartz's eight-step scenario building model and Schoemaker's ten-step model, Dator's scenario approach of four archetypes, and its modifications, for example, Bezold's aspirational futures, Slaughter, Hayward and Voros' integral framework for scenario development, List's flexible scenario network mapping (SNM) approach

[51]. Since there are so many procedures, models, and approaches to develop different types of scenarios, some FL researchers call this field "methodological chaos" [51].

In general, it is customary to distinguish three schools of approaches for the development of scenarios, which are associated with two geographical centres, the UK/USA and France [45]. Thus, the first two are attributed to 'the American' centre of scenario planning, and the last one to 'the French' (further in detail, based on research [51]):

1. 'Qualitative' school of intuitive logic or "Shell approach" [52]. A method based on the fact that the decisions made are complexly dependent on the interrelationships of economic, political, social, resource and environmental factors. The resulting scenarios are sequences of events built with regard to causal processes and decision points. There are many models for building scenarios of intuitive logic, but the SRI methodology is most often used.
2. 'Quantitative' school of probabilistic modified trends (PMT-school), which uses trend impact analysis and cross impact analysis. These matrix methods include probabilistic modifications of extrapolated trends. The most popular procedures and applications in this area are, already mentioned, IFS of the Battelle Memorial Institute and SMIC by "La Prospective", as well as INTERAX by Enzer from the Center for Future Research of the University of Southern California.
3. The French school "La Prospective", which is an integration of two previous schools and based on four concepts, 'base' (analysis and scanning of the current state of the object), 'external context' (study of the environment of the system, that is, external social, economic, environmental, political, national, and international factors of influence), 'progression' (simulation based on the dynamic 'base' and constraints of the 'external context') and 'images' of the future [53]. Michel Godet developed a mathematical and computer-based probabilistic approach for generating scenarios, the best-known implementations of which are now widely used as web applications, MORPHOL and SMIC PROB-EXPERT.

The most popular quantitative methods used in scenario development include INTERAX, SMIC, IFS, TIA and fuzzy cognitive maps [51].

It should be noted that the further into the future the foresight study tries to "look", the less useful quantitative procedures are and the more preference is given to qualitative FL procedures. However, any FL project or, more narrowly, the foresight process is built on a flexible combination of widely used, specialized and unique techniques [54], such as network analysis and the network format of data presentation in a particular system foresight [55] or the construction and analysis of families of scenarios based on many previous FL-projects of other researchers using a unique qualitative iterative method [56], or the use of global databases [57], or the creation of participatory perspective analysis for the co-development of scenarios [58].

6 Foresight Studies in the Countries of the World and in Ukraine

Activities in the field of foresight and, in most cases, the use of the scenario approach in foresight studies differ in actions, coverage and level in countries of the world. For example, Great Britain (Foresight Programme), Singapore (Risk Assessment and Horizon Scanning system), the Netherlands (Horizon Scan Project), Finland (Finnsight forum), and other countries and the European Union have achieved significant development in the field of national strategic foresight [59–64].

As for Ukraine, in 2012, experts of the WikiCityNomica, the organizing committee of the Human Capital Forum, and the Kyiv Business School conducted a foresight study of "Human Capital of Ukraine 2025" with the involvement of a wide range of experts and entrepreneurs [65]. The researchers attempted to determine the main trends in the transformation of Ukraine's human capital and developed the four most likely development scenarios ("Totalitarian corporation", "Mafia", "GM society" and "Kaleidoscope"), as well as strategic project initiatives that can become a factor of change [66].

In 2014, at the World Economic Forum in Davos, with the support of financial and industrial companies of Ukraine (Group DF, SCM and Smart-Holding), the study "Scenarios of the economic development of Ukraine" for the period until 2030 was presented [67]. As part of this strategic foresight, three alternative scenarios ("Starting a virtuous cycle", "Back to the future", and "Lost in stagnation") were developed based on such factors as the institutional environment and the external economic situation. The ways of their development and the consequences of implementation were determined. The study covered three stages—synthesis of opinions of interested persons (about three hundred experts), scenarios development and discussion of the consequences of scenarios and actions necessary for the implementation of a certain scenario.

The World Data Center for Geoinformatics and Sustainable Development based on the Igor Sikorsky Kyiv Polytechnic Institute was created in 2006, which continued and expanded the FL research [68], in particular, based on the modelling of sustainable development processes, and performed the following foresight studies:

- Foresight of Ukrainian Economy: mid-term (2015–2020) and long-term (2020–2030) time horizons [69];
- Foresight and construction of the strategies of socio-economic development of Ukraine on mid-term (up to 2020) and long-term (up to 2030) time horizons [70];
- Foresight 2018: systemic world conflicts and global forecast for XXI century [71], and others.

A key component of research [68–71] was scenario modelling, along with the use of SWOT analysis and the Delphi method. For example, within the framework of "Foresight-2016" [70], eight scenarios of the socio-economic development of Ukraine until 2030 (Fig. 5) were specified using the methodology of scenario modelling and SWOT analysis, and an expert study of the socio-economic segment of

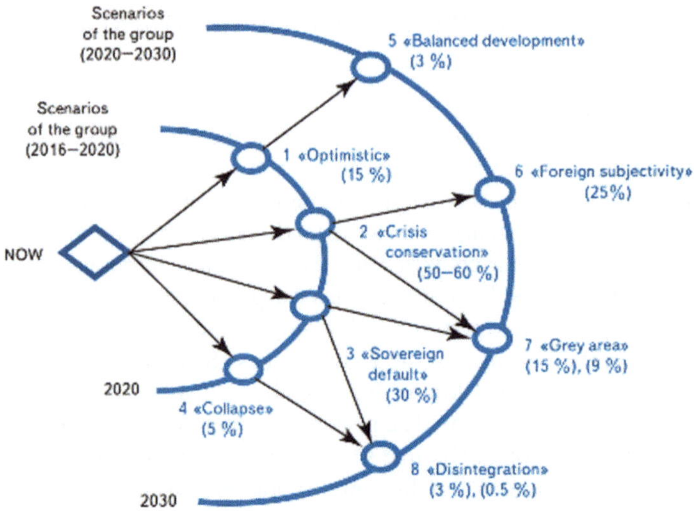

Fig. 5 Scenarios of social and economic development of Ukraine on the mid-term (up to 2020) and long-term (up to 2030) time horizons

society's development was carried out, the presence of human capital capable of carrying out preferred transformations, and fifty main actions of the government were formed in the form of a strategy for socio-economic development in the medium and long term. The advantage of these studies is comprehensiveness from the point of view of the three-pillar model of sustainable development, systematicity and repeatability.

However, it is not so much the scenario modelling or foresight research based on it that is crucial but the implementation of the research results and their influence on the further development of the evaluated system. Thus, in 2012, Dirk Meissner [72] attempted to analyze the impact of national foresight studies in OECD countries and the European Research Area on the development of national innovation systems and concluded that the contribution of foresight to the development and change of innovation systems is significant.

The COVID-19 pandemic has demonstrated the need for FLA tools usage for better preparedness, coordination and response to future infectious threats. Thus, in 2022, the World Health Organization launched its 1st foresight report, "Imagining the Future of Pandemics and Epidemics" [73]. Since the beginning of the pandemic, a lot of FL studies different in scale and techniques were conducted in this field. The coronavirus pandemic development scenarios developed by the scientific team of the COVID-19 Foresight project of the World Data Center for Geoinformatics and Sustainable Development should be mentioned [74]. These scenarios helped answer the questions of how the Ukrainians are going to live under the pandemic conditions during the months to come, how the world and Ukraine will change after the pandemic is over, and when can it happen. Since the time that the COVID-19 pandemic began, the COVID-19 Foresight project presented short-term COVID-19 forecasts and about twenty foresight studies, “Impact on economy and society” [75], “The middle phase of development” [76], “Transition to the phase of pandemic attenuation” [77], “Fourth stage of the quarantine measures weakening” [78], “Exacerbation during the adaptive quarantine” [79], “The rise of the pandemic at the school year beginning” [80], “The World transformation after the COVID-19 pandemic, European context” [81], “Analysis of the vaccination impact on the pandemic attenuation in Ukraine and the World” [82], “COVID-19 waves analysis caused by Delta and Omicron variants of SARS-CoV-2” [83], and others. The COVID-19 Foresight project team focuses on predicting the spread of the COVID-19 pandemic using neural networks: back-propagation neural network, long-term memory neural network, multilayer perceptron-type neural network. Figure 6 shows an example of using a forward perceptron neural network to predict the spread of the Delta virus SARS-CoV-2 in August-December 2021.

7 The Prospects for the Use of Artificial Intelligence for Forward Looking Research

As can be seen from the previous case of the COVID-19 pandemic (e.g. Fig. 6), the use of artificial intelligence for foresight and scenario development looks promising [84, 85], from machine learning to natural language processing applications for policy-and decision-makers.

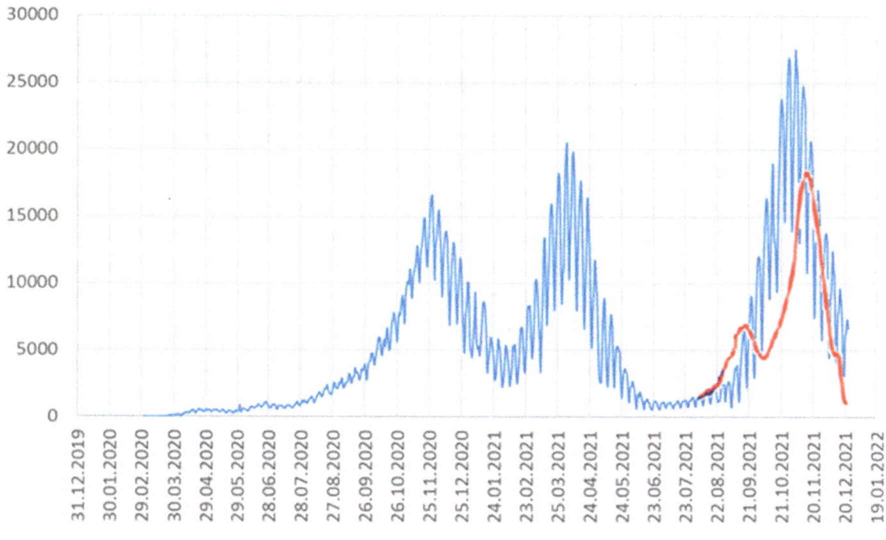

Fig. 6 Comparison of real number of infected (blue color) with predicted by a forward perceptron neural network distribution (red color) of the Delta variant of the SARS-CoV-2 virus in August-December 2021

The Kyiv School of mathematicians and system analytics considers artificial intelligence (AI) as a branch of computer linguistics and informatics. AI deals with the formalization of problems and tasks similar to human actions. At the same time, this powerful tool should not replace a human in thinking and creativity but only serve a human. This Ukrainian scientific school carries out a cycle of fundamental theoretical and applied research in the field of the theory of artificial and computational intelligence, neural networks, and their practical applications. The main results of these studies are presented in a series of monographs [86–89] and analytical reports of the World Data Center for Geoinformatics and Sustainable Development, located at Kyiv Polytechnic Institute. The Kyiv scientific school concentrates its efforts on the artificial intelligence area, which is commonly called computational intelligence (CI). This is a set of methods and software tools that are designed to solve problems using formal apparatus and logic of human mental activity, namely qualitative and intuitive approaches, creativity, fuzzy logical conclusion, self-learning, classification, pattern recognition, etc.

The structure of CI consists of two components: technologies (neural networks; fuzzy logic systems; fuzzy neural networks, evolutionary modelling) and methods and algorithms (learning methods; self-learning methods; methods of self-organization; genetic algorithms; swarm and ant algorithms). The applied tasks of CI include forecasting; classification and recognition of images; cluster analysis; data mining. Considerable attention of The Kyiv Scientific School of mathematicians and system analytics is paid to wide practical application and improvement of the classic back propagation neural network, recurrent neural networks with feedback (by Hopfield and Hamming), and neural networks with Kohonen self-organization; to fuzzy logic systems and fuzzy neural networks.

The use of hybrid neural networks [88] for solving practical problems is very wide, from the task of classification in medicine [90] to intelligent system of distribution of road traffic flows. In context of FLA the task of forecasting and decision-making is an important, e.g. in the conditions of attacks on the civilian population and fires (evacuation plan). One of the cases is the problem of optimal exit of people from complex buildings (big offices, multi-story shopping and entertainment centres).

One of the practical problems that scientists of the School solved in the field of modern economics was the problem of forecasting the risk of bankruptcy of corporations in conditions of uncertainty. Fuzzy neural networks were used to solve this problem. The results of predicting bankruptcy risk using different methods for 26 companies that later went bankrupt show that the classic methods of Altman and Nedosekin gave a prediction accuracy of 61% and 81%, respectively, and fuzzy neural networks—90%.

8 Conclusion

In this paper, an attempt was made to show the place of scenario modelling among the methods of FL studies. Some classifications of scenario types are presented, from simple quantitative-qualitative to six subtypes and van Notten's typology. The unification problem of approaches to scenario planning and the development of different types of scenarios is highlighted based on Bishop and Hines's categorisation of scenario development procedures.

Attention is paid to the experience of using scenarios in national foresight studies with a focus on Ukraine's background, the World Data Center for Geoinformatics and Sustainable Development in particular. Scenario modelling, SWOT analysis and the Delphi method in the context of sustainability concept are key tools for foresight studies of the Center in areas of the socio-economic development of Ukraine, world conflicts and global development processes. The contribution of scenarios to the development of the government's actions is shown. The COVID-19 Foresight project team of the World Data Center for Geoinformatics and Sustainable Development also has wide experience in foresight studies with the use of neural networks.

Prospects for the use of artificial intelligence in the field of FLA are considered. The latest achievements of the Kyiv School of mathematicians and system analytics in the field of computational intelligence are presented and their application for solving practical problems, including cases of forecasting in the conditions of attacks on the civilian population and forecasting the risk of bankruptcy of corporations in conditions of uncertainty.

References

1. Rohrbeck, R., Battistella, C., Huizingh, E.: Corporate foresight: an emerging field with a rich tradition. Technol. Forecast. Soc. Chang. **101**(12), 1–9 (2015). https://doi.org/10.1016/j.techfore.2015.11.002
2. Shell International B.V.: Scenarios: An Explorer's Guide (2008). https://www.shell.com/energy-and-innovation/the-energy-future/scenarios/what-are-the-previous-shell-scenarios/new-lenses-on-the-future.html
3. IPCC.: Climate Change (2014): Synthesis Report. Contribution of Working Groups I, II and III to the Fifth Assessment Report of the Intergovernmental Panel on Climate Change [Core Writing Team, Pachauri, R.K., Meyer, L.A. (eds.)], pp. 151. IPCC, Geneva, Switzerland (2014). ISBN 978-92-9169-143-2
4. Destatte, P.: Foresight: a major tool in tackling sustainable development. Technol. Forecast. Soc. Chang. **77**, 1575–1587 (2010). https://doi.org/10.1016/j.techfore.2010.07.005
5. Kuribayashi, M., Hayashi, K., Akaike, S.: A proposal of a new foresight platform considering of sustainable development goals. Eur. J. Futur. Res. **6**, 4 (2018). https://doi.org/10.1007/s40309-017-0130-8
6. Woodhill, J., Hasnain, S.: Ä framework for understanding foresight and scenario analysis, working draft. Foresight4Food Initiative by OSF (2019). https://www.foresight4food.net/wp-content/uploads/2019/05/Foresight-Approach_May_2019.pdf

7. Ramirez, R., Wilkinson, A.: Strategic Reframing: The Oxford Scenario Planning Approach, p. 280. Oxford University Press (2016). https://doi.org/10.1093/acprof:oso/9780198745693.001.0001
8. Foresight4Food (2019). https://www.foresight4food.net/
9. Gibson, E., Daim, T., Garces, E., Dabic, M.: Technology foresight: a bibliometric analysis to identify leading and emerging methods. Foresight STI Gov. **12**(1), 6–24 (2018). https://doi.org/10.17323/2500-2597.2018.1.6.24
10. Miller, R.: Chapter 5. Futures studies, scenarios, and the "possibility-space" approach. In: Think Scenarios, Rethink Education. OECD (2006). https://www.oecd.org/site/schoolingfortomorrowknowledgebase/futuresthinking/scenarios/37246742.pdf. https://doi.org/10.1787/9789264023642-en
11. Bishop, P., Hines, A., Collins, T.: The current state of scenario development: an overview of techniques. Foresight J. Futur. Stud. **9** (2007). https://doi.org/10.1108/14636680710727516
12. van Woensel, L., Vrščaj, D.: Towards scientific foresight in the European parliament: in-depth analysis. Eur. Parliam. Res. Serv. (2015). https://doi.org/10.2861/367909
13. Fergnani, A.: Mapping futures studies scholarship from 1968 to present: a bibliometric review of thematic clusters, research trends, and research gaps. Futur. **105**, 104–123 (2019). https://doi.org/10.1016/j.futures.2018.09.007
14. Da Costa, O., Warnke, P., Cagnin, C., Scapolo, F.: The impact of foresight on policy-making: insights from the FORLEARN mutual learning process. Technol. Anal. & Strat. Manag. **20**, 369–387 (2008). https://doi.org/10.1080/09537320802000146
15. European Foresight Platform. http://www.foresight-platform.eu/
16. Popper, R., Amanatidou, E., Jones, B., Teichler, T.: Forward looking activities (FLA) mapping towards a fully-fledged FLA mapping system. EFP Consortium (2012). https://doi.org/10.13140/RG.2.2.13017.67687
17. van Rij, V.: New emerging issues and wild cards as future shakers and shapers. Foresight STI Gov. **6**(1), 67–89 (2012). https://doi.org/10.1007/978-1-4614-5215-7
18. Giesecke, S., Crehan, P., Elkins, S. (eds.): The European Foresight Monitoring Network, Collection of EFMN Briefs, Part 1 (2008). https://doi.org/10.2777/17436
19. Heiets, V.M., Aleksandrova, V.P., Skrypnychenko, M.I., Fedulova, L.I., Naumovets, A.H., et al. (2007). Zvedenyi prohnoz naukovo-tekhnolohichnoho ta innovatsiinoho rozvytku Ukrainy na naiblyzhchi 5 rokiv ta nastupne desiatylittia. K.: Feniks, 152. ISBN 978-966-651-472-4
20. Popper, R.: 1st EFP mapping report: practical guide to mapping forward-looking activities (FLA) practices, players and outcomes towards a fully-fledged futures mapping. EFP consortium (2011). http://www.foresight-platform.eu/wp-content/uploads/2011/01/EFP-mapping-report-1.pdf
21. The European Foresight Monitoring Network Briefs. http://foresight-platform.eu/briefs-resources/
22. Martins, M.K., Bodo, B.: Deliverable D5.5: Raw Materials Foresight Guide. Mineral Intelligence Capacity Analysis (MICA Project) (2017). http://www.mica-project.eu/wp-content/uploads/2016/03/D5.5_Raw-Materials-Foresight-Guide.pdf
23. Falck, W.E., Tiess, G., Murguia, D., Machacek, E., Domenech, T., Hamadová, B.: Deliverable D5.1: Raw Materials Intelligence Tools and Methods. Mineral Intelligence Capacity Analysis (MICA Project) (2017). http://mica.eurogeosurveys.org/wp-content/uploads/2017/02/D5.1_Raw-Material-Intelligence-Methods-and-Tools.pdf
24. Bengston, D.N.: The futures wheel: a method for exploring the implications of social-ecological change. Soc. & Nat. Resour. **29**, 374–379 (2015). https://doi.org/10.1080/08941920.2015.1054980
25. Inayatullah, S.: Causal layered analysis: theory, historical context, and case studies. In: Inayatullah, S. (ed.) The Causal Layered Analysis (CLA) Reader: Theory and Case Studies of an Integrative and Transformative Methodology, pp. 8–49. Tamkang University Press, Taipei, Taiwan (2004)
26. Inayatullah, S.: Six pillars: futures thinking for transforming. Foresight. **10**(1), 4–21 (2008). https://doi.org/10.1108/14636680810855991

27. UNDP.: Foresight: The Manual (2014). https://www.undp.org/content/dam/undp/library/capacity-development/English/Singapore%20Centre/GCPSE_ForesightManual_online.pdf
28. The Millennium Project. http://www.millennium-project.org/
29. Olavarrieta, C., Glenn, J.C., Gordon, T.J. (eds.): Futures. The Millennium Project. http://107.22.164.43/millennium/FUTURES.html
30. Glenn, J.C., Gordon, T.J. (eds.): Futures Research Methodology Version 3.0. The Millennium Project. http://107.22.164.43/millennium/FRM-V3.html
31. Glenn, J.C., Florescu, E.: State of the Future. V. 19.0. The Millennium Project Team. The Millennium Project. http://107.22.164.43/millennium/2017SOF.html
32. Global Futures Intelligence System (GFIS). http://107.22.164.43/millennium/GFIS.html
33. Giaoutzi, M., Sapio, B.: Recent Developments in Foresight Methodologies, p. 310. Springer, US (2013). https://doi.org/10.1007/978-1-4614-5215-7
34. EEA.: Knowledge base for forward-looking information and services (FLIS) (2011). https://doi.org/10.2800/63246
35. Ahamer, G.: Forward looking needs systematised megatrends in suitable granularity. Campus Wide Inf. Syst. **31**, 81–199 (2014). https://doi.org/10.1108/CWIS-09-2013-0044
36. Reshetnyak, O.: Choice of foresight methods to substantiate directions of scientific development. Mod. Econ. (18), 166–173 (2019). https://doi.org/10.31521/modecon.V18(2019)-25
37. UN CSTD.: Report of the Secretary-General on Strategic foresight for the post-2015 development agenda (2015). https://doi.org/10.13140/RG.2.1.3880.8488
38. Popper, R.: "How are foresight methods selected?. Foresight. **10**(6), 62–89 (2008). https://doi.org/10.1108/14636680810918586
39. La Prospective. http://en.laprospective.fr
40. Futuribles International takes the torch from the CAP to manage and disseminate free online software developed by Michel Godet. La Prospective (2018). http://en.laprospective.fr/dyn/anglais/news/free-online-software-now-handled-by-futuribles-international.pdf
41. Keller, J., von der Gracht, H.: The influence of information and communication technology (ICT) on future foresight processes—results from a Delphi survey. Technol. Forecast. Soc. Chang. **85**, 81–92 (2013). https://doi.org/10.1016/j.techfore.2013.07.010
42. Raford, N.: Online foresight platforms: evidence for their impact on scenario planning & strategic foresight. Technol. Forecast. Soc. Chang. **97**, (2014). https://doi.org/10.1016/j.techfore.2014.03.008
43. Kyzym, M.O., Heiman, O.A.: Stsenarne modeliuvannia rozvyku sotsialno-ekonomichnykh system: napriamky, osoblyvosti ta mekhanizmy. Rehionalna ekonomika **4**, 16–23 (2009)
44. Ash, N., Blanco, H., Brown, C., Garcia, K., Tomich, T., Vira, B.: Ecosystems and Human Well-Being: A Manual for Assessment Practitioners, p. 288. Island Press (2010). ISBN-10: 1597267104
45. Rialland, A., Wold, K.: Future Studies, Foresight and Scenarios as basis for better strategic decisions, NTNU, IGLO-MP2020 project, WP1.5. MARINTEK (2009). http://www.forschungsnetzwerk.at/downloadpub/IGLO_WP2009-10_Scenarios.pdf
46. Börjeson, L., Höjer, M., Dreborg, K.-H., Ekvall, T., Finnveden, G.: Scenario types and techniques: towards a user's guide. Futur. **38**, 723–739 (2006). https://doi.org/10.1016/j.futures.2005.12.002
47. Woodhill, J., Zurek, M., Laanouni, F., Soubry, B.: Foresight4Food working paper (2017). https://www.foresight4food.net/wp-content/uploads/2018/03/Foresight-Background-Paper-Final.pdf
48. van Notten, P.: Chapter 4. Scenario development: a typology of approaches. In: Think Scenarios, Rethink Education. OECD (2006). ISBN: 926402364X
49. Minkkinen, M., Auffermann, B., Ahokas, I.: Six foresight frames: classifying policy foresight processes in foresight systems according to perceived unpredictability and pursued change. Technol. Forecast. Soc. Chang. **149**, 119753 (2019). https://doi.org/10.1016/j.techfore.2019.119753
50. Peter, M.K., Jarratt, D.G.: The practice of foresight in long-term planning. Technol. Forecast. Soc. Chang. **97**, 49–61 (2015). https://doi.org/10.1016/j.techfore.2013.12.004

51. Amer, M., Daim, T., Jetter, A.: A review of scenario planning. Futur. **46**, 23–40 (2013). https://doi.org/10.1016/j.futures.2012.10.003
52. Wayland, R.: Three senses of paradigm in scenario methodology: a preliminary framework and systematic approach for using intuitive logics scenarios to change mental models and improve strategic decision-making in situations of discontinuity. Technol. Forecast. Soc. Chang. **146**, 504–516 (2019). https://doi.org/10.1016/j.techfore.2017.09.005
53. Godet, M.: Creating Futures Scenario Planning as a Strategic Management Tool, 2nd edn, p. 349. Economica, Paris (2006). ISBN-10: 2717852441
54. Böhme, K., Holstein, F., Wergles, N., Ulied, A., Blosca, O., Nogera, L., Guevara, M., Kruljac, D., Spiekermann, K., Kluge, L., Sessa, C., Enei, R., Faberi, S.: Possible European Territorial Futures. Final Report. Volume C. Guidelines to Territorial Foresight, Version 21/02/2018, ESPON, p. 98 (2017). https://www.espon.eu/sites/default/files/attachments/vol
55. Nugroho, Y., Saritas, O.: Incorporating network perspectives in foresight: a methodological proposal. Foresight. **11**(6), 21–41 (2009). https://doi.org/10.1108/14636680911004948
56. Lacroix, D., Laurent, L., Menthière, N., Schmitt, B., Béthinger, A., David, B., Didier, C., Châtelet, J.: Multiple visions of the future and major environmental scenarios. Technol. Forecast. Soc. Chang. **144**, 93–102 (2019). https://doi.org/10.1016/j.techfore.2019.03.017
57. Ahamer, G.: Applying global databases to foresight for energy and land use: the GCDB method. Foresight STI Gov. **12**(4), 46–61 (2018). https://doi.org/10.17323/2500-2597.2018.4.46.61_
58. Bourgeois, R., Penunia, E., Bisht, S., Boruk, D.: Foresight for all: co-elaborative scenario building and empowerment. Technol. Forecast. Soc. Chang. **124**, 178–188 (2017). https://doi.org/10.1016/j.techfore.2017.04.018
59. Habegger, B.: Strategic foresight in public policy: reviewing the experiences of the UK, Singapore, and the Netherlands. Futur. **42**, 49–58 (2010). https://doi.org/10.1016/j.futures.2009.08.002
60. Könnölä, T., Brummer, V., Salo, A.: Diversity in foresight: insights from the fostering of innovation ideas. Technol. Forecast. Soc. Chang. **74**(5), 608–626 (2007). https://doi.org/10.1016/j.techfore.2006.11.003
61. Ejdys, J., Gudanowska, A., Halicka, K., Kononiuk, A., Magruk, A., Nazarko, J., Nazarko, Ł., Szpilko, D., Widelska, U.: Foresight in higher education institutions: evidence from Poland. Foresight STI Gov. **13**(1), 77–89 (2019). https://doi.org/10.17323/2500-2597.2019.1.77.89
62. Pouris, A., Raphasha, P.: Priorities setting with foresight in South Africa. Foresight STI Gov. **9**(3), 66–79 (2015). https://doi.org/10.17323/1995-459x.2015.3.66.79
63. Berze, O.: Mapping Foresight Practices Worldwide, Discussion Paper. http://projects.mcrit.com/esponfutures/documents/International%20Studies/Ottilia%20Berze_Mapping%20Foresight%20Practices%20Worldwide.pdf
64. Finnsight (2019). https://www.businessfinland.fi/ajankohtaista/tapahtumat/2019/finnsight-2019/
65. Shevchenko, L.S.: Tekhnolohichnyi forsait u konteksti innovatsiinoho rozvytku Ukrainy. Current issues of intellectual property and innovative development : Materials of the II Int. science and practice conf. (Kharkiv, 21 March 2014), pp. 149–153 (2014)
66. Pekar, V., Pjestjernikov, J.: Ljuds'kyj kapital Ukrai'ny 2025. Pidsumky Forsajtu. [Human capital of Ukraine 2025. Results of Forsyth] (2012). http://wikicitynomica.org/future/lyudskiy-kapital-ukraini-2025-pidsumki-forsaytu.html
67. Eide, E.B., Rösler, P.: Scenarios for Ukraine. Reforming institutions, strengthening the economy after the crisis. World Scenario Series. WEF, Geneva (2014). http://www3.weforum.org/docs/WEF_ScenariosSeries_Ukraine_Report_2014.pdf
68. The Sustainable Development Global Simulation: Quality of Life and Security of World Population (2005–2007/2008). The project research supervisors M.Z. Zgurovsky, A.D. Gvishiani. - K.: Publishing House "Polytekhnika", p. 336 (2008)
69. Foresight of Ukrainian Economy: mid-term (2015–2020) and long-term (2020–2030) time horizons. Scientific advisor of the project acad. of NAS of Ukraine M. Zgurovsky. International Council for Science (ICSU) ; Committee for the System Analysis of the Presidium of NAS of Ukraine; National Technical University of Ukraine "Kyiv Polytechnic Institute" ; Institute

for Applied System Analysis of NAS of Ukraine and MES of Ukraine ; World Data Center for Geoinformatics and Sustainable Development. 2nd edn. Kyiv : NTUU "KPI", Publishing House "Polytechnica", p.136 (2016). ISBN 978-966-622-750-1
70. Foresight and construction of the strategies of socio-economic development of Ukraine on mid-term (up to 2020) and long-term (up to 2030) time horizons. Scientific advisor of the project acad. of NAS of Ukraine M. Zgurovsky. International Council for Science (ICSU); Committee for the System Analysis of the Presidium of NAS of Ukraine; National Technical University of Ukraine "Igor Sikorsky Kyiv Polytechnic Institute"; Institute for Applied System Analysis of MES of Ukraine and NAS of Ukraine; World Data Center for Geoinformatics and Sustainable Development; Agrarian Superstate Foundation. 2nd edn. Kyiv : NTUU "Igor Sikorsky KPI", Publishing House "Polytechnica", p. 184 (2016). ISBN 978-966-622-783-9
71. Foresight 2018: systemic world conflicts and global forecast for XXI century. International Council for Science etc.; Scientific Supervisor M. Zgurovsky. K. : NTUU "Igor Sikorsky Kyiv Polytechnic Institute", p. 226 (2018). ISBN 978-966-622-878-2
72. Meissner, D.: Results and impact of national foresight-studies. Futur. **44**, 905–913 (2012). https://doi.org/10.1016/j.futures.2012.07.010
73. Imagining the future of pandemics and epidemics: a 2022 perspective. WHO, Epidemic and pandemic foresight initiative (2022). https://pandemic-foresight.who.int/publications/i/item/9789240052093
74. The COVID-19 Foresight project. The World Data Center for Geoinformatics and Sustainable Development (2020). http://wdc.org.ua/en/covid19-project
75. Foresight COVID-19: impact on economy and society. WDC-Ukraine (2020). http://wdc.org.ua/en/node/190017
76. Foresight COVID-19: the middle phase of development. WDC-Ukraine (2020). http://wdc.org.ua/en/covid19-ua
77. Foresight COVID-19: transition to the phase of pandemic attenuation. WDC-Ukraine (2020). http://wdc.org.ua/en/covid19-attenuation
78. Foresight COVID-19: fourth stage of the quarantine measures weakening. WDC-Ukraine (2020). http://wdc.org.ua/en/covid19-fourth-stage-quarantine-weakening
79. Foresight COVID-19: exacerbation during the adaptive quarantine. WDC-Ukraine (2020). http://wdc.org.ua/en/covid19-exacerbation-during-adaptive-quarantine
80. Foresight COVID-19: the rise of the pandemic at the school year beginning. WDC-Ukraine (2020). http://wdc.org.ua/en/covid19-school-year-beginning
81. Foresight COVID-19: the World transformation after the COVID-19 pandemic, European context. WDC-Ukraine (2020). http://wdc.org.ua/en/covid19-transformation-after-pandemic-europe
82. Foresight COVID-19: analysis of the vaccination impact on the pandemic attenuation in Ukraine and the World. WDC-Ukraine (2020). http://wdc.org.ua/en/covid19-vaccination-impact-on-attenuation
83. Foresight COVID-19: COVID-19 waves analysis caused by Delta and Omicron variants of SARS-CoV-2. WDC-Ukraine (2020). http://wdc.org.ua/en/covid19-delta-omicron-spread
84. Spaniol, M.J., Rowland, N.J.: AI-assisted scenario generation for strategic planning. Futur. & Foresight Sci. e148, (2023). https://doi.org/10.1002/ffo2.148
85. Gruetzemacher, R., Whittlestone, J.: The transformative potential of artificial intelligence. Futur. **135**, (2021). https://doi.org/10.1016/j.futures.2021.102884
86. Zgurovsky, M.Z., Zaychenko, Y.P.: The Fundamentals of Computational Intelligence: System Approach. Springer International Publishing, Cham (2017). https://doi.org/10.1007/978-3-319-35162-9
87. Zgurovsky, M.Z., Zaychenko, Y.P.: Big Data: Conceptual Analysis and Applications. Springer (2019). https://doi.org/10.1007/978-3-030-14298-8
88. Zgurovsky, M.Z., Sineglazov, V.M., Chumachenko, E.I.: Artificial Intelligence Systems Based on Hybrid Neural Networks : Theory and Applications. Springer International Publishing, Imprint Springer, Cham (2021). https://doi.org/10.1007/978-3-030-48453-8

89. Zaychenko, Y., Zgurovsky, M.: Banks bankruptcy risk forecasting with application of computational intelligence methods. In: IEEE 2019 14th International Scientific and Technical Conference on Computer Sciences and Information Technologies, CSIT 2019—Proceedings (2019). https://doi.org/10.1109/STC-CSIT.2019.8929872
90. Zaychenko, Y., Hamidov, G.: Hybrid fuzzy CNN network in the problem of medical images classification and diagnostics. In: Advances in Intelligent Systems and Computing (2020). https://doi.org/10.1007/978-3-030-32456-8_95

Assessing the Development of Energy Innovations and Its Impact on the Sustainable Development of Countries

Maryna Kravchenko, Olena Trofymenko, Kateryna Kopishynska, and Ivan Pyshnograiev

Abstract The article presents an analysis of the impact of energy innovations on the sustainable development of 26 countries, including the EU countries, USA, China, and Ukraine. The study of the essence of energy innovations provided an opportunity to determine their key types, which mainly include digital technologies, such as the Internet of Things, Use of Artificial Intelligence, Automation in the energy sector, Cloud computing, Blockchain technology. One of the methodologies used to assess the level of energy innovation development in countries is the Global Energy Innovation Index. However, the limitation of the available results of the calculation of this index for a significant period of time determined the use of one of the indicators included in it (Patents on environmental technologies) to determine the impact on the sustainable development of countries (where the key indicator for SDG 7: Affordable and Clean Energy is Energy intensity of GDP). The use of correlation analysis for countries based on certain indicators showed the presence of a strong inverse relationship between the indicators, i.e., an increase in the number of energy innovations leads to a decrease in the energy intensity of GDP, i.e., to an improvement in the country's sustainable development. With the help of cluster analysis, five clusters of countries were determined according to the level of intensity of development of energy innovations and the energy intensity of GDP, and the key measures taken by countries to implement plans for achieving the goals of sustainable development were analyzed.

M. Kravchenko (✉) · O. Trofymenko · K. Kopishynska · I. Pyshnograiev
Igor Sikorsky Kyiv Polytechnic Institute, Kyiv, Ukraine
e-mail: marina.kravchenko.kpi@gmail.com

O. Trofymenko
e-mail: olena.trofymenko@lll.kpi.ua

K. Kopishynska
e-mail: kopishynska@ukr.net

M. Zgurovsky and N. Pankratova (eds.), *System Analysis and Artificial Intelligence*, Studies in Computational Intelligence 1107,
https://doi.org/10.1007/978-3-031-37450-0_24

1 Introduction

Energy permeates all spheres of social life, the sphere of industry and the sphere of services, is a component of industrial revolutions, technological systems, sustainable development, etc. The innovative development of the energy sector will significantly contribute to the socio-economic development of society in the conditions of the Sixth Technological Order, the Post-Industrial Era, the Knowledge Economy, the Fourth Industrial Revolution (Industry 4.0), taking into account the provisions of sustainable development, resource-saving, frugal energy supply, and preservation of the environment for future generations. According to the main world priorities, the acceleration achievement goals for sustainable development until 2030 approved at the UN Summit, in particular, SDG 7: Affordable and Clean Energy for the achievement of which the introduction and development of energy innovations is an urgent need. It is the development of energy innovations that forms the basis for the implementation of the Paris Climate Agreement, which was adopted as part of the United Nations Framework Convention on Climate Change (UNFCCC) to regulate measures to reduce carbon dioxide emissions and will form the agenda for environmental preservation in the third decade of the 21st century.

In order to ensure the goals of sustainable development in 2019, the European Union agreed on a plan – The National Energy and Climate (ENCP), which, within the limits of the agreed needs, each EU country must form and implement for the period from 2021 to 2030, which is necessary to achieve the common goals reducing emissions for EU needs. The main areas to which this plan belongs correspond to such areas as: decarbonization; energy efficiency; energy security; internal energy market; research, innovation and competitiveness. The full-scale war has an extremely negative impact on the energy system of Europe, provoking a significant increase in the prices of energy resources, the need for a quick replacement for their main supplier, and endangering the achievements of previous years in the direction of decarbonization of the economy. This poses modern risks to energy and economic security, which requires the formation of new approaches and mechanisms for its provision, however, on the basis of previous achievements in the field of energy innovations, it is possible to form the main directions of ensuring the sustainable development of countries, which is of additional importance in the conditions of global transformations of the energy market.

Today, there are different views of scientists on the definition of the concept and components of the development of energy innovations. We agree with K. Gallagher, J. Holdren, and A. Sagar [1], that innovation in the field of energy technologies is a set of processes leading to new or improved energy reduce the economic, environmental or political costs associated with the supply and use of energy technologies that can increase energy resources; improve the quality of energy services reduce the economic, environmental or political costs associated with the supply and use of energy. R. Margolis [2] emphasizes that energy innovation is a set of processes for the development and improvement of technologies and equipment in the field of energy, which can exist either in a new form or improve and refine previously existing technologies. Researchers A. Grubler and Ch. Wilson apply a systemic approach to innovations in energy, where the innovative system of technologies in the field of energy is the analytical basis for the application of energy innovations, which defines innovative stages, processes and drivers for their provision, and identifies the sphere of four components: knowledge, actors and institutions, use of resources and technologies. A systematic approach to the structuring of energy innovations was carried out in the studies of Ch. Freeman and C. Peres [4]. A. Grubler and Ch. Wilson [3], Ch. Freeman and C. Peres [4] applies a systemic approach to innovations in energy.

According to the Annual Review of the Environment and Resources, the energy technology innovation system is the application of a systems perspective on energy technology innovation, which includes all aspects of energy systems (demand and supply), all stages of the development and technology cycle, all innovation processes, feedback connections, participants, institutions and networks. Approaches to determining the impact of energy innovations on individual processes of sustainable development are of considerable interest. Scientists G. Noja, M. Cristea, M. Panait, S. Trif, and C. Ponea [5] examine the role of energy innovations, digital technological transformation, and environmental performance in enhancing the sustainable economic development of EU countries. The approach of T. Berling, I. Surwillo, and V. Slakaityte [6] to analysis highlights how the study of energy innovation in the Baltic Sea Region can contribute to the conversation between critical security studies and science and technology studies, which deserves attention. Ch. Smith and D. Hart offered to evaluate the Global Energy Innovation Index (GEII), which provides a multifaceted assessment of national contributions to the global energy innovation system [6, 7].

The purpose of the article is to assess the measure of the impact of energy innovations on the sustainable development of countries according to the indicators included in GEII and to systematize the countries according to the main indicators.

2 Trends and Dynamics in the Development of Energy Innovations: Global Energy Innovation Index

In recent years, digital technologies have increasingly spread to production and production processes at a global level. Technologies in the areas of the Internet of Things, Big Data, Robotics, Blockchain technology, Sensors, Artificial intelligence, Augmented reality and Rapid prototyping technologies have moved into the manufacturing industry. Today, goods are designed, produced and consumed thanks to these technologies, and these technologies drive the development of new business models, services and consumer behavior. At the same time, the transition to Industry 4.0 is taking place simultaneously with such global problems as climate change, food shortages, limited energy resources, water scarcity, biodiversity loss, and such global problems of socio-economic systems as population growth, urbanization and mass migrations, new and ongoing conflicts, pandemics, etc.

It is appropriate to highlight five main digital technologies that are transforming the energy sector [8]:

- The Internet of Things can improve the efficiency of energy production, sale, and distribution. According to forecasted data, the value growth of the IoT in the global energy market in the period 2020–2025 will be about 15 billion USD: from 20.2 billion USD in 2020 to 35.2 billion USD in 2025;
- Use of Artificial Intelligence and advanced analytics in the energy sector, which involves the management of smart grids that ensure the intelligent flow of energy and data between the energy supplier and the consumer;
- Automation in the energy sector in particular involves automating repetitive tasks such as confirming meter readings, invoicing, canceling payments, and complaint management can be automated using robotic process automation in energy trading;
- Combining the energy sector with cloud computing used in the management of energy production, distribution, and supply. Cloud computing applications can enable collaboration and improve the visibility of financial and operational data across the network;
- Blockchain technology that leverages transparent peer-to-peer energy trading and can track the source of renewable energy and record the carbon footprint of the various parties involved in the network. Blockchain can store excess energy through smart meters, and algorithms can automatically match buyers and sellers of that energy [8].

The high speed of development of innovative technologies and approaches to innovative activities in the field of energy, constant changes in the external environment determine the need to define methodological approaches to the assessment of innovative development of the energy sector in order to make management decisions at the state level based on the results of the assessment, to determine an effective strategy for the innovative development of energy and form appropriate mechanisms for ensuring the sustainable development of the energy industry. Today approaches to assessing the overall level of development of innovations in the energy sector are already beginning, which is a positive trend for further global development. In our opinion, the methodology from the Information Technology and Innovation Foundation [7], based on which the Global Energy Innovation Index (GEII) is defined, is interesting. The GEII rests at its base on 20 indicators, which are aggregated into 10 functional categories and 3 subindices (Fig. 1).

Thus, the Knowledge Development and Diffusion subindex includes the following categories and indicators: Public Investments in Low-Carbon Energy R&D; Knowledge Generation (number of publications, share of highly cited publications); Invention (development and diffusion, attraction and absorption).

The Entrepreneurial Experimentation and Market Formation subindex includes such categories and indicators as Demonstration (public investment, number of projects); Entrepreneurial Ecosystem (early-stage venture capital investments, number of high-impact start-ups, number of successful companies exits); Industry and

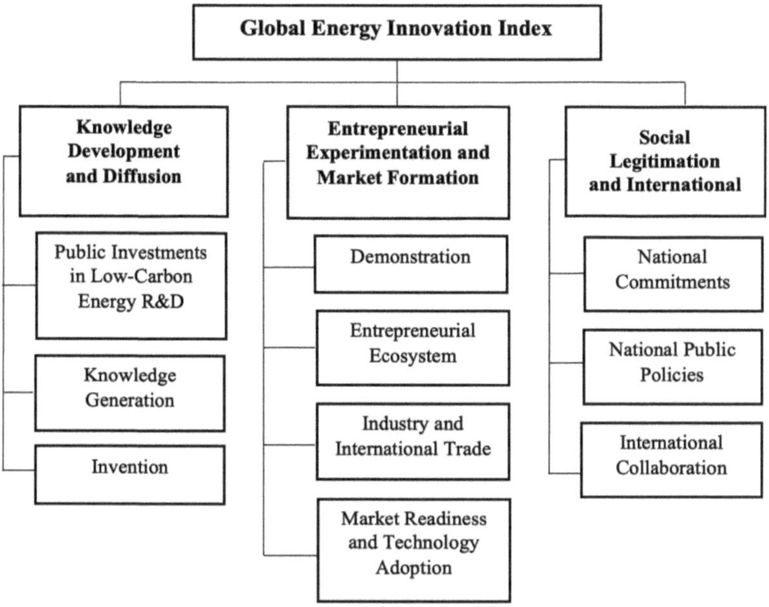

Fig. 1 Subindices and functional categories of the Global Energy Innovation Index (compiled by the authors based on [7])

International Trade (exports); Market Readiness and Technology Adoption (energy efficiency, clean energy consumption).

The Social Legitimation and International Collaboration subindex includes such categories and indicators as National Commitments (global climate action agenda, domestic clean energy innovation agenda); National Public Policies (standards and regulations, effective carbon rates); International Collaboration (public R&D funding, knowledge development); Technology Development (co-inventions).

The GEII weights the Knowledge Development and Diffusion subindex at 40.0%, the Entrepreneurial Experimentation and Market Formation subindex—at 40.0%, and the Social Legitimation and International Collaboration subindex—at 20.0%.

The results of the assessment of the Global Energy Innovation Index of some countries for 2021 are shown in Fig. 2. According to the results of the evaluation of the Global Energy Innovation Index, it can be proven that the top ten of the rating are mainly EU countries. This is explained by the significant efforts of the European Union to comply with the Green Course and the active transition to the use of alternative energy sources. At the same time, countries such as the USA and China ranked 16th and 24th, respectively out of 34 countries included in the ranking.

In the development of the innovation system in the energy sector, the focus on greening and decarbonization is decisive which is one of the main tasks for the implementation of the global goals of sustainable development. For example, in November 2018, the European Commission presented a long-term strategic concept

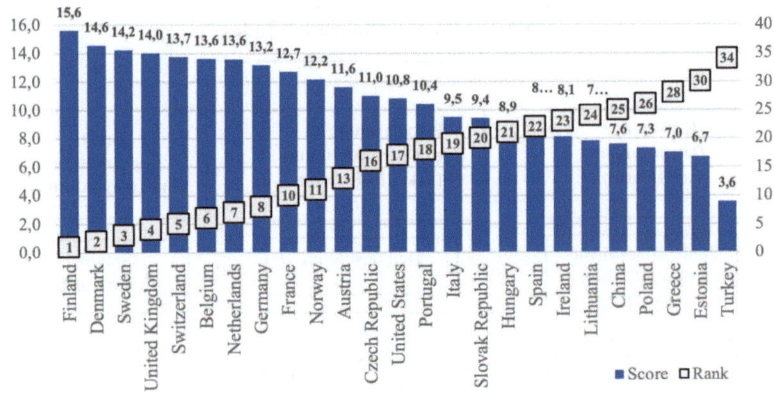

Fig. 2 Ranking of some countries based on the results of the Global Energy Innovation Index, 2021 (Data source [7])

for reducing greenhouse gas emissions, which determined how Europe can pave the way to a climate-neutral economy with net-zero greenhouse gas emissions by 2050. It contains seven main strategic components: maximization of energy efficiency; maximum implementation of renewable energy sources (RES) and electrification; transition to ecologically clean transport; introduction of circular economy (closed cycle economy); development of "smart" networks and communications; expansion of bioenergy and natural carbon absorption; absorption of the remaining CO2 emissions due to carbon absorption and storage technologies [9].

3 Determination of the Impact of Energy Innovations on the Sustainable Development of Countries

The use of the global energy innovation index as a factor influencing sustainable development is significantly limited due to the lack of sufficient data, because currently the authors of the methodology presented its calculation for only a few years. Therefore, as an indicator reflecting the level of development of energy innovations, we chose number of patents on environmental technologies, which shows the patent applications filed under the PCT for the country. It combines the number of patents on environmental technologies in the following areas: climate change mitigation technologies related to buildings; climate change mitigation technologies related to energy generation, transmission or distribution; capture, storage, sequestration or disposal of greenhouse gases; environmental management; climate change mitigation technologies related to transportation; climate change mitigation technologies in the production or processing of goods; climate change mitigation technologies related to wastewater treatment or waste management. All of these components directly or indirectly affect the development of innovations in the energy sector.

According to the data in the Table 1, it can be noted that the USA and China have the largest number of patents on environmental technologies among the countries of the world, and Germany and France among the European countries. At the same time, countries such as Estonia, Lithuania, Greece and the Slovak Republic have the least number of patented technologies in this field.

One of the key indicators of achieving the SDG 7 is the Energy intensity of GDP (primary energy consumption per unit of GDP). For the selected countries, we will consider the dynamics of changes in the indicator (Table 2).

Table 1 Number of patents on environment technologies, 2015–2019, number (Data source [10])

Country	2015	2016	2017	2018	2019
Austria	212,0	203,5	226,5	234,0	241,7
Belgium	147,9	200,7	174,1	172,9	137,6
Czech Republic	18,4	24,7	27,3	27,8	25,4
Denmark	266,5	332,1	333,0	365,8	372,0
Estonia	3,0	6,0	7,6	8,2	2,2
Finland	189,4	207,7	215,0	195,5	215,5
France	1 201,7	1 027,9	1 069,2	1 166,0	998,8
Germany	2 558,8	2 735,7	2 863,0	2 777,0	2 514,6
Greece	12,2	17,8	12,9	19,9	10,8
Hungary	34,7	23,1	30,4	35,2	26,7
Ireland	44,9	42,3	25,6	34,8	34,8
Italy	394,5	425,6	379,6	385,5	390,9
Lithuania	4,8	2,9	4,9	1,4	1,5
Netherlands	436,0	422,2	359,5	391,9	343,5
Norway	98,4	116,4	134,1	139,4	129,2
Poland	78,4	43,2	45,9	36,5	42,6
Portugal	34,0	32,6	26,5	21,0	24,5
Slovak Republic	13,0	15,8	9,3	11,0	6,0
Spain	248,1	215,4	246,3	235,1	236,9
Sweden	402,6	447,1	429,6	463,8	434,4
Switzerland	188,0	225,3	222,7	240,4	247,4
Turkey	98,0	121,3	127,0	129,7	123,7
United Kingdom	828,9	769,6	748,1	670,7	756,1
United States	5 650,9	5 740,0	5 631,0	5 206,1	4 754,4
China	2 434,4	3 658,9	4 222,0	4 691,9	5 214,3
Ukraine	16,8	14,8	29,3	27,5	27,8

Table 2 Energy intensity of GDP for the countries of the world, 2015–2019, kWh per USD (Calculated from the data source [11])

Country	2015	2016	2017	2018	2019	Δ2019/2000, %
Austria	0,92	0,89	0,88	0,81	0,81	51,5
Belgium	1,29	1,31	1,27	1,21	1,17	55,2
Czech Republic	1,32	1,22	1,18	1,11	1,01	64,5
Denmark	0,72	0,68	0,63	0,60	0,56	63,8
Estonia	1,78	1,71	1,76	1,62	1,21	71,2
Finland	1,41	1,35	1,25	1,22	1,14	55,9
France	1,03	0,96	0,92	0,89	0,80	58,9
Germany	0,97	0,92	0,89	0,83	0,77	56,6
Greece	1,06	1,02	0,86	1,00	0,93	47,3
Hungary	0,95	0,93	0,93	0,87	0,81	64,8
Ireland	0,54	0,53	0,49	0,45	0,42	71,9
Italy	0,81	0,75	0,73	0,71	0,67	51,1
Lithuania	0,76	0,73	0,72	0,69	0,63	76,2
Netherlands	1,15	1,12	1,05	0,99	0,94	52,4
Norway	1,75	1,80	1,65	1,48	1,39	60,3
Poland	1,09	1,09	1,06	1,00	0,89	64,2
Portugal	0,95	0,95	0,89	0,86	0,76	49,4
Slovak Republic	1,13	1,14	1,19	1,12	1,02	71,4
Spain	0,97	0,92	0,87	0,86	0,78	55,1
Sweden	1,31	1,23	1,22	1,15	1,12	54,0
Switzerland	0,61	0,55	0,54	0,53	0,53	61,5
Turkey	0,79	0,80	0,79	0,77	0,79	44,2
United Kingdom	0,81	0,77	0,73	0,71	0,66	61,5
United States	1,41	1,38	1,33	1,30	1,24	52,0
China	1,98	1,92	1,87	1,77	1,71	46,8
Ukraine	2,30	2,21	1,94	1,90	1,71	77,4

For all the studied countries, a steady trend towards a decrease in the level of energy intensity of GDP can be observed. Over the past 20 years, the highest rate of decrease in Energy intensity of GDP was observed in Ukraine (77.4%), Ireland (71.9%), Slovak Republic (71.4%) and Estonia (71.2%). Today, as a result of the global energy crisis caused by the Russian Federation's full-scale military invasion of Ukraine, reducing record high consumer bills and ensuring reliable access to supply is a central political and economic imperative for almost all governments. According to experts from the International Energy Agency [12], while countries have many ways to deal with the current crisis, focusing on energy efficiency measures is the main answer to achieving affordability goals at the same time, security of supply and climate. This indicates the need for the development of energy innovations to ensure energy efficiency goals.

In order to determine the impact of Energy Innovations on the Energy Intensity of the country's GDP, as a key indicator of the SDG 7, we will establish the existence of a correlation between the specified indicators. Consider the auto- and cross-correlation function for the indicators Patents on environmental technologies (x) and Energy intensity of GDP (y) for the period 2000–2019. Table 3 shows the correlation value and the corresponding lag. If lag k is indicated in the table, then the maximum value of correlation by modulus is observed between $x(t+k)$ and $y(t)$.

The specified countries have a fairly small number of patents on environmental technologies, which was already mentioned above, so for them the specified indicator does not have a significant impact on the energy intensity of GDP.

Table 3 The maximum values of the auto-and cross-correlation function of the countries

Country	Lag (k)	Correlation
Austria	0	–0,904
Belgium	0	–0,872
Czech Republic	0	–0,744
Denmark	0	–0,963
Estonia	1	–0,550
Finland	0	–0,853
France	0	–0,891
Germany	0	–0,835
Greece	–1	–0,467
Hungary	0	–0,669
Ireland	0	–0,740
Italy	0	–0,907
Lithuania	–2	–0,482
Netherlands	0	–0,859
Norway	0	–0,769
Poland	0	–0,822
Portugal	0	–0,813
Slovak Republic	1	–0,643
Spain	0	–0,843
Sweden	0	–0,946
Switzerland	0	–0,815
Turkey	0	–0,874
United Kingdom	0	–0,861
United States	0	–0,849
China	0	–0,902
Ukraine	0	–0,782

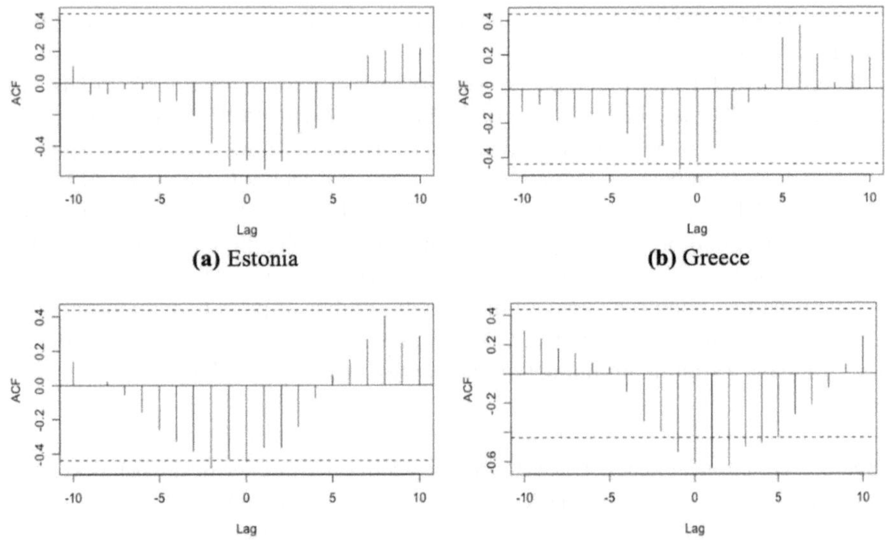

(a) Estonia (b) Greece

(c) Lithuania (d) Slovak Republic

Fig. 3 Value of auto-and cross-correlation function for countries

As we can see from the table, the greatest impact of Patents on environmental technologies is observed in the same year, but there are exceptions for which there is a lag of impact, as well as the correlation coefficient by module is less than others. Let us present the auto-and cross-correlation function for Estonia (Fig. 3a), Greece (Fig. 3b), Lithuania (Fig. 3c) and Slovakia (Fig. 3d).

According to the results of the correlation analysis, it can be noted that for most of the studied countries there is a strong inverse relationship between the Patents on environmental technologies indicator and the Energy intensity of GDP. This indicates that with the increase in the number of Patents on environmental technologies, that is, energy innovations, the energy intensity of GDP decreases, which is a positive result.

4 Clustering of Countries According to the Main Indicators

For further analysis of countries, we will use cluster analysis. To carry out clustering, in addition to the previously studied indicators of Energy intensity of GDP and Patents on environmental technologies per capita of countries, data of countries according to the indicator of Total energy supply, measured in Megajoules per GDP, based on 2017 USD, were used.

In order to bring the data to one dimension, the min-max normalization method was chosen according to the formula (1):

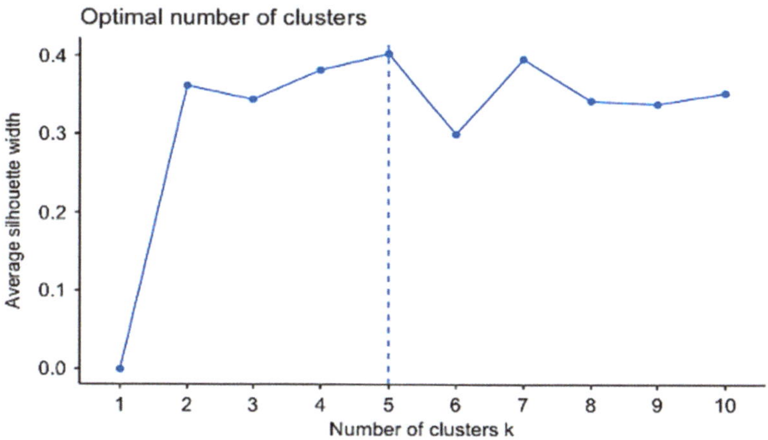

Fig. 4 Criterion for choosing the number of clusters

$$X^i_{j,norm} = \frac{X^i_j - \min_j X^i_j}{\max_j X^i_j - \min_j X^i_j}, \quad (1)$$

where i—number of the indicator, j—number of the country, X^i_j—value of the ith indicator for the jth country.

For cluster analysis, it will be used the k-means method [13]. It is an iterative method that forms clusters in which the sum of the Euclidean distances from the center to the elements of the clusters is minimal.

The next step is choosing the required number of clusters. Figure 4 shows the dependence of the silhouette coefficient on the number of clusters. Its larger value indicates a greater degree of difference of the obtained clusters. Thus, we will use five clusters for further analysis.

The results of clustering and the formed clusters of countries are presented in the Table 4. According to the calculations, an association of countries was formed into five clusters, as well as individual countries:

- Cluster 1 includes such countries as Belgium, the Czech Republic, Estonia, the Slovak Republic, and the United States;
- Cluster 2 includes the largest number of countries, in particular: France, Greece, Hungary, Ireland, Italy, Lithuania, Netherlands, Poland, Portugal, Spain, Turkey, United Kingdom;
- Cluster 3 includes countries such as Austria, Denmark, Germany, and Switzerland.
- Cluster 4 includes China and Ukraine;
- Cluster 5 is formed from Finland, Norway, and Sweden.

The preliminary analysis of the correlation between the data sets Number of patents on environmental technologies, Total energy supply per GDP, and Energy

Table 4 The maximum values of the auto-and cross-correlation function of the countries

Country	Total energy supply per GDP, MJ per 2017 USD	Energy intensity of GDP, kWh per USD	Patents on environmental technologies, number per 1 billion people	Cluster
Austria	2,82	0,81	27,22	3
Belgium	3,87	1,17	11,98	1
Czech Republic	4,12	1,01	2,38	1
Denmark	2,00	0,56	63,98	3
Estonia	4,49	1,21	1,66	1
Finland	5,19	1,14	39,03	5
France	3,29	0,80	14,82	2
Germany	2,76	0,77	30,26	3
Greece	2,90	0,93	1,01	2
Hungary	3,51	0,81	2,73	2
Ireland	1,32	0,42	7,05	2
Italy	2,45	0,67	6,54	2
Lithuania	3,08	0,63	0,54	2
Netherlands	3,05	0,94	19,80	2
Norway	3,33	1,39	24,16	5
Poland	3,42	0,89	1,12	2
Portugal	2,54	0,76	2,38	2
Slovak Republic	4,11	1,02	1,10	1
Spain	2,64	0,78	5,03	2
Sweden	3,80	1,12	42,26	5
Switzerland	1,68	0,53	28,85	3
Turkey	2,61	0,79	1,48	2
United Kingdom	2,30	0,66	11,31	2
United States	4,51	1,24	14,48	1
China	6,31	1,71	3,70	4
Ukraine	6,95	1,71	0,63	4

intensity of GDP showed a significant and inverse correlation, which is explained by the implementation of the principles of the Knowledge economy to support energy innovations in the energy sector to obtain economic, environmental and social profit.

For a deeper study of this correlation dependence, these two indicators will be compared. Within the formed clusters, we will compare the Number of patents on environmental technologies per unit of GDP, and the Total energy supply per GDP. The growth of this ratio indicates an increase in the productivity of the use of patents. That is, we have a general conclusion that the increase in the number of patents on environmental technologies ensures an increase in the level of energy efficiency. To increase the order of the obtained values of the ratio of the Number of patents on environmental technologies per unit of GDP, and the Total energy supply per GDP multiplied by 1e+10 and the dimension of the indicator is the Number patents on environmental technologies in relation to the energy used in Megajoules.

In the countries of Cluster 1, the Total energy supply per GDP ratio is observed in the range from 3.9 to 4.5, which is a unifying feature for these countries. Belgium ranks first in terms of energy intensity (3.87), in second place is Slovakia (4.11), followed by the Czech Republic (4.12), Estonia (4.49), and the United States (4.51). At the same time, according to the ratio of the Number of patents on environmental technologies per unit of GDP, and the Total energy supply per GDP, the first place has Belgium with a value of 0.6, the second place has the USA with a value of 0.51, followed by the Czech Republic (0.14), Estonia (0.10) and Slovakia (0.08). The energy market in Belgium is changing rapidly, among the influencing factors we can single out the processes of deregulation and liberalization ongoing in the EU; needs for the development of renewable energy sources; changing the structure of energy distribution in the country. Belgium has many energy-intensive companies, and energy prices and labor costs are rising faster than in other European countries. It is worth noting that Belgium is the leader in the volume of frozen Russian assets—50 billion euros. The cluster also includes a leading country on the international scene—the USA, which uses mainly fossil fuels and supports innovative trends in energy. The USA is a leader in the production and supply of energy and is also one of the largest consumers of energy in the world. Specifically, the USA is the leading energy exporter in 2019–2021. Total USA energy exports exceeded total energy imports by approximately 3.82 quadrillion British thermal units (BTUs) in 2021, the largest margin in USA history. Growing consumer demand and world-class innovation, combined with a competitive workforce and supply chain in certain industry segments, make the USA energy industry competitive in the $6 trillion global energy market. In general, Cluster 1 is characterized by an energy-intensive economy, however, according to the indicators of the countries, it shows tendencies towards a decrease in energy intensity. In particular, over the past 20 years, the highest rate of decrease in the energy intensity of GDP was observed among the countries of the cluster in the Slovak Republic (71.42%) and Estonia (71.22%). In accordance with the energy-intensive economy of the cluster countries, it is important to continue the implementation of the policy of increasing energy efficiency, which is a prerequisite for achieving the goals of sustainable development. Taking into account the fact that in Cluster No. 1 are concentrated powerful countries-leaders in supporting Ukraine in the fight against Russian aggression—this also forms the basis for ensuring post-war recovery and modernization on the basis of a "green" economy for both Ukraine and partner countries in the direction of the implementation of common projects in the energy sector and the development of energy innovations.

The largest Cluster 2 is characterized by the Total energy supply per GDP ratio ranging from 1.32 to 3.51. According to this indicator, Ireland stands out, which has a minimum value of 1.32, while all other countries have an average value of about 3. This indicates the implementation of energy efficiency measures, in addition to the fact that it also has a positive effect on the environment, because according to calculations experts, thanks to the use of energy efficiency and renewable energy sources, Ireland manages to avoid more than 6 million tons of CO2 emissions annually [14]. According to the ratio of the Number of patents on environmental technologies per unit of GDP, the Total energy supply per GDP, the first place in cluster has the Netherlands with a value of 1.14, the second place has the United Kingdom with a value of 1.04, the third place has France with a value of 0.98, followed by Italy (0.63), Ireland (0.62), Spain (0.47), Portugal (0.27), Hungary (0.24), Turkey (0.2), Greece (0.12), Poland (0.1), Lithuania (0.05). Among the main energy innovations, for which the Netherlands is the leader in the number of patents, are hydrogen technologies. During the period from 2011 to 2020, 283 patents of Netherlands companies accounted for innovations in the production of hydrogen. Accordingly, the Netherlands is in the top three in the number of patents in the field of hydrogen technologies in Europe [15]. At the same time, the Energy intensity of the GDP of the Netherlands is not in the first place among the countries in Cluster 2, this indicates that it is also necessary to take into account the real application of technologies and inventions in production, which will testify to the practical implementation of registered patents. In general, this cluster can be characterized as energy-dependent with a priority for implementing energy innovations for sustainable development, combining economic and political efforts within the framework of cooperation of the cluster countries, fulfilling the tasks of Goal 17, and achieving other sustainable development goals.

Austria, Denmark, Germany, and Switzerland are included in Cluster 3. This Cluster is characterized by a relatively low level of energy intensity in relation to other clusters—the Total energy supply per GDP ratio is observed in the range from 1.68 to 2.82, which indicates a relatively high energy efficiency of the formed Cluster. The leader in the Cluster according to this indicator is Switzerland. Switzerland is actively implementing energy innovations. In particular, Renewable energy share in final energy demand increased by 28.5% between 2009 and 2019, and in 2021 total installed capacity of renewable power was 17.7 GW, which allows Switzerland to generate more than 95,0% of its energy needs from renewables. Financial support for renewable energy is being introduced, in 2021 Switzerland allocated USD 521 million for solar rebates [16]. At the same time, in terms of Energy intensity of GDP, Denmark is in second place in this cluster and is 2 MJ per USD of GDP, however, according to the ratio of the Number of patents on environmental technologies per unit of GDP, and the Total energy supply per GDP, Denmark has a value of 5.63 and is the leader among all the studied countries in five clusters. Today, 50,0% of electricity in Denmark is produced on the basis of wind and solar energy, bioenergy technologies are developing. By 2030, the country has set a goal that the electricity system in Denmark will be completely independent of fossil fuels [17]. The Danish companies Vestas and Siemens are among the world's leading innovators in the field of wind energy. In 2021, Denmark registered 551 "green" patents, which was 93 "green" patents per million inhabitants and cements Denmark's status as one of the leaders in this field. A major part of the favorable tax climate in Denmark is the corporate tax rate of 22.0%. Denmark's track record in successfully stimulating environmental innovation is worth considering, with the Danish government introducing a permanent research and development deduction of 130.0% to promote commercial research and accelerate environmental investment. In addition, if the R&D expenses lead to tax losses, companies are entitled to a cash refund of 22% of the losses related to these expenses - up to a maximum tax value of 25 billion DKK. Also, companies are entitled to a full deduction of patents and know-how in the year of an acquisition. In Austria and Germany, the ratio of the Number of patents on environmental technologies per unit of GDP, and the Total energy supply per GDP are 1.72 and 2.03, respectively, which also indicates high positions. According to the indicated indicators, this cluster can be considered as an example to follow for achieving the goals of sustainable development in the energy sector, the development of the high-tech industry sector, and high economic indicators, which together determine a high level of quality of life in this cluster on the basis of sustainable development.

Cluster 4 includes only two countries—China and Ukraine—and is characterized by a high level of energy intensity, which indicates high energy dependence and the need for energy innovations. However, the difference in the ratio of the Number of patents on environmental technologies per unit of GDP, and the Total energy supply per GDP is 29,0% in favor of China (0.37 in China and 0.07 in Ukraine). This shows that China does not use its patents effectively enough in production, because in terms of the Number of patents on environmental technologies, this indicator has the highest indicator among all the countries studied. Indirectly, this may indicate that some of the patents are for sale. Although both countries have a high level of energy intensity of GDP, China has shown almost exponential growth in the ratio of indicators over the past 10 years. Taking into account the latest indicators, the full-scale war of the Russian Federation in Ukraine led to interruptions in the functioning of the country's energy system, and the damage caused to energy infrastructure facilities, according to experts, amounts to more than 8.1 billion dollars. The issue of energy efficiency is particularly acute today in Ukraine in the midst of Russia's massive missile attacks on the energy infrastructure, growing energy and economic crises. Considering that Ukraine and China are in the same cluster, this also shows the potential of joint cooperation to achieve the goals of sustainable development. At the same time, China has long demonstrated neutrality in its position regarding the war in Ukraine and the provision of sanctions for the aggressor country Russia. This is a threat at the global level, as well as a significant obstacle to sustainable development.

Cluster 5 includes Finland, Norway and Sweden. The Total energy supply per GDP ratio is 5.19 in Finland, 3.33 in Norway, and 3.80 in Sweden. At the same time, the ratio of the Number of patents on environmental technologies per unit of GDP, and the Total energy supply per GDP is 1.55, 1.12 and 2.1, respectively, which is a relatively high value in relation to most clusters. According to these indicators, Sweden looks the most effective. In 2012, Sweden achieved its 2020 target of 50.0% renewable energy in its energy supply, and plans to use 100.0% renewable energy by 2040 and achieve net carbon emissions by 2045. It is interesting to consider the experience of the development of energy innovations in Norway. The Norwegian government has developed a strategy for carbon capture and storage (CCS), which aims to identify measures to promote technology development and reduce CCS costs. Part of the CCS strategy is the continued support of CLIMIT, the national program for research, development and demonstration of CCS technologies for both power plants and industry. The program covers areas from basic research to innovative projects and demonstrations of CCS technology, Norwegian Research Centers for Climate Energy and international cooperation in the field of innovation, identified as a priority for the Norwegian government's investment in R&D [18]. Continuation of the policy of supporting energy innovations will ensure further progress in achieving the goals of sustainable development. It is important for other countries to adopt the experience of these countries regarding the implementation of policies for sustainable development.

Figure 5 shows a visualization of the comparison indicator of the Number of patents on environmental technologies per unit of GDP and the Total energy supply per GDP for the five identified clusters.

Among energy innovations, digitalization is at the heart of supporting the development of clean energy. However, despite the significant number of benefits from the implementation of digital technologies, energy companies face various obstacles and challenges during digital transformation, which, among other things, are related to the specifics of the industry itself and established management practices.

Factors that complicate the process of implementing digital technologies by energy companies include the following [19]:

- Physical orientation. The energy business is sensitive to the laws of physics-the geophysics of oil and gas deposits, the quantum physics of solar energy, the hydrodynamics of wind, the thermodynamics of fossil energy, or the electromagnetics of energy transmission;
- Health and safety risks. Energy is a powerful commodity: it supports our daily lives, but it is potentially dangerous if left untreated. Given the inherent risks, energy companies are under constant regulatory scrutiny, are risk-averse and attempt to control it through detailed and rigorous processes;
- An engineer-driven culture. In a physically oriented, highly regulated sector, engineers are privileged. Oil, gas and energy companies are dominated by current and former senior level engineers whose management style involves detailed planning, finding the perfect solution, and prioritizing thorough analysis and process over quick judgment and flexibility;
- Strong dependence on third parties. The operation of energy companies depends on an extensive and fragmented supply chain. Energy industry puts supplier collaboration at the heart of operations;
- Long career of managers and their narrow influence. Many energy executives have been with the same company for at least 30 years, have been rewarded not for innovation, but for caution and adherence to tradition, so they focus more on surviving business cycles than driving permanent change;
- Global legal and operational environment. The legal and operational environment varies considerably from country to country. Labor resources differ in capabilities, reliability, size and cost. Supply chains vary in maturity. All of this in combination significantly complicates the implementation of digital technologies [19].

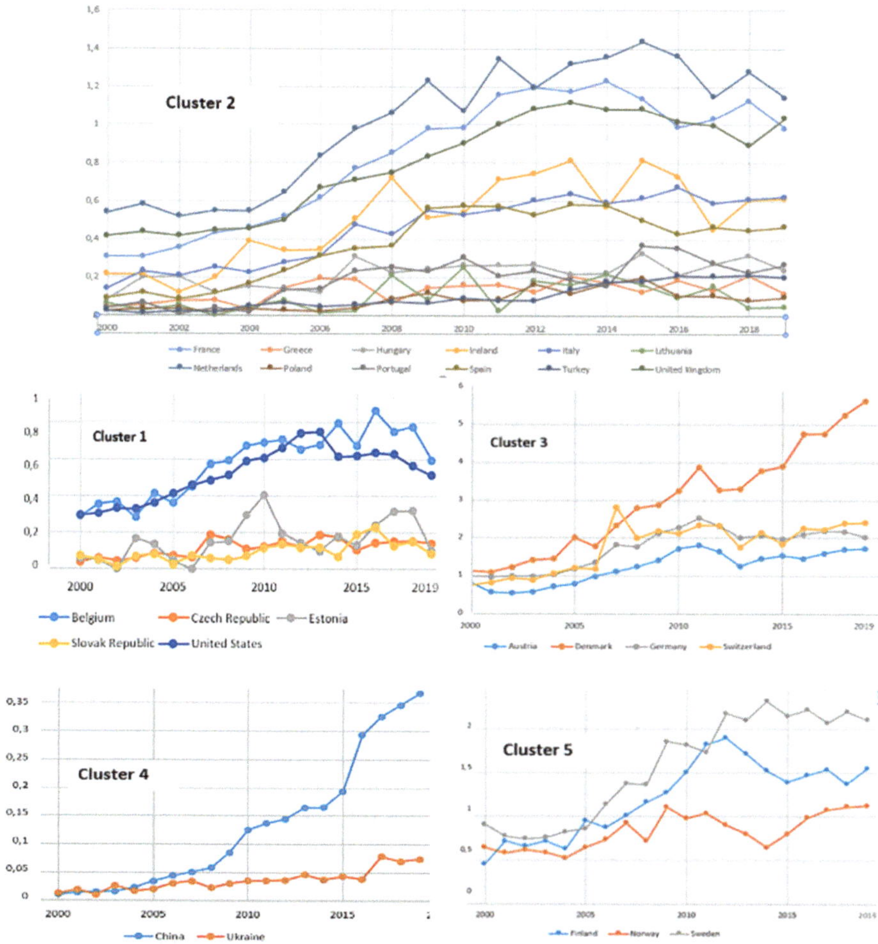

Fig. 5 The value of the indicator of the Number of patents on environmental technologies per unit of GDP and the Total energy supply per GDP in the countries of the five identified clusters, numbers of patents per MJ

Therefore, the efforts of countries should be directed not only to the development of energy innovations to achieve the established goals of sustainable development, but also to overcoming resistance to the implementation of such innovations and avoiding possible risks associated with their implementation and further use.

5 Conclusion

The orientation to achieve the goals of sustainable development, as well as the desire to achieve climate neutrality by 2050, determined the need to assess the impact of energy innovations on the sustainable development of countries, with special importance on Goal 7: Affordable and clean energy.

1. A decrease in the level of energy intensity of GDP is characteristic of all the studied countries. Over the past 20 years, the highest rate of decrease in energy intensity of GDP was observed in Ukraine (77.44%), Ireland (71.92%), Slovak Republic (71.42%), and Estonia (71.22%). However, these countries had a fairly high level of energy intensity during the studied period, therefore, the continuation of the policy of increasing energy efficiency in these countries is a mandatory condition for achieving the goals of sustainable development.
2. According to the Patents on environmental technologies indicator, the leaders among the studied countries are the USA, China, Germany, and France, while only China has seen a more than two-fold increase in the indicator over the past 5 years. This may indicate, among other things, the likely lower cost of conducting research and developing relevant technologies, and as a result, their transfer to other countries.
3. The results of the correlation analysis made it possible to establish a strong inverse relationship between the Patents on environmental technologies indicator and the energy intensity of GDP, and to testify that as the number of energy innovations increases, the energy intensity of GDP decreases, which is a positive result.
4. Clustering of countries according to the indicators of Energy intensity of GDP, Patents on environmental technologies, and Total energy supply per GDP showed the presence of 5 Clusters. The best ratio of indicators Number of patents on environmental technologies per unit of GDP and the Total energy supply per GDP was demonstrated by the countries of Clusters 3 and 5. The leader among the countries of all five Clusters is Denmark, where today 50% of electricity is produced from alternative sources. As for the countries that demonstrate the lowest indicators in terms of the ratio of the studied indicators, these are the countries of Cluster 4—China and Ukraine. So far, both countries have a high level of energy intensity of GDP, but China has shown almost exponential growth in the ratio of indicators over the past 10 years.

In general, countries from Clusters 3 and 5 are recommended to continue the policy of supporting energy innovations, which will ensure further progress in achieving the goals of sustainable development. Other countries need to adopt the experience of these countries regarding the implementation of policies in the field of sustainable development.

References

1. Gallagher, K., Holdren, J., Sagar, A.: Energy-technology innovation. Annu. Rev. Environ. Resour. **31**(1), 193–237 (2006). https://doi.org/10.1146/annurev.energy.30.050504.144321
2. Margolis, R.: Understanding Technological Innovation in the Energy Sector : The Case of Photovoltaics. Princeton University (2002). https://www.belfercenter.org/publication/understanding-technological-innovation-energy-sector-case-photovoltaics
3. Grubler, A., Wilson, C.: Energy Technology Innovation: Learning from Historical Successes and Failures. Cambridge University Press, Cambridge (2013). https://doi.org/10.1017/CBO9781139150880
4. Freeman, C., Perez, C.: Structural crises of adjustment, business cycles and investment behaviour. Struct. Cris. Adjust. **130**, 38–67 (1988). https://carlotaperez.org/wp-content/downloads/publications/theoretical-framework/StructuralCrisesOfAdjustment.pdf
5. Noja, G., Cristea, M., Panait, M., Trif, S., Ponea, C.: The impact of energy innovations and environmental performance on the sustainable development of the EU countries in a globalized digital economy. Front. Environ. Sci. **10**, 934404 (2022). https://doi.org/10.3389/fenvs.2022.934404
6. Berling, T., Surwillo, I., Slakaityte, V.: Energy security innovation in the Baltic sea region: competing visions of technopolitical orders. Geopolit. (2022). https://doi.org/10.1080/14650045.2022.2131546
7. Smith, C., Hart, D.: The 2021 Global Energy Innovation Index : National contributions to the global clean energy innovation system. Information Technology and Innovation Foundation (2021). https://www2.itif.org/2021-global-energy-innovation-index.pdf
8. Javaid, S.: Top 5 Digital technologies transforming the energy sector. AImultiple (2022). https://research.aimultiple.com/digital-transformation-in-energy-industry
9. Ministry of Environmental Protection and Natural Resources of Ukraine: Green Concept (2021). https://mepr.gov.ua/
10. Organisation for Economic Co-operation and Development: Patents on environment technologies (indicator) (2023). https://doi.org/10.1787/fff120f8-en
11. World Bank: World Bank Open Data (2022). https://data.worldbank.org
12. International Energy Agency: Energy efficiency: The first fuel of a sustainable global energy system (2022). https://www.iea.org/topics/energy-efficiency
13. Hartigan, J., Wong, M.: Algorithm AS 136: a K-means clustering algorithm. Appl. Stat. **28**, 100–108 (1979). https://doi.org/10.2307/2346830
14. Sustainable Energy Authority of Ireland: Ireland's Energy Targets (2016). https://www.seai.ie/publications/Ireland___s-Energy-Targets-Progress-Ambition-and-Impacts.pdf
15. Agro-chemistry: Netherlands global leader in hydrogen patents (2023). https://www.agro-chemistry.com/news/netherlands-global-leader-in-hydrogen-patents
16. Renewables 2022 : Global Status Report. Switzerland Factsheet (2022). https://www.ren21.net/wp-content/uploads/2019/05/GSR2022_Fact_Sheet_Switzerland.pdf
17. Denmark official website: Denmark is a laboratory for green solutions (2022). https://denmark.dk/innovation-and-design/green-solutions
18. Ministry of Petroleum and Energy of Norway: The Government's carbon capture and storage strategy (2022). https://www.regjeringen.no/en/topics/energy/carbon-capture-and-storage/the-governments-carbon-capture-and-storage-strategy/id2353948
19. Booth, A., Patel, N., Smith, M.: Digital transformation in energy: Achieving escape velocity. Mckinsey (2020). https://www.mckinsey.com/industries/oil-and-gas/our-insights/digital-transformation-in-energy-achieving-escape-velocity#download

Studies of the Intercivilization Fault Level Dynamics

Michael Zgurovsky, Maryna Kravchenko, and Ivan Pyshnograiev

Abstract The dynamics of rifts between civilizations, which are considered as one of the important factors in the emergence and course of modern world conflicts that have a civilizational character and as a result change the vector of global development, are studied. A system of indicators and a mathematical model for determining the size of faults in ethno-cultural, economic and social dimensions are proposed. The main global trends in the development of fault lines are identified, the Slavic—Eastern Orthodox civilization and the place of Ukraine in it are considered.

1 Introduction

At the end of the 20th century, the american political scientist S. Huntington expressed several non-trivial but convincing theses about the change in the world system after the end of the Cold War [1]. In the research he comes to the conclusion that after the collapse of the Soviet Union, unions and blocs, which were previously linked by a common ideology, began to be replaced by civilizational groups based on people's belonging to a common culture, religion and value system, and the ideological confrontation that took place during of the Cold War, turns into a confrontation of civilizations.

After the full-scale invasion of Russia on Ukraine S. Huntington's theory of civilizational confrontations received a new round of attention, but in a different context. It was addressed either with cautious support: in the context of the fact that Russia seeks a clash of civilizations, and the task of the West is to prevent this [2]; or openly criticizing: in the context of the fact that the author's theory is refuted by the war of countries that divide Orthodoxy to this day were considered civilizationally similar, which Ukraine and Russia are, and Putin's attempt to restore "Great Russia" [3, 4].

M. Zgurovsky · M. Kravchenko (✉) · I. Pyshnograiev
Igor Sikorsky Kyiv Polytechnic Institute, Kyiv, Ukraine
e-mail: marina.kravchenko.kpi@gmail.com

M. Zgurovsky
e-mail: mzz@kpi.ua

M. Zgurovsky and N. Pankratova (eds.), *System Analysis and Artificial Intelligence*, Studies in Computational Intelligence 1107,
https://doi.org/10.1007/978-3-031-37450-0_25

In 1996, S. Huntington distinguished eight basic civilizations. In [5], the authors refined the ethno-cultural civilizational distribution of countries based on the method of fuzzy clusters and identified thirteen civilizations based on data during 2008–2018. A mathematical model of faults between civilizations has been developed, which is based on the calculation of quantitative characteristics of fault lines using group expert evaluations. Among other things, the increase in the number of civilizations identified in 1996 and in 2018 may be evidence of civilizational shifts accompanying the transition from one civilization to another. The obtained results made it possible to follow the regularities and characteristics of the faults, but in a general sense they are not complete enough for a detailed analysis of the dynamics.

The main goal of this research is the development and testing of a mathematical model of the quantitative measurement of such faults for the analysis of the dynamics of their changes and the impact on global security processes using multidimensional statistical analysis. This approach will make it possible to consider civilizations comprehensively according to several groups of parameters, as well as to determine their homogeneity.

2 Data Framework and Methodology

A detailed list of thirteen civilizations, which will be used in the study, is presented in Table 1.

To model the magnitude of faults, we will define three groups of indicators:

- ethno-cultural—reflecting the cultural, value, religious and political historically formed results of joint life activities of representatives of a certain civilization and their originality;
- economic—determining the level of industrial development and other economic results of the activities of representatives of a certain civilization;
- social—demonstrating the social and public foundations and features of a certain civilization and the level of its social development.

The measurement was carried out by finding integral indicators for a group of countries belonging to a certain civilization in the time range of 1995–2021.

2.1 Formation of a Group of Ethno-Cultural Indicators

A group of criteria defined by experts in [5] was taken as the basis for the formation of a set of indicators, in accordance with which a group of indicators was selected.

Criterion 1. Individual freedom in society determines the degree of freedom of movement, personal life, formation of social groups, etc. The Personal Freedom Index, which contains 44 indicators and is part of the Human Freedom Index, was used to quantify the criterion. It is calculated by The Cato Institute [6].

Table 1 The list of civilizations [5]

Civilization	Countries
Muslim—Arabic	Afghanistan, Algeria, Bahrain, Bangladesh, Bosnia-Herzegovina, Burkina Faso, Cabo Verde, Chad, Egypt, Iran, Iraq, Jordan, Kuwait, Lebanon, Libya, Maldives, Mali, Mauritania, Morocco, Niger, Oman, Pakistan, Qatar, Saudi Arabia, Senegal, Somalia, State of Palestine, Sudan, Syria, Tajikistan, Timor-Leste, Tunisia, United Arab Emirates, Western Sahara, Yemen (North Yemen)
Muslim—Turkic	Albania, Azerbaijan, Kazakhstan, Kyrgyzstan, Turkey, Turkmenistan, Uzbekistan
Western—European	Andorra, Australia, Austria, Belgium, Christmas Island, Cocos (Keeling) Islands, Cook Islands, Denmark, Estonia, Faeroe Islands, Finland, France, Germany, Gibraltar, Guernsey, Iceland, Ireland, Israel, Italy, land Islands, Latvia, Liechtenstein, Lithuania, Luxembourg, Malta, Monaco, Netherlands, New Zealand, Norway, Portugal, Romania, Samoa, San Marino, Solomon Islands, Spain, St. Martin (French part), Sweden, Switzerland, Tuvalu, United Kingdom, Vanuatu
African	Angola, Benin, Botswana, British Indian Ocean Territory, Burundi, Cameroon, Central African Republic, Comoros, DR Congo (Zaire), Congo, Cote d'Ivoire, Djibouti, Equatorial Guinea, Eritrea, Ethiopia, French Southern Territories, Gabon, Gambia, Ghana, Guinea, Guinea-Bissau, Kenya, Lesotho, Madagascar (Malagasy), Malawi, Mayotte, Mozambique, Namibia, Nigeria, Runion, Rwanda, Saint Helena, Sao Tome and Principe, Sierra Leone, South Africa, South Sudan, Kingdom of eSwatini (Swaziland), Tanzania, Togo, Uganda, Zambia, Zimbabwe
Latin American	Japan, Anguilla, Antigua and Barbuda, Argentina, Bahamas, The, Barbados, Belize, Bermuda, Bolivia, Bonaire, Sint Eustatius and Saba, Brazil, Cayman Islands, Chile, Colombia, Costa Rica, Cuba, Curacao, Dominica, Dominican Republic, Ecuador, El Salvador, Falkland Islands (Malvinas), French Guiana, Grenada, Guadeloupe, Guatemala, Guyana, Haiti, Honduras, Jamaica, Martinique, Mexico, Montserrat, Nicaragua, Panama, Paraguay, Peru, Puerto Rico, Saint Barthlemy, Sint Maarten (Dutch part), South Georgia and the South Sandwich Islands, St. Kitts and Nevis, St. Lucia, St. Vincent and the Grenadines, Suriname, Trinidad and Tobago, Turks and Caicos Islands, Uruguay, Venezuela
Slavic—Eastern orthodox	Armenia, Belarus, Georgia, Moldova, Russia, Ukraine
Western—North American	Aruba, Canada, Greenland, Kiribati, Liberia, Marshall Islands, Micronesia, Fed. Sts., Nauru, Palau, Papua New Guinea, Philippines, Seychelles, Tonga, United States of America
Hindu	Vietnam, Bhutan, Cambodia (Kampuchea), Fiji, India, Mauritius, Myanmar (Burma), Nepal, Sri Lanka
Muslim—Malayan	Brunei Darussalam, Indonesia, Malaysia
Slavic—Central-Eastern orthodox	Bulgaria, Cyprus, Greece, Hungary, Macedonia, Montenegro, Serbia
Confucian	China, Hong Kong SAR, Korea, Dem. Rep., Korea, Rep., Laos, Macao SAR, China, Mongolia, Singapore, Taiwan, Province of China, Thailand, Vietnam
Slavic—Western catholic	Croatia, Czech Republic, Poland, Slovak Republic, Slovenia
Japanese	Japan

Criterion 2. The status of women in society reflects differences in the status of women and men in society. The Gender Inequality Index calculated by the United Nations Development Program [7] was used for evaluation. It shows the degree of gender inequality using three dimensions: reproductive health, empowerment and the labor market.

Criterion 3. Penetration of religion in people's lives shows how much society respects the right to practice religion of one's own choice, as well as how much religion influences social and political processes. As a quantitative assessment of the criterion, the Religion Index was taken, which determines the role of religious institutions in people's lives and is part of the Human Freedom Index [6].

Criterion 4. Ethnic homogeneity—in this case, it is defined as the probability that two randomly selected individuals of the country will be representatives of different ethnic groups, and is estimated using the Ethnic Fractionalization Index [8].

Criterion 5. Civilization's receptivity to other cultures reflects the degree of immutability of traditions and worldview of society. Criterion 6. Traditionalism of culture and thinking is measured by the traditions and worldview of society changing. For a generalized quantitative assessment of these criteria, the Social Globalization Index was used, which is calculated by The Swiss Institute of Technology in Zurich and takes into account the degree of informational, social and cultural exchange between countries [9].

Criterion 7. Radicalism of political life—how stable the political course and safe political life of society are. The criterion is measured using the Political Stability and Absence of Violence/Terrorism Index [10], which is actually defined by the World Bank as the probability of political instability or politically motivated violence based on a set of 9 indicators.

Thus, a system of indicators was formed to assess the level of ethno-cultural differences (Table 2).

2.2 *Formation of the Group of Economic and Social Indicators*

In previous studies, the identification of civilizations by social and economic indicators was not carried out. Therefore, we propose to form a set of social and economic indicators on the basis of the economic and social dimensions of sustainable development, which is carried out by the World Data Center for Geoinformatics and Sustainable Development in Ukraine [11].

The indicators chosen to characterize the economic level of civilization [12] are presented in Table 3.

This set of indicators includes the various economic results of the functioning of a group of countries belonging to a certain civilization: the level of production, international trade, investments and reserves. In order to take into account, the

Table 2 System of criteria characterizing ethno-cultural differences between civilizations

The name of the criterion	Code	Weight	Indicator
Individual freedom in society	C^1	0,231	Personal freedom index
The status of women in society	C^2	0,077	Gender inequality index
Penetration of religion in people's lives	C^3	0,231	Religion index
Ethnic homogeneity	C^4	0,077	Ethnic fractionalization index
Susceptibility of civilization to other cultures. Traditionalism of culture and thinking	C^5	0,153	Social globalization index
Radicalism of political life	C^6	0,231	Political stability and absence of violence/Terrorism index

Table 3 System of indicators characterizing economic differences between civilizations

The name of the indicator	Code	Weight	Units
Agriculture, forestry, and fishing, value added	E^1	0,118	Current USD
Industry (including construction), value added	E^2	0,118	Current USD
Gross domestic product	E^3	0,176	Current USD
Foreign direct investment, net	E^4	0,118	BoP, current USD
Total reserves	E^5	0,176	Includes gold, current USD
Exports of goods and services	E^6	0,118	Current USD
Imports of goods and services	E^7	0,118	Current USD
Market capitalization of listed domestic companies	E^8	0,059	Current USD

Table 4 System of indicators characterizing social differences between civilizations

The name of the indicator	Code	Weight	Units
Hospital beds	S^1	0,059	Per 1 000 people
The infant mortality rate	S^2	0,118	Per 1 000 live births
Total life expectancy at birth	S^3	0,176	Years
Government expenditure on education, total	S^4	0,118	% of GDP
Literacy rate, adult total	S^5	0,059	% of people ages >15
School enrollment, primary	S^6	0,118	% gross
Corruption perception index	S^7	0,176	Coef.
Income inequality coefficient	S^8	0,176	Coef.

different potential capabilities of civilizations depending on their size, the indicators are reduced to values per capita.

Indicators [12, 13], that were chosen to characterize the social level of civilization, are presented in Table 4.

This set of indicators takes into account the level of health, education, social security, manifestation of corruption and social inequality.

2.3 *Formalization of Civilization Faults*

According to the available data sets, the period 1995–2021 is considered in the study. The smallest object for measurement is a country.

Let every country X_i at each moment of time is characterized by three vectors:

$$X_i(t) = \{C_i(t), E_i(t), S_i(t)\}, \tag{1}$$

where i is a country number, $t = \overline{1995..2021}$, and

$$C_i(t) = (C_i^j(t))_{j=\overline{1..6}}, E_i(t) = (E_i^j(t))_{j=\overline{1..8}}, S_i(t) = (S_i^j(t))_{j=\overline{1..8}}. \tag{2}$$

The set of countries is divided into 13 subsets that correspond to the civilizations highlighted in [5]. Let's define the characteristic of civilization, as

$$V_k(t) = \{\frac{1}{card(I_k)} \sum_{i \in I_k} C_i(t), \frac{1}{card(I_k)} \sum_{i \in I_k} E_i(t), \frac{1}{card(I_k)} \sum_{i \in I_k} S_i(t)\}, \quad (3)$$

where k is a civilization number, I_k is a set of indices of the countries included in $k - th$ civilization.

To avoid data heterogeneity, all indicators are normalized according to the minmax formula:

$$P_{i,norm} = \frac{P^i - \min_i P_i}{\max_i P_i - \min_i P_i}, \quad (4)$$

where P_i is a specific value of a certain parameter, i is a parameter value number in the entire set.

To determine the degree of difference of objects, we will choose the weighted Euclidean distance in the space of selected indicators separately for cultural, economic and social dimensions. Thus, the difference of objects will be characterized by the following vector:

$$Df_{o1,o2}(t) = (Dc_{o1,o2}(t), De_{o1,o2}(t), Ds_{o1,o2}(t)) =$$

$$= (\sqrt{\sum_{j=1}^{6} \sigma_j (C_{o1}^j(t) - C_{o2}^j(t)}, \sqrt{\sum_{j=1}^{8} \phi_j (E_{o1}^j(t) - E_{o2}^j(t)}, \sqrt{\sum_{j=1}^{8} \omega_j (S_{o1}^j(t) - S_{o2}^j(t))}, \quad (5)$$

where $o1$, $o2$ are object numbers being compared, σ_j, ϕ_j, ω_j are weight coefficients of indicators of cultural, economic and social dimensions, respectively. Objects can be understood as both civilizations and individual countries.

Let the vector $Df_{o1,o2}(t)$ characterizes the degree of ethno-cultural, economic, and social fault lines, and we introduce the integral degree of the fault line as the Minkowski norm:

$$D_{o1,o2}(t) = ((Dc_{o1,o2}(t))^3 + (De_{o1,o2}(t))^3 + (Ds_{o1,o2}(t))^3)^{\frac{1}{3}}. \quad (6)$$

3 Analysis of the Intercivilization Fault Level Dynamics

The results characterizing the ethno-cultural, economic and social dimensions of the fault lines between civilizations were obtained.

Figure 1 presents the sum of ethno-cultural distances between civilizations for the 1995–2021. As we can see, the global trend of ethno-cultural differences tends to be cyclical with peaks in 2001 and 2015. From 2019, the size of the fault began to increase, with a projected peak in 2023–2024.

Table 5 presents the distances between civilizations in the ethno-cultural dimension in 2021. The biggest difference, and therefore possible conflict with others, in Japanese, African, Muslim—Arabic and Western—European civilizations. This

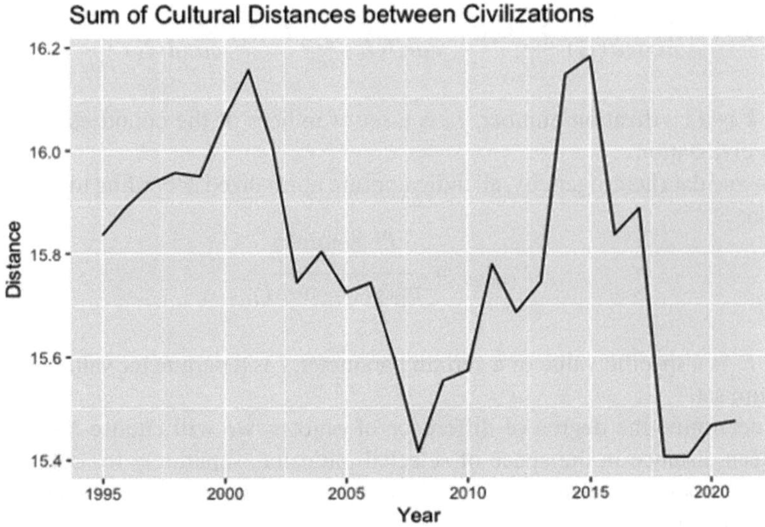

Fig. 1 An integral measure of the ethno-cultural dimension of the rifts between civilizations for 1995–2021

Table 5 Measure of the ethno-cultural dimension of rifts between civilizations, 2021: 1 African; 2 Confucian; 3 Hindu; 4 Japanese; 5 Latin American; 6 Muslim—Arabic; 7 Muslim—Malayan; 8 Muslim—Turkic; 9 Slavic—Western catholic; 10 Slavic—Central-Eastern orthodox; 11 Slavic—Eastern orthodox; 12 Western—European; 13 Western—North American

	1	2	3	4	5	6	7	8	9	10	11	12	13	Confl. level
1		0,211	0,099	0,384	0,190	0,146	0,211	0,184	0,327	0,273	0,200	0,353	0,189	2,769
2	0,211		0,119	0,238	0,140	0,196	0,162	0,102	0,198	0,134	0,109	0,227	0,179	2,016
3	0,099	0,119		0,306	0,125	0,137	0,142	0,112	0,248	0,187	0,122	0,276	0,137	2,009
4	0,384	0,238	0,306		0,202	0,414	0,358	0,322	0,083	0,144	0,264	0,078	0,239	3,031
5	0,190	0,140	0,125	0,202		0,249	0,223	0,194	0,146	0,115	0,149	0,172	0,076	1,981
6	0,146	0,196	0,137	0,414	0,249		0,156	0,113	0,357	0,283	0,169	0,388	0,270	2,879
7	0,211	0,162	0,142	0,358	0,223	0,156		0,122	0,294	0,228	0,166	0,320	0,221	2,603
8	0,184	0,102	0,112	0,322	0,194	0,113	0,122		0,269	0,192	0,087	0,299	0,226	2,221
9	0,327	0,198	0,248	0,083	0,146	0,357	0,294	0,269		0,080	0,205	0,034	0,178	2,419
10	0,273	0,134	0,187	0,144	0,115	0,283	0,228	0,192	0,080		0,128	0,111	0,160	2,034
11	0,200	0,109	0,122	0,264	0,149	0,169	0,166	0,087	0,205	0,128		0,236	0,193	2,028
12	0,353	0,227	0,276	0,078	0,172	0,388	0,320	0,299	0,034	0,111	0,236		0,195	2,690
13	0,189	0,179	0,137	0,239	0,076	0,270	0,221	0,226	0,178	0,160	0,193	0,195		2,263

Table 6 Changes in the measure of the ethno-cultural dimension of faults between civilizations (largest and smallest values), 1995–2021

Civilization 1	Civilization 2	$Dc_{o1,o2}(t)$	Value (%)
Muslim—Turkic	Slavic—Eastern orthodox	0,045	+67,40
African	Hindu	0,033	+38,97
Muslim—Arabic	Hindu	0,041	+36,79
Muslim—Turkic	Hindu	0,023	+23,16
African	Slavic—Eastern orthodox	0,041	+22,51
Hindu	Slavic—Central-Eastern orthodox	–0,051	–22,71
Slavic—Central-Eastern orthodox	Slavic—Western catholic	–0,022	–24,97
Latin American	Hindu	–0,052	–32,82
Western—North American	Hindu	–0,061	–34,40
Western—European	Slavic—Western catholic	–0,013	–38,62

is explained by the difference in cultural values in society. Japanese civilization has always been separated from the rest of the world, which influenced the formation of its ethno-cultural features, which have a significant contribution even today. Western—European civilization is characterized by the most developed democracy, freedom, a diversity of cultures and political stability. African civilization is characterized by significant cultural diversity with a small level of political stability and gender equality, and Muslim—Arabic—by a significant influence of religious institutions.

Table 6 presents the five pairs of civilizations that increased the cultural gap the most, as well as the five pairs that decreased it the most. The Slavic-Eastern orthodox civilization increases the cultural gap with Muslim and African countries. Also, Muslim civilizations are becoming more distant from Hindu civilization, which may indicate tensions in the region.

Figure 2 presents the sum of economic distances between civilizations for 1995–2021. During the period under study, there is a trend towards increasing differences in the economic development of civilizations. The integral value of the economical measurement of faults increased by 1.847 times. However, it should be noted that as a result of the cumulative manifestation of economic differences, civilization is inferior to ethno-cultural ones by almost 4 times. This testifies to the accelerated course of the processes of economic globalization, compared to the ethno-cultural one, and is largely determined by the digitalization of economic processes, which leads to the merging of individual national markets into a single world market.

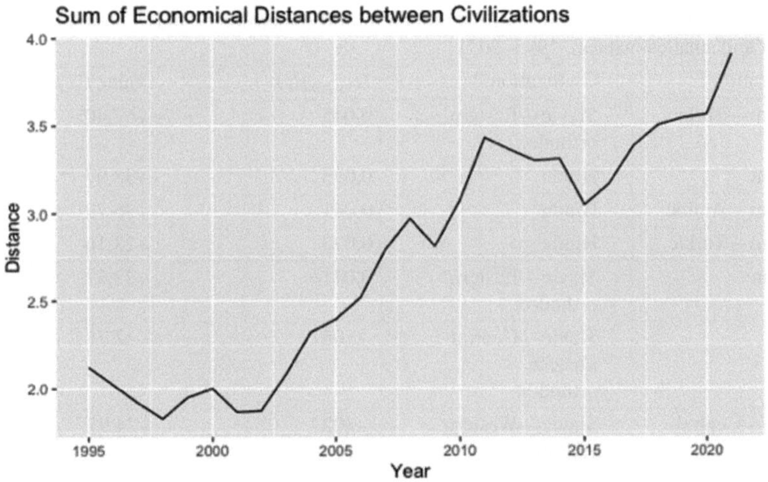

Fig. 2 An integral measure of the economic dimension of faults between civilizations, 1995–2021

Table 7 The measure of the economic dimension of the fault lines between civilizations, 2021: 1 African; 2 Confucian; 3 Hindu; 4 Japanese; 5 Latin American; 6 Muslim—Arabic; 7 Muslim—Malayan; 8 Muslim—Turkic; 9 Slavic—Western catholic; 10 Slavic—Central-Eastern orthodox; 11 Slavic—Eastern orthodox; 12 Western—European; 13 Western—North American

	1	2	3	4	5	6	7	8	9	10	11	12	13	Confl. level
1		0,114	0,007	0,080	0,029	0,018	0,037	0,013	0,052	0,032	0,012	0,121	0,049	0,563
2	0,114		0,110	0,073	0,096	0,098	0,088	0,108	0,078	0,092	0,106	0,078	0,088	1,129
3	0,007	0,110		0,076	0,025	0,014	0,032	0,007	0,048	0,028	0,007	0,116	0,043	0,513
4	0,080	0,073	0,076		0,055	0,064	0,051	0,072	0,042	0,055	0,070	0,060	0,052	0,748
5	0,029	0,096	0,025	0,055		0,016	0,015	0,021	0,026	0,007	0,019	0,093	0,028	0,430
6	0,018	0,098	0,014	0,064	0,016		0,022	0,014	0,036	0,019	0,011	0,105	0,037	0,438
7	0,037	0,088	0,032	0,051	0,015	0,022		0,028	0,020	0,011	0,026	0,085	0,023	0,483
8	0,013	0,108	0,007	0,072	0,021	0,014	0,028		0,044	0,023	0,004	0,112	0,037	0,512
9	0,052	0,078	0,048	0,042	0,026	0,036	0,020	0,044		0,022	0,042	0,070	0,033	0,424
10	0,032	0,092	0,028	0,055	0,007	0,019	0,011	0,023	0,022		0,021	0,089	0,025	0,464
11	0,012	0,106	0,007	0,070	0,019	0,011	0,026	0,004	0,042	0,021		0,110	0,037	1,122
12	0,121	0,078	0,116	0,060	0,093	0,105	0,085	0,112	0,070	0,089	0,110		0,083	0,534
13	0,049	0,088	0,043	0,052	0,028	0,037	0,023	0,037	0,033	0,025	0,037	0,083		0,454

Table 7 presents the distances between civilizations in economic terms in 2021. The biggest difference, and therefore possible conflict with others, in Confucian, Western—European and Japanese civilizations, which is definitely explained by the dynamics of rapid economic development of their countries.

Due to the positive global trend of the economic dimension of faults the growing distance between civilizations is natural. The five pairs of civilizations that,

Table 8 Changes in the measure of the economic dimension of faults between civilizations (largest and smallest values), 1995–2021

Civilization 1	Civilization 2	$De_{o1,o2}(t)$	Value (%)
Latin American	Confucian	+0,082	+177,23
Muslim—Malayan	Confucian	+0,073	+164,25
Slavic—Central-Eastern orthodox	Slavic—Western catholic	+0,017	+152,08
Slavic—Central-Eastern orthodox	Confucian	+0,072	+148,32
Muslim—Arabic	Confucian	+0,074	+147,58
Slavic—Central-Eastern orthodox	Japanese	–0,018	-31,40
Muslim—Arabic	Slavic—Eastern orthodox	–0,004	–33,45
Muslim—Malayan	Japanese	–0,019	–36,34
Latin American	Slavic—Central-Eastern orthodox	–0,004	-50,21
Slavic—Western catholic	Japanese	–0,032	–61,44

according to the results of simulations for the period 1995–2021, increased the economic gap the most, as well as the five pairs that decreased it the most, are presented in Table 8. In this sense, two trends can be determined: the rapidly growing distance of Confucian civilization from others, as well as the minor declining distance of Japanese civilization from others.

Figure 3 presents the sum of social distances between civilizations for 1995–2021. During the studied period, there is a tendency to decrease differences in the social development of civilizations. The integral value of the fault measurement decreased by 1.227 times.

Table 9 presents the distances between civilizations in the social dimension in 2021. The biggest difference, and therefore a possible conflict with others, in Japanese, African and Western—European civilizations. This is explained by the difference in their social processes in society. Japanese and Western—European civilizations have always been characterized by a high degree of social development, while African civilization mostly shows the opposite trend.

Due to the negative global trend in the social dimension of faults, only a few pairs of civilizations have increased distances since 1995 by more than 1% of the average distance. These pairs, as well as those that have become the closest, are presented in the Table 10. It can be seen that social rifts have grown between European civilizations, and the Confucian and Muslim-Arabic differences with other civilizations

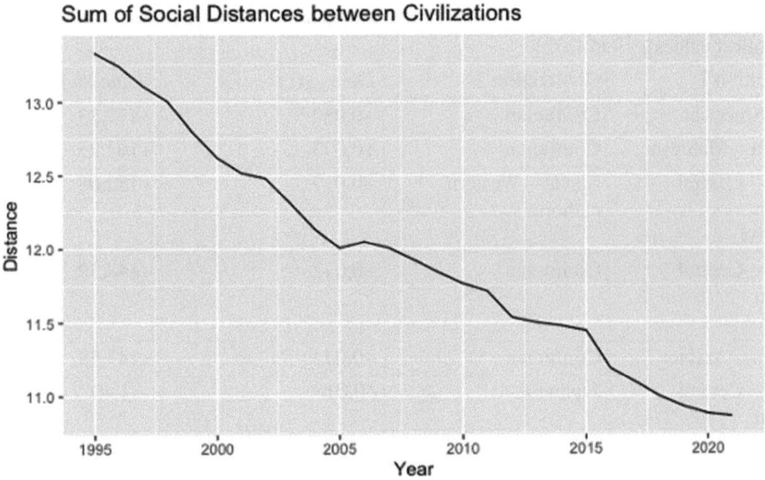

Fig. 3 An integral measure of the social dimension of the rifts between civilizations, 1995–2021

Table 9 Measure of the social dimension of rifts between civilizations, 2021: 1 African; 2 Confucian; 3 Hindu; 4 Japanese; 5 Latin American; 6 Muslim—Arabic; 7 Muslim—Malayan; 8 Muslim—Turkic; 9 Slavic—Western catholic; 10 Slavic—Central-Eastern orthodox; 11 Slavic—Eastern orthodox; 12 Western—European; 13 Western—North American

	1	2	3	4	5	6	7	8	9	10	11	12	13	Confl. level
1		0,195	0,134	0,333	0,126	0,121	0,154	0,162	0,264	0,192	0,210	0,277	0,161	2,328
2	0,195		0,091	0,158	0,124	0,122	0,071	0,108	0,097	0,036	0,089	0,116	0,081	1,288
3	0,134	0,091		0,235	0,121	0,051	0,078	0,085	0,137	0,086	0,100	0,165	0,064	1,349
4	0,333	0,158	0,235		0,256	0,268	0,209	0,254	0,156	0,182	0,204	0,120	0,205	2,580
5	0,126	0,124	0,121	0,256		0,135	0,077	0,144	0,212	0,126	0,179	0,205	0,124	1,829
6	0,121	0,122	0,051	0,268	0,135		0,109	0,078	0,163	0,109	0,114	0,206	0,097	1,571
7	0,154	0,071	0,078	0,209	0,077	0,109		0,116	0,146	0,074	0,126	0,142	0,070	1,373
8	0,162	0,108	0,085	0,254	0,144	0,078	0,116		0,141	0,082	0,071	0,207	0,118	1,566
9	0,264	0,097	0,137	0,156	0,212	0,163	0,146	0,141		0,092	0,081	0,107	0,123	1,718
10	0,192	0,036	0,086	0,182	0,126	0,109	0,074	0,082	0,092		0,067	0,136	0,085	1,267
11	0,210	0,089	0,100	0,204	0,179	0,114	0,126	0,071	0,081	0,067		0,164	0,111	1,516
12	0,277	0,116	0,165	0,120	0,205	0,206	0,142	0,207	0,107	0,136	0,164		0,128	1,974
13	0,161	0,081	0,064	0,205	0,124	0,097	0,070	0,118	0,123	0,085	0,111	0,128		1,366

are also increasing. The social connection between the Muslim-Arabic and Muslim-Turkic civilizations, as well as between the Hindu and other civilizations, became even closer.

The integral value of faults in the world as a whole is slightly decreasing—by 7% as of 2021 (Fig. 4). In general, this indicates the growing processes of globalization,

Table 10 Changes in the measure of the social dimension of faults between civilizations (largest and smallest values), 1995–2021

Civilization 1	Civilization 2	$Ds_{o1,o2}(t)$	Value (%)
Slavic—Central-Eastern orthodox	Slavic—Western catholic	+0,015	+18,49
Western—European	Slavic—Western catholic	+0,009	+8,94
Muslim—Arabic	Hindu	+0,004	+8,37
Western—European	Slavic—Central-Eastern orthodox	+0,009	+7,03
Western—North American	Confucian	+0,005	+6,01
Muslim—Arabic	Confucian	+0,003	+2,85
Western—European	Japanese	+0,002	+1,57
Slavic—Eastern orthodox	Hindu	–0,058	–46,18
African	Latin American	–0,090	–50,55
Hindu	Slavic—Central-Eastern orthodox	–0,063	–53,10
Muslim—Arabic	Muslim—Turkic	–0,056	–56,29
Slavic—Central-Eastern orthodox	Confucian	–0,030	–73,41

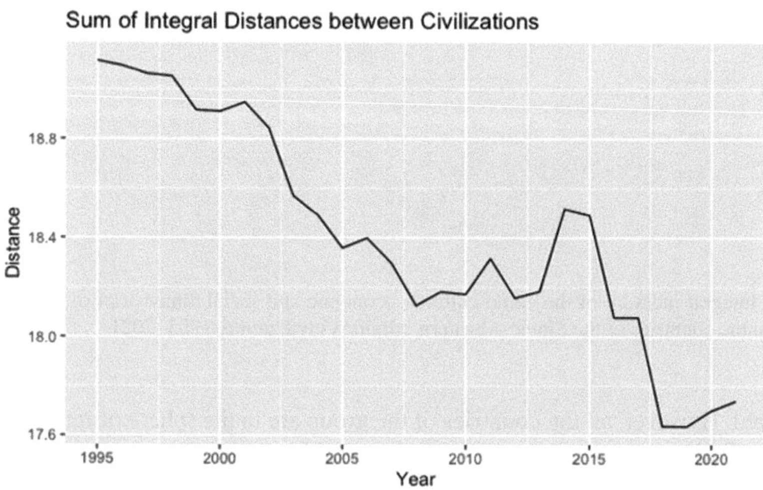

Fig. 4 An integral measure of faults between civilizations, 1995–2021

which continue to spread throughout the world. In the Table 11 shows the civilizations with the greatest changes in distances.

Let's consider the Slavic—Eastern Orthodox civilization, which includes Ukraine, Belarus, Armenia, Georgia, Moldova and the Russian Federation. Figure 5 presents the sums of the distances between countries according to fault measurements. During the studied period, there is a trend of growing cultural distance in this group of countries. Also, until 2013–2014, the growth of the economic and social gap was

Table 11 Change in the measurement of faults between civilizations (largest and smallest values), 1995–2021

Civilization 1	Civilization 2	$D_{o1,o2}(t)$	Value (%)
Muslim—Arabic	Hindu	+0,039	+34,43
Muslim—Turkic	Slavic—Eastern orthodox	+0,017	+20,63
Muslim—Turkic	Confucian	+0,022	+16,86
Western—European	Slavic—Western catholic	+0,014	+12,78
Slavic—Eastern orthodox	Confucian	+0,014	+11,21
African	Latin American	–0,053	–22,33
Muslim—Arabic	Muslim—Turkic	–0,031	–23,02
Hindu	Slavic—Central-Eastern orthodox	–0,063	–26,78
Western—North American	Hindu	–0,064	–34,63
Latin American	Hindu	–0,068	–35,76

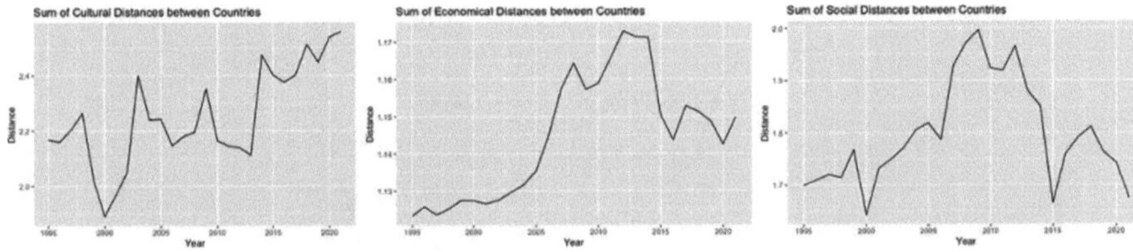

Fig. 5 Integral measure of the ethno-cultural, economic and social dimensions of the fault lines between the countries of the Slavic—Eastern orthodox civilization, 1995–2021

observed. However, all the countries of the group are in the sphere of interests of the Russian Federation, which by its actions was able to stop the increase in economic and social distances.

It is also necessary to understand the place of Ukraine in civilization (Fig. 6). In fact, the peak of Ukraine's cultural, economic and social differences occurred in 2010–2014. Thus, the actions of the Russian Federation as an aggressor not only directly affected the social and economic development of the country, but also stopped the processes of differentiation of countries within the boundaries of civilization.

S. Huntington assumed that Ukraine could split politically into halves. He also explained how the confrontation of civilizations takes place: defending one's borders, one's language and one's social structure, which is followed in Russia's war with Ukraine. From the beginning of its independence, Ukraine has been increasing its differences with other countries of the Slavic—Eastern Orthodox civilization, which

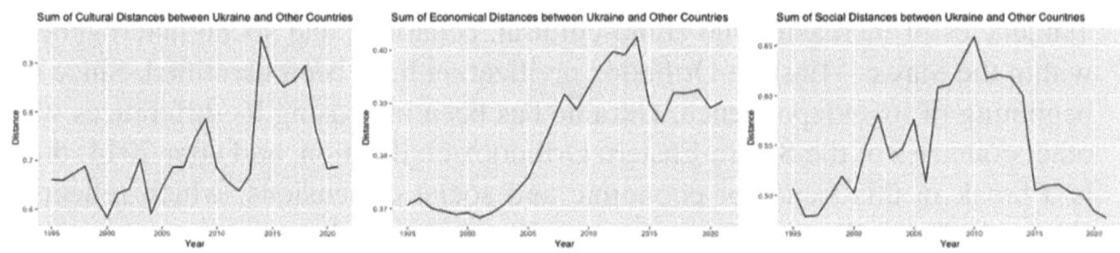

Fig. 6 Integral measure of the ethno-cultural, economic and social dimensions of the rifts between Ukraine and the countries of the Slavic—Eastern orthodox civilization for 1995–2021

could not happen according to the political paradigm of Russia, and this led to the beginning of military confrontation, and then to a full-scale invasion.

The only thing that cannot be explained in this war with the help of S. Huntington's theory is that the war is between two countries that are in demographic stagnation, and not as a result of the clash of two overpopulated civilizations.

4 Conclusion

The study examines the dynamics of faults between civilizations, which are considered as one of the important factors in the emergence and course of modern world conflicts that have a civilizational character and, as a result, change the vector of global development.

1. A system of indicators and a mathematical model are proposed for determining the ethno-cultural, economic and social dimension of faults between civilizations in dynamics. Quantitative values of fault values for 1995–2021 were determined, conflict levels of civilizations were calculated.
2. A cyclical change in the integral value of the ethno-cultural dimension of fault lines in combination with a significant trend component was revealed, which generally indicates the aspiration of each civilization to preserve its inherent ethno-cultural features and their self-identification.
3. In combination with the deepening of cultural differences, an additional threat may be the increasing trend of the economic dimension of the rifts between civilizations, as the unfair distribution of material goods has historically served as the cause of conflict situations. The fact that the measure of the economic dimension of faults is inferior to the cultural one by almost 4 times, indicates the accelerated course of the processes of economic globalization, compared to the ethno-cultural one.
4. The fact that, according to the results of modeling, the social dimension of the faults between civilizations is significantly decreasing over time, also clearly indicates the acceleration of globalization processes. This phenomenon is natural due to the fact that access to information, goods and services is facilitated, the standard of living of people generally increases, etc.

5. Tendencies of increasing the ethno-cultural, economic and social heterogeneity within the Slavic—Eastern Orthodox civilization have been identified. Since the beginning of its independence, Ukraine has been increasing its differences with other countries of the Slavic-Eastern Orthodox civilization, and after 2014, there is a break in this trend for economic and social dimensions, which reflects a change in the civilizational paradigm of Ukraine's development.

References

1. Huntington, S.: The clash of civilizations?. Foreign Aff. **72**(3), 22–49 (1993). https://doi.org/10.2307/20045621
2. Meaney, T.: Putin Wants a Clash of Civilizations. Is "The West" Falling for It?. The New York Times (2022). https://www.nytimes.com/2022/03/11/opinion/nato-russia-the-west-ukraine.html?searchResultPosition=1
3. Caldwell, C.H.: After the clash of civilizations. Compact. (2022)
4. Roy, O.: Ukraine and the Clash of Civilization Theory. Robert Schuman Centre for Advanced Studies of the European University Institute (2022). https://www.eui.eu/news-hub?id=ukraine-and-the-clash-of-civilisation-theory-an-interview-with-oliver-roy
5. Zgurovsky, M., Kravchenko, M., Pyshnograiev, I., Perestyuk, M.: Modeling of the intercivilization fault effect on the conflict intensity throughout the world. Syst. Res. & Inf. Technol. **4**, 7–26 (2021). https://doi.org/10.20535/SRIT.2308-8893.2021.4.01
6. Human Freedom Index. The Cato Institute and the Fraser Institute (2022). https://www.cato.org/human-freedom-index/2022
7. Gender Inequality Index. United Nations Development Programme (2021). https://hdr.undp.org/data-center/thematic-composite-indices/gender-inequality-index#/indicies/GII
8. Fearon, J.: Ethnic and cultural diversity by country. J. Econ. Growth. **8**, 195–222 (2023). https://doi.org/10.1023/A:1024419522867
9. Savina, G., Haelg, F., Potrafke, N., Sturm, J.-E.: The KOF globalisation index—revisited. Rev. Int. Organ. **14**(3–7), 543–574 (2019). https://doi.org/10.1007/s11558-019-09344-2
10. The Worldwide Governance Indicators (WGI) Project. The World Bank (2022). http://info.worldbank.org/governance/wgi
11. Zgurovsky, M., Yefremov, K., Pyshnograiev, I., Boldak, A., Dzhygyrey, I.: Quality and security of life: a cross-country analysis. In: 2022 IEEE 3rd International Conference on System Analysis & Intelligent Computing (SAIC), pp. 1–5 (2022). https://doi.org/10.1109/SAIC57818.2022.9923006
12. World Bank Open Data. The World Bank (2022). https://data.worldbank.org
13. Corruption Perception Index. Transparency International (2022). https://www.transparency.org/en/cpi/2022

Exploring the Vulnerability of Social Media for Crowdsourced Intelligence Under a False Flag

Illia Varzhanskyi

Abstract This research examines hybrid intelligence-gathering techniques that incorporate elements of human intelligence (HUMINT) and open-source intelligence (OSINT) in today's information environment. It draws attention to how disinformation techniques, such as fake pages and personal accounts on social media, can be automated within a short time to collect and process large amounts of eyewitness information to increase the effectiveness of OSINT. The research introduces the concept of crowdsourced intelligence under a false flag as one of the vectors of active OSINT. The risks associated with disguised intelligence gathering through communication with enemy civilians and military personnel on social media are identified. Examples of such cases from the history of the Russian-Ukrainian war since 2014 are disclosed. The research also discusses the new challenges arising in countering information operations. The research concludes by suggesting possible methods to detect and disrupt such operations, thereby providing protection against unwanted infiltration and intelligence gathering on social media. A mathematical apparatus has been developed based on the data collected on alleged Russian information operations in the Telegram messenger, which allows estimating the probability that an account is a sybil bot. Some scenarios of applying systems for automated detection of information operations in messengers are also proposed.

Keywords OSINT · Crowdsourced intelligence · False flag · Reflexive control · Sybil bots · Information operations

I. Varzhanskyi (✉)
National Technical University of Ukraine "Igor Sikorsky Kyiv Polytechnic Institute", Kyiv 03056, Ukraine
e-mail: i.varzhanskyi@sms.ed.ac.uk

M. Zgurovsky and N. Pankratova (eds.), *System Analysis and Artificial Intelligence*, Studies in Computational Intelligence 1107,
https://doi.org/10.1007/978-3-031-37450-0_26

1 Introduction

The development of social media has led to many data of interest to the Intelligence Community being made publicly available. It, in turn, has influenced the creation of new Open Source Intelligence (OSINT) techniques adapted for use in the new environment. Some researchers even distinguish this area as a specific type of OSINT called SOCINT. For example, social media accounts can be used to identify the location, interests and activities of specific users with minimal risk and resources. Most current Internet information-gathering techniques, such as web scraping and social network analysis, can be easily automated. It allows large amounts of information to be collected and processed at short notice, increasing research efficiency and allowing researchers to respond quickly to changes in the area under study. The vast amount of information that reaches social media makes it possible to use OSINT to obtain even sensitive information. For example, it is known that publishing anonymous statistics of popular jogging routes from fintech trackers has inadvertently revealed the location of military bases and even the identities of military personnel [1]. Another example: during the preparation and conduct of combat operations, it is photos and videos of the movement of military equipment, firing and locations of hits from ordinary social media users that are an essential source of intelligence data [2]. Tasks such as missile strike corrections, for example, require a high speed of decision-making and, therefore, intelligence acquisition. Therefore, OSINT is one alternative for operational intelligence in the deep rear.

However, public outreach and the criminalization of the dissemination of information that can be exploited by the enemy bring rewards for reducing these phenomena during a military conflict. Adversaries, therefore, have to resort to what is known as active OSINT. Active OSINT involves interacting with a source to obtain information unavailable through passive analysis, such as using social engineering techniques to subscribe to a closed social networking page. One type of active OSINT, combining OSINT and HUMINT methodologically, is termed crowdsourced intelligence. This technique derives from the basic principles of operational intelligence, which have been based primarily on eyewitness testimony throughout the history of law enforcement. In turn, the development of social networks allows witnesses to be interviewed on a larger scale and at a higher speed. Whereas the survey is carried out through a thematically motivated community, each member can also survey their circle of contacts, increasing their reach, or use their OSINT skills to search in more detail in an area that they know. At the same time, however, the risk of disguised intelligence gathering through social networking with civilians and even enemy soldiers increases. In this research, we look specifically at scenarios of false-flag intelligence gathering in social media, using examples from the history of the Russian-Ukrainian war since 2014 and possible methods to detect and disrupt such operations.

2 Classification of "Crowdsourced Intelligence Under a False Flag" Information Operations

In general, social media intelligence-gathering methods can be classified into two categories: personal and mass communications. Personal communications involve masquerading as individual accounts of people who may ask questions, ask for help, or assert something to evoke certain emotions in the audience and obtain credible information as a rebuttal. The use of this technology by the Russian Federation in 2022 was reported by the State Service for Special Communications and Information Protection of Ukraine [3]. In the comments of Facebook groups and open Telegram chats, as well as in the comments of the accounts of opinion leaders that relate to the cities shelled by the Russian Armed Forces, the same type of questions from fake accounts began to appear en masse, which mainly concerned the degree of damage from the bombing and the exact coordinates of the hits, probably, to adjust the fire. Since the start of full-scale hostilities in February 2022, users themselves have been actively sharing personal photos and videos of missile strikes. But the Ministry of Defence of Ukraine has conducted broad information and explanatory campaign among the population about the inadmissibility of public disclosure of the exact objects of destruction, the location of air defence systems, and other possible military targets [4]. Therefore, in order to encourage Ukrainian users to disclose information, some typical and template questions with emotional overtones were framed as follows: "What do you have there, Ivano-Frankivsk? Although I'm far from home, my heart hurts with every alarm". Another type of questions concerned open routes to the cities of the Kyiv region, for example: how to get to Kyiv, who can give a ride with a baby to Kyiv, how to get from Brovary to Boryspil. The Office of the National Security and Defense Council of Ukraine officially warned about the mass appearance of such messages in the first days of a full-scale invasion [5]. When reaching large audiences in hundreds of open WhatsApp, Viber, Telegram, and Facebook chat, the likelihood of obtaining useful intelligence information increases, and the data processing is automated. We do not know conclusively whether comments are manually disseminated through fake accounts or whether it is also automatic, but until recently, the deterrent to endlessly scaling up this approach was the need to constantly change the wording of the issue to avoid monotony. Unfortunately, using the latest autoregressive generative language models like GPT-4 can effectively address this challenge, automatically generating relatively compelling social media messages with the right tone, style and other parameters. The automatic distribution of the posts and the collection of the information received in response makes it possible to reach all public chat rooms around the clock to overcome the limited capacity of local moderation. A modern variation of the traditional means of HUMINT—information gathering through dating sites—has also become widespread. Depending on some circumstances, military personnel in correspondence may give away information about their psycho-emotional state in the unit, their location, and other, no less sensitive, information. Thus, the only relevant countermeasure remains to raise public awareness and spread skills to identify fake accounts by their key attributes, which will be discussed in detail in the next section.

Mass methods of gathering intelligence through social media do not involve creating fake identities, but specific Internet resources that aggregate the necessary information. The first version of such technology is the setting up and sharing of open chat rooms to exchange information about military activities, emergencies, accessible routes for travelling through the territory where the fighting is taking place, etc. According to Russian information security specialists, this idea has already been developed in Ukrainian active measures [6]. Some "pseudo-activists" themselves create mobilization chats in the Telegram messenger for different cities in the Russian Federation. Using database leaks from Russian government agencies and private companies, it becomes possible to identify Telegram accounts that joined to chats and thus identify Russian military personnel and their relatives. If the standard measures to counter such information operations include complaints to group administrators about suspicious questions, then chat control allows you not only to ignore warnings but also to block and delete all messages from users who have discovered the fraud, which increases the duration of the operation. Also, the authority of the "group administrator" allows you to effectively implement individual HUMINT activities regarding users in the chat.

The second option is the creation of improvised means of mass communication in social media, such as Telegram channels. This format has significant potential to gain an audience by publishing high-profile materials, such as photos and videos of shelling, damage after shelling, movement of military equipment, and air defence operations. Since, during martial law, reliable Internet resources refrain from publishing such materials or censor them heavily, the format of anonymous Telegram channels makes it possible to gather an audience around impressive "forbidden" footage. At the same time as the audience grows, i.e. the channel's popularity grows, its coverage among the potential authors of the videos increases. The Telegram Channel Administrators recruited or controlled by the adversary may not always be able to publish the material received. What matters is the inbound flow of eyewitness accounts, which grows in proportion to the size of the audience. Maintaining a popular channel on military topics under a false flag allows targeted information-gathering activities to be carried out regularly, such as announcing the collection of information on problems involving material support and mismanagement, ostensibly to centralize it to the top political and military leadership or the military prosecutor's office. The primary target audience of such operations are relatives of military personnel who, on the one hand, stay in touch with them and are aware of the main problems but, on the other hand, have more time to use social media and are less attentive to information security.

A third variant of mass information-gathering technology through social networks is the creation of false-flag websites to obtain information about dead, wounded and captured military personnel. As a rule, an argument is made on such a site about the inadmissibility of publishing lists in the public domain. Therefore, any interested parties are required to fill in an online form with the personal data of the wanted military personnel. Because of filling in the online form, the user may receive a universal answer, as if "such a person is not in the lists", but in reality, the data obtained is collected for intelligence purposes. Links to such sites are distributed through fake accounts in comments under topical social media posts.

A fourth variant of the technology is the distribution through social networks of links to specially created video games for mobile devices that induce users to take pictures of critical infrastructure. In 2022, the Ukrainian Security Service claimed to have discovered a mobile app that lured users with geolocation and photos of strategically vital facilities in the form of a game [7]. Players took pictures of the terrain, including military and critical infrastructure in various locations, while traversing the route. All geolocation information of the quest participants was uploaded to the database of the game application. According to the investigation, an IT company controlled by Russian special services had access to the application's administration.

In tasks that involve covertly imposing disadvantageous solutions on an adversary, there is one significant methodological problem. Since in this research, we consider the experience of the Russian-Ukrainian conflict, to describe this problem, we will use the Soviet theory of reflexive control, which is popular among Russian military commanders [8–17]. Thus, if reflexive control is the transfer to the adversaries of the reasons for their decision, then the rank of reflexive control in this context, is a quantitative assessment of the level of the structural complexity of the operation of reflexive control. For example, publishing information on behalf of fake accounts intended to cause panic in the civilian population is a reflexive control of the first rank. Exposing an enemy agent and passing disinformation (chickenfeed) through him is a reflexive control of the second rank. Hypothetically, the index of the rank is not limited. Nevertheless, many scientific works show that, as a practical matter, the stable implementation of reflexive control of the third rank and above is highly complicated [18]. The only scientific way of determining the rank of an adversary's reflexivity is based on the principles of fuzzy logic, particularly the accumulation of a specific database of cases of reflexive control in the past and a rough estimate of the probability of using a particular rank of reflexivity depending on the frequency of its use in the past [19]. The widespread use of fake accounts, news, websites and other tools of reflexive control employed by the Russian security services, which peaked in February 2022 after the full-scale invasion began, could not go unnoticed by the Ukrainian and international communities. Thanks to the information work of Ukrainian state authorities, experts and journalists, the population was informed promptly about the many forms and methods of disinformation and the enemy's reconnaissance activities in social media, including those mentioned above. After an initial jump in vigilance that translated into, to some extent, paranoid sentiment among the general population, the average citizen's reflection rank came closer to the first—an awareness of themselves as a possible target of manipulation. Thus, the increased awareness of the population quickly enough led to a drop in the effectiveness of the widespread methods of active OSINT under a false flag (first-rank reflexive control technologies). This can be seen in more active moderation of comments by social media administrators, exposure of fake accounts and news by ordinary users, and increased vigilance in face-to-face interactions when dating online. On the other hand, receiving timely information about enemy intelligence-gathering regarding a specific target is often a significant outcome, which can sometimes even predetermine the course of combat operations. For example, if the information is received that fake messenger accounts are currently trying to find out which road from one city to another is free to pass through, it can be concluded that a sabotage and reconnaissance group is planning to move in. By applying second-rank reflexive control in this situation, counter-intelligence agencies can realistically "answer" the questions posed by advising a specific "free" road. If the adversaries do not have the sufficient reflexive rank to recognize disinformation or other channels for verifying information, they may trust this data and take advantage of it, which should lead to their safe neutralization. As a result, templated intelligence work through open sources can be ineffective and even dangerous for its initiator. Meanwhile, successful second-rate reflective management of this type requires implementing a system to monitor social media and automatically detect anomalies that could indicate the conduct of in-person reconnaissance activities or the distribution of links to resources for mass variants of crowdsourced intelligence under a false flag. In the next section, we look at tools to detect attempts at such social media intelligence operations.

3 Automatic Detection of Crowdsourced Intelligence Under a False Flag in Social Networks

Automated detection of reconnaissance operations on social media is a significant information security challenge. Collecting sensitive information under a false flag can threaten national security as well as individuals and businesses in peacetime, serving as a preparatory stage for cyberattacks or other criminal activities. Recognition of informational operation can be accomplished by algorithms that analyse text and graph data related to user activity on social media. However, traditional methods of detecting information operations are not always suitable for detecting intelligence gathering. For example, the method proposed by D. Lande [20] is based on a scheme to spread disinformation through multiple low-rated sources, but it is rarely used for crowdsourced intelligence. Since the bulk of the resources through which mass information is acquired are resources with static content, automating their detection may be limited to monitoring indexed links and web page content for keyword sets related to critical infrastructure or other likely adversary targets. Analysing and identifying the owners of all resources that actively inform on military and special topics is an ongoing task for national security counterintelligence units. Therefore, in this section, we will focus on identifying false-flag information operations of the personality type, typically conducted from fake personal accounts.

Automatic detection of fake social media accounts is the process of using computer algorithms, including machine-learning techniques, to determine whether a particular social media account is real or fake. Such algorithms are usually based on various account parameters analysis, such as activity, post content, frequency of posts, type of followers and many others. Social networks and messengers have their mechanisms to identify bots, such as those based on measuring the speed of actions (the number of actions in a certain amount of time) of a user or their audience [21]. Such methods can produce false positive results. However, they cannot detect fake accounts run by real people, including those belonging to so-called troll farms. Therefore, it is essential to use several methods and algorithms to detect fake accounts as accurately as possible. We will give some examples of advanced algorithms for such a task:

1. Classification of accounts based on statistically normal attributes such as posting frequency, subscribers and other metrics. Abnormal account performance may indicate that the account is being used for informational operations.
2. Link analysis—in social networks where information on user interactions is open, fake accounts can be used as part of "troll farms" for mutual promotion. Thus, a massive network can be established from its connections by detecting one or more fake accounts. In addition, accounts created at the same time with similar characteristics may indicate the creation of groups of accounts for intelligence gathering.
3. Tone analysis algorithms can detect the emotional colouring of posts and comments. It can help identify accounts that systematically write posts with a strong emotional tone, which could indicate they are being used for provocation or intelligence gathering.
4. Geolocation analysis—using automated OSINT methods for some social media and messenger accounts, using leak databases, it is possible to identify the geographical origin of an account, such as registering using a country phone number different from the one declared.
5. Message content metadata analysis—identifying frequency anomalies of keyword mentions that relate to specific locations or critical infrastructure facilities.
6. Cluster content analysis—cluster analysis can be used to group accounts with similar message topics, keywords and behaviours that are related to intelligence activities.
7. Deep learning—using neural networks such as convolutional and recurrent neural networks, models can be trained to analyze different aspects of an account, including the text and images it uses. It can help identify accounts that use sophisticated methods of manipulation and misinformation.

The problem of numerous bots in popular social networks is well known. According to Facebook, 3.8% of all accounts on the social media were fake in 2020, a number that has more than doubled since 2015 [22]. In a trial in 2022, data was made public that experts estimate the number of fake accounts on Twitter to be between 5 and 11% [23]. To combat the problem, social networking developers are implementing new mechanisms to identify fake accounts. For example, Facebook's Deep Entity Classification identifies fake accounts based on 20,000 attributes, such as age or gender distribution in a friends list. External researchers have significantly fewer indicators available for parsing, but there are examples of deep learning systems that classify bots on Facebook with 98% accuracy using just 11 indicators: profile picture, work place, education, place of residence, relationship status, check in, life events, introduction, number of mutual friends, number of pages liked, number of groups joined [24]. On Instagram, modern algorithms also show up to 98% accuracy, but using 15 indicators: username length, full name length, availability of a profile picture, biography length, the number of followers of the account, the number of users which the account follows, prevalence of the number of accounts being followed over that of followed by, the number of shared posts, belonging to any business, setting the profile as private, evidence of being verified by Instagram, availability of the channel, an established link between the account and an external URL, evidence of the number of highlights being pinned to the account, an established link between the account and a Facebook profile [25].

A team of researchers from Indiana University has developed an algorithm for detecting bots on Twitter that is 99% accurate in an experimental environment [26]. This algorithm, called Botometr, is implemented as an online platform, so you can check the activity of individual profiles manually or connect an API to your own system.

According to a survey by the Kyiv International Institute of Sociology, Telegram messenger has become the main source of news for Ukrainians in a full-scale war [27]. It was indicated by 65.7% of respondents. Therefore, in this research, we propose a metric for identifying sybil bots (fake identities, as opposed to fake spam accounts) specifically in the Telegram messenger. The difficulty in identifying bots in this messenger is that it is positioned as private and has many privacy settings. It is difficult to identify social connections of a certain account, and all the information a user can provide about themselves is their name, username, account description (biography) of 70 characters, and profile picture, not counting their phone number, which by default privacy settings hide from third-party users.

To assess the probability that a Telegram account is fake, we need to get the following information about the account: account id, username, number of profile pictures, dates of profile pictures installation, copy of profile pictures, presence of Premium status, dates and content of user's public comments. From this data, we can calculate the indicators that, given the appropriate weighting factors, will determine the likelihood that an account is fake. Obtaining data for statistical analysis of sybil bots in the Telegram messenger can be a difficult task, as those accounts that an adversary can unambiguously recognize as fake are typically deleted in chats under their control. To roughly assess the significance of some of the criteria, we prepared a sample of 109 accounts that joined to Ukrainian right-wing radicals' Telegram chats and wrote only those messages, the content of which fully corresponded to the prevalent thesis of Russian military propaganda. Thus, it can be assumed that the vast majority of these users were aware that a constructive discussion was impossible due to the inherent difference of opinion, which means that their goal could be defined as trolling in the broadest sense of the term. We also selected 100 accounts of ordinary participants in these Telegram chats for comparison. The data collected is sufficient to construct several hypotheses about the possible indicators that it is reasonable to collect to establish the likelihood that an account is fake.

The X_1 indicator is the newness of the account. In order to approximate the date of creation of a Telegram account, it is necessary to make calculations based on the account id, which is an ordinal number (assigned sequentially). If you get information about the id of new Telegram accounts at least once a month (e.g. by creating accounts with this frequency yourself), it will be possible to find the closest value of the control id to the id under investigation, thus approximately establishing the month of creation. Let:

A is the investigated id,

$B = \{a_1, a_2, \ldots, a_k\}$ is the list of control id, where k is the total number of control id,

$\Delta(a_1, a_2)$ is the absolute value of the difference between id a_1 and a_2.

Then the closest id to the control id can be found by the formula:

$$A_x = A_{\arg\min}(\Delta(A, a_i)), \text{ where } i = 1, 2, \ldots, k. \quad (1)$$

The month of creation of A_x will approximately correspond to the month of creation of account A. To increase accuracy, you can calculate the approximate number of accounts created between the two control id's in a month. Let a_{x-1} and a_{x+1} be the control id between the creation dates of account A. The number of accounts created between the closest control id will be $\Delta(a_{x-1}, a_{x+1})$. The number of days from the creation of account a_{x-1} to the creation of investigated A can then be calculated using the formula:

$$\frac{\Delta(a_{x-1}, a_{x+1})}{\Delta(A, a_{x-1})} \quad (2)$$

By adding this number of days to the date of creation of the a_{x-1}, we get the approximate date of creation of the investigated account—D_1.

The ratio of an account's approximate creation date to its oldest public post detected is an X_1 indicator—the sooner the account owner started posting public comments after the account was created, the more likely it is a sybil bot (Fig. 1). Since the creation date estimates are approximate, we included an additional factor

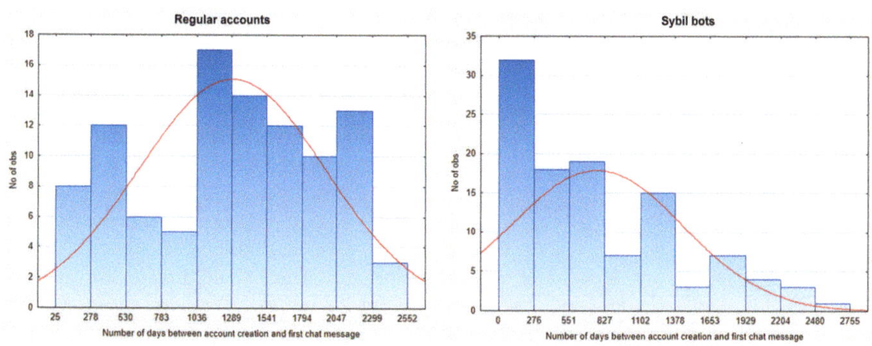

Fig. 1 Comparison of the interval between account creation and first comment in the group for regular accounts and sybil bots

in this indicator: the time between the date of adding the first profile picture to the account and the first message in the chat. Using these two criteria, we built a logistic regression model which, on average, reliably identifies 68.85% of the fake accounts in our sample, with statistical significance $p = 0.00006$. Let D_2 be the date the first profile picture was added to the account, and D_3 be the date the first post was written in the group under study, then:

$$X_1 = \frac{e^{(-1.2437-0.0002\Delta(D_3,D_2)+0.0013\Delta(D_3,D_1))}}{1+e^{(-1.2437-0.0002\Delta(D_3,D_2)+0.0013\Delta(D_3,D_1))}} \quad (3)$$

The X_2 indicator based on the number of profile pictures. As a rule, the fewer profile pictures a user has uploaded, the more likely the account is fake. The logistic model based on this assumption is highly sensitive to bots (83.5%) but leads to a large number of false positives (73%). One should also consider that privacy settings can disable the visibility of account profile pictures for external users, which expands the expected number of legitimate users without a visible profile picture. This indicator should only be considered as an auxiliary if it is not possible to calculate X_3.

The X_3 indicator is the concentration of profile picture installations. A low number or lack of profile pictures makes even inexperienced users suspicious, so in an attempt to increase the credibility of a fake account, due to the rush, creators may often install 3–5 profile pictures in one day, usually right when the account is created (Fig. 2).

Let $d_1, d_2, d_3, \ldots, d_n$ be the dates of adding all profile pictures of the account, where n is the total number of profile pictures of the account. The indicator function I is equal to 1, when the time between dates of inserting consecutive profile pictures is less than 2 days. The total number of times an account profile picture was changed before two days:

$$C = \sum_{i=1}^{n-1} I(|d_{i+1} - d_i| < 2) \quad (4)$$

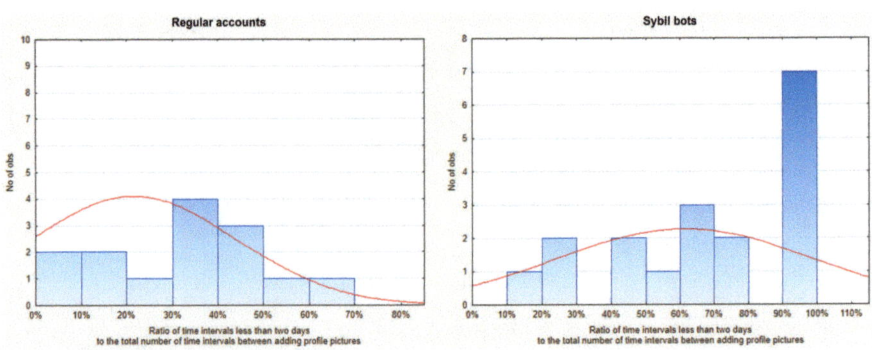

Fig. 2 Comparison of the number of profile image updates within a short timeframe for regular accounts and sybil bots

Then the logistic regression based on the ratio of time intervals less than two days to the total number of time intervals between adding profile pictures is:

$$X_3 = \frac{e^{\left(1.6720106916364 - 3.997801394842 \frac{c}{n}\right)}}{1 + e^{\left(1.6720106916364 - 3.997801394842 \frac{c}{n}\right)}} \quad (5)$$

This regression coefficient has statistical significance $p = 0.00009$ and allows us to identify fake accounts with 75% accuracy. It should be noted that despite the significant accuracy in this sample, this indicator detects unprofessional sybil bots. This indicator can be further developed by refining the model to account for the variance of gaps between profile picture setups. Low variance scores would indicate that profile pictures were uploaded in "batch" at narrow intervals, which is not natural behaviour for average users.

The X_4 indicator is the presence of paid premium Telegram status. This indicator also distinguishes sybil bots from fake spam accounts. At the time of writing, the cost of 1 month of "premium" Telegram subscription for different countries ranged from 3.99 USD to 5.99 USD. At an average expense of 0.50 USD per fake account, a troll farm can create at least nine regular bots instead of one "premium" fake account. Thus, unlike spambots, which generate some profit for their creators by increasing conversions, paying for "premium" subscriptions for sybil bots seems irrational. It should also be noted that existing Telegram subscription payment mechanisms require troll farm organizers to implement additional measures to ensure the secrecy of the origin of funds and increase risks in this area. Therefore, the presence of "premium" status is a significant indicator, which most often indicates that the account does not belong to the category of sybil bots. However, sometimes, when the information operation requires a high audience trust level, fake accounts may also apply for Premium status. According to Telegram, one million people out of the messenger's 700 million users bought a premium subscription in December 2022 [28]. At the time of research, we were unable to find a single sybil bot with Premium status, so a mathematical model for this indicator can be built after the sample is expanded. Next, we propose a few more indicators, the statistical significance of which is difficult to judge due to the insufficient sample size.

The X_5 indicator is the replacement of some Cyrillic letters in the text with Latin ones, or the use of numbers instead of letters. This outdated technique of evading spam detection is used when automatic keyword moderation works in a chat room or the identical text is sent to many Internet resources simultaneously to increase its technical "uniqueness". It is rarely used in modern information-psychological operations, but is a strong indicator of an illegitimate account.

The X_6 indicator is the degree of compression of the profile picture. The presence of visible artefacts on the profile picture, which affect its aesthetic qualities, indicates the creator's carelessness in preparing the fake account, since such photos are usually not original, but have been compressed in the process of re-saving several times. There are several algorithms to no-reference JPEG image quality (the format in which profile pictures are stored on Telegram servers), the most common of which are BRISQUE (Blind/Referenceless Image Spatial Quality Evaluator) and BLIINDS (Blind Image Integrity Notator using DCT Statistics).

The X_7 indicator is the number of reverse image search results for an profile picture image on search engines. The reverse image search technology, which is present in Google, Yandex, TinEye and several other search engines, suggests the degree of originality of the image. The majority of current users, though not all, prefer to set their own images on their profile pictures, or at least not very popular images. The higher the value of similar images detected in the reverse search, the more likely it is that the account was created algorithmically or without enthusiasm, i.e. fake.

The X_8 indicator is resorting to alternative means of generating portraits for profile pictures. There is another way to stream profile pictures in addition to using existing images from search engines. Since a person's face profile picture inspires the most trust, in order to make it impossible to detect "identity theft" using image reverse search technology, photos are generated using generative-adversarial neural networks. While studying information-psychological operations in the framework of the Russian-Ukrainian conflict, we have not encountered the use of in-house designed neural networks for this task. The Nvidia StyleGAN neural network is most often used for mass generation. The photos it produces are fairly uniform in composition, and the cropping of portraits is not exactly typical of conventional profile pictures. Therefore, when creating algorithms for detecting fake photos generated by StyleGAN neural network, we can focus not so much on identifying generation artefacts, which is a time-consuming task for machine learning, but rather on comparing the content to a typical compositional pattern. Thus, the use of neural network generated portraits as profile pictures is also often a sign of a fake account.

Several other indicators were also examined and dismissed as statistically insignificant, in particular the length of the biography (description) of the Telegram account and the language used in the messages. Although 96% of the users who belonged to the group of "trolls" in our sample used Russian, it was also actively used by regular chat participants, including simultaneously with Ukrainian and in a mixed form of social dialect, which makes it difficult to determine the leading language by software tools. Separately, we should consider the "hidden" account indicator, which reflects the privacy setting in the Telegram messenger that prevents the author of a message from being identified when it is forwarded by another user. With a statistical significance of $p = 0.003$, this criterion allows us to identify 60% of fake accounts, as they are significantly less likely to include this privacy setting. We chose not to include this indicator, primarily because it is easy to falsify, but also because the use of "hidden" mode is beneficial for the application of sybil bots. Therefore, there is reason to believe that this pattern may soon change.

Once the sample is expanded, weight coefficients for creating a general logistic regression model and classifying Telegram accounts into categories real or fake can be assigned. Decision support systems based on the hypotheses outlined here can be augmented with keyword filters and advanced language models to highlight narratives in order to optimize further manual verification of identified accounts by an analyst.

The system for detecting fake accounts in the Telegram messenger can be used to solve many practical problems, both for individual users and organizations. Examples of such tasks include: protecting users from fraud and spam, preventing the spread of misinformation and fake news, detecting attempts to manipulate public opinion, protecting users' privacy and security by identifying and blocking accounts using an identity other than one's own. For companies and brands, the system can provide protection against negative reviews and comments from fake accounts, as well as help in the fight against competitors resorting to unfair promotional methods. For government agencies and law enforcement agencies, such a system can help to reveal and stop illegal activities involving the use of fake accounts to distribute extremist material, conduct cyber espionage, in particular, crowdsourced intelligence under a false flag, or organize illegal mass events. In the field of academic research, the system represents a valuable instrument for analyzing social networks and public opinion, identifying anomalies in users' behavior that may indicate attempts at fraudulent impersonating citizens.

4 Conclusions

The study examined different information operations conducted to gather intelligence on social media under false flag. We have presented a classification of the main types of such operations: personal and mass communications. Examples of false flag intelligence data gathering through personal communications include comments in groups and channels, the use of dating sites. Collecting information through mass communications includes the creation of fake groups and channels, websites, video games. An adversary's use of crowdsourced intelligence can pose a serious threat to national security, as it can be an intermediate stage of both scientific and technical or political intelligence, as well as preparations for sabotage or missile and bomb attacks. Understanding the vulnerability of social media users to such threats is an important step in developing strategies and technologies to combat information and psychological operations on the Internet.

This research proposes a mathematical apparatus for detecting fake accounts in the Telegram messenger, which can be used for timely detection and prevention of false-flag information operations. The following main indicators of a fake Telegram account were identified: a short interval between creating an account and writing the first comment in the group, as well as updating profile images within a short timeframe. Also, the application of the developed method will make it possible to implement complex multi-stage operations to identify sabotage and reconnaissance groups or directions of enemy troop strikes. It should be noted that during large-scale combat operations when the warring parties can lose control over large territories in a short period of time, a certain number of disoriented enemy soldiers or units emerge. They constitute a distinct group of individuals who may use messengers and social media through fake accounts to infiltrate the rear to avoid detection and internment.

Understanding the vulnerability of social media users to such threats is an important step in developing strategies and technologies to combat information and psychological operations on the Internet. In the future, the results of this research could be adapted for machine learning and used to develop comprehensive systems to detect and prevent false-flag attacks in the digital space.

References

1. Gritten, D.: Strava app Flaw Revealed Runs of Israeli Officials at Secret Bases. https://www.bbc.com/news/world-middle-east-61879383 (2022). Last accessed 30 Apr 2023
2. Detector. media: Maliar on the Danger of Open Sources During war: The Enemy is Studying them, this is a Very Serious Challenge. https://detector.media/infospace/article/208601/2023-03-03-malyar-pro-nebezpeku-vidkrytykh-dzherel-pid-chas-viyny-vorog-ikh-vyvchaie-tse-duzhe-seryoznyy-vyklyk (2022). Last accessed 30 Apr 2023
3. State Special Communications Service of Ukraine: The Center for Countering Disinformation at the National Security and Defense Council of Ukraine Reports on a new Type of Information Threat!. https://www.facebook.com/dsszzi/posts/pfbid0vh9Zz2GGSk4j9xjmRKdgPQ2qnuZusaqhwzANcmhpS8bEZFWYNgVEuJPscMySK13yl (2022). Last accessed 30 Apr 2023
4. 24tv.ua: The Ministry of Defense Once Again Urged not to Disseminate Information About the Place Where Enemy Shells hit. https://24tv.ua/ru/minoborony-eshhe-raz-prizvali-ne-rasprostranjat-informaciju-o_n1905558 (2022). Last accessed 30 Apr 2023
5. National Security and Defense Council of Ukraine: Official Statement of the NSDC Apparatus of Ukraine. https://www.rnbo.gov.ua/ua/Diialnist/5286.html (2022). Last accessed 30 Apr 2023
6. XakNet Team: Pseudo-activists Create Mobilization Chats in Telegram. https://t.me/xaknet_team/373 (2022). Last accessed 30 Apr 2023
7. Security Service of Ukraine: The SSU has Exposed the Russian Special Services for Using Smartphone Games to Recruit Ukrainian Children. https://ssu.gov.ua/novyny/sbu-vykryla-spetssluzhby-rf-na-vykorystanni-smartfonihor-dlia-verbuvannia-ukrainskykh-ditei (2022). Last accessed 30 Apr 2023
8. Ionov, M.: On the methods of influencing an opponent's decisions. Mil. Thought. **12** (1971)
9. Ionov, M.: On the impact on the enemy in anti-aircraft combat. Vestnik PVO. **1** (1979)
10. Ionov, M.: On reflective enemy control in a military conflict. Mil. Thought. **1** (1995)
11. Druzhinin, V., Kontorov, D.: Voprosi voennoi sistemotehniki, Vojennoe Izdateltsvo (1976)
12. Tarakanov, K.: Matematyka y vooruzhennaia borba (1974)
13. Leonenko S.: Refleksyvnoe upravlenye protyvnykom, Armeiskyi sbornyk, **8** (1995)
14. Turko, N., Modestov, S.: Refleksyvnoe upravlenye razvytyem stratehycheskykh syl, kak mekhanyzm sovremennoi heopolytyky. In: Otchet o konferentsyy "Systems Analysis on the Threshold of the 21 Century: Theory and Practice" (1996)
15. Komov, S.: Forms and methods of information warfare. military theory and practice. Mil. Thought. **4** (1997)
16. Chausov, F.: Osnovu refleksyvnoho upravlenyia protyvnykom, Morskoi sbornyk, **1** (1999)
17. Chausov, F.: Nekotorue podkhodu k sovershenstvovanyiu systemu upravlenyia voiskamy (sylamy) novoho oblyka, Morskoi sbornyk, **3** (2011)
18. Novikov, D., Chkhartishvili, A.: Reflexion and Control: Mathematical Models. Publishing House of Physical and Mathematical Literature, pp. 194–195 (2013)
19. Mints, A., Schumann, A., Kamyshnykova, E.: Stakeholders rank of reflexion diagnostics in a corporate social responsibility system. Econ. Ann. XXI. **181**(1–2), 97 (2020). https://doi.org/10.21003/ea.v181-08

20. Dodonov, A., Lande, D., Tsyganok, V., Andreichuk, O., Kadenko, S., Grayvoronskaya, A.: Recognition of Information Operations, pp. 201 (2017)
21. Facebook: Verifying the Identity of People Behind High-Reach Profiles. https://about.fb.com/news/2020/05/id-verification-high-reach-profiles (2020). Last accessed 30 Apr 2023
22. Hao, K.: How Facebook Uses Machine Learning to Detect Fake Accounts. https://www.technologyreview.com/2020/03/04/905551/how-facebook-uses-machine-learning-to-detect-fake-accounts/(2020). Last accessed 30 Apr 2023
23. Reuters: Twitter Lawyer Tells Court Musk has not Backed up Claims of Fake Accounts. https://www.reuters.com/markets/deals/elon-musks-deposition-twitter-litigation-rescheduled-oct-6-7-2022-09-27/ (2022). Last accessed 30 Apr 2023
24. Albayati, M., Altamimi, A.: An empirical study for detecting fake Facebook profiles using supervised mining techniques. Inform. **43**(1) (2019). https://doi.org/10.31449/inf.v43i1.2319
25. Sheikhi, S.: An efficient method for detection of fake accounts on the Instagram platform. Revue d'Intelligence Artificielle **34**(4) (2020). https://doi.org/10.18280/ria.340407
26. Yang, K.C., Ferrara, E., Menczer, F.: Botometer 101: social bot practicum for computational social scientists. J. Comput. Soc. Sci.**5** (2022). https://doi.org/10.1007/s42001-022-00177-5
27. OPORA: Media Consumption of Ukrainians in Conditions of Full-scale war. OPORA Surv. https://www.oporaua.org/report/polit_ad/24068-mediaspozhivannia-ukrayintsiv-v-umovakh-povnomasshtabnoyi-viini-opituvannia-opori (2022). Last accessed 30 Apr 2023
28. Singh, M.: Telegram Premium Tops 1 Million Subscribers. https://techcrunch.com/2022/12/06/telegram-premium-tops-1-million-subscribers/ (2022). Last accessed 30 Apr 2023

MIX
Papier aus verantwortungsvollen Quellen
Paper from responsible sources
FSC® C105338

If you have any concerns about our products,
you can contact us on
ProductSafety@springernature.com

In case Publisher is established outside the EU,
the EU authorized representative is:
Springer Nature Customer Service Center GmbH
Europaplatz 3, 69115 Heidelberg, Germany

Printed by Libri Plureos GmbH
in Hamburg, Germany